How can the interior of the Sun, white dwarfs and other stars be studied by stellar seismology? What can Doppler imaging tell us about high-degree pulsations? What impact are CCD and infrared observations having on extending the Cepheid and RR Lyrae distance scale? And how are the other classes of pulsators providing independent checks of the distance scale? These and many other critical questions are answered in this timely review of the dramatic advances made in pulsating star research in the last decade.

This survey collects together more than thirty comprehensive reviews and over one hundred summaries of research papers from the 139th IAU Colloquium, held in Victoria, British Columbia. Together these cover all aspects of recent developments in the field of variable star research and preview some of the exciting advances anticipated for the next decade. This volume provides an essential review for graduate students and researchers.

New Perspectives on Stellar Pulsation and Pulsating Variable Stars

IAU Astronomical Union
Union Astronomique International

The following Colloquia of the International Astronomical Union are published for the Union by Cambridge University Press.

82. Cepheids. *Edited by Barry F. Madore.* 0 521 30091 6. 1985

91. History of Oriental Astronomy. *Edited by G. Swarup, A. K. Bag and K. S. Shukla.* 0 521 34659 2. 1987

92. Physics of Be Stars. *Edited by A. Slettebak and T. P. Snow.* 0 521 33078 5. 1987

101. Supernova Remnants and the Interstellar Medium. *Edited by R. S. Roger and T. L. Landecker.* 0 521 35062 X. 1988

105. The Teaching of Astronomy. *Edited by Jay M. Pasachoff and John R. Percy.* 0 521 35331 9. 1990

106. Evolution of Peculiar Red Giant Stars. *Edited by Hollis Johnson and Ben Zuckerman.* 0 521 36617 8. 1989

111. The Use of Pulsating Stars in Fundamental Problems of Astronomy. *Edited by Edward G. Schmidt.* 0 521 37023 X. 1989

136. Stella Photometry – Current Techniques and Future Developments. *Edited by C. J. Butler and I. Elliott.* 0 521 41866 6. 1993

139. Stellar Pulsation and Pulsating Variable Stars. *Edited by J. Nemec and J. Matthews.* 0 521 44382 2. 1993

New Perspectives on Stellar Pulsation and Pulsating Variable Stars

Proceedings of IAU Colloquium No. 139
Victoria, British Columbia
15–18 July 1992

Edited by

James M. Nemec
University of Washington

and

Jaymie M. Matthews
University of British Columbia

Published by the Press Syndicate of the University of Cambridge
The Pitt Building, Trumpington Street, Cambridge CB2 1RP
40 West 20th Street, New York, NY 10011–4211, USA
10 Stamford Road, Oakleigh, Melbourne 3166, Australia

© Cambridge University Press 1993

First published 1993

Printed in Great Britain at the University Press, Cambridge

A catalogue record for this book is available from the British Library

Library of Congress cataloguing in publication data available

ISBN 0 521 44382 2 hardback

CONTENTS

Group photograph xvi
List of Participants xix
Preface xxiii

I. VARIABLE STARS AS DISTANCE INDICATORS

1. RR Lyrae Stars

The Oosterhoff Period Effect and Age of the Galactic Globular Cluster System
    Allan Sandage 3

RR Lyrae Stars in the Magellanic Clouds
    Alistair R. Walker 15

The Infrared Period-2.2μm Magnitude Relation for RR Lyrae stars
    A.J. Longmore 21

Infrared Period-Luminosity Relations of RR Lyrae Stars in M5 and M15
    T. Liu & K. A. Janes 30

Period-Luminosity-Metal Abundance Relations for Pop II Variable Stars
    James M. Nemec & Thomas E. Lutz 31

2. Cepheids

Cepheids in IC 4182, Calibration of SN Ia 1937c and the Hubble Constant
    A. Saha *et al.* 53

Recent Improvements to the Cepheid Distance Scale
    Wendy L. Freedman & Barry F. Madore 61

Calibration of the Cepheid Distance Scale
    Wolfgang P. Gieren & Pascal Fouqué 72

Cepheids and Long-Period Variables in Virgo Cluster Galaxies
    M.J. Pierce *et al.* 81

The Calibration of Colours and Luminosities for Classical Cepheid Variables
    D.G. Turner 90

I-Band Cepheid Distance to NGC 6822
    Myung Gyoon Lee, Wendy L. Freedman & Barry F. Madore 91

I-Band Cepheid Distance to WLM
    Myung Gyoon Lee, Wendy L. Freedman & Barry F. Madore 92

A Preliminary Distance to the SMC by the Surface Brightness Technique
    Thomas G. Barnes III, Thomas J. Moffett & Wolfgang P. Gieren 93

An Adjustment to Cepheid Distances Using Model Atmospheres
    Robert Hindsley & R.A. Bell 94

Galactic Cepheid Kinematics as a Probe of Large Scale
Non-Axisymmetry of the Galaxy
    J. A. R. Caldwell *et al.* 95

3. Long Period Variable Stars

An Optical Period-Luminosity Relation for Long Period Variables
    M. J. Pierce & D.R. Crabtree 102

Implication of a $P - L$ relation for Mira Variable Stars
    Hiromoto Shibahashi 103

Studies of Large-Amplitude δ Scuti variables
    W.J.F. Wilson, E.F. Milone & D.J.I. Fry 104

II. STELLAR SEISMOLOGY

1. White Dwarfs

White Dwarf and Pre-White Dwarf Oscillations
    Arthur N. Cox 107

An Example Demonstrating the Potential for Asteroseismology
of DB White Dwarf Stars
    P. A. Bradley & M. A. Wood 116

The Internal Structure of White Dwarf Stars Using the
Whole Earth Telescope
    P. A. Bradley 117

White Dwarf Seismology at the Université de Montréal
    G. Fontaine *et al.* 120

The Effect of a Vertical Magnetic Field on the Periods of Trapped
g-Modes in White Dwarfs
    Bradley W. Carroll 121

2. Rapidly oscillating Ap stars

Seismology of pulsating Ap stars:
Results from the past decade, prospects for the next
    Jaymie M. Matthews 122

Magnetic Fields of Rapidly Oscillating Ap Stars
    G. Mathys 132

Chaos in Pulsating Variable Stars: Preliminary Analysis of Photometric
Photometry and Observational Constraints of Detection
    T.J. Kreidl 133

Pulsation of Rotating Magnetic Stars
    H. Shibahashi & M. Takata 134

3. δ Scuti Stars

Nonradial Pulsation among δ Scuti Stars
    Michel Breger 135

The Pulsation Characteristics of HD 93044
    Zong-li Liu & Zhi-ping Li 144
Mode determination by Fourier analysis of line profile variations:
Application to the δ Scuti star τ Peg
    E.J. Kennelly et al. 147
Fourier Analysis of Line-Profile Variations: Toward Stellar m–ν Diagrams?
    W.J. Merryfield & E.J. Kennelly 148
Frequency Analysis of Multiperiodic δ Scuti Stars
    L. Mantegazza, E. Poretti & F.M. Zerbi 149
Search for a Secondary Frequency in the Large-Amplitude
δ Scuti Star CY Aqr
    C. Coates et al. 150

4. The Sun

Interpretations of Solar Oscillations
    Arthur N. Cox 151
A Striking Similarity Between the Sun, Binaries and RR Lyrae stars
in globular clusters
    Valery A. Kotov 160

III. BEYOND THE CLASSICAL INSTABILITY STRIP

1. The β Cephei and B/Be stars

New Opacities and the β Cephei Stars
    L. A. Balona 163
Focal Points in Contemporary β Cephei Star Research
    C. Sterken 171
Slowly Pulsating B Stars
    C. Waelkens 180
Line-Profile Variations of Rotating Pulsating Stars
    C. Aerts & M. De Pauw 182
Did β Canis Majoris Quit Pulsating?
    Andrew P. Odell & Robert D. Watson 183
β Cephei Pulsation Anomalies: Potential New Windows into the Instabilities
and Evolution of Early B Stars
    Bruce A. Goldberg et al. 184
Beta Cephei: A Magnetic Be star?
    H.F. Henrichs et al. 186
An Extraordinary Early-Type Eclipsing Binary
    L. A. Balona & J. Cuypers 188
HR 8762: Low-Amplitude Photometric Variation in a Pre-Shell Phase
    S. González-Bedolla et al. 189

Time-Series Spectroscopy of ζ Ophiuchi
  A.H.N. Reid et al. 190

2. Miras and Long Period Variables

Pulsation and Mass Loss in LPVs
  George H. Bowen 191
Long Period Variables in the Magellanic Clouds and the Galaxy
  S. M. G. Hughes 192
Properties of Mira Photospheres
  M. Scholz 201
Pulsation Properties of Hydrodynamic Models with Dimensional Analysis
  Toshiki Aikawa 204
Dust Induced Dynamics of Circumstellar Shells around LPVs
  A. Gauger, A. J. Fleischer & E. Sedlmayr 206
NLTE Synthetic Spectra of Mira-Type Variable Stars
  Donald G. Luttermoser, George H. Bowen & Lee Anne Wilson 207
Nonlinear Models of Miras Including Time-Dependent Convection
  Dale A. Ostlie & Arthur N. Cox 208
The Evolution of Hα Profiles in S-type Mira Stars
  A.W. Woodsworth 209
Variable UV Line Emission in S Carina: Miras do not fear change
  E.W. Brugel, R. Davis & J. Bookbinder 210

3. R CrB, RV Tauri and H-deficient stars

Pulsations and Declines of R CrB stars
  P.L. Cottrell & W.A. Lawson 212
Recent Insights into R CrB Stars from Recent UV and Visible Observations
  Geoffrey C. Clayton & Barbara A. Whitney 214
A Spectroscopic Study of R CrB Stars in the Galaxy and the LMC
  Karen Pollard, P.L. Cottrell & W.A. Lawson 215
A Photometric and Spectroscopic Study of Southern RV Tauri Stars
  Karen Pollard et al. 216
Hydrodynamic Models of Radially Pulsating Hot Extreme Helium Stars
  Yu. A. Fadeyev 217

IV. THEORETICAL BREAKTHROUGHS

1. The New Opacities and their Impact on Pulsation Theory

OPAL Opacities
  F. J. Rogers & C.A. Iglesias 221
The Opacity Project
  M. J. Seaton 231

Contents

The Bump Cepheid Mass Discrepancy Laid to Rest (?)
P. Moskalik & J. R. Buchler 237
Comparative Pulsation Calculations with OP and OPAL Opacities
S.M. Kanbur & N. R. Simon 240
Masses of Oosterhoff I and II RR Lyrae Stars
Arthur N. Cox 241
Radiation Hydrodynamics in Pulsating Stars
Michael U. Feuchtinger & E. A. Dorfi 242

2. Diffusion and convection

Element Diffusion in Pulsating Variable Stars
Joyce Ann Guzik 243
Convection in RR Lyrae Stars
R. F. Stellingwerf & G. Bono 252
Convection and the Bump Cepheid Resonance
Arthur N. Cox 261
A Survey of RR Lyrae Models
G. Bono & R.F. Stellingwerf 262

3. Nonadiabaticity and nonlinearity

Higher Vibrational Modes in RR Lyrae Stars
S. A. Glasner & J. R. Buchler 263
A Full-Amplitude Nonlinear Model for RR Lyrae:
Pulsations, Shock Waves and Hα Peculiarities
Andrew Fokin 265
Nonlinear Radiative RR Lyrae Models: Search for Double-Mode Behaviour
Joyce A. Guzik & Arthur N. Cox 266

4. Pulsation models

Modelling Cepheids and RR Lyrae stars
G. Kovács 267
Double-Mode RR Lyrae Models
G. Bono & R.F. Stellingwerf 275
A Survey of BL Herculis-type models
J. Robert Buchler & Pawel Moskalik 277
Theoretical Implications of Triple-Mode RR Lyrae Pulsations
Géza Kovács & J. R. Buchler 278
On the Explanation of the Sandage Effect
M. Catelan 280
Hydrogen Emission Lines from Extended Pulsating Atmospheres
P. de Laverny & C. Magnan 281

Propagation of Radial Pulsation Modes in the Outer Atmosphere of
Arcturus: First results
    M. Cuntz 283
Pulsations of Proto-Giant-Planets
    Gunther Wuchterl 284

5. Evolution and the effects of metallicity

Classical Models of Cepheid Stars
    C. Chiosi 285
Evolutionary Models of RR Lyrae Stars
    Young-Wook Lee 294
The Red Edge of the Cepheid Instability Strip
    Yan Li 304
The Cepheid Instability 'Wedge'
    Siobahn M. Morgan 307
Time Dependent Convection and the Pulsations of Polaris
    Siobhan M. Morgan & Arthur N. Cox 308
Stellar Structure and RR Lyrae Masses
    Ben Dorman 309
Mass-Loss During the RR Lyrae Phase of the HB
    Rebecca A. Koopmann et al. 312

V. WINDOWS ON THE INSTABILITY STRIP

1. Photometry

The Masses and Luminosities of Globular Cluster RRc Stars
    Norman R. Simon & Christine M. Clement 315
Masses of c-type RR Lyrae Variables in Globular Clusters
    Carla Cacciari & A. Bruzzi 324
CCD photometry of RR Lyrae stars in M3
    Carla Cacciari et al. 325
Period Shifts in RR Lyrae Stars
    J. Fernley 332
New Results on Field and Cluster RR Lyrae Stars
    G. Clementini et al. 335
Search for Variables in the Globular Clusters NGC 6544 and NGC 6642
    Martha L. Hazen 337
CCD Photometry of RR Lyrae Stars in NGC 6388 and M15
    N. A. Silbermann, H. A. Smith & M. Bolte 338
Amplitude of RR Lyrae Star Light Curves: Comparison between
Observations and One-Zone Model Predictions
    E. Antonello & S. Cernuti 339

Pulsation Variables in the AF Stars of the Case Low-Dispersion Survey
    T. D. Kinman 340

Fundamental Mode and First Overtone Mode Cepheids in the SMC
    H. A. Smith et al. 349

Resonance Effects in Fundamental Mode, First-Overtone, and
Double-Mode Cepheids
    Elio Antonello 357

Variable Stars in the Sculptor Dwarf Galaxy
    C. G. Goldsmith 358

Cepheids in Magellanic Cloud Clusters
    D. L. Welch, M. Mateo & E. Olszewski 359

CCD Observations of Forgotten Cepheids
    David L. DuPuy et al. 368

CCD Photometry of Faint Cepheids
    Arne A. Henden 369

First Overtone Pulsators Among Cepheids
    L. Mantegazza & E. Poretti 370

On Period Ratios and Resonance Sequences
    J. O. Petersen 371

Harmonics and Coupling-Terms in the Pulsation of the Double-Mode
Cepheid TU Cas
    L. Szabados 372

The Periods of X Cyg, T Mon, Y Oph, S Vul and SV Vul
    A.M. Heiser 373

2. Spectroscopy and Radial Velocities

A Spectroscopic Abundance Study of Dwarf Cepheid V1719 Cygni
    Chulhee Kim & Kozo Sadakane 374

Asymmetry of Metallic Spectral Lines in Cepheids
    Michael Albrow & Peter Cottrell 375

RR Lyrae: Shock Waves and Atmospheric Motions
    Agnès Lebre 376

Spectrophotometric Determination of Effective Temperatures
for Six Short-Period Cepheids
    E. Antonello, S. Fossati & L. Mantegazza 377

Spectroscopic Study of Cepheids in ω Cen
    Guillermo Gonzalez & George Wallerstein 378

A photometric and Spectrographic Study of SX Phe
    Chulhee Kim et al. 380

Investigation of the Double-Mode Cepheid TU Cas: Atmospheric
Parameters and Chemical Composition
    S. M. Andrievsky et al. 382

The HF Precise Radial Velocity Programme at DAO
    S. Yang et al. 383
The Radial Velocity Variability of Yellow Giant Stars
    A.M. Larson et al. 384
Extremely Low Amplitude Cepheids
    R. Paul Butler 385
New Radial Velocities and the Barnes-Evans Method
    Robert Hindsley & R.A. Bell 386

3. Other Spectral Regions and Pulsators in Binary Systems

Ultraviolet Studies of Cepheids
    Erika Böhm-Vitense 387
X-ray Sources in δ Scuti stars: an Ultraviolet Study of 71 Tau
    L.E. Pasinetti Fracassini et al. 396
A New Pulsating White Dwarf Seen By the Wide-Field Camera?
    D. Wonnacott 397
Recent Results on Binary Cepheids
    Nancy Remage Evans 398
New Ways of Revealing Cepheid Binary Systems
    L. Szabados 406
Orbital Solution and Physical Parameters of the Binary Cepheid AW Per
    J. Vinko 408

4. Long-term behaviour and evolution

The Periods of RR Lyrae
    Arthur N. Cox 409
The Blazhko Effect in RR Lyrae
    Terry J. Teays 410
Moving Through the Instability Strip
    Emilia Pisani Belserene 419
RR Lyrae Variables in the Second-Parameter Globular Cluster NGC 7006
    Amelia Wehlau & James M. Nemec 420
The m_1 Index in RR Lyrae Stars
    Eloy Rodríguez, Angel Rolland & Pilar López de Coca 421
BV Photometry of V9 in the Globular Cluster 47 Tuc
    Michael Corwin & Bruce Carney 421
The Stability of Cepheid Lightcurves
    J. D. Fernie 422
Period Variation of the Pop.II Cepheid AU Peg
    J. Vinko 423

Long-Term Brightness Changes in Cool Pulsating Variable Stars
  John R. Percy 425
Period Changes in the SX Phoenicis Star CY Aqr
  Michael D. Joner & John M. Powell 428
Strömgren Photometry of the Dwarf Cepheids DY Her and BP Peg
  J. H. Peña, R. Peniche & R. Garrido 429
28 And: Simultaneous Strömgren photometry
  Eloy Rodríguez *et al.* 430
ρ Pup: A Monoperiodic Radially Pulsating δ Scuti Star
  Eloy Rodríguez, Angel Rolland & Pilar López de Coca 430

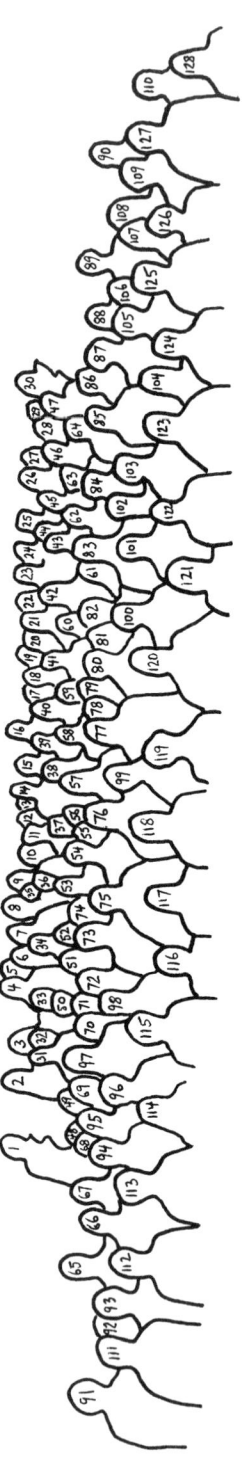

#	Name	#	Name	#	Name
1	P Garnavich	27	F Rogers	53	J Mould
2	A Larson	28	J Guzik	54	J-W Pel
3	E Böhm-Vitense	29	N Baker	55	J Graham
4	H Henrichs	30	C Davis	56	unidentified
5	D Zurek	31	T Kennelly	57	P Butler
6	E Rodríguez	32	N Evans	58	A Longmore
7	P Brassard	33	Y-W Lee	59	M Corwin
8	F Wesemael	34	S González-Bedolla	60	M Albrow
9	P Bergeron	35	C Kim	61	K Pollard
10	N Simon	36	R Stellingwerf	62	P Durrell
11	G Clementini	37	G Bono	63	M Catelan
12	A Glasner	38	T Liu	64	A Sandage
13	G Kovács	39	L Balona	65	A Cox
14	A Walker	40	B Pedersen	66	J Vinko
15	A Saha	41	unidentified	67	D Fernie
16	R Buchler	42	P Moskalik	68	L Szabados
17	T Aikawa	43	J Petersen	69	W Merryfield
18	M Breger	44	A Wehlau	70	H Shibahashi
19	unidentified	45	E Antonello	71	G Gonzalez
20	Z-P Li	46	G Clayton	72	M Scholz
21	J Fernley	47	L Mantegazza	73	C Goldsmith
22	unidentified	48	D Hartwick	74	T Kinman
23	W Wehlau	49	M Hazen	75	J Matthews
24	C Iglesias	50	C Koenig	76	A Odell
25	D Luttermoser	51	M Griscom	77	T Kreidl
26	R Crowe	52	C Cacciari	78	D Turner

#	Name	#	Name
79	M Joner	105	J Caldwell
80	C Chiosi	106	J Cuypers
81	A Heiser	107	P de Laverny
82	B Carroll	108	C Aerts
83	D Ostlie	109	unidentified
84	G Bowen	110	R Sreenivasan
85	G Mathys	111	J Nemec
86	E Belserene	112	D Welch
87	C Waelkens	113	S Yang
88	A Hearn	114	R Davis
89	P Bradley	115	S Morgan
90	D Dupuy	116	R Peniche
91	A McWilliam	117	J Peña
92	J Cox	118	S Kanbur
93	P Cottrell	119	D Wonnacott
94	J Percy	120	B Martin
95	C Sterken	121	B Dorman
96	E Brugel	122	T Teays
97	H Smith	123	W Gieren
98	N Silbermann	124	M Feuchtinger
99	A Reid	125	G Wuchterl
100	C Aikman	126	A Lebre
101	A Henden	127	T Lutz
102	M Cuntz	128	B Goldberg
103	Y Li		
104	S Höfner		

LIST OF PARTICIPANTS

Name	Country	e-Mail Address
Conny Aerts	Belgium	kgafa01@cc2.kuleuven.ac.be
Toshiki Aikawa	Japan	aikawa@izcc.tohoku-gakuin.ac.jp
Chris Aikman	Canada	aikman@dao.nrc.ca
Michael Albrow	New Zealand	phys170@csc.canterbury.ac.nz
(S. Andrievsky)	Ukraine	
Elio Antonello	Italy	antonello@astmim.astro.it
Norman Baker	USA	nbaker@lulu.phys.columbia.edu
Luis Balona	South Africa	lab@saao.ac.za
Lee Belserene	USA	emilia@sequim.wa.com
Pierre Bergeron	Canada	bergeron@astro.umontreal.ca
Erika Böhm-Vitense	USA	erica@phast.phys.washington.edu
Giuseppe Bono	Italy	bono@astrts.astro.it
George Bowen	USA	s1.ghb@isumvs
Paul Bradley	USA	bradley@astro.as.utexas.edu
Pierre Brassard	Canada	brassard@astro.umontreal.ca
Michel Breger	Austria	breger@avia.una.ac.at
Edward Brugel	USA	brugel@puppis.colorado.edu
Robert Buchler	USA	buchler@ufhepa.phys.ufl.edu
Paul Butler	USA	paul@astro.umd.edu
Carla Cacciari	Italy	cacciari@astbo3.cineca.it
John Caldwell	South Africa	jac@saao.ac.za
Brad Carroll	USA	bcarroll@cc.weber.edu
Joe Catanzarite	USA	jhc@lynx.ociw.edu
Márcio Catelan	Brazil	marcio@iag.usp.ansp.br
Cesare Chiosi	Italy	39003::chiosi (span)
Geoff Clayton	USA	clayton@hyades.dnet.nasa.gov
Gisella Clementini	Italy	gisella@alma02.cineca.it
Mike Corwin	USA	corwin@unccvax.uncc.edu
Peter Cottrell	New Zealand	phys130@canterbury.ac.nz
Arthur Cox	USA	anc@beta.lanl.gov
Dennis Crabtree	Canada	crabtree@dao.nrc.ca
Rick Crowe	USA	crowe@uhunix.uhcc.hawaii.edu
Manfred Cuntz	USA	cuntz@hao.ucar.edu
Jan Cuypers	Belgium	jan@astro.oma.be
Cecil Davis	USA	cgj@corrtex.lanl.gov
Rebecca Davis	USA	rdavis@puppis.colorado.edu
Patrick de Laverny	France	delaverny@friap51.bitnet
Ben Dorman	USA	bd4r@borealis.astro.virginia.edu
David Dupuy	USA	fpydupuy%faculty%vmi@ist.vmi.edu
Patrick Durrell	Canada	durrell@physun.physics.mcmaster.ca
Nancy Evans	Canada	evans@nereid.sal.ists.ca

Name	Country	e-Mail Address
(Yu. Fadeyev)	Russia	iaas@node.ias.msk.su
Donald Fernie	Canada	fernie@centaur.astro.utoronto.ca
John Fernley	Spain	jaf@vilspa.span
Michael Feuchtinger	Austria	fm@avia.una.ac.at
(Andrew Fokin)	Russia	demos!node.ias.msk.su!iaas@fuug.fi
Gilles Fontaine	Canada	fontaine@astro.umontreal.ca
Wendy Freedman	USA	wendy@ociw5.caltech.edu
Andreas Gauger	Germany	andy0434@w415zrz.physik.tu-berlin.de
Wolfgang Gieren	Chile	wgieren@astro.puc.cl
Ami Glasner	Israel	glasner@hujivms.huji.ac.il
Chris Goldsmith	Canada	goldsmit@aries.yorku.ca
Bruce Goldberg	USA	fax:818-393-6962
Guillermo Gonzalez	USA	gonzalez@phast.phys.washington.edu
S. Gonzalez-Bedolla	Mexico	FAX: 525-5483712
John A. Graham	USA	graham%dtm.span@sds.sdsc.edu
Matthew Griscom	USA	griscom@raven.phys.washington.edu
Joyce Guzik	USA	joy@beta.lanl.gov
F.D.A. Hartwick	Canada	hartwick@uvphys.bitnet
Martha Hazen	USA	martha@cfa3.bitnet
A.G. Hearn	Netherlands	ahearn@fys.ruu.nl
Arnold Heiser	USA	heisera@ctrvax.vanderbilt.edu
Arne Henden	USA	henden@mps.ohio-state.edu
Huib Henrichs	Netherlands	huib@astro.uva.nl
Robert Hindsley	USA	rbh@phobos.usno.navy.mil
Susanne Höfner	Austria	hoefner@avia.una.ac.at
Shaun Hughes	USA	smh@deimos.caltech.edu
Carlos Iglesias	USA	opal@ocfmail.ocf.llnl.gov
Michael Joner	USA	jonerm@astro.byu.edu
Shashi Kanbur	USA	skanbur@unlcdc2
Ted Kennelly	Canada	usershoe@mtsg.ubc.ca
Chulhee Kim	Korea	snu00200@krsnucc1.bitnet
Tom Kinman	USA	tkinman@noao.edu
Chris Koenig	Canada	koenig@geop.ubc.ca
Valery Kotov	Crimea/USA	vahe@corona.stanford.edu
Géza Kovács	Hungary/USA	kovacs@ufhepa.phys.ufl.edu
Tobias Kreidl	USA	tjk@lowell.edu
Ana Larson	Canada	larson@otter.phys.uvic.ca
Agnès Lebre	France	lebre@frip51.bitnet
Young-Wook Lee	USA	lee@yalastro.bitnet
Yan Li	China	fax:011-86-871-71845
Zhi-ping Li	China	bmabao@ica.beijing.canet.cn
Tim Liu	USA	liu@astro.ucla.edu
Andrew Longmore	UK	ajl@starlink.roe.ac.uk
Donald Luttermoser	USA	lutter@fester.physics.iastate.edu
Thomas Lutz	USA	tlutz@beta.math.wsu.edu
Luciano Mantegazza	Italy	mantegazza@pavia.infn.it

List of Participants

Name	Country	e-Mail Address
Brian Martin	Canada	tomt@ualtamts
Philippe Mathias	France	obshpa::mathias
Gautier Mathys	Switzerland	mathys@cgeuge54.bitnet
Jaymie Matthews	Canada	matthews@geop.ubc.ca
Andrew McWilliam	USA	andy@lynx.ociw.edu
William Merryfield	Canada	userpgnz@ubcmtsg.bitnet
Siobahn Morgan	USA	morgans@iscsvax.uni.edu
Pawel Moskalik	Poland	moskalik@pine.circa.ufl.edu
Jeremy Mould	USA	jrm@deimos.caltech.edu
James Nemec	USA	nemec@phast.phys.washington.edu
Andrew Odell	Australia	odell@nauvax.bitnet
Dale Ostlie	USA	dostlie@cc.weber.edu
Brian Pedersen	Canada	brian@orca.phys.uvic.ca
Jan-Willem Pel	Netherlands	pel@hrdksw5
José Peña	Mexico	rpeniche@alfa.astroscu.unam.mx
Rosario Peniche	Mexico	rpeniche@alfa.astroscu.unam.mx
John Percy	Canada	percy@utorphys.bitnet
Jörgen Petersen	Denmark	oz%cuobs.dk@sds.sdsc.edu
Michael Pierce	USA	mpierce@noao.edu
Karen Pollard	New Zealand	k.pollard@canterbury.ac.nz
Chris Pritchet	Canada	pritchet@clam.phys.uvic.ca
Andy Reid	UK	ahnr@starlink.ucl.ac.uk
Eloy Rodríguez	Spain	eloy%16488.span@sds.sdsc.edu
Forrest Rogers	USA	opal@ocfmail.ocf.llnl.gov
Abhijit Saha	USA	Saha@stsci.edu
Allan Sandage	USA	
Michael Scholz	Germany	b15@dhdurz2.bitnet
Michael Seaton	UK	mjs@starlink.ucl.ac.uk
H. Shibahashi	Japan	shibahashi@dept.astron.s.u-tokyo.ac.jp
Nancy Silbermann	USA	silbermann@msupa.bitnet
Norman Simon	USA	phys020@unlcdc2.bitnet
Horace Smith	USA	smith@msupa.bitnet
Ranga Sreenivasan	Canada	sreenivasan@uncamult.bitnet
Robert Stellingwerf	USA	rfs@demos.lanl.gov
Chris Sterken	Belgium	csterken@bbrbfu60.bitnet
Fritz Swenson	Canada	swenson@otter.phys.uvic.ca
László Szabados	Hungary	h1036ibv%ella.uucp@mail.germany.eu.net
Terry Teays	USA	teays%iuesoc.span@draco.tuc.noao.edu
John Telting	Netherlands	john@astro.uva.nl
David Turner	Canada	turner@husky1.stmarys.ca
Don VandenBerg	Canada	davb@uvvm.uvic.ca
Jozsef Vinko	Hungary	h2674sza%ella.uucp@mail.germany.eu.net

Name	Country	e-Mail Address
Chris Waelkens	Belgium	fgafa01@cc1.kuleuven.ac.be
Alistair Walker	USA/Chile	awalker@noao.edu
Amelia Wehlau	Canada	a447@nve.uwo.ca
William Wehlau	Canada	whwehlau@uwovax.uwo.ca
Doug Welch	Canada	welch@physun.physics.mcmaster.ca
François Wesemael	Canada	wesemael@astro.umontreal.ca
William Wilson	Canada	phone:(403)220-6088
Dave Wonnacott	UK	dww@astrophysics.starlink.rutherford.ac.uk
Andy Woodsworth	Canada	wdswrth@dao.nrc.ca
Gunther Wuchterl	Germany	wcah%ds0rus1i.bitnet@cunyvm.cuny.edu
Stephenson Yang	Canada	yang@orca.phys.uvic.ca
Dave Zurek	Canada	zurek@otter.phys.uvic.ca

PREFACE

In July 1992 International Astronomical Union (IAU) Colloquium 139, "New Perspectives on Stellar Pulsation and Pulsating Variable Stars", was held at the Victoria Conference Centre in Victoria, British Columbia, Canada. The four-day colloquium was attended by approximately 150 astronomers from over 20 countries. The proceedings of the conference are contained in this book. The five chapters reflect the main topics that were discussed: variable stars as distance indicators, stellar seismology, variable stars beyond the classical Cepheid instability strip, theoretical breakthroughs, and new windows on the instability strip. Each of these research areas has undergone dramatic changes in the last decade. The approximately 30 review papers and 100 shorter contributions describe the most recent developments, and give a preview of some of the advances expected in the decade to come.

The scientific organizing committee consisted of M. Breger (Austria), A. Cox (USA), M. Feast (South Africa), W. Gieren (Chile), M. Jerzykiewics (Poland), J. Lub (Netherlands), J. Nemec (USA, co-chair), A. Renzini (Italy), H. Shibahashi (Japan), P. Smeyers (Belgium), A. Walker (Chile), D. Welch (Canada, co-chair) and P. Wood (Australia). The local organizing committee consisted of A.F. Linnell Nemec (Canada), J. Nemec (USA, co-chair), C. Pritchet (Canada), R. Sreenivasan (Canada, co-chair) and G. Wallerstein (USA).

The meeting was co-sponsored by five IAU commissions: No.27 (Variable Stars); No.30 (Radial Velocities); No.35 (Stellar Constitution); No.36 (Stellar Atmospheres); No.37 (Star Clusters & Associations). We gratefully acknowledge this financial support and thank J. Bergeron (IAU General Secretary) and the respective commission presidents (at the time) for their help: M. Breger, D.W. Latham, A. Maeder, D.F. Gray and G.L.H. Harris. We also thank several organizations that were instrumental in providing financial and other support: the Department of Physics, University of Calgary (Calgary, Alberta); the Canadian Institute for Theoretical Astrophysics (CITA); the Dominion Astrophysical Observatory of the Herzberg Institute of Astrophysics, National Research Council of Canada; International Statistics & Research Corporation (Brentwood Bay, British Columbia); the Department of Physics & Astronomy, McMaster University; the Natural Sciences & Engineering Research Council (NSERC); the Department of Physics & Astronomy, University of Victoria (Victoria, British Columbia); the Department of Astronomy, University of Toronto (Toronto, Ontario); the Program in Astronomy of the Department of Mathematics, Washington State University (Pullman, Washington); and the Department of Astronomy, University of Washington (Seattle, Washington). We also acknowledge the volunteered time and talents of M. Chapin, C. Koenig (University of British Columbia), E. Chapin and other individuals and organizations who helped with the meeting.

Invited talks were given by L. Balona, E. Böhm-Vitense, G. Bowen, M. Breger, C. Cacciari, J. Caldwell, C. Chiosi, A. Cox, N. Evans, W. Freedman, W. Gieren, J. Guzik, S. Hughes, Y.-W. Lee, A. Longmore, J. Matthews, J. Nemec, M. Pierce,

F. Rogers, A. Saha, A. Sandage, M. Seaton, N. Simon, H. Smith, R. Stellingwerf, C. Sterken, T. Teays, A. Walker and D. Welch. The chairs of the various sessions were: R . Buchler, C. Cacciari, G.Clayton, A. Cox, J.Cuypers, W. Freedman, T. Kreidl, J. Mould and J. Nemec. Poster papers were on display throughout the meeting, and were discussed in a special session chaired by J. Matthews.

On Tuesday (July 14) a welcoming reception was held at the Victoria Conference Center, with a delightful concert of light classical musical provided by the Juan de Fuca Chamber Orchestra. A banquet was held on Thursday evening at the Chantecler restaurant in Royal Oak, the highlight of which was an entertaining speech by D. Fernie. On Friday night a tour of the Dominion Astrophysical Observatory was given by C. Aikman, followed by a visit to the Butchart Gardens on the Saanich Peninsula.

As the editing of these proceedings was near an end we learned of the death of Helen Sawyer Hogg, a pioneer in the study of globular cluster variable stars. This book is dedicated to her memory.

J.M. Nemec (Seattle, Washington, USA)
J.M. Matthews (Vancouver, British Columbia, Canada)
March 1993

Variable Stars as Distance Indicators

The Oosterhoff Period Effect and the Age of the Galactic Globular Cluster System

Allan Sandage

The Observatories of the Carnegie Institution of Washington
813 Santa Barbara Street, Pasadena, CA 91101, USA

Abstract

The Oosterhoff division of globular clusters into two dichotomous mean period groups is a result of the variation with metallicity of the combined effects of (1) a mean increase in period with decreasing metallicity, and (2) the change of globular cluster horizontal branch (HB) morphology from the M3 to the M13 HB type within the instability strip in the metallicity range of [Fe/H] between −1.7 and −1.9. A new representation of the Oosterhoff period effect showing this property is made from the individual cluster data in Figure 1. The relation between period and metallicity for cluster and for field RR Lyraes at the blue fundamental edge of the instability strip in the HR diagram as read from this figure is

$$\log P_{ab} = -0.122(\pm 0.02)([\text{Fe/H}]) - 0.500(\pm 0.01)$$

using the metallicity scale of Butler.

The high slope coefficient is consistent with the extant models of the HB when they are read at the varing temperature of the fundamental blue edge given by equation (3) of the text. Most of the current literature treats only the constant temperature condition, which is manifestly incorrect. It is this temperature effect that reconciles the observations and the models.

A new calibration of the absolute magnitudes of RR Lyrae stars as a function of metallicity, combined with new oxygen enhanced isochrones for globular clusters (Bergbusch & VandenBerg 1992) reduces the age of the Galactic globular cluster system to 14.1 ± 0.3 Gyr (internal error). The resulting lower age of the universe which, when combined with a Hubble constant near 50 km s^{-1} Mpc^{-1} determined from type I supernovae, shows that the cosmological expansion has been decelerated by an amount consistent with the closure density, permitting $\Omega \sim 1$ now from the timing test.

1. Introduction

The Oosterhoff (1939, 1944) separation of globular clusters into two RR Lyrae period groups and the subsequent discovery that this is also a separation by metallicity (Arp 1955, Kinman 1959) is generally believed to hold a deep clue to the absolute magnitudes of RR Lyrae stars as a function of metallicity. From that, the age of the galactic globular clusters follows, leading to the age of the universe, T_0, which, by the timing test comparing T_0 with H_0 gives the deceleration parameter, q_0 from which the eschotology of the cosmos follows.

Recent advances in an understanding of the Oosterhoff effect suggest that such a recalibration of $M(\mathrm{RR}) = f([\mathrm{Fe/H}])$ can now be made. The method and the result are summarized in this report. The data, precepts, and analysis are set out elsewhere (Sandage 1993, hereafter S93).

In outline, the RR Lyrae luminosity calibration is made through the pulsation equation that relates period, temperature, and mass with luminosity. The first three parameters have been determined as a function of [Fe/H] along the fundamental blue edge (hereafter FBE) of the instability strip. The absolute magnitude calibration of the RR Lyrae stars at this edge is then calculated from these parameters.

2. The Period-Metallicity Relation at the Fundamental Blue Edge

Figure 1 shows the distribution of observed periods of type RR*ab* variables in globular clusters of various metallicities. The period data are taken from the Third Catalogue of Variable Stars in Globular Clusters (Sawyer Hogg 1973), updated with more recent data from the literature. The metallicity scale is that of Zinn & West (1984).

The envelope line defining the shortest period for a given [Fe/H] has been determined from the data in Figure 1, corrected to a constant sample size for the different clusters and then combined with similar data for the field RR Lyrae stars in the sample of Blanco (1992). The field star data are based on the metallicity scale of Butler (1975) which is more metal rich by 0.21 dex at a given period than the scale of Zinn & West.

The equation of the adopted line for the FBE, corrected for sample richness and reduced to the metallicity scale of Butler, is

$$\log P = -0.122(\pm 0.02)([\mathrm{Fe/H}]) - 0.500(\pm 0.01). \tag{1}$$

This is the envelope line drawn in Figure 1 near the shortest period for a given metallicity but on the metallicity scale of Zinn & West.

Figure 1 also shows that the Oosterhoff dichotomy (*i.e.*, into two period groups) is a masquerade of the near discontinuity between [Fe/H] of -1.7 and -1.9. This discontinuity throws the periods for variables that exist in clusters such as M13, M2, M22, NGC 4833, NGC 6809, etc. into the long period group, making an apparent dichotomy. The behaviour is due to the extreme blueness of the HB of many clusters in this metallicity range. Their HB are so blue that they generally miss the instability strip. The variables that do occur have longer periods than they would on the ZAHB because of elevated evolutionary tracks, as was first evident in M13 (Sandage 1970), generalized by Lee, Demarque, and Zinn (1990, hereafter LDZ). This explanation of the Oosterhoff dichotomy was first made by Renzini (1983), Castellani (1983), Lee

(1990, 1993), and we suspect others.

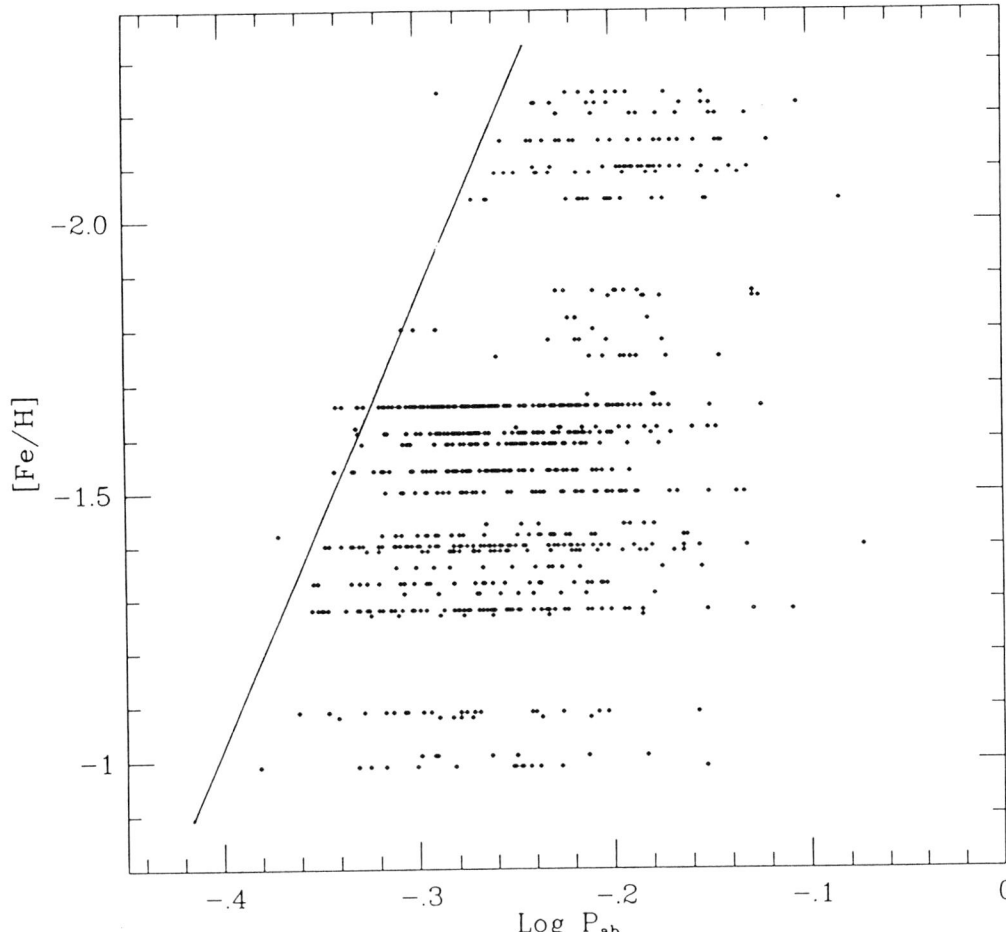

Figure 1. Distribution of observed RR Lyrae periods with metallicity (scale of Zinn & West) in Galactic globular clusters. The envelope line, reduced to the metallicity scale of Butler (1975), has the slope of equation (1) of the text.

3. Temperature of the Fundamental Blue Edge as f([Fe/H])

Figure 2 shows the variation of the highest temperature at a given metallicity based on the field star sample from Blanco. His new $E(B-V)$ values, applied to $(B-V)_{\rm mag}$ values from the general catalog of RR Lyrae star parameters by Nikolov, Buchantsova & Frolov (1984), give unreddened colors. The temperatures are calculated from

$$\log T_e = (0.049[{\rm Fe/H}] - 0.288)(B-V)^0_{\rm mag} - 0.002[{\rm Fe/H}] + 3.931, \qquad (2)$$

based on the atmospheric models of Bell (Sandage 1990, Figure 2) but made cooler by $\Delta \log T_e = 0.010$ based on models by Kurucz as used by Carney, Storm & Jones (1992, their equation 13).

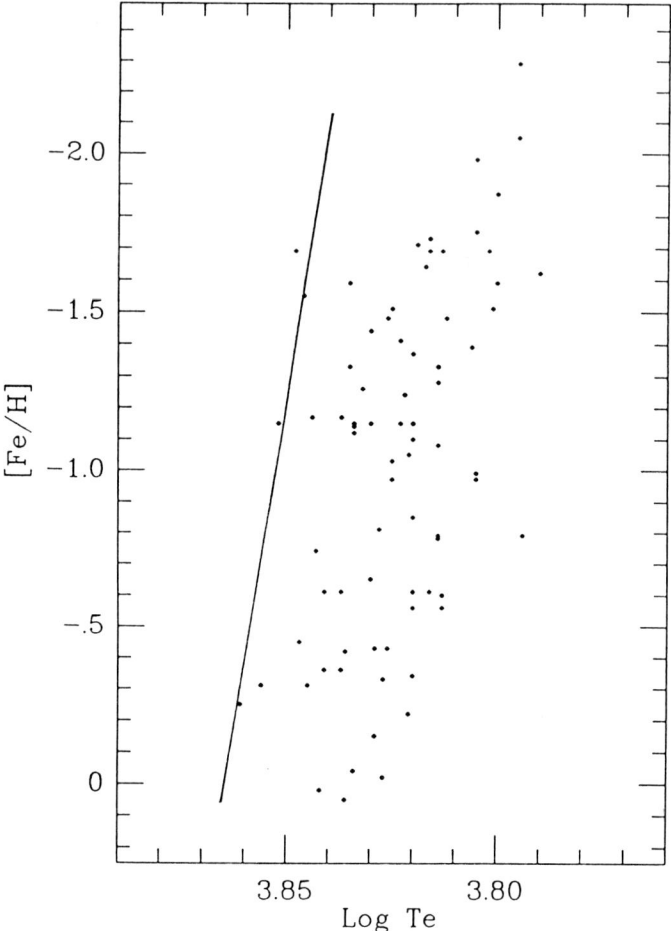

Figure 2. Adopted temperature-metallicity correlation for the fundamental blue edge of the RR Lyrae instability strip. Individual points are from the field star sample of Blanco. The blue envelope is equation (3) of the text.

It has been shown (Preston 1961; Sandage 1990) that the proper color to use for RR Lyrae stars is $(B-V)_{\mathrm{mag}}$, i.e. integrated over the color curve in magnitudes. It is also shown elsewhere (S93) that the criticism of this type of color by Carney, Storm & Jones (1992) is not established.

The equation of the line of bluest temperature in Figure 2, taken to define the fundamental blue edge, is

$$\log T_e = 0.012(\pm 0.01)[\text{Fe/H}] + 3.865(\pm 0.01). \tag{3}$$

4. The Mass-Metallicity Relation at the Fundamental Blue Edge

It can be shown from the new calculation of masses for double mode RR Lyrae stars by Kovacs *et al.* (1992, Table 2) using the new Livermore opacities (Iglesias & Rogers 1991; Rogers & Iglesias 1992) that the slope of the mass-metallicity relation for the FBE (double mode variables are on the FBE) is $d\log\mathcal{M}/d[\text{Fe/H}] = -0.058$. This also agrees with Dorman's (1992) new oxygen enhanced HB models for the age zero HB at the blue fundamental edge.

The sense of the correlation of mass with metallicity is the same as for the standard cononical HB models of Sweigart & Gross (1976) and of Sweigart, Renzini & Tornambe (1987), but the slope is smaller by about a factor of two. Calculations by Cox (1992), also using the new opacities, give similar results.

The zero point of the mass has been taken from the Dorman models. This zero point has the appropriate property that the masses so determined are consistent with the known distribution of white dwarf masses (but note that these are well known only for population I white dwarfs, Weidemann 1990).

The adopted mass-metallicity relation is

$$\log \mathcal{M} = -0.058[\text{Fe/H}] - 0.288(\pm 0.01). \tag{4}$$

From the way we have interpolated in the Dorman tables, equation (4) refers to the fundamental blue edge at the ZAHB.

5. Exploring the Parameter Space of the Pulsation Equation

The pulsation equation of van Albada & Baker (1973) is

$$\log L = 1.19 \log P + 0.81 \log \mathcal{M} + 4.143 \log T_e - 13.687. \tag{5}$$

with similar equations by Iben (1971) and by Cox (1987). Clearly, in investigations of the slope dependences of L, $\mathcal{M}$, P, and T_e on metallicity, the relation $d\log L/d[\text{Fe/H}] = 1.19\, d\log P/d[\text{Fe/H}] + 0.81\, d\log \mathcal{M}/d[\text{Fe/H}] + 4.143\, d\log T_e/d[\text{Fe/H}]$ must hold. Fixing the first term on the right by equation (1) permits an exploration of the permitted variations of the remaining slope coefficients of L, $\mathcal{M}$, and temperature with [Fe/H].

Explorations in the relevant parameter space relating P, L, $\mathcal{M}$, and T_e using the slope coefficients in equations (1) - (4) shows (S93) that the best value for $d\log L/d[\text{Fe/H}]$ is -0.143 ± 0.05, or $dM_{\text{bol}}/d[\text{Fe/H}] = 0.36 \pm 0.12$. It is certain that $L(\text{RR})$ must depend on metallicity to explain the Oosterhoff-Arp period variation with metallicity for cluster and for field RR Lyraes (Preston 1959). The introduction of the variation of temperature in the explanation of the Oosterhoff-Arp-Preston slope in equation (1) reconciles now the observations and the models. This was first understood for the RRc variables by Simon & Clement (1993) via a different route.

6. The RR Lyrae Absolute Magnitude Calibration

With the slope coefficients and the absolute zero points of the P, T_e, and $\mathcal{M}$ dependences on metallicity fixed by equations (1)-(4), equation (5) gives

$$M_{\text{bol}} = 0.36(\pm 0.12)[\text{Fe/H}] + 1.00(\pm 0.13). \tag{6}$$

The error on the slope and the error on the zero point are coupled, and only one of the values must be used to assess the overall accuracy of the calibration.

The result in M_V is shown in Figure 3 where equation (6) is compared with the summary given elsewhere (Sandage & Cacciari 1990) of the results of the Baade-Wesselink method obtained by (1) Liu & Janes (1990) whose pioneering analysis of the observations first established beyond doubt that L and [Fe/H] are correlated, (2) by Cacciari et al. (1988), (3) by Jones, Carney & Latham (1988), and by others cited in the aforementioned summary.

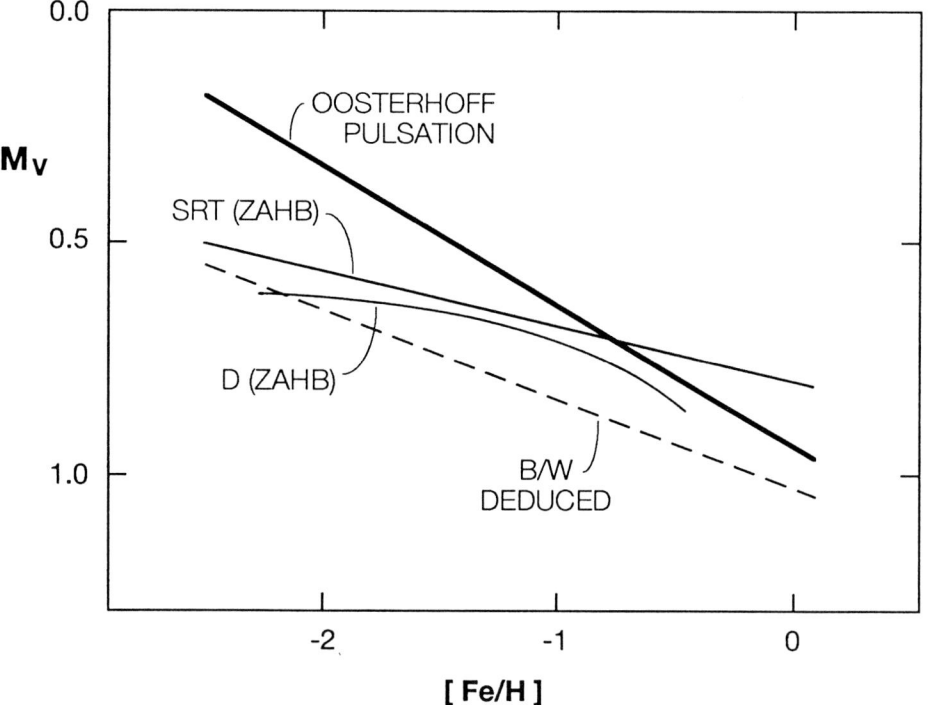

Figure 3. Adopted calibration of the RR Lyraes derived here (labeled "pulsation") compared with other calibrations.

The predictions of the zero age horizontal branch models of Sweigart, Renzini &

Tornambe (1987) and of Dorman (1992) are also shown in Figure 3. The predictions from the evolved HB models of Lee, Demarque & Zinn (1990) are similar to the SRT and D curves at high metallicities but are brighter by ~0.2 mag at lower metallicities, agreeing well with equation (6).

Our final calibration is

$$M_V = 0.30[\text{Fe/H}] + 0.94, \qquad (7)$$

based on the bolometric correction of $BC = 0.06[\text{Fe/H}] + 0.06$, used to convert M_{bol} to M_V for the SRT and D curves in Fig. 3. By the method of derivation, this equation applies to the FBE. Because the HB in V is nearly flat through the instability strip, it should apply closely to most RR Lyrae stars in globular clusters.

Support for this increase in the absolute luminosity of RR Lyrae stars by ~0.2 mag is available from (1) the difference in distance modulus by this amount in galaxies where both classical cepheids and RR Lyrae stars are observed (Saha *et al.* 1992 for a summary), and (2) the difference in the globular cluster luminosity function between the Galaxy and M31 (Racine & Harris 1992) based on the older RR Lyrae absolute magnitude calibration (*i.e.*, B/W in Figure 3). The zero point of the cluster luminosity function in the Galaxy must be made ~0.2 mag brighter than in M31 using the cepheid distance to M31.

7. Age of the Galactic Globular Cluster System

New cluster ages can be calculated (S93) once this RR Lyrae calibration is used to determine the main sequence turnoff luminosity from the observed magnitude difference between the HB and the turnoff. Interpolation in the recent isochrone tables for oxygen enhanced main sequence termination models by Bergbusch & VandenBerg (1992) gives the age dating equation

$$\log T(\text{Gyr}) = 0.39 M_{\text{bol}}(\text{TO}) - 0.10[\text{Fe/H}] + 8.441. \qquad (8)$$

The metallicity dependence here is smaller than for the non-oxygen enhanced models used by Sandage & Caccari (1990, their equations 13-15) based on Ciardullo & Demarque 1977) and VandenBerg (1983). This is because the oxygen abundance is 250 times higher than for Fe. An increased oxygen abundance increases the effective "metal" abundance.

The ages calculated from equation (8) are shown in Figure 4 for 24 Galactic globular clusters. The mean age is 14.1 Gyr, with a formal error of ±0.3. The one sigma range is ±1.4 Gyr. The actual range of error for the age of any given cluster is of course larger. Note that an error of only 0.1 mag in the combined steps needed to determine $M_{\text{bol}}(\text{TO})$ gives a 10% error in the age. Most of the spread in the points in Figure 4 is, then, due to this random uncertainty in the absolute magnitude of the main sequence termination point, although some of the spread has recently been documented as real (*e.g.* Stetson *et al.* 1989).

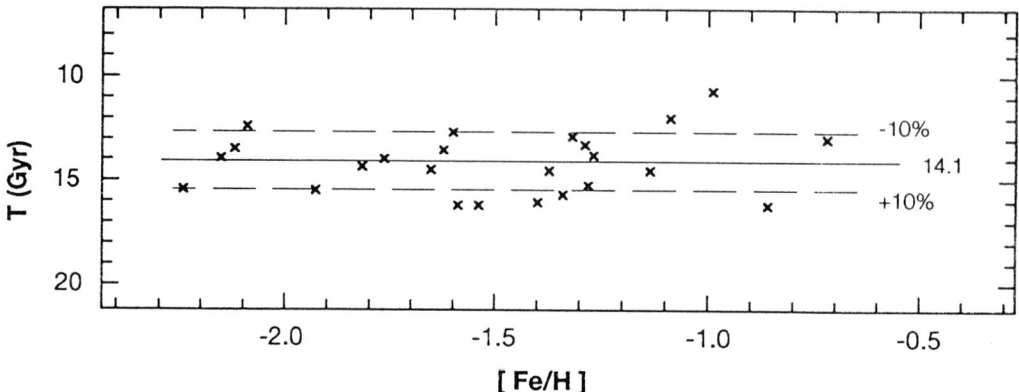

Figure 4. Age of 24 Galactic globular clusters based on the RR Lyrae calibration derived here and on the oxygen enhanced main sequence turnoff models of Bergbusch & VandenBerg (1992).

Nevertheless, it is incorrect to take the lower envelope of Figure 4 to define the oldest globular clusters in the Galaxy. The statistical error of that envelope is the same as the ridge line. In view of the uncertainty in each of the points by at least 10%, the ridge line gives the correct statistical estimate for the age of the globular cluster system by this method.

8. The Age of the Universe and the Value of the Cosmological Omega

The age of the universe is the age of the globular cluster system plus the time for galaxy formation after the initial singularity. No primeval galaxies have yet been found but many theories require the epoch of galaxy formation to be at redshifts of at least $z = 10$. The look-back time here is less than 5% of the present age of the universe.

An observational way to estimate the galaxy gestation time is to note that the look-back time to the quasar with the presently known largest redshift near $z = 5$ (Schneider, Schmidt, & Gunn 1991) is 93% of the age of the universe in a $q_0 = 1/2$ model. (A table of look-back times for various q_0 values is given elsewhere, Sandage 1961b). If this were to be the time of formation of galaxies, then this would be the same as the age of the globular clusters on the almost certain proposition that globular clusters are the first objects to form in the Galaxy. The age of the universe would then be $14.1 + 0.96 = 15$ Gyr with an uncertainty of perhaps ± 2 Gyr.

Combining such age estimates with the inverse Hubble constant gives the estimate of the deceleration, and hence a direct determination of Ω shown in Figure 5. The values of H_0 here are based on the calibration of the absolute magnitude of the prototype supernova 1937C via Cepheids in IC 4182 (Sandage et al. 1993). Each quoted possibility is within the limits posed by the several precepts of the experiment.

The curve gives the $H_0 T_0$ time-scale ratios for various Ω values ($=2q_0$ if $\Omega = 0$) tabulated elsewhere (Sandage 1961a, Table 8). From the ratio of the times as $H_0 T_0 = 0.66 \pm 0.15$ for the cross and 0.69 ± 0.17 for the triangle, calculated from the adopted T_0 and H_0^{-1} values, the formal solution is $\Omega = 1(+2, -0.7)$. The ranges are shown as arrows along the abscissa.

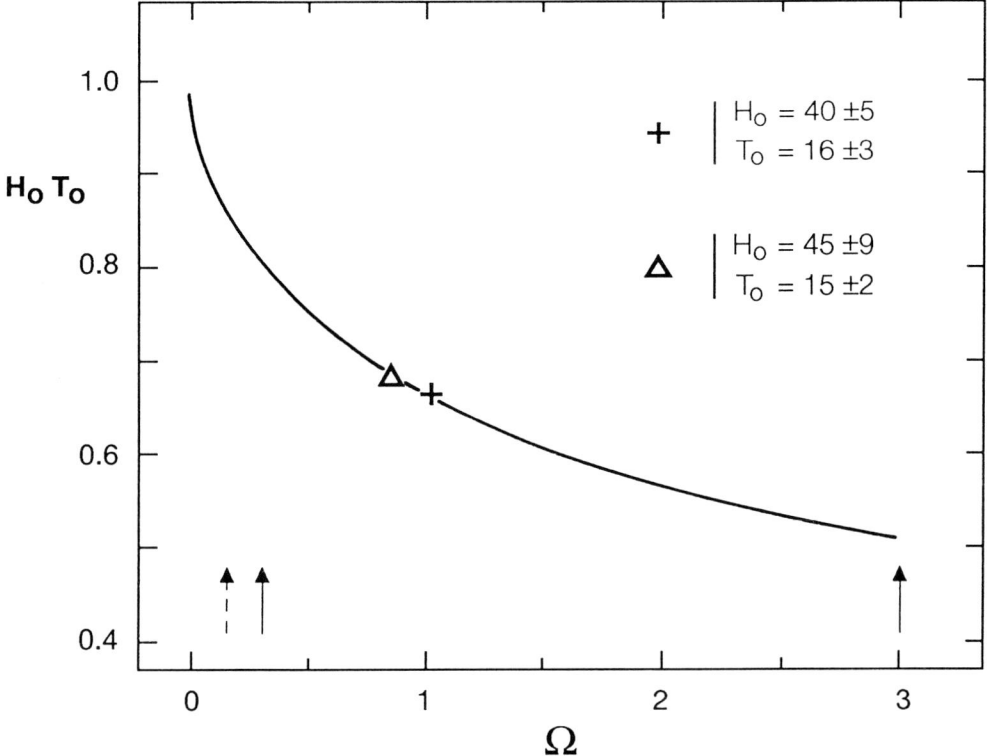

Figure 5. The ratio of the age of the universe to the the inverse Hubble constant as a function of Ω. The two inferred "observed" points are shown as cross and triangle.

The conclusion is that the new calibration of the absolute magnitude of RR Lyrae stars at ~ 0.2 mag brighter than previous calibrations at [Fe/H] ~ -1.8, together with the new oxygen enhanced main-sequence-termination isochrones, reduce the ages of the Galactic globular clusters to 14 Gyr. The new value of H_0 via type I supernovae gives a Hubble time near 24 Gyr. If true, there has been a deceleration of the expansion sufficient to accomodate a closure density, giving $\Omega \sim 1$ directly from the timing argument.

It is a pleasure to acknowledge the hospitality of G. A. Tammann and the astronomers at the Astronomisches Institut der Universität Basel where this report was written during an extended visit in the summer of 1992 after the Victoria meeting. Thanks are also extended to K. Cardy in Pasadena and D. Cerrito in Basel for producing the diagrams, and to James Nemec for assistance in preparing the camera-ready manuscript.

References:

Arp, H. C. 1955, AJ, 60, 317
Bergbusch, P. A., & VandenBerg, D. A. 1992, ApJS, 81, 163
Blanco, V. 1992, AJ, 104, 734
Butler, D. 1975, ApJ, 200, 68
Castellani, V. 1983, Mem. Soc. Astr. Ital., 54, 141
Carney, B. W., Storm, J., & Jones, R. V. 1992, ApJ, 386, 663
Ciardullo, R. B., & Demarque, P. 1977, Trans. Yale Obs., Vol. 33
Cacciari, C., Clementini, G., Prevot, L., & Buser, R. 1988, A&A, 209, 141
Cox, A. N. 1987 in IAU Colloq. 95, Second Conference on Faint Blue Stars, ed. A.G.D. Philip, D.S. Hayes, & J.W. Liebert (Schenectady: Davis), p 161
—— 1992, private communication
Dorman, B. 1992, ApJS, 81, 221
Iben, I. 1971, PASP, 83, 697
Iglesias, C. A. & Rogers, F. J. 1991, ApJL, 371, 73
Jones, R. V., Carney, B. W., & Latham, D. W. 1988, ApJ, 332, 206
Kinman, T. D. 1959, MNRAS, 119, 538
Kovacs, G., Buchler, J. R., Marom, A., Iglesias, C. A., & Rogers, F. J., 1992, A&A, 259, L46
Lee, Y.-W. 1990, PhD thesis, Yale Univ.
—— 1993, these proceedings, pp. 285-294.
Lee, Y.-W., Demarque, P., & Zinn, R. 1990, ApJ, 350, 155
Liu, T., & Janes, K. A. 1990, ApJ, 354, 273
Nikolov, N., Buchantsova, N. & Frolov, M. 1984, Catalog of RR Lyrae Photometric Properties; Astron. Council, USSR Acad Sci., Sofia.
Oosterhoff, P. Th. 1939, Observatory, 62, 104
—— 1944, Bull. Astr. Inst. Netherlands, 10, 55
Preston. G. W. 1959, ApJ, 130, 507
—— 1961, ApJ, 133, 29
Racine, R., & Harris, W. E. 1992, AJ, 104, 1068
Renzini, A. 1983, Mem. Soc. Astr. Ital., 54, 335
Rogers, F. J., & Iglesias, C. A. 1992, ApJS, 79, 507
Saha, A., Freedman, W. L., Hoessel, J. G., & Mossman, A. E. 1992, AJ., 104, 1072
Sandage, A. 1961a, ApJ, 333, 355
—— 1961b, ApJ, 334, 916

—— 1990, ApJ, 350, 603
—— 1993, AJ, submitted (S93)
Sandage, A. & Cacciari, C. 1990, ApJ, 350, 645
Sandage, A., Saha, A., Tammann, G. A., Panagia, N., & Macchetto, D. 1993, ApJL, in press
Sawyer Hogg, H. 1939, David Dunlop Obs. Pub., 1, No. 4
—— 1955, ibid, 2, No. 4
—— 1973, ibid, 3, No. 6
Schneider, D. P., Schmidt, M., & Gunn, J. E. 1991, AJ, 102, 837
Simon, N. R. & Clement, C. M. 1993, these proceedings, pp. 304-312.
Stetson, P., VandenBerg, D. A., Bolte, M., Hesser, J. E., & Smith, G. H. 1989, AJ, 97, 1360
Sweigart, A. V., & Gross, P. G., 1976, ApJS, 32, 367
Sweigart, A., Renzini, A., & Tornambe, A, 1987, ApJ, 312, 762
van Albada, T. S., & Baker, N. 1973, ApJ, 185, 477
VandenBerg, D. A. 1983, ApJS, 51, 29
Weidemann, V. 1990, ARA&A, 28, 103
Zinn, R. & West, M. J. 1984, ApJS, 55, 45

DISCUSSION

C.CACCIARI: RR Lyrae masses of about 0.8 solar masses, as recently found using the Lawrence-Livermore opacities, are significantly larger than the values found by stellar evolution for HB stars, and are a problem for the interpretation of HB morphologies. Could you please comment on that?

A.SANDAGE: You refer to the fact that the mass at the tip of the RGB is also ~0.8 solar masses, leaving no room for mass loss to populate the HB along its length? Due to the apparently high sensitivity of the mass to [Fe/H] I could believe that the exact values of calculated HB masses are uncertain, but that the trend in [Fe/H] might be believed – but I agree it is a real problem if the calculated mass of 0.8 solar masses is set in stone. The Dorman model masses used in the final version of his paper are lower than I had in the verbal presentation, and should be fine for this problem.

Y.-W. LEE: The crucial point of the Oosterhoff dichotomy is why two clusters of more or less the same [Fe/H] (e.g., M2 and M3) have such a large difference in $<P_{ab}>$, and hence produce the gap in $<P_{ab}>$. That this can be explained by the blueward shift of the HB morphology with decreasing [Fe/H] and accompanying evolutionary effect has been announced by me two years ago in Bologna (see also Lee in this volume). Renzini and Castellani noted that the non-monotonic behavior of the HB morphology with decreasing [Fe/H] has something to do with the Oosterhoff effect, but they did not mention specifically the importance of the redward evolution in the way explained by Lee and LDZ. Hence they do not deserve the credit on that point, and also they did not explain the origin of the Oosterhoff gap.

A. SANDAGE: Thank you. It is clear that your evolutionary ideas for M13 type, extreme blue HB clusters have such high $\log < P_{ab} >$ values even in the range [Fe/H] between -1.5 and -1.8. This is the prime reason for the dichotomy into the 0.65 OoII bin and why that bin is so sharp. But there is no LDZ evolution in the OoI bin where both red and blue parts of the HB are on the ZAHB. The density distribution through the instability strip must also be taken into account in any detailed discussion of why the $\log < P_{ab} > = $ f([Fe/H]) plot shows two nearly vertical distributions (at $< P_{ab} >$ near 0.55 and 0.65 day) rather than the continuous $\log P_{ab} = $ f([Fe/H]) shown by the blue fundamental edge plots both for the cluster data and for the field RR Lyrae stars (Blanco's new $E(B-V)$ and ΔS data).

B.DORMAN: (question concerning masses of metal-poor HB stars)

A. SANDAGE: For [Fe/H] $= -2.26$ one obtains 0.80 solar masses at 15 Gyr, and 0.84 at 12 Gyr. Masses of 0.80 solar masses are thus possible, but only if there is little mass loss. The inference of no mass loss is at least "uncomfortable". However, for example, the M15 HB morphology seems to exclude the scaled solar models which give this mass in the instability strip. In this case you must have masses of between 0.70 and 0.75 solar masses. So there is still a "mass problem" if Cox's mass value continues to stand. Note that the above-quoted masses are completely independent of CNO abundance.

Of course there must be mass loss from the top of the AGB to the ZAHB to populate the HB along its length. I believe one cannot believe the absolute values of Mass(RR) to within say $\Delta \log \mathcal{M} \sim 0.1$, from theoretical HB models (various calculations based on different opacities [Kovacs, Cox, Livermore vs. Los Alamos, etc]), but one can believe (I think) required *trends* of mass = f([Fe/H]).

RR Lyraes in the Magellanic Clouds

Alistair R. Walker

Cerro Tololo Inter-American Observatory, Casilla 603, La Serena, Chile

Abstract

Recent work on the Magellanic Cloud RR Lyrae stars is reviewed. The absolute magnitudes of LMC RR Lyraes, when calibrated from a distance modulus of 18.5 mag, disagrees with the Galactic calibration. The revised distance scale makes distances greater, and ages younger, within our galaxy. Field star studies show that the "halo" population of the LMC is very similar to that of our own Galaxy outside of the solar circle. This result supports a Searle and Zinn model of galaxy formation.

1. Introduction

RR Lyraes in the Magellanic Clouds (MC) are important for a number of reasons. Perhaps the two most important are as follows: RR Lyraes are important standard candles, and via their calibration of the luminosity of the main-sequence turnoff of Galactic globular clusters, they allow ages to be determined for the oldest recognizable component of our Galaxy. The MC, in particular the LMC, is the best place for comparing the RR Lyrae luminosity calibration with those for other standard candles. Secondly, the field RR Lyraes map the distribution of the oldest MC population, and together with the oldest LMC clusters we can compare their properties with the halo stars and clusters of our own galaxy. This can give important clues as to how the Galaxy and the MC originally formed, and together with abundances and kinematics, can map out the chemical and dynamic evolution in the first few Gyr after formation.

In this review I concentrate on progress made since my review of the same subject at IAU Symposium No. 148 (Walker 1991).

2. RR Lyrae absolute magnitudes

RR Lyraes are known in seven LMC clusters (NGC 1466, 1786, 1835, 1841, 2210, 2257, and Reticulum). Walker (1992a) summarizes his CCD photometry for 182 variables in these clusters, from which mean magnitudes, metal abundances, and reddenings are derived. These results are summarized in Table 1, together with data from various studies of the field RR Lyraes. We tabulate (1) location of the field, (2) distance in degrees of the LMC clusters from the LMC center, from Suntzeff et al. (1992), (3) metallicity, except where noted, determined from the RR Lyraes themselves, (4) reddening, generally a mean of several methods, (5) number of RR Lyraes, (6) intensity mean V magnitude, (7) and (8) refer to the references and notes attached to the table. For NGC 1466, NGC 1841, and Reticulum color-magnitude

Table 1.

Summary of the Photometry of Magellanic Cloud RR Lyrae Variables

Location (1)	Radius (2)	[Fe/H] (3)	E(B-V) (4)	N(RR) (5)	<V> (6)	Ref (7)	Notes (8)
LMC							
NGC 1466 cluster	8.4	-1.8	0.09	38	19.33	W5	
NGC 1786 cluster	2.5	-2.3	0.07	9	19.27	WM2	1
NGC 1835 cluster	1.4	-1.8	0.13	33	19.37	W6	
NGC 1841 cluster	14.9	-2.2	0.18	22	19.31	W3	
NGC 2210 cluster	4.4	-1.9	0.06	9	19.12	W1	
NGC 2210 cluster	4.4		0.08	10	19.19	HN	2,3
NGC 2257 cluster	8.4	-1.8	0.04	39	19.03	W2	
Reticulum cluster	11.4	-1.7	0.03	32	19.07	W4	
NGC 1466 field				3	19.5	K	4
NGC 1783 field		-1.6:	0.06:	73	19.2	G2	5
NGC 1841 field				2	19.35	K	4
NGC 2210 field		-1.8	0.08	52	19.26	HN	
NGC 2257 field			0.04	22	19.30	HNU	
NGC 2257 field		-1.8	0.04	9	19.20	W2	
Reticulum field				1	19.1	K	4
SMC							
NGC 121 cluster		-1.4:	0.04	4	19.59	WM1	6
NGC 121 field				75	19.6	G1	4
NGC 361 field		-1.5:	0.06	42	19.5	S	4,7

References: G1 = Graham (1975), G2 = Graham (1977), HN = Hazen and Nemec (1992), HNU = Hesser, Nemec and Ugarte (1976), K = Kinman et al. 1991, S = Smith et al. 1992, W1 = Walker (1985), W2 = Walker (1989), W3 = Walker (1990), W4 = Walker (1992b), W5 = Walker (1992c), W6 = Walker (1992d), WM1 = Walker and Mack (1988a), WM2 = Walker and Mack (1988b).

Notes: (1) Suntzeff et al. (1992) tabulate [Fe/H] = -1.87 from spectroscopy of RGB stars. For all other clusters the RR Lyrae-derived [Fe/H] values agree with those from spectroscopy of RGB stars to within 0.1 dex.
(2) V32 which has <B>=20.32 has not been included in the mean.
(3) Another 33 cluster variables have been discovered (Nemec and Hazen, in prep.)
(4) <V> derived from <B>, not directly measured.
(5) G77 V magnitude scale found to be correct by Blanco and Blanco (1986), but B requires a correction of -0.2 mag.
(6) [Fe/H] from Stryker, Da Costa and Mould (1985).
(7) Of the 42 stars, 22 have periods determined.

diagrams extending to the main-sequence turnoff were prepared by combining several of the frames used for measuring the variables. All the clusters are metal poor, with mean $[Fe/H] = -1.9$. If a distance modulus of 18.5 mag for the LMC is assumed, then the RR Lyraes have mean $< M_V > = 0.45$. Since the old clusters which contain the RR Lyraes are fairly evenly distributed across the face of the LMC, correcting individual distances for some assumed LMC geometry has very little effect on the mean. Given that the LMC modulus quoted above is probably reliable to within 0.15 mag, the relatively bright absolute magnitude found for the RR Lyraes is not consistent with the galactic RR Lyrae calibration (from statistical parallaxes and Baade-Wesselink analyses) by some 0.3 mag., especially now that the slope of the RR Lyrae magnitude-metallicity relation is generally accepted to be in the range 0.15-0.20. Since the galactic RR Lyraes, via the globular clusters, are used to calibrate both distances and ages within our galaxy, the new LMC calibration will increase distances (eg R_0 of 8.5 kpc rather than 8.0 kpc) and reduce ages (the oldest globular clusters are now no more than 15 Gyr old, fitting VandenBerg and Bell (1985) isochrones to the main-sequence turnoff) within our Galaxy. Some support for this brighter calibration is provided by the LMC Mira variables, for which the zeropoint can be calibrated in the same way as for the RR Lyraes. These in turn calibrate the metal rich galactic globular clusters which happen to contain Mira variables. This calibration agrees with the new RR Lyrae calibration if the slope of the magnitude-metallicity relation is 0.17.

3. Field RR Lyraes in the LMC

Studies of field RR Lyraes in the LMC are in almost all cases based upon photographic surveys. For the more recent work the photometric calibrations are based on CCD sequences thus giving confidence that no gross systematic errors exist in the magnitude scales. In addition, much of the later work is based on CTIO 4m plates rather than 1.5m plates. The plate limit of the 1.5m telescope is such that the earlier surveys (eg Graham 1975, 1977) are seriously incomplete for RRc stars.

Hazen and Nemec (1992) have found 52 new field RR Lyraes near the LMC cluster NGC 2210. The mean period $< P_{ab} > = 0.576 \pm 0.057$ day lies between that of Oosterhof groups I and II in our galaxy, and is also similar to that found for LMC field RR Lyraes near the LMC clusters NGC 1783 and 2257. The mean metallicity is about $[Fe/H] = -1.8$, within the errors similar to the mean metallicity found for the old clusters. There is some evidence for a gradient in abundance between the (inner) NGC 1783 field, via the NGC 2210 field to the outer NGC 2257 field, based on mean periods and period-amplitude plots. Given that the NGC 1783 field results (Graham 1977) are seriously incomplete for low amplitude stars, this result must be treated with caution.

Kinman et al. (1991) found four new type-ab field RR Lyraes in the direction of NGC 1466, and a possible field variable near Reticulum. They then fitted exponential and King models to all available LMC field RRab results, and calculated a central surface density of about 200 stars per square degree. By associating each RR Lyrae

(RRab and RRc) with a halo globular cluster population of absolute visual magnitude -4.74 (Suntzeff, Kinman and Kraft 1991), and assuming a mass-to-light ratio of 1.6, they found that the field halo stars make up about 2% of the mass of the LMC, a figure close to that for our Galaxy. In addition, they found that period-frequency distributions are similar for the LMC field variables and those in the outer halo of the Galaxy. Finally, the surface density ratio of the older long-period variables to RR Lyraes (old-disk to halo) is within a factor of two of that in our own Galaxy at the solar circle. All these results suggest that the efficiency of the first few Gyr of star formation in the Galaxy and the LMC was comparable.

4. New Field Star Surveys

Two new surveys which are discovering many more RR Lyraes are underway. That of N. Reid and W. Freedman uses U.K. Schmidt 6x6 degree plates analysed by COSMOS to search for variable stars in the LMC. Although many RR Lyrae variables are being found, follow-up CCD observations centered on some of the variables has resulted in the discovery of extra candidate RR Lyraes, showing that the COSMOS/UKST survey significantly underestimates the true RR Lyrae density. From their results it appears that earlier surveys also underestimate the density of RR Lyraes, perhaps by as much as a factor of two. A paper on the variables in a field centered on NGC 2210 is near completion.

H. Smith and collaborators are studying a 1x1.3 degree SMC field near NGC 361. Their work uses CTIO 1.5m plates taken by J. Graham, on which many short-period Cepheids and RR Lyraes have been discovered (Smith et al. 1992). The surface density of the RR Lyraes is comparable with that found for the outlying NGC 121 field by Graham (1975), showing that SMC RR Lyraes are not strongly centrally concentrated. In addition, the period-amplitude relation is much the same as found for the stars in the NGC 121 field. Further observations (in B and V) are scheduled using the CTIO Curtis-Schmidt telescope using a CCD as detector.

5. Population studies

Suntzeff et al. (1992) have made a critical comparison between the LMC Population II field stars and clusters, and those in our own Galaxy. They find that the mean cluster metallicity, the absolute magnitude distribution of the clusters, and the relative numbers of RR Lyraes per unit cluster luminosity are very similar to the Galactic globular cluster population outside of the solar circle. In addition, they calculate that the total luminosities in both clusters and field stars scale as the total luminosities of the LMC and Galaxy. They conclude that the evidence strongly supports the Searle and Zinn (1978) idea that the Galactic halo originated from LMC-sized units. Further evidence for the Searle and Zinn scenario comes from ages determined from HB (horizontal branch) type, $(B-R)/(B+V+R)$ where B, V and R are the numbers of blue HB stars, RR Lyraes and red HB stars respectively. Plots as a function of metallicity, with fiducials from HB models, for various subgroups of HB stars show that there is an age gradient within our Galaxy, with the Galactic Bulge population the oldest (Lee 1992). Using the same method, Walker (1992c) finds that in the mean the LMC clusters for which the HB type can be calculated are one Gyr younger than the Galactic globular clusters outside the solar circle.

References:

Blanco, V.M., and Blanco, B.M. 1986, PASP, 98, 1162

Graham, J.A. 1975, PASP, 87, 641

Graham, J.A. 1977, PASP, 89, 425

Hazen, M.L., and Nemec, J.M. 1992, AJ, 104, 111

Hesser, J.E., Nemec, J.M., Ugarte, P. 1976, ApJS, 32, 283

Kinman, T.D., Stryker, L.L., Hesser, J.E., Graham, J.A., Walker, A.R., Hazen, M.L., and Nemec, J.M. 1991, AJ, 103, 1279

Lee, Y.-W. 1992, PASP, in press

Lee, Y.-W., and Demarque, P. 1990, ApJS, 73, 709

Searle, L., and Zinn, R. 1978, ApJ, 225, 357

Smith, H.A., Silbermann, N.A., Baird, S.R. and Graham, J.A. 1192, PASP, in press

Stryker, L.L., Da Costa, G.S., and Mould, J.R. 1985, ApJ, 298, 544

Suntzeff, N.B., Kinman, T.D., and Kraft, R.P. 1991, ApJ, 367, 528

Suntzeff, N.B., Schommer, R.A., Olszewski, E.W., and Walker, A.R. 1992, AJ, in press

VandenBerg, D.A., and Bell, R.A. 1985, ApJS, 58, 561

Walker, A.R. 1985, MNRAS, 212, 343

Walker, A.R. 1989, AJ, 98, 2086

Walker, A.R. 1990, AJ, 100, 1532

Walker, A.R. 1991, in: *The Magellanic Clouds*, IAU Symposium No. 108, ed. R. Haynes and D. Milne, Kluwer, Dordrecht, p307.

Walker, A.R. 1992a, ApJ, 390, L81

Walker, A.R. 1992b, AJ, 103, 1166

Walker, A.R. 1992c, AJ, in press

Walker, A.R. 1992d, in preparation

Walker, A.R., and Mack, P.M., 1988a, AJ, 96, 872

Walker, A.R., and Mack, P.M., 1988b, AJ, 96, 1362

DISCUSSION

Y.-W. LEE: I think there are two reasons why you obtained a small LMC distance modulus from your RR Lyrae analyses, compared to that of 18.5 from the Cepheids: (1). Some LMC clusters have blue HB's, and hence evolutionary effects must be taken into account (0.1 mag.). (2). The slope of the magnitude-metallicty relation which you adopted is slightly small compared to that from my models and Baade-Wesselink measurements (0.19). At $[Fe/H] = -2.0$ this difference alone would create 0.1 mag difference. A combination of (1) and (2) would explain most of the difference you suggested.

A. R. WALKER: With regard to point (1), the Lee and Demarque (1990) HB evolutionary tracks show that HB stars evolving redwards from BHB stars spend only a small fraction of their HB lifetime within the instability strip. All the LMC clusters under consideration have rather more RR Lyraes and RHB stars than expected if these stars were all highly evolved. Thus in the mean, the expected increase in brightness for the RR Lyrae sample under consideration is expected to be small, certainly much less than 0.1 mag. See also the contribution by M. Catelan (this conference). For (2), the LMC clusters containing RR Lyraes have mean metallicity 0.7 dex more metal poor than the Galactic field RR Lyraes used in the statistical parallax analyses. Increasing the slope of the magnitude-metallicity relation by 0.04 is not going to make a significant difference to the discrepancy.

The Infrared Period - 2.2μm Magnitude Relation for RR Lyrae Stars

A.J. Longmore

Royal Observatory, Blackford Hill, Edinburgh, EH9 3HJ, Scotland.

Abstract

The observational and theoretical basis for the $\log P$ vs. K-mag relation is reviewed. The observed gradient in all well observed globular clusters agrees well with the theoretical prediction. An indistinguishable result is found for field RR Lyrae stars whose absolute magnitudes have been determined from infrared Baade-Wesselink analyses. A full reference list is given for the source of these magnitudes. Application of the results to find the distance to globular clusters and the Galactic Centre is discussed.

1. Introduction

At the last pulsation meeting, in Bologna, I concluded my presentation by saying 'If one can even begin to realise the full potential of the contribution of IR photometry to the subject matter of this conference, it will not be possible to summarise it in 20 minutes at the next Pulsation Meeting.' This meeting's organisers clearly remembered these words for two years and have consequently allowed 25 minutes for this review! To meet the time requirement, I will concentrate on the use of K (2.2μm) magnitude observations, with special reference to the log(Period) vs K-mag relation. Although there are several variations on the IR photometric system (see Bessell & Brett 1988 for one excellent comparison) there is very little difference between K filters. Typical systematic corrections between systems are less than 0.02 mag, with negligible colour terms for stars hotter than 4500 K. Therefore all K magnitudes here are taken as published. Another bonus for 2.2μm observations is the reduced effect of dust extinction: $A_K \sim 0.1 A_V$.

Although my first 2.2μm observations of RR Lyrae stars were made in about 1980 the full picture of their usefulness did not emerge until more extensive observations in 1983/84. These were published in 1985 (Longmore *et al.*, the first Baade-Wesselink (BW) analysis using infrared photometry, which with other BW analysis of dwarf Cepheids constituted the subject of J.Fernley's thesis) and 1986 (Longmore, Fernley & Jameson [LFJ], demonstrating the $\log P$ vs K-mag relation). The 1985 BW analysis of VY Ser had a fortuitous timing coincidence. Despite their use of an excellent radial velocity curve, Carney & Latham (1984) were unable to find a solution for the radius of VY Ser with optical photometry alone. Finding a solution by combining optical and IR photometry and using the Carney & Latham velocities immediately demonstrated the 'infrared advantage'. This review traces the development of the $\log P$ vs K-mag relation (section 2), summarizes the latest BW results (section 3) and discusses additional applications of them (section 4).

2. The log(Period) versus K-mag relation

A simple way to think of the $\log P$ vs K-mag relation is as a K-mag, radius relation. Period is then introduced via the period, mean-density relation, which is strongly radius dependent. From the definition of bolometric luminosity $\Delta M_{bol} = 10 \Delta \log T_e - 5 \Delta \log R$ (all symbols have their usual meanings). Empirical (eg. Carney 1983) and model atmosphere (eg. R.Kurucz, 1992 private communication) calculations indicate a tight T_e-$(V-K)$ relation of the form

$$\log T_e = -0.112(M_V - M_K) + 3.934 \qquad (1)$$

Therefore, noting that for RR Lyrae stars $M_{bol} \sim M_V$, it follows that $\Delta M_K \sim 0.10$ $\Delta M_{bol} - 4.5 \Delta \log R$. The $\log P$ vs K-mag relation itself is easily derived from the pulsation equation (van Albada & Baker 1971, Cox 1988):

$$\log P = -0.68 \log(M/M_\odot) - 0.336 M_{bol} - 3.48 \log T_e + 13.09, \qquad (2)$$

where, from the R. Kurucz models

$$M_{bol} = M_K - 7.560 \log T_e + 29.846. \qquad (3)$$

From (1), (2) & (3)

$$\log P = -0.441 M_K + 0.105 M_V - 0.68 \log(M/M_\odot) - 0.635, \qquad (4)$$

$$M_K = -2.27 \log P + 0.24 M_V - 1.54 \log(M/M_\odot). \qquad (5)$$

Equations (4) and (5) show clearly the relatively small scatter likely to be introduced by differences in the visual magnitudes of the stars. The small mass range exhibited by RR Lyrae stars implies ± 0.02 mag effects on M_K within clusters and ± 0.07 mag between clusters from the $\log(M/M_\odot)$ term, although the absolute value of RR Lyrae masses is still uncertain (Cox 1991, Kovacs et al. 1992, Simon & Clement 1993). Fig. 1 in Fernley et al. (1987) illustrates particularly the relevance of the $\log P$ vs K-mag relation to horizontal branch (HB) morphology.

On a sound basis theoretically, the $\log P$ vs K-mag relation is also well established observationally. It was found in all three clusters observed by LFJ, who pointed out that the relation could be used as a distance indicator if 'normalised' to a fixed period, chosen to be 0.5 days ($\log P = -0.3$). Results for eight clusters (M3, M4, M5, M15, M107, NGC 3201, NGC 5466 and ω Cen) were published by Longmore et al. (1990). Distances to these clusters were derived, all eight demonstrating a $\log P$ vs K-mag relation with the same gradients within the observational errors. Figures 1(a) and 1(b) show two of the five best established results, including some unpublished data for M3 and M4. Using these data for M3, Buckley *et al.* (1991) examined the K residuals. Despite the size of the residual being close to that expected simply from the scatter in the mean of approximately six random phase observations, they found a correlation between the residual and the V mag height ΔV above the zero age horizontal branch (ZAHB) as determined by Sandage (1990). The reason for this correlation is still being explored. Proving that the relation is not a function of observer, T.Liu (1992 private communication) has recently observed M5 and M15 in detail. Table 1 lists the gradients found for the five best observed clusters (six or more observations per star, ≥ 25 stars per cluster). The mean K residual is about 0.03 mag for all five clusters.

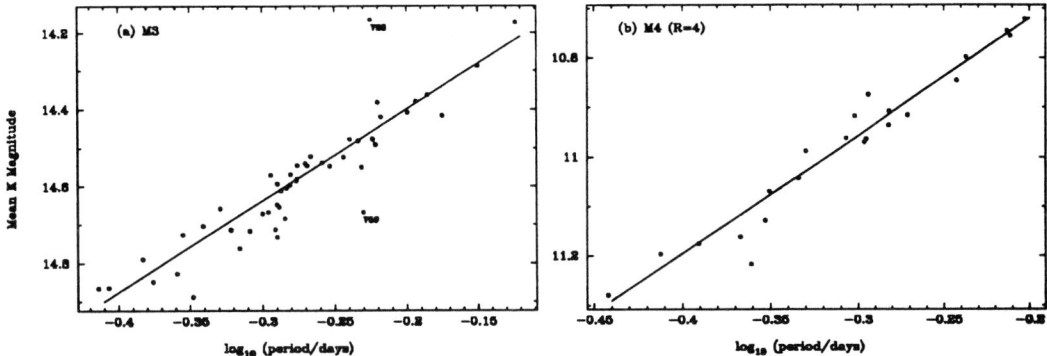

Figs. 1 (left) and **2** (right). The $\log P$ vs K-mag relation for M3 (left) & M4 (right). A fixed slope of -2.38, the mean value from Table 1, is drawn on each plot. M3-V23 is a cluster non-member, and M3-V59 suffers from contamination by a nearby star(s).

Table 1. Gradients of the five best determined $\log P$ vs K-mag relations in globular clusters.

Cluster	Gradient	Reference
ω Cen	-2.28	Longmore et al 1990
M3	-2.34	Buckley et al 1991
M4	-2.33	" "
M5	-2.42	T.Liu, private communication, 1992
M15	-2.46	" "

Fig. 3 (from Buckley *et al.* 1991) shows theoretical evolution away from the ZAHB tracks in the $\log P$ vs K-mag plane for stars of 0.68, 0.76 and $0.80 M_\odot$. For simplicity all pulsations are assumed to be in the fundamental mode. They are derived using the Lee & Demarque (1990) HB models and Equations (2) and (5) and indicate well the small effect that evolution has in this plane.

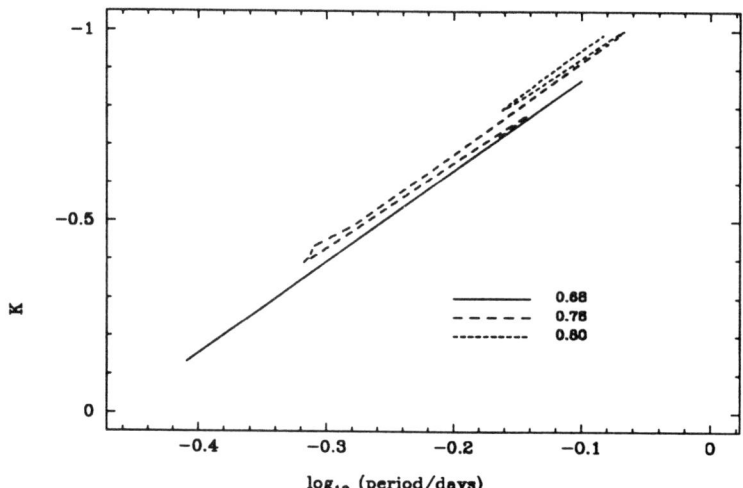

Fig. 3. Theoretical evolutionary tracks in the $\log P$ vs K-mag plane for horizontal branch stars of mass 0.68, 0.76 and 0.80 $M_\odot$.

LFJ also suggested that the nature of the $\log P$ vs K-mag relation should be insensitive to metallicity, a great advantage for a distance indicator. This is substantiated by the tight relationship in ω Cen, despite a range in metallicities of the individual stars $-2.2 \leq$ [Fe/H] ≤ -0.5. No correlation between metallicity and residual was found by Longmore et al. (1990).

3. Baade-Wesselink analysis

Since 1985, three groups have been primarily responsible for developing, in parallel, the infrared version of the BW analysis. Two main variations of methodology have emerged: the infrared-flux method (eg. Fernley et al. 1989, Skillen et al. 1989, Fernley et al. 1990a,b), and the surface-brightness method (Jones et al. 1987a,b, 1988a,b; Jones 1988; Liu & Janes 1989, 1990a). Moffett (1988) reviewed these techniques. Independent non-IR BW analyses have continued (Burki & Meylan 1986, Cacciari et al. 1989). Of particular note is the BW inversion technique introduced by Simon (1987, 1989). Despite the extremely intensive observational requirements of this type of programme (full phase coverage for highly accurate optical and IR light curves and radial velocities good to 1-2 km/s) the technique has now been applied to the much fainter RR Lyrae stars in nearby globular clusters (M4, Liu & Janes 1990b; M5 & M92, Storm et al. 1991, Storm 1991, Storm et al. 1992; M5, Cohen & Matthews 1992a,1992b, J.Cohen 1992 private communication). 29 different field stars and 13 different globular cluster RR Lyraes have been measured, six of the field stars by more than one group. Some of the globular cluster data are not yet fully analyzed.

The following list summarises the literature to date:
- Carney, Jones, Latham, Kurucz & Storm. The Baade-Wesselink Method and the Distance to RR Lyrae Stars, Papers I-VIII. Nine field stars;
- Fernley, Skillen, Longmore, Jameson, Marang, Kilkenny, Lynas-Gray & Stobie. The Absolute Magnitudes of RR Lyrae Stars, Papers I-V. 10 field stars;
- Liu and Janes. The Luminosity Scale of RR Lyrae Stars with the Baade-Wesselink Method, Papers I-III. 13 field stars, four globular cluster stars;
- Cacciari, Clementini & Fernley. Three field stars
- Storm, 1991, three globular cluster stars;
- Storm et al. - relevant observational data papers;
- Cohen & Matthews. Six globular cluster stars.

There are at least three parameters which are not dealt with uniformly between the groups: (a) correction for the projection factor used to convert observed to true radial velocity (values from 1.30 to 1.36 have been used); (b) the zero point of the surface brightness method (Carney's group use one 0.04 mag brighter than that used by Liu & Janes); (c) the optimum phase range to use. (a) is known to depend on the spectroscopic dispersion used (Parsons 1972) but would greatly benefit from a re-analysis specifically for RR Lyrae stars, using modern model atmospheres and advanced simulation techniques. (b) reflects the problem common to both the techniques mentioned - the conversion from colour and flux to temperature. Both methods need to invoke model atmospheres at some stage. A weakness is that these are static models. (c) is a problem because shocks generated near phase 0.95 distort the colours around maximum light. Most of these and other residual uncertainties in the BW method would be resolved if suitable non-static model atmospheres could be constructed.

Now that there is a significant body of data it is useful to collate all the results. This has been done by Jones et al. (1992), Carney et al. (1992, CSJ), Skillen et al. (1992) and Cacciari et al. (1992). CSJ give an especially detailed discussion, also applying the unified results to comparisons of globular cluster distances and ages. They find ages ≥ 14 Gyr, depending on the assumed [O/Fe] ratio. Using the BW M_V results they determine that $M_V = 0.16[\text{Fe/H}] + 1.02$ (see Fig.1 in their paper).

The existence of an independently determined $\log P$ vs K-mag relation can be used to test the relative accuracy of individual BW results. Fig.3 of CSJ shows that all stars (field and globular cluster) lie on a relation $M_K = -2.33 \log P - 0.88$ within the observational errors. Fig. 4 below is an up-dated version of that figure including the most recent results. Four stars (SS Leo, BB Pup, AR Per & V445 Oph) are omitted from the figure and subsequent analyses because of reddening or other uncertainties mentioned in the respective original papers. The linear regression on 25 field stars gives $M_K = -2.56 \log P - 0.98$ with a mean residual of $0^\text{m}.09$ (well within the quoted errors of each of the BW analyses). To test for the metallicity dependence I also carried out a multiple linear regression, finding

$$M_K = -2.38 \log P + 0.04[\text{Fe/H}] - 0.88 \tag{6}$$

The $\log P$ coefficient agrees precisely with the mean gradient in Table 1, while the

[Fe/H] term is in the same sense as, but significantly smaller than, that for M_V vs [Fe/H]. However, the correlation coefficient and the residuals were essentially unchanged by including the extra variable. Slightly modifying the sample could significantly change the coefficients. The reason for this indeterminism can be seen from Fig.5 - there is a strong $\log P$-[Fe/H] relation for this sample of stars. Such a selection effect is difficult to overcome as it is a property of the field RR Lyrae population. The argument needs to be inverted: from Table 1, -2.38 is the correct gradient for $\log P$ vs K so the best estimate for the metallicity term is 0.04. This result can only be considered indicative at present.

Using equation (6) and the five best-determined globular cluster $\log P$ vs K-mag relations, distances can be derived that are almost independent of errors in reddening determinations (Table 2). For explanation of the different M_V values see Longmore et al. (1990).

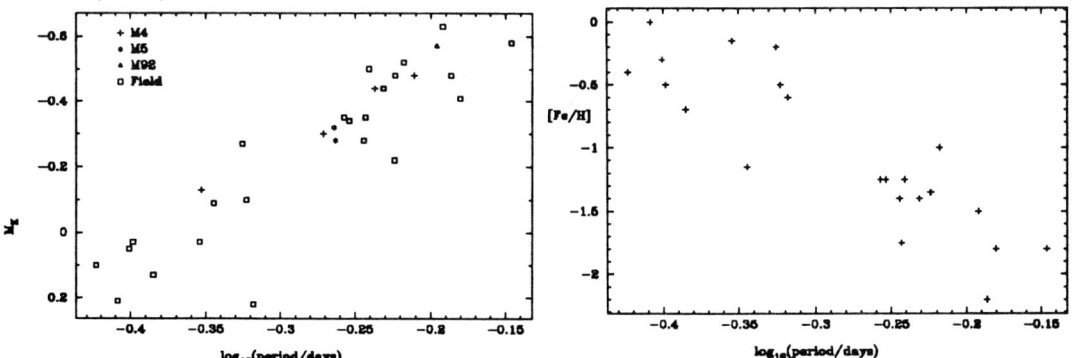

Fig. 4 (left) $\log P$ vs M_K from BW analysis of field and cluster RR Lyrae stars.
Fig. 5 (right) $\log P$ vs [Fe/H] for field RR Lyraes in Fig.4.

Table 2. Distances and HB absolute magnitudes for five globular clusters.

Cluster	$(M-M)_0$ (from field star BW)	$(M-M)_0$ (BW direct)	M_V (HB)	M_V (RR)
M3	14.88		0.72	0.78
M4 *	11.21	11.19	0.73	0.71
M5	14.21	14.19	0.83	0.79
M15	14.95		0.52	0.59
ω Cen	13.54		0.62	0.67
(M92)		(14.49)		

* E(B-V) = 0.37, R = 3.8

4. Other applications of RR Lyrae IR photometry.

4.1 Distance to the Galactic Centre

Fernley et al. (1987) used $H(1.65\mu m)$ photometry, assumed $(H-K) \sim 0.03$ for RR Lyrae stars, and employed the $\log P$ vs K-mag relation to find the distance to the Galactic Centre. They measured the RR Lyrae stars in Plaut's (1973b and references therein, Oort & Plaut 1975) $l=0$ deg, $b=-12$ deg field. They found $R(0) = 8$ Kpc; re-calibrating using eqn.(6) and assuming [Fe/H] = -1.0 (Walker & Terndrup 1991) a revised distance of 7.5 ± 0.6 kpc is found. Walker (1992) deduced $R(0) = 8$-8.5 Kpc from RR Lyrae calibrations based on LMC distance indicators, including Cepheids. This difference implies that only about 15% mag uncertainty remains between RR Lyrae and Cepheid scales, which is almost within the errors of the IR BW analysis alone. B.Carney (private communication) has completed $2.2\mu m$ observations of RR Lyraes in Baade's window. We can look forward to seeing the result of the distance calculation from this data set.

4.2 RR Lyrae Temperatures and Globular Cluster Colour-Magnitude Diagrams

For well-established reasons, $V-K$ colours of well observed RR Lyrae stars give a better RR Lyrae relative temperature determination than $B-V$. Tighter temperature - amplitude relations, for example, confirm this. However, even using the new Kurucz models, $(V-K)$ gives lower mean RR Lyrae temperatures than $B-V$ by ~ 170 K at $\log g =2.7$. Fig.6 (from Dixon 1991) shows a $V,V-K$ diagram of M4. Although marginal, there is an indication that $V-K$ for the variables (triangles) is displaced redward in $V-K$ compared with the HB non-variables. If the effect is real it is still unexplained but could account for the temperature difference noted above.

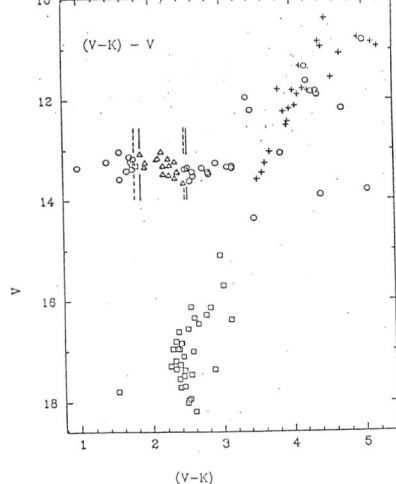

Fig.6. An optical - infrared CMD (V vs $V-K$) of the globular cluster M4, from Dixon (1991). The vertical solid lines indicate the probable range in (V-K) of the RR Lyrae stars while the vertical dashed lines indicate similarly for the non-variable HB stars.

5. Summary

1. This work is observationally highly intensive!
2. Baade-Wesselink M_K residuals from the $\log P$ vs K-mag relation are all within observational errors.
3. The $\log P$ vs K-mag relation is essentially identical between field and cluster RR Lyraestars, in gradient and absolute calibration.
4. The absolute calibration implies globular cluster ages greater than 14 Gyr.
5. RR Lyrae temperatures are not yet satisfactorily determined.
6. Walker's (1992) RR Lyrae calibration using the LMC may be a problem for the BW results, but only at the ~15% level.

References:

Bessell, M.J. & Brett, J.M., 1988, PASP, 100, 1134.
Buckley, D.R.V., Longmore, A.J. & Dixon,R.I.J., 1992, 'New Results on Standard Candles', ed. F.Caputo, in press.
Burki,G. & Meylan,G., 1986, Astron.Astrophys. 156,131.
Cacciari,C., Clementini,G., Fernley,J.A., 1992, ApJ, 396,219.
Cacciari,C., Clementini,G. & Buser,R., 1989, AAp 209, 154.
Carney,B.W., 1983, AJ, 88, 623.
Carney,B.W & Latham,D.W., 1984, Astrophys.J. 278,241 (paper I).
Carney,B.W., Storm,J., Tramell,S.R. & Jones,R.V., 1992, PASP, 104, 44.
Carney,B.W., Storm,J. & Jones,R.V., 1992, Astrophys.J. 386,663 (paper VIII) (CSJ).
Cohen,J.G. & Matthews,K., 1992a, PASP, 104, 1205.
Cohen,J.G. & Matthews,K., 1992b, PASP, in press
Cox,A.N., 1987, The Second Conference on Faint Blue Stars, I.A.U. Colloq. 95 (ed. Davis-Philip,A.G., Dayes,D.S. & Liebert,J.), p.161.
Cox, A.N., 1991, ApJ, 381, L71.
Dixon,R.I.J.D.A.T., 1991, Ph.D.Thesis, University of Edinburgh.
Fernley,J.A, Longmore,A.J., Jameson,R.F., Watson,F. & Wesselink, T., 1987, MNRAS, 226, 927.
Fernley,J.A., Lynas-Gray,A.,Skillen,I.,Jameson,R., Marang,F.,Kilkenny,D. & Longmore, A. 1989, MNRAS, 236, 447 (Paper I).
Fernley,J.A., Skillen,I., Jameson,R. & Longmore,A. 1990a, MNRAS, 242, 68 (Paper III).
Fernley,J.A., Skillen,I., Jameson,R.F., Barnes,T.G, Kilkenny,D. & Hill,G. 1990b, MNRAS, 247, 287 (Paper IV)
Jones,R.V. 1988, ApJ, 326, 305 (Paper IV).
Jones,R.V., Carney,B.W. & Latham,D.W. 1988a, ApJ, 326,312 (Paper V).
Jones, R.V., Carney,B.W. & Latham,D.W. 1988b, ApJ, 332, 206 (Paper VI).
Jones,R.V., Carney,B., Latham,D. & Kurucz,R. 1987a, ApJ, 312, 254 (Paper II).
Jones,R.V., Carney,B., Latham,D. & Kurucz,R. 1987b, ApJ, 314, 604 (Paper III).
Jones,R.V., Carney,B.W, Storm,J. & Latham,D.W., 1992, ApJ, 386, 646 (Paper VII).
Kovacs, G., Buchler, J.R., Marom, A., Iglesias, C.A, Rogers, F.J., 1992, AAp, 259, L46.

Lee,Y.-W. & Demarque,P., 1990, Astrophys.J.Suppl. 73,709.
Liu,T. & Janes,K.A., 1989, ApJS, 69, 593 (Paper I).
Liu,T. & Janes,K.A., 1990a, ApJ, 354, 273 (Paper II).
Liu,T. & Janes,K.A., 1990b, ApJ, 360, 561 (Paper III).
Longmore,A.J., Fernley,J.A., Sherrington,M.R. & Frank,J., 1985, MNRAS, 216, 873.
Longmore,A.J., Fernley,J.A. & Jameson,R.F., 1986, MNRAS, 220, 279.
Longmore, A.J., Dixon,R., Skillen,I., Jameson,R., Fernley,J., 1990, MNRAS, 247, 684.
Moffett,T.J., 1989, The Use of Pulsating Stars in Fundamental Problems in Astronomy, IAU Coll.111,p191. (ed. Schmidt,E.G.).
Oort,J.H. & Plaut,L., 1975, AAp, 41, 71.
Parsons, S.B., 1972, ApJ, 174, 57.
Plaut,L. 1973, AApS, 12, 351.
Sandage, A.R., 1990, ApJ, 350, 603.
Simon, N.R., 1987, PASP, 99, 868.
Simon, N.R., 1989, MNRAS, 237, 163.
Simon, N.R. & Clement, C.M. 1993, these proceedings, pp.304-312.
Skillen,I., Fernley,J., Stobie,R., Marang,F. & Jameson,R. 1992 (preprint) (paper V).
Skillen,I.,Fernley,J.,Jameson,R.,Lynas-Gray,A. & Longmore,A. 1989, MNRAS, 241, 281.
Storm,J., Carney,B.W. & Latham,D.W., 1992, PASP, 104, 159.
Storm,J., 1991, in *New Results on Standard Candles*, ed. F.Caputo, in press.
Storm,J., Carney,B.W. & Beck,J.A., 1991, PASP, 103, 1264.
van Albada,T.S & Baker, N., 1971, ApJ, 169, 311.
Walker, A.R. & Terndrup,D.M., 1991, ApJ, 378, 119.
Walker, A.R., 1992, ApJ, 390, L81.

DISCUSSION

D.HARTWICK: How do the RR Lyrae distance moduli for globular clusters compare with those from main sequence fitting?

A.LONGMORE: They are comparable, given the uncertainties. Depending on the Lutz-Kelker corrections, the RR Lyrae calibrations may give distance moduli up to $\sim 0^m.15$ closer. I consider this to be within the overlap of likely systematic errors.

N.SIMON: Could you convert your M_K vs $\log P$ fit into a M_V vs $\log P$ fit?

A.LONGMORE: It could be done indirectly, if the temperatures of the stars are known independently. M_V could then be calculated because $(V-K)$ is very well correlated with temperature. Alternatively, if individual masses are known or assumed, equation (6) would yield M_V.

A.SANDAGE: (in reply to a question from the floor on the large scatter of M_V = f(Period) compared with the small scatter for M_K = f(P): it is expected that M_V(RR) = f(P) has great scatter because the CMD in V is flat through the instability strip, yet in a given cluster there is a large spread in period because the instability strip has finite width. Therefore M_V does not change but P does (e.g., from $0^d.4$ to $0^d.7$ in M3). But in K, the HB is NOT horizontal, therefore there must be a much tighter relation in M_K = f(P) than in M_V = f(P).

Infrared Period-Luminosity Relations of RR Lyraes in M5 and M15

T. Liu[1], K. A. Janes[2]

[1]*University of California at Los Angeles,* [2]*Boston University*

Abstract

We have carried out K band photometric observations of RR Lyrae stars in two globular clusters which both have large populations of RR Lyraes but different characteristics: the moderately metal-rich cluster M5 ([Fe/H] = -1.40) and the metal-poor one M15 ([Fe/H] = -2.15). The purpose is to accurately calibrate the linear relationship between RR Lyrae infrared (K) absolute magnitudes and their periods that has been confirmed by recent Baade-Wesselink type studies of RR Lyraes and IR photometry of cluster variables. A total of 47 RR Lyraes in M15 was observed and each has more than 8 measures on the average, which allows the accurate determination of a mean K magnitude for each star. In M5 44 stars have been observed with each RR Lyrae having 4 measurements. Our preliminary results show that the RR Lyrae infrared period–luminosity relations for the two clusters have roughly the same slope, despite the fact that they have a large metallicity difference. This suggests that the metallicity effect on the $\langle M_K \rangle$–$\log P$ relation is indeed small as one would expect. The M5 and M15 RR Lyrae IR photometry gives a reliable determination for the slope of the infrared period–luminosity relation because of the large number of stars measured. A well-calibrated $\langle M_K \rangle$–$\log P$ relation will be very useful in distance determinations to heavily reddened star regions such as the Galactic center and globular clusters in the Galactic bulge.

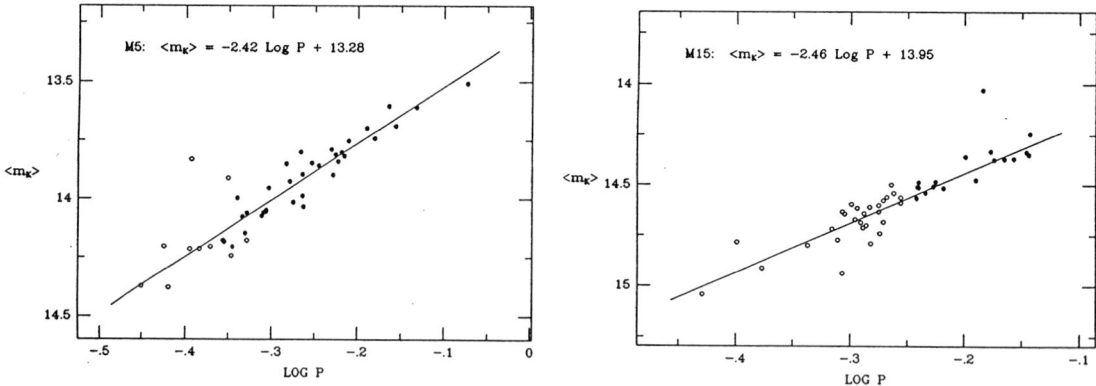

Figure 1 $\langle m_K \rangle$ vs. $\log P$ for RR Lyrae stars in M5 and M15. Filled circles are ab-type variables, while open circles represent c-type stars. The solid lines are linear least-squares fit to the data.

Period-Luminosity-Metal Abundance Relations for Population II Variable Stars

James M. Nemec[1,2] & Thomas E. Lutz[2]

[1] *Astronomy Department, University of Washington, Seattle WA 98195 USA*
[2] *Program in Astronomy, Washington State University, Pullman WA 99164 USA*

Abstract

New period-luminosity-metallicity (P-L-[Fe/H]) relationships for Pop. II Cepheids, RR Lyrae stars, anomalous Cepheids and SX Phe (variable-blue straggler) stars are presented. These were computed by fitting regression lines to observed pulsation periods and mean B, V, K magnitudes for over 1200 stars in $\sim$40 stellar systems. The stars were assumed to be pulsating in either the fundamental (F) or first-overtone (H) modes (excluding double-mode and other multi-periodic variables). Eight P-L-[Fe/H] relationships (one for each of the two pulsation modes for the four kinds of stars) were simultaneously fitted for each filter. After accounting for the metal abundance differences, the slopes of the P-L relations were tested for departures from equality. The results are consistent with the assumption that, for each kind of star, the relations for the F and H stars are vertically offset, with a family of lines corresponding to the different metallicities. In the case of the globular cluster Cepheids, the available B, V data support Arp's 1955 contention that the Cepheids are oscillating in the F and H modes; moreover, the majority of the short-period Cepheids (BL Her stars) appear to be first-overtone pulsators, while most of the Cepheids with periods between 10 and 30 days (W Vir stars) appear to be fundamental-mode pulsators. For the RR Lyrae stars, the slopes of the P-L-[Fe/H] relations in B, V and K show a clear trend with filter type, namely, the absolute values of the slopes increase from B to K. Finally, for the SX Phe stars the differences between the P-L-[Fe/H] relations in B and V for the F and H stars are found to be consistent with the known period-ratio for the double-mode star SX Phe.

1. Introduction

Several kinds of variable stars occupy the Pop. II Cepheid instability strip: Pop. II Cepheids (P2Cs), anomalous Cepheids (ACs), RR Lyrae stars and SX Phe (variable blue straggler) stars. By far, most of these are RR Lyrae stars (see Sawyer Hogg 1973). All four kinds of stars are also found throughout our Galaxy's halo and nuclear bulge, in nearby dwarf galaxies, and in other more distant galaxies. Since the stars appear to obey P-L relations which can be used to derive distances, it is surprising that, with the exception of the RR Lyrae stars, Pop. II variables are rarely used as distance indicators. This is in part because there are only a few P2Cs, ACs or

SX Phe stars in any given cluster, and thus determination of the P-L relations is very uncertain. In addition much of the available photometry is limited to photographic B magnitudes, and there is a dependence of the luminosity level of the RR Lyrae stars on [Fe/H] and evolutionary state.

The aim of the present paper is to show that by analyzing all four kinds of Pop. II variables simultaneously, and by taking into account reddening corrections, metallicity differences, and the evolutionary states of the RR Lyrae stars, it is possible to derive a consistent set of P-L-[Fe/H] relations for the B, V and K passbands, from which more reliable distance estimates for globular clusters and nearby dwarf galaxies can be obtained. The estimated P-L-[Fe/H] relations are obtained by a linear least squares regression analysis with indicator variables, or equivalently, an analysis of covariance (see §2). In general, this method can be used to carry out a simultaneous analysis of as many different types of stars, with different pulsation modes, in different clusters, and through various filter types, as the data allow. Estimated uncertainties for the model parameters are also calculated. This method of pooling the data has not previously been used to derive P-L-[Fe/H] relations, and has potential for improving other distance estimation methods, e.g., Tully-Fisher, Faber-Jackson, etc.

To illustrate the power of the method, the mean photometric B magnitudes of the pulsating variable stars in many globular clusters and dwarf galaxies are used in §3 to calibrate the B-passband P-L-[Fe/H] relations for the four kinds of stars. In §§4-7 we discuss the results of a simultaneous multi-wavelength analysis for the RR Lyrae stars, the Pop. II Cepheids, the anomalous Cepheids and the SX Phe stars.

2. Methodology

Each of the eight types of stars is assumed to follow a linear absolute magnitude-period-metallicity relation in each passband. For observations through the B filter,

$$M_{Bijk} = a_{Bj} + b_{Bj} \log P_{ijk} + c_B [\text{Fe/H}]_k + \epsilon_{Bijk}.$$

The absolute B magnitude and the corresponding pulsation period (log base 10) are denoted M_{Bijk} and $\log P_{ijk}$, respectively, where the subscripts i, j and k represent the 'star', 'type' and 'cluster'. For the cases considered here, the type index ($j = 1, 2, \ldots, 8$) identifies the four kinds of stars and the two pulsation modes. The quantity $[\text{Fe/H}]_k$ represents the mean metal abundance of the stars in an individual stellar cluster (k) or system. Of course, there will always be some unexplained variability due to random measurement error, represented by the term ϵ_{Bijk} in the model. The apparent B magnitude is given by

$$m_{Bijk} = a_{Bj} + b_{Bj} \log P_{ijk} + c_B [\text{Fe/H}]_k + d_k + A_{Bk} + \epsilon_{Bijk},$$

where we have transformed from absolute to apparent magnitude by adding to both sides of the equation the true distance modulus of the cluster, d_k, and the interstellar extinction in B for cluster k, A_{Bk}. The latter is equal to $R_B\, \text{E}_k(B-V)$, where R_B is the ratio of total to selective extinction for the B passband, and $\text{E}_k(B-V)$ is the

B–V reddening for cluster k. Similar equations could be written for other filters. We will be discussing, in addition to the B passband, the V and K passbands.

The coefficients a_{Bj}, b_{Bj}, d_k and ϵ_{Bijk} were estimated by first removing the effects of differences in the metal abundances and reddenings to obtain de-reddened, metallicity-corrected magnitudes:

$$m'_{Bijk} = m_{Bijk} - c_B[\text{Fe/H}]_k - A_B.$$

These 'corrected' magnitudes were then fitted to the model

$$m'_{Bijk} = a_{Bj} + b_{Bj}\log P_{ijk} + d_k + \epsilon_{Bijk}$$

and the coefficients were estimated by ordinary unweighted least squares. The calculations were made using **PROC GLM** in the **SAS** statistical package.

To estimate the true distance moduli, values for the metal abundances and extinction coefficients were assumed. For the three filters, the metallicity coefficients were set equal to the values $c_B=0.35$, $c_V=0.32$ and $c_K=0.06$ (see Zinn 1985b; Longmore et al. 1990, hereafter L90; and Longmore 1993) for all four kinds of variable stars (P2Cs, RR Lyraes, ACs and SX Phe stars) – this assumption should be closely examined when more information becomes available. The interstellar extinction coefficients were assumed to be the standard values $R_B=4.1$, $R_V=3.1$ and $R_K=0.35$, for all but three of the program clusters, NGC 3201, NGC 6121(=M4) and N6171(=M107). These systems are known for their high and variable star-to-star extinctions, and, in the case of M4 and M107 there is also the possibility of abnormal ratios of total to selective extinction. To correct NGC 3201 the Cacciari (1984) internal extinction values were adopted, and for M4 and M107 we used $R_V=3.8$ rather than 3.1 (as recommended by L90).

Student's t-test was used to test whether the lines for the fundamental and first-overtone pulsators are parallel for each filter. Failure to find a significant difference was taken to mean that the first-overtone and fundamental mode stars can be combined into a single line, or equivalently, the first-overtone pulsators can have their periods "fundamentalized" and then combined with the fundamental mode pulsators to form a single line (as L90 and other before them have assumed without testing). For the RR Lyrae stars tests were also made to see whether there was a vertical separation at a given period between the ab- and c-type lines for the B and V P-L relations. Finally we tested whether the slopes of the P-L lines were zero for each mode. This test is of interest only for the B and V regressions for the RR Lyrae stars since it is common practice to draw a horizontal line through the RR Lyrae stars in a cluster (i.e., assume a zero slope) and then to make a vertical shift to the RR Lyrae stars in another stellar system thereby establishing the distance of the latter.

The strength of our analysis method is that any number of different wavelength intervals, different types of stars, different clusters, etc., can be analyzed using this method. The distance scale is determined using not just the RR Lyrae stars but all of the available instability strip variables. The derived P-L-[Fe/H] relations are estimated in a consistent manner, and estimates of the uncertainties in the slopes, zero points, and differences in distance moduli, can be derived. The method also has the advantage that one can determine the slope for each P-L-[Fe/H] relation using only a small number of stars of each type in any one star system. Thus a substantially larger number of stars is available for examining each P-L relation than would normally be available.

Finally, a comment about notation. In the subsequent sections, instead of using the subscript j to represent the star type, we shall note explicitly in parentheses the kind of star and the pulsation mode. For example, the absolute B magnitude for a fundamental mode SX Phe star will be denoted $M_B(\text{SX},\text{F})$. We will also adopt the usual convention of denoting fundamental-mode pulsation periods by P_0 and first-overtone pulsation periods by P_1.

3. P-L Relations (in B) for RR Lyrae stars, Pop. II Cepheids, Anomalous Cepheids and SX Phe Stars

Fig. 1 shows the apparent B magnitudes versus the pulsation periods for a sample of 710 variable stars (RR Lyraes, P2Cs, ACs and SX Phe variables) in 28 stellar systems. Most of the systems contain two or more of the four kinds of variable stars (e.g., RR Lyrae stars and P2Cs; or, RR Lyraes, ACs and SX Phe's; etc.).

Fig. 2 shows the data with the eight fitted P-L relations for the F and H modes for the four different kinds of variable stars. The overall scatter of the observations about all four lines is $\sigma_B = 0.11$ mag. In this figure the B magnitudes have been adjusted by taking the Ursa Minor dwarf galaxy, whose RR Lyrae stars have an overall mean magnitude $B=20.16$ mag, as the zero-point cluster. The B mags for the other systems have been adjusted. It is apparent that the slopes of the P-L relations for the two pulsation modes for the stars of a given kind are approximately parallel. This hypothesis was later tested and found to be a reasonable assumption to within the measuring errors. No metallicity corrections were made, but this did not matter since we were initially interested in the slopes of the P-L relations and not absolute distances.

Although the results of this analysis are encouraging, it does not include the V photometry that is available for all four kinds of stars, or the recently obtained K photometry for approximately 200 RR Lyrae stars. By including the V and K photometry it should be possible to increase the accuracy of the derived distance estimates. In fact, the K-band photometry, although limited to a handful of globular clusters, provides some of the most reliable distance information because of the reduced effects of the reddening and metallicity corrections.

For the subsequent analyses we have taken the globular cluster M15 as the zero point cluster, for several reasons: (1) M15 contains over 100 well-studied RR Lyrae stars with varying amounts of U, B, V and K photometry (Sandage, Katem & Sandage 1981; L90); (2) its color-magnitude diagram is well-defined from the tip of the giant branch to below the main-sequence turnoff (Sandage 1970; Fahlman, Richer & VandenBerg 1985); (3) the mean metal abundance for M15 is very low and well-

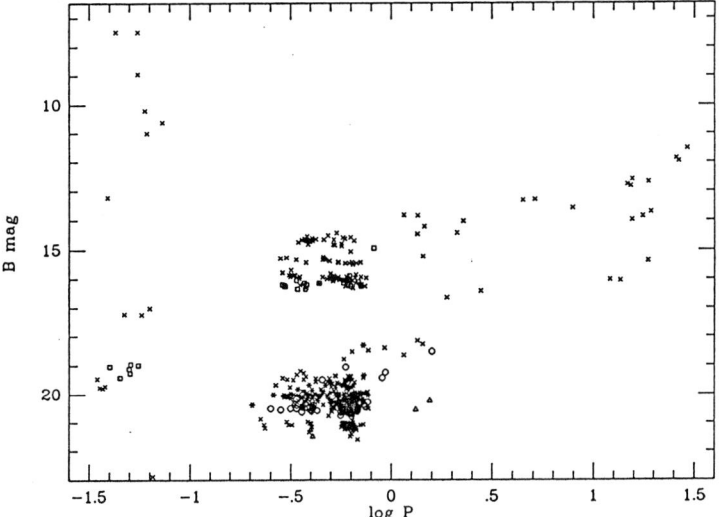

Fig. 1. Period, apparent B magnitude diagram for the program stars. The stars with periods shorter than $\log P = -1$ are SX Phe variables, those with $-0.8 \leq P \leq -0.1$ are mostly RR Lyrae stars (with some possible ACs included), and the remainder are either ACs or P2Cs.

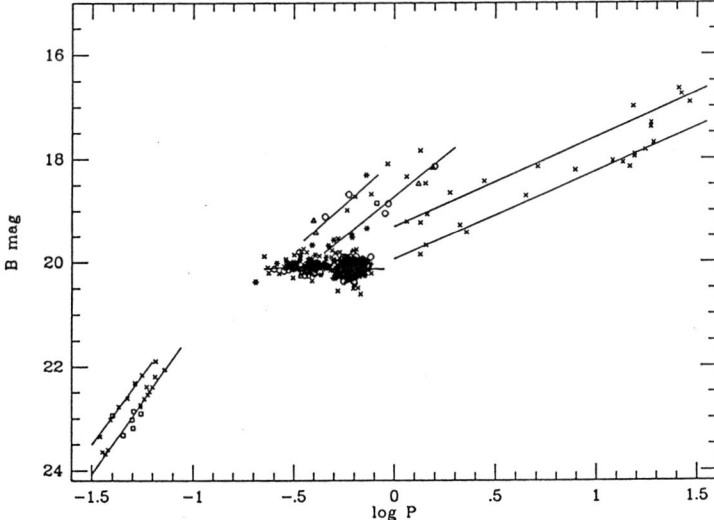

Fig. 2. Period, adjusted B magnitude diagram and P-L relations for the four different kinds of variable stars. The magnitudes have been de-reddened and shifted to the distance of Ursa Minor. Note that the slopes of the P-L relations for the assumed F and H modes for each kind of variable star are approximately parallel. Subsequent t-tests showed that the assumption of parallel slopes is reasonable and cannot be rejected.

established at [Fe/H]=−2.15 (Zinn & West 1984); furthermore, there is no evidence of star-to-star [Fe/H] variations (Sandage & Katem 1977); (4) the vertical height of the M15 horizontal branch is among the narrowest of any globular cluster (Sandage 1990); (5) the reddening is well established at $E(B-V) = 0.10 \pm 0.02$; (6) the true distance modulus of M15 is reasonably-well established at $(m-M)_0 = 15.03 \pm 0.05$. The existence of the U, B, V and K photometry of the RR Lyrae stars (L90), and the deep CCD photometry to the level of the main sequence, means that we can relate the variable and non-variable stars; and finally, (7) M15 contains more than 12 double-mode RR Lyrae (RRd) stars whose masses are known (see Petersen 1992), and which permit a tie-in to the theoretical models of the horizontal branch stars.

In the sections that follow we discuss our results for the RR Lyrae stars (§4), the P2Cs (§5), the ACs (§6) and the SX Phe stars (§7). The stars are discussed separately because the photometry available and the issues of interest are different for the four kinds of stars. The magnitude offsets permit us to derive absolute magnitudes to all the stars. Using the relations calibrated by the globular cluster stars it was also possible to derive absolute magnitudes for several field stars (see §7). Our complete results will be presented in detail elsewhere.

4. P-L-[Fe/H] Relations in B, V and K for RR Lyrae Stars

RR Lyrae stars have traditionally been used to derive distance estimates for Pop. II stellar systems. Because they are common in globular clusters, and their P-L relations in B and V are approximately flat, one had, traditionally, only to assume a constant mean luminosity for the RR Lyrae stars in the systems of interest (typically M_B=0.80 and M_V=0.50), make a reddening correction, and use the difference in the mean magnitudes to obtain a measure of the relative distances. However, it is now clear that this approach is inadequate.

The absolute magnitudes of RR Lyrae stars almost certainly depend on their evolutionary states, which in turn depend on their ages and metal abundances. In a given cluster the mean absolute magnitude of the RR Lyraes appears to correlate with the mean metallicity of the stars, in the sense that the more metal-poor (Oo II) RR Lyraes are more luminous (Sandage 1957, 1990) and more massive (see Simon & Clement 1993). Sandage (1970) argued that the few RR Lyrae stars in M13 are extra-luminous; and more recently, Lee, Demarque & Zinn (1990) suggest that the RR Lyrae stars in the most metal-poor systems are in advanced post-ZAHB evolutionary states and are more luminous than the metal-rich (Oo I) RR Lyrae stars (see Lee 1993). Thus, using the 'lower envelope' of the theoretical ZAHB rather than the 'evolutionary mean level' (see Lee 1990, Catelan 1992) would cause the distances for the parent clusters to be underestimated. Furthermore, in addition to systematic effects there are also random effects in that individual stars are found in evolutionary states different from the bulk of the RR Lyrae stars. (The cluster M15 harbors several such stars – Sandage, Katem & Sandage 1981). Thus, if RR Lyrae stars are to be used as Pop. II distance indicators then these systematic and random effects must be taken into account. Finally, the derived distances must be in accord with distance estimates based on main sequence fitting techniques (see Buonanno et al. 1989, and Lee 1993).

We have followed L90, who used the reddening-corrected K mag at $\log P=-0.3$ as their distance indicator, and used the reddening-corrected B, V and K magnitudes at $\log P=-0.3$ as our distance indicators. Thus, our best estimate of the true distance modulus to a particular system is the average of the B, V and K estimates. This assumes that each of the stars in a particular stellar system has the same metal abundance, distance and interstellar extinction (except for the cases noted above), namely, the mean values for the cluster, regardless of the type of variable star. The luminosity, [Fe/H] dependence was assumed to be the same for all four kinds of variables, and identical to the slope found by L90 for the RRab stars at $\log P=-0.3$:

$$M_B(\text{RR}ab, -0.3) = 0.35[\text{Fe/H}] + 1.53,$$
$$M_V(\text{RR}ab, -0.3) = 0.32[\text{Fe/H}] + 1.19,$$
$$M_K(\text{RR}ab, -0.3) = 0.06[\text{Fe/H}] - 0.24.$$

Table 1 gives the number of RR Lyrae stars (and other kinds of stars) for which B, V, K photometry, and B–V and V–K colors are available (the list of references to be published later). The numbers are summarized separately for the fundamental (F) and first-overtone (H) pulsators.

Table 1

	Number of Stars				
	B	V	K	$B-V$	$V-K$
TYPE					
Anomalous Cepheids (F)	13	1	0	1	0
Anomalous Cepheids (H)	15	5	0	5	0
Pop. II Cepheids (F)	19	14	0	14	0
Pop. II Cepheids (H)	22	18	0	18	0
RR Lyrae stars (F)	743	414	152	411	101
RR Lyrae stars (H)	369	256	51	256	41
SX Phe stars (F)	25	27	0	25	0
SX Phe stars (H)	9	9	0	9	0
TOTAL	1215	744	203	739	142

Fig. 3 shows the period-'adjusted magnitude' diagrams in B (top panel), V (middle panel) and K (bottom panel) for the 203 RR Lyrae stars that presently have K photometry. The adjusted magnitudes were reddening corrected, normalized to the metallicity of M15, and shifted to the distance of M15. Also shown are the P-L relations computed using the analysis of covariance technique described in §2. These were

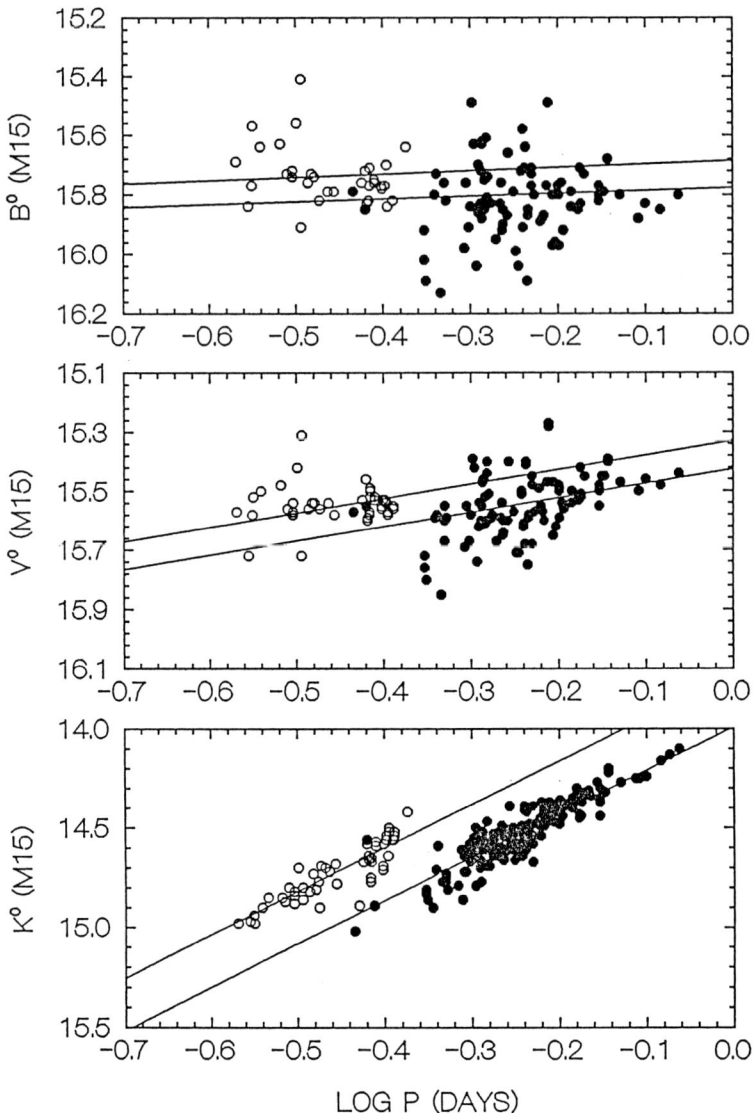

Fig. 3. Period, adjusted magnitude diagrams and *P-L* relations in *B* (top panel), *V* (middle panel), and *K* (bottom panel) for those globular cluster RR Lyrae stars with *K* photometry. The magnitudes have been de-reddened, adjusted to the metallicity of M15, and shifted to the distance of M15. The RR*ab* stars are shown as solid circles, and the RR*c* stars are shown as open circles. The scales are unequal.

derived by fitting regression lines to the observed pulsation periods and reddening-corrected, metallicity adjusted magnitudes shifted to the distance of M15.

No evidence was found to suggest that the slopes of the P-L relations in B, V or K for the F and H pulsators are not parallel. The obvious result that for the RR Lyrae stars the K intercepts are different was recovered. Since these tests do not rule out using a single slope for the P-L relation in each passband, we subsequently assumed parallel slopes for the F and H stars through all three filters.

Under this assumption, the derived P-L-[Fe/H] relations in B, V, and K for the RR Lyrae stars are:

$$M_B(\text{RRab}) = 1.51 - 0.06(\pm 0.09)\log P_0 + 0.35[\text{Fe/H}],$$
$$M_B(\text{RRc}) = 1.44 - 0.06(\pm 0.09)\log P_1 + 0.35[\text{Fe/H}],$$
$$M_V(\text{RRab}) = 1.09 - 0.44(\pm 0.11)\log P_0 + 0.32[\text{Fe/H}],$$
$$M_V(\text{RRc}) = 1.01 - 0.44(\pm 0.11)\log P_1 + 0.32[\text{Fe/H}],$$
$$M_K(\text{RRab}) = -0.94 - 2.32(\pm 0.11)\log P_0 + 0.06[\text{Fe/H}],$$
$$M_K(\text{RRc}) = -1.23 - 2.32(\pm 0.11)\log P_1 + 0.06[\text{Fe/H}].$$

Of interest here is the statistically significant increase in the slopes of the P-L relations in going from the blue to the infrared. The existence of a non-zero slope for the V photometry also means that one should be cautious when dealing with a small number of RR Lyrae stars in using a simple mean V magnitude in estimating distances. Notice that, since much of the same data were analyzed and the M_K-[Fe/H] relation of L90 was assumed, the $M_K(\text{RRab})$ relation is in good agreement with equation (9) of L90 (see also Liu & Janes 1993, and equation 6 of Longmore 1993).

5. Pop. II Cepheids: Fundamental and First-Overtone Pulsators?

In Arp's (1955) analysis of the P-L relations for globular cluster Cepheids (in m_{pg} and m_{pv}) he concluded that these stars pulsate in the fundamental and first-overtone modes (and possibly in higher pulsation modes), and that the stars of each mode follow separate P-L relations offset from and parallel to each other. Since the early 1970's this hypothesis has been largely ignored in favor of the assumption that all P2Cs are F pulsators obeying a single P-L relation. For example, based on 17 stars in the globular clusters ω Cen, M2, M13 and M14, Demers & Harris (1974) obtained the single line $M_V = -1.59 \log P - 0.08$, with a residual scatter of ± 0.21 mag. More recently, Harris (1985) assumed the Demers & Harris relation for stars with periods less than 12.5 days, and adopted a steeper relation $M_V = -3.6 \log P + 2.2$ for longer period stars. In both papers the scatter about the P-L relation is as much as 0.3 mag at a given period.

Fig. 4 shows the period, adjusted magnitude diagrams in V (top panel) and in B (middle panel) for the P2Cs. Also shown are the derived P-L relations. For a given filter common slopes were assumed in the calculation for the F and H modes, after no evidence was found that the slopes of the P-L relations are not parallel.

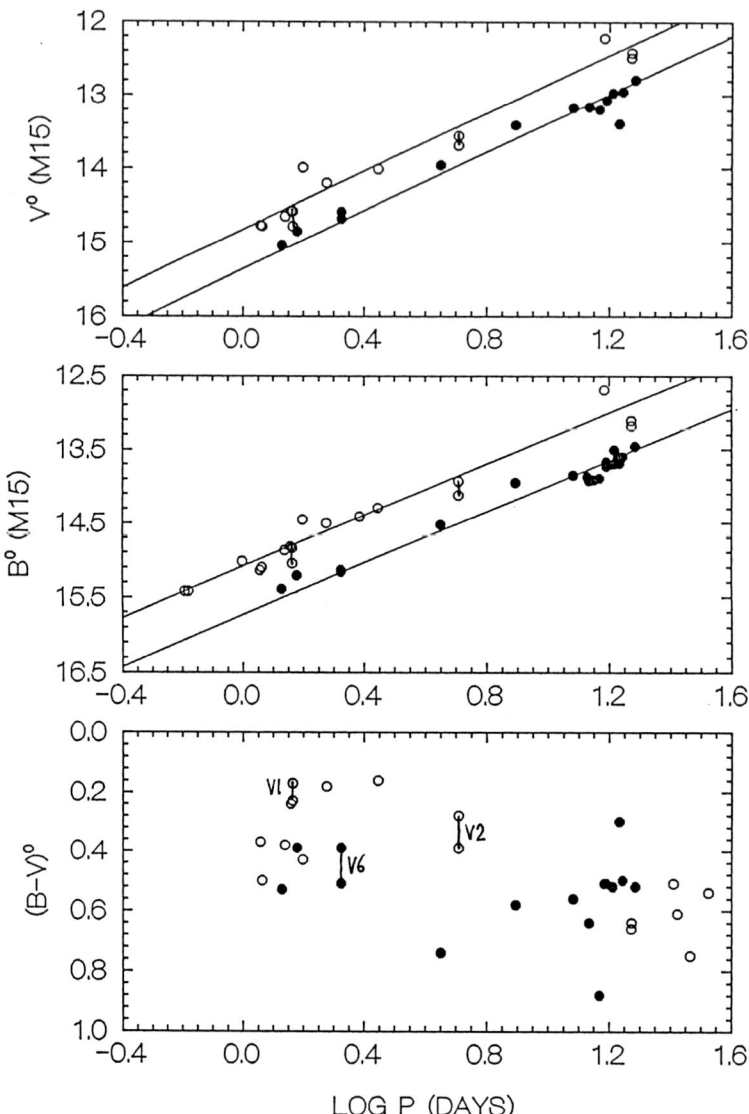

Fig. 4. Period, adjusted magnitude diagrams and *P-L* relations in *V* (top panel) and in *B* (middle panel) for the globular cluster Cepheids. The magnitudes have been de-reddened and metallicity corrected and shifted to the distance of M15. At a given period the assumed F pulsators (solid circles) are seen to be, on average, less luminous than the assumed H pulsators (open circles). The *P-C* diagram (bottom panel) shows that the short-period Pop. II Cepheids (the BL Her stars) are systematically bluer than the long-period Cepheids (W Vir stars), and that at a given pulsation period the first-overtone stars are systematically bluer than the fundamental-mode stars.

The lines have slopes of -1.72 (B) and -1.96 (V). The derived B slopes were found to lie midway between the corresponding slopes shown in Fig. 2. The new slope is steeper than the Demers-Harris slope, in accord with Harris' modification for the longer period stars; however, in our model there is no need for a change in slope at mid-periods.

The bottom panel of Fig. 4 shows the period-color (P-C) diagram for the P2Cs. Clearly the longer-period (W Vir) Cepheids are redder on average than the shorter-period (BL Her) stars, a result seen previously in Fig. 1 of Harris (1985). A linear P-C relation through all the points was derived by Harris: $(B-V) = 0.275 + 0.206 \log P$. In addition to supporting this result, the bottom panel of Fig. 4 suggests that the first-overtone BL Her stars are bluer than the fundamental-mode BL Her stars, paralleling the situation for the slightly lower-luminosity RR Lyrae variables. At longer periods and higher luminosities the discrimination between F and H stars is reduced and there is no obvious color dependence.

Fig. 5 shows the period, absolute magnitude diagrams in V (top panel) and in B (bottom panel) for the P2Cs. The P-L-[Fe/H] families of lines (not plotted) are given by:

$$M_B(\text{P2C}, \text{F}) = 1.42 - 1.72(\pm 0.05) \log P_0 + 0.35 [\text{Fe/H}],$$
$$M_B(\text{P2C}, \text{H}) = 0.77 - 1.72(\pm 0.05) \log P_1 + 0.35 [\text{Fe/H}],$$
$$M_V(\text{P2C}, \text{F}) = 0.97 - 1.96(\pm 0.05) \log P_0 + 0.32 [\text{Fe/H}],$$
$$M_V(\text{P2C}, \text{H}) = 0.44 - 1.96(\pm 0.05) \log P_1 + 0.32 [\text{Fe/H}].$$

Because the [Fe/H] range of the Cepheids is only from -1.39 (for M14) to -2.15 (for M15) the scatter at a given period is not much larger than that seen in Fig. 4.

The success with which the P-L relations in Figs. 2 and 4 fit the P2C data suggest that Arp's assumption of two or more distinct P-L relations for P2Cs, with the brighter stars at a given period being first-overtone pulsators and the fainter stars being fundamental-mode pulsators, is correct. If P2Cs do pulsate in the F and H modes there would be an appealing symmetry with the well-known *ab*- and *c*-type RR Lyrae stars, with the ACs and SMC Cepheids that almost certainly follow two parallel lines (see §6), with the SX Phe stars that obey two P-L relations (see §7), as well as with the Pop.I Cepheids that are now being discovered in young Magellanic Cloud clusters (see Welch *et al.* 1993).

Arp's hypothesis also explains a number of outstanding puzzles, including the problem of the Cepheids V1, V2 and V6 in M13 (labelled in the bottom panel of Fig.4 – the observations of Pike & Meston 1977, and of Demers & Harris 1974, were treated separately and both sets of data have been plotted). These stars were studied extensively by Böhm-Vitense *et al.* (1974). They noted that, while V6 has a reddish color and an effective temperature consistent with pulsation in the F mode, the very blue colors and early spectral types (A2 - Joy 1949) of V1 and V2 are anomalously bluer than the theoretical fundamental-mode blue edges for various assumed helium abundances. Böhm-Vitense *et al.* argue that if the relatively high temperatures of V1 and V2 are to be believed, then both stars must have very low masses (~ 0.3-0.4

solar masses) and relatively high envelope helium abundances (∼0.6-0.7). Our fits suggest that V1 and V2 lie on the brighter of the two P2C P-L lines plotted in Fig. 4, and that V6 lies on the lower of the two lines, and thus V1 and V2 are H pulsators while V6 is an F pulsator. This being the case there is no longer the need to suggest unusual mases or abundances.

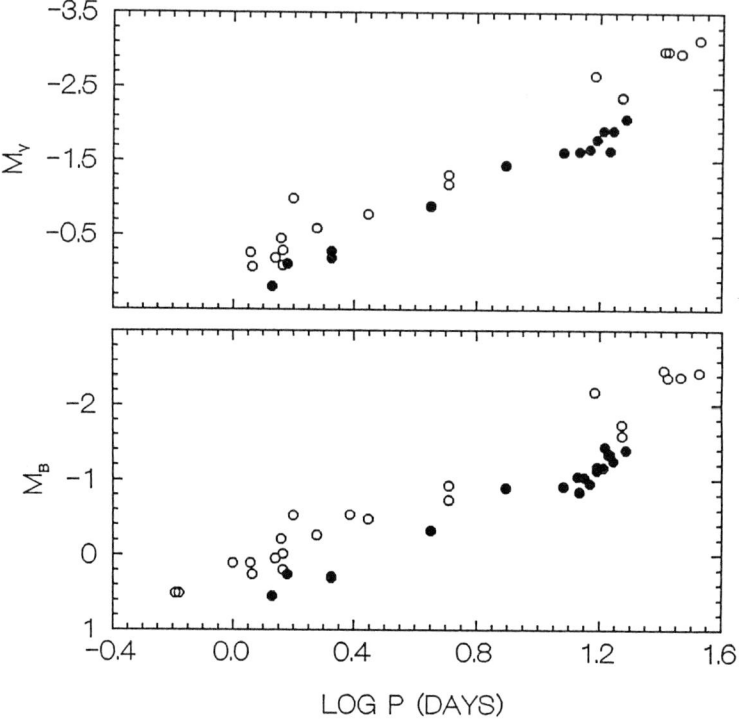

Fig. 5. Period, absolute magnitude diagrams for the globular cluster Cepheids.

If the assignment of pulsation modes based on the P-L diagram is valid, then one might expect there to be some evidence of the different pulsation modes in the light amplitudes and shapes of the light curves. This is certainly the case for the RR Lyrae stars, where the period-amplitude (P-A) diagram shows well-defined differences and correlations, with the ab-type RR Lyrae stars tending to have larger amplitudes than the c-type stars (except at small amplitudes where the two amplitude distributions overlap). Furthermore, for a given metallicity, the amplitudes of the ab-type stars exhibit a linear trend with $\log P$, in the sense that the shorter-P stars have larger amplitudes than the longer-P stars. The lower metallicity stars exhibit the same overall amplitude trends as do higher metallicity stars but with the $\log P$-A relation shifted toward longer periods at a given amplitude (temperature).

A more quantitative use of light curve morphology for mode assignment might be possible if results of Fourier decomposition analyses were used (see Simon & Lee

1981, Antonello et al. 1986, and Petersen 1993). Unfortunately, most P2Cs lack well-defined light curves based on modern observations and hence such analyses are not presently possible. Alternatively, one could use period-amplitude diagrams to sort F and H stars, and the periods of double-mode P2Cs to establish better the separation of the two parallel lines.

Until such analyses have been carried out, we propose (based on the P-L, P-C and P-A diagrams) the following mode assignments: N2419-V18(H); ωCen-V1(H), V29(F), V43(H), V48(F), V60(H), V61(F), V92(F); M3-V154(H); M5-V42(H), V84(H); M80-V1(F); M10-V1(F), V2(H), V3(F); M13-V1(H), V2(H), V6(F); M19-V1(F), V2(F), V3(F), V4(H); M14-V1(F), V2(F), V7(F), V17(F), V76(H); M28-V4(F), V5(H), V9(H), V22(H); N6752-V1(H); M15-V1(H), V72(H), V86(F); and M2-V1(F), V5(F), V6(F).

6. P-L-[Fe/H] Relations in B and V for Anomalous Cepheids

Anomalous Cepheids have been extensively reviewed by Zinn (1985a), DaCosta (1988) and Nemec (1989). Most of the ∼50 ACs that are known are located in seven nearby dwarf galaxies (see Table IX of Nemec, Wehlau & Oliveira 1988, hereafter NWO). In addition, at least one such star is known in the galactic globular cluster NGC 5466 (Zinn & King 1982), and several ACs have recently been identified in the Magellanic Cloud globular cluster NGC 1786 (Walker & Mack 1988). At a given luminosity ACs tend to have shorter periods than the Cepheids found in globular clusters, which can be explained if the stars are more massive than the P2Cs. The leading explanation for the ACs is that they are coalesced binary systems above the horizontal branch in the Cepheid instability strip and that they are possibly related to the blue stragglers found in the same systems.

NWO showed that ACs obey two distinct linear P-L relations, one for the F mode pulsators and one for the H pulsators. They determined the relations by shifting the mean magnitudes of the RR Lyrae stars in the individual clusters to the mean apparent magnitude for the Ursa Minor RR Lyrae stars, at B =20.16 mag. The absolute magnitudes were then found to be in the range $-1.5 \leq M_B \leq 0.4$ by assuming that the mean absolute magnitude for the UMi RR Lyrae stars is $M_B = 0.80$. The resulting P-L relations seemed to diverge.

Fig. 6 shows our period, adjusted magnitude diagrams in B (middle panel) and in V (upper panel), and the new P-L solutions. The top panel shows that in V only the P-L relation for the first-overtone ACs is available. In contrast to the result of NWO, no evidence for a lack of parallelism for the F and H pulsators (cf. Fig.6, and Fig.13 of NWO) was found for the slopes of the P-L relations in either B or V. This change was brought about in part by the existence of the new Goldsmith (1993) Sculptor RR Lyrae star photometry that constrains the luminosities of the Sculptor ACs and has resulted in revised mode classifications for V25 and V26 from F to H.

Fig. 7 shows the period, absolute magnitude diagrams in V (top panel) and in B (bottom panel) for the ACs. The large scatter at a given period is due to the large range in metallicity of the stars, from approximately −0.60 for the SMC field stars,

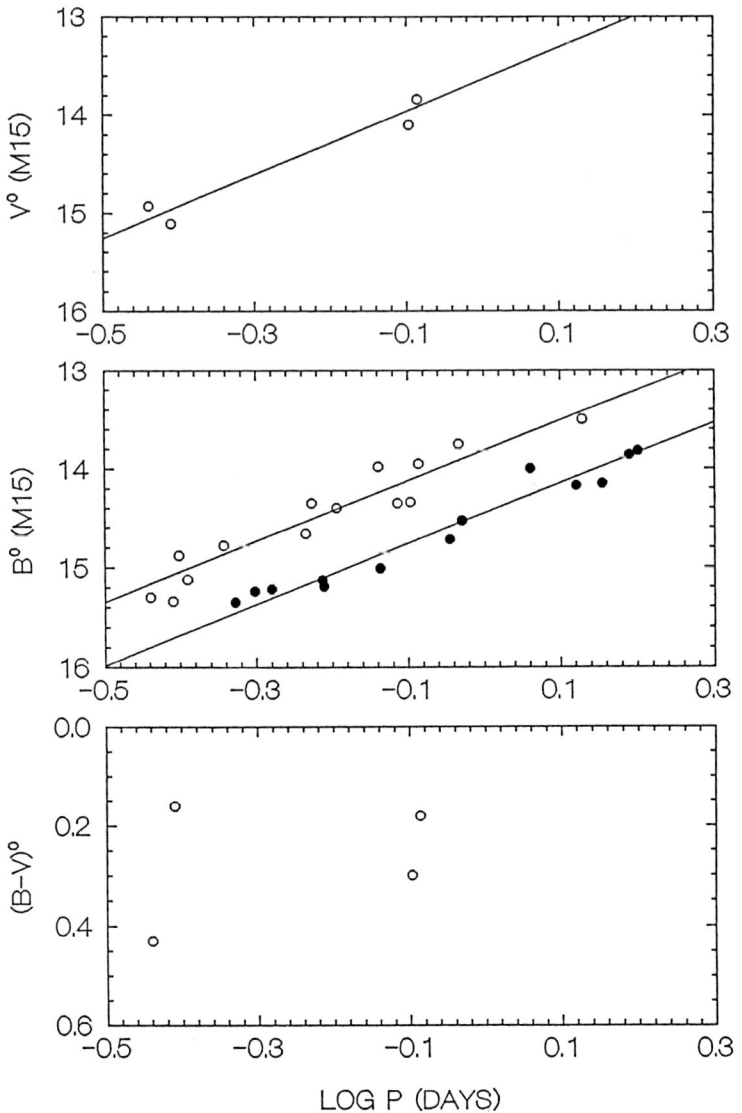

Fig. 6. Period, adjusted magnitude diagrams in V (top panel) and in B (middle panel) for the anomalous Cepheids. The assumed F and H pulsators are show as solid circles and open circles, respectively. The photometry has been reddening corrected, adjusted to the metallicity of M15, and shifted to the distance of M15. Parallel slopes were assumed for the two modes. Clearly more photometry is needed before anything substantive can be said about possible period-color (bottom panel) relations.

to -2.02 for the Ursa Minor ACs. The derived P-L-[Fe/H] relations are as follows:

$$M_B(\text{AC}, \text{F}) = +0.18 - 3.04(\pm 0.16)\log P_0 + 0.35[\text{Fe/H}],$$
$$M_B(\text{AC}, \text{H}) = -0.44 - 3.04(\pm 0.16)\log P_1 + 0.35[\text{Fe/H}],$$
$$M_V(\text{AC}, \text{H}) = -0.63 - 3.09(\pm 0.40)\log P_1 + 0.32[\text{Fe/H}].$$

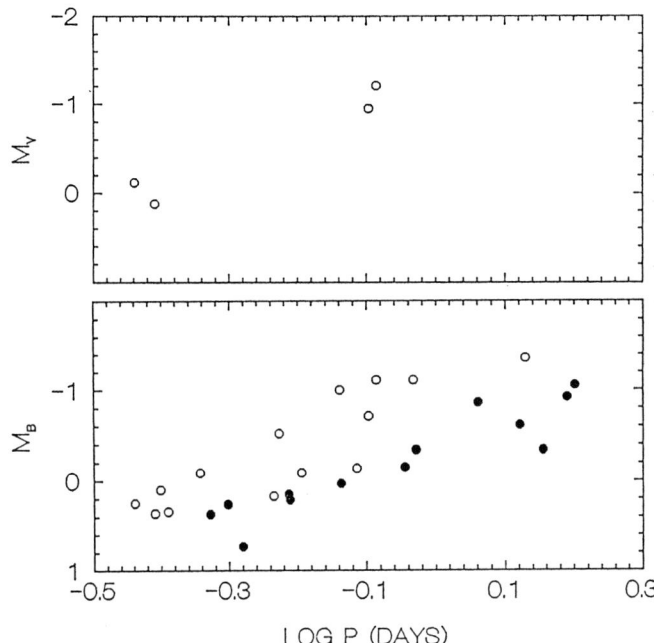

Fig. 7. Period, absolute magnitude diagrams for the anomalous Cepheids. The large scatter at a given period seen in the lower panel is due mainly to the different metallicities of the program stars.

It has long been thought that the Cepheids in the Small Magellanic Cloud (SMC) may be related to the anomalous Cepheids found in the dwarf galaxies. Recently, Smith *et al.* (1992, hereafter SSBG) published the results of an extensive investigation of the variable stars in a NE region of the SMC. They present photographic B photometry and derive the P-L characteristics for a large number of short-period Cepheids. The clean division of the SMC Cepheids into two groups with nearly parallel $\log P$-M_B relations, just as the ACs studied here, is of considerable interest. The P-L relations derived by SSBG for the F and H pulsators in the SMC were found to be:

$$B(\text{SMC}, \text{F}) = -2.71(\pm 0.09)\log P_0 + 18.29(\pm 0.04),$$
$$B(\text{SMC}, \text{H}) = -2.98(\pm 0.17)\log P_1 + 17.61(\pm 0.04).$$

Since the separation of the two lines at $\log P = -0.1$ (the center of the range in $\log P$ for the ACs) is 0.65 mag, compared with the 0.66 mag for the ACs found here, and since the slopes derived by SSBG, -2.71 and -2.98, are very close to the slope -3.04 ± 0.16 derived for the ACs, our results for the ACs appear to be consistent with those of SSBG for the SMC Cepheids. At the very least, the identification of the two modes with F and H pulsation seems secure. To settle the matter it would be useful to have metal abundances for the SMC Cepheids and for the ACs to remove the confounding effects of the metallicities on the distance determinations so that a direct comparison can be made.

7. P-L-[Fe/H] relations in B and V for SX Phe Variables

Over 30 SX Phe stars are now known in several globular clusters: three in ω Cen (Jörgensen & Hansen 1984), six in NGC 5466 (Mateo et al. 1988; 1993, in preparation), five in NGC 5053 (Nemec 1989; Nemec et al. 1993, in preparation), one in M71 (Hodder et al. 1992), one (candidate) star in M3 (DaCosta 1988), and eight which were recently discovered in NGC 4372 (Kaluzny & Krzeminski 1992 preprint). In addition five field SX Phe stars (BL Cam, DY Peg, SX Phe, CY Aqr and KZ Hya) are also known (see Nemec & Mateo 1990). Fig. 7 of Nemec (1989) shows an early $\log P$-M_V diagram that included most of these stars, with the P-L relations for the suspected F and H stars drawn through the V data available at that time. Also shown was a period-amplitude ($\log P$-A_V) diagram showing that the visual amplitudes of SX Phe stars tend to increase with lengthening pulsation period.

In **Fig. 8** period, adjusted magnitude diagrams for the SX Phe stars are plotted in V (top panel) and in B (middle panel), and in the bottom panel we plot the period-color diagram. To obtain estimates of the P-L relations, two cases, both of which assumed parallel slopes, were analyzed. The upper diagrams show the least squares lines derived *without* the RR Lyrae stars in the globular clusters NGC 5053, NGC 5466 and ω Cen. The reason for their exclusion was that the present photometry of the RR Lyrae stars in these systems is very uncertain. In all three panels of Fig. 8 the well-known double-mode star SX Phe has been plotted twice, once for the first-overtone component, and once for the fundamental-mode component, with the two points connected by a line. It is encouraging that its accurately-known period ratio, $P_1/P_0 = 0.7782$, is very close in value to the horizontal separations of the F and H lines (*i.e.*, period ratios, P_1/P_0) of 0.81 (B) and 0.83 (V). The differences between the observed and predicted period ratios can be attributed to measurement errors and the small sample sizes (particularly for the H pulsators). Other field SX Phe stars are also plotted in the diagrams, at their fitted positions on the lines.

The derived P-L-[Fe/H] equations for the case where the RR Lyrae stars in NGC 5053, NGC 5466 and ω Cen were excluded, are given by

$$M_B(\text{SX}, \text{F}) = -3.99 - 5.44(\pm 0.75)\log P_0 + 0.35[\text{Fe/H}],$$
$$M_B(\text{SX}, \text{H}) = -4.53 - 5.44(\pm 0.75)\log P_1 + 0.35[\text{Fe/H}],$$
$$M_V(\text{SX}, \text{F}) = -4.10 - 5.42(\pm 0.70)\log P_0 + 0.32[\text{Fe/H}],$$
$$M_V(\text{SX}, \text{H}) = -4.60 - 5.42(\pm 0.70)\log P_1 + 0.32[\text{Fe/H}].$$

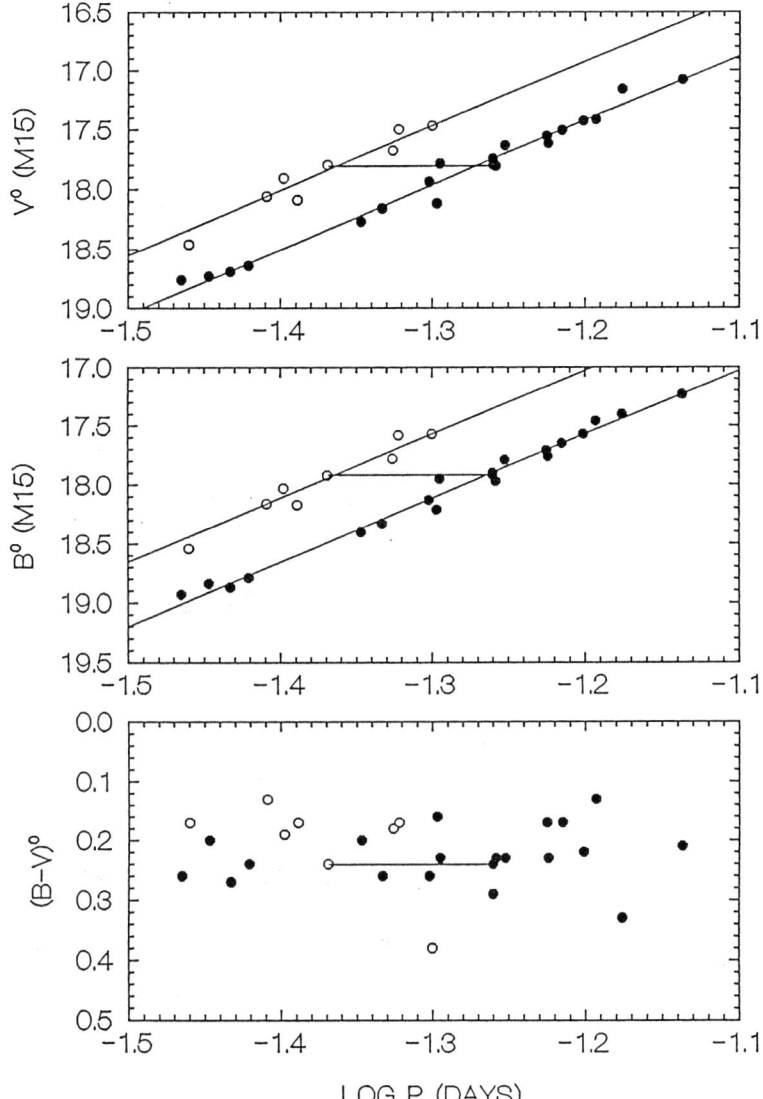

Fig. 8. Period, adjusted magnitude diagrams and P-L relations in V (top panel) and in B (middle panel) for the SX Phe stars. The assumed fundamental-mode (F) pulsators are plotted as solid circles, and the assumed first-overtone (H) pulsators are represented by open circles. Both periods for SX Phe are shown, connected by horizontal lines. The photometry has been reddening corrected, adjusted for metallicity differences, and shifted to the distance of the globular cluster M15. In computing the P-L relations parallel slopes were assumed for the two modes. The P-C diagram (bottom panel) shows that at a given period the H pulsators tend to be bluer than the F pulsators (an exception being H1 in M71 at $\log P = -1.3$).

These are shown in **Fig. 9**, where we have plotted the corresponding period, absolute magnitude diagrams in V (top panel) and in B (bottom panel).

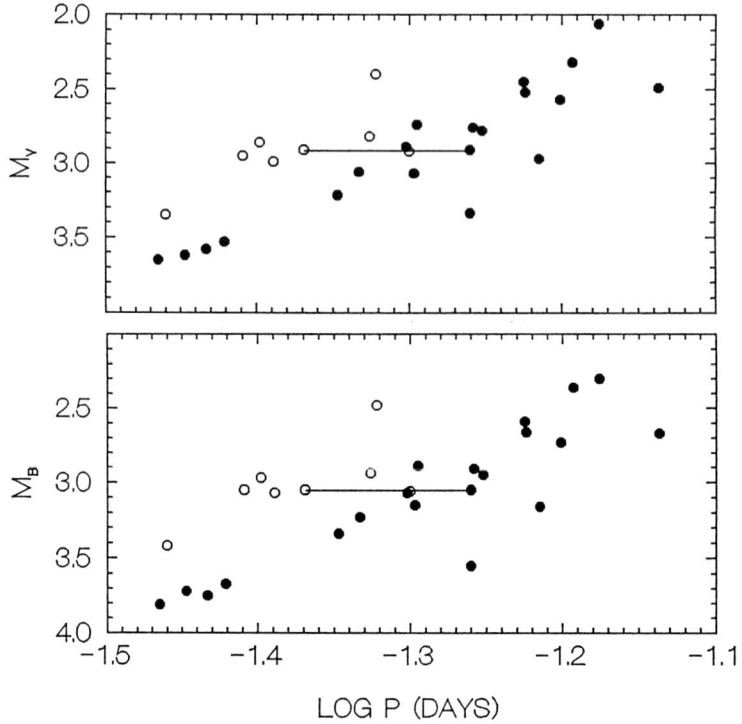

Fig. 9. Period, absolute magnitude diagrams for the SX Phe stars. The large scatter at a given period seen at the longer periods is mainly due to the metal richness of the SX Phe stars in the field compared with those in the globular clusters.

When the RR Lyrae stars in ω Cen, NGC 5466 and NGC 5053 were included in the analysis the family of P-L-[Fe/H] lines was found to be:

$$M_B(\text{SX},\text{F}) = -1.85 - 4.44(\pm 0.74)\log P_0 + 0.35[\text{Fe/H}],$$
$$M_B(\text{SX},\text{H}) = -2.27 - 4.44(\pm 0.74)\log P_1 + 0.35[\text{Fe/H}],$$
$$M_V(\text{SX},\text{F}) = -1.71 - 4.19(\pm 0.70)\log P_0 + 0.32[\text{Fe/H}],$$
$$M_V(\text{SX},\text{H}) = -2.05 - 4.19(\pm 0.70)\log P_1 + 0.32[\text{Fe/H}].$$

The slopes in this case agree at the 0.05 level of significance with the preceding estimates, although the scatter about the P-L lines is somewhat greater as expected. If the photometry of the NGC 5053, NGC 5466 and ω Cen RR Lyrae stars were more reliable this would be the preferred solution. However, with the available data we can only conclude that the slope is approximately −5.0 ±0.5. The results of the photometry projects underway by H.Harris *et al.* for the RR Lyrae stars in NGC 5466, and by Corwin *et al.* for the RR Lyrae stars in NGC 5053 are eagerly awaited.

Fig. 8 and Fig. 9 also show the field SX Phe stars BL Cam, KZ Hya and DY Peg (all are on the P-L lines in the upper two panels of Fig. 8). Assuming that the metallicities, reddenings, and pulsation modes for these stars are as given by Nemec & Mateo (1990), it is possible to compute their absolute magnitudes using the above equations. For example, for SX Phe with [Fe/H]=−1.70 and P_0=0.055 day, the second set of equations gives M_B=2.7 ± 0.6 and M_V=2.6 ± 0.5. It will be of considerable interest when the zero-points of the P-L-[Fe/H] relations are set using the accurate parallax for SX Phe obtained by HIPPARCOS.

8. Summary

Period-luminosity-metallicity relations in B, V and K have been derived for RR Lyrae stars, anomalous Cepheids, Pop. II Cepheids and SX Phe stars in globular clusters. For all four kinds of stars the assumption that the stars pulsate either in the fundamental or first-overtone modes appears to be reasonable (obvious exceptions are the double-mode stars and the Blazhko RR Lyrae stars). Statistical tests show that the hypohesis of equal slopes for the F and H P-L relations in each filter cannot be rejected. For the P2Cs it was found that the majority of the shorter-period Cepheids pulsate in the first-overtone mode, a result anticipated by Arp in 1955 but largely ignored since then. The displacements of the parallel P-L-[Fe/H] lines for the RR Lyrae and SX Phe stars are in agreement with the well-known period ratios for double-mode RR Lyrae and SX Phe stars (0.746 and 0.778, respectively). This agreement lends support to the method of mode assignment based on the P-L diagram. It was also found that the SMC Cepheids studied by Smith et al. (1992) are similar to the anomalous Cepheids, with comparable slopes and separation for the P-L relations; however, we cannot be sure that they are the same kind of star. Finally, we note that the period-amplitude (P-A) relations for the SX Phe and P2Cs are somewhat unusual because the H stars tend to have, at a given period, larger amplitudes than the F pulsators. Complete details of our analyses will be presented elsewhere.

In the future, P-L-[Fe/H] relations should provide more accurate absolute magnitudes for the individual stars and estimates of relative distances for the parent stellar systems. It would be highly desirable to identify double-mode P2Cs and ACs, since these will serve to confirm the separations between the P-L-[Fe/H] lines for the ACs and P2Cs. It will also be useful to obtain improved photometry (and more K band photometry) that will be useful for refining the results presented here, as well as for Fourier decomposition purposes. It would also be valuable to have a theoretical explanation for the observed period-amplitude relations.

We thank Amanda F. Linnell Nemec for useful discussions and assistance with the SAS package and ANCOVA methods, and acknowledge Erika Böhm-Vitense, Robert Buchler, Guillermo Gonzalez, Hugh Harris, Young-Wook Lee, Pawel Moskalik, Allan Sandage, Norman Simon and George Wallerstein for useful discussions and preprints. J.M.N. gratefully acknowledges an NSERC operating grant and an adjunct professorship at York University.

References:

Antonello, E., Broglia, P., Conconi, P. & Mantegazza, L. 1986, A&A, 169, 122.
Arp, H.C. 1955, AJ, 60, 1.
Böhm-Vitense, E. 1988, ApJ, 324, L27.
Böhm-Vitense, E., Szkody, P., Wallerstein, G. & Iben, I. 1974, ApJ, 194, 125.
Buonanno, R., Corsi, C.E., Fusi Pecci, F. 1989, A&A, 216, 80.
Buonanno, R., Cacciari, C., Corsi, C.E., Fusi Pecci, F. 1990, A&A, preprint.
Cacciari, C. 1984, AJ, 89, 231.
Catelan, M. 1993, these proceedings, p.271.
Da Costa, G. 1988, In *The Harlow Shapley Symposium on Globular Cluster Systems in Galaxies, IAU Symp 126*, ed. J.Grindlay & A.G.D.Philip, Dordrecht:Reidel, p.217.
Demers, S. & Harris, W.E. 1974, AJ, 79, 627.
Fahlman, G.G., Richer, H.B. and VandenBerg, D.A. 1985, ApJ, 58, 25.
Fernie, D. 1992, AJ, 103, 1647.
Goldsmith, C. 1993, these proceedings, p.347
Harris, H.C. 1985, In *Cepheids: Theory and Observations* ed. B.F.Madore, Cambridge University Press, p.232.
Hodder, P.J.C., Nemec, J.M., Richer, H.B. & Fahlman, G.G. 1992, AJ, 103, 460.
Jörgensen, H. & Hansen, L. 1984, A&A, 133, 165.
Joy, A. H. 1949, ApJ, 110, 105.
Lawson, C. L. & Hanson, R. J. *Solving Least Squares Problems* (Prentice-Hall, Englewood Cliffs, New Jersey, 1974).
Lee, Y.-W. 1990, Ph.D. Thesis, Yale University.
Lee, Y.-W. 1993, these proceedings, pp. 285-294.
Lee, Y.-W., Demarque, P. & Zinn, R. 1990, ApJ, 350, 155.
Liu, T. & Janes, K.A. 1993, these proceedings, p. 28.
Longmore, A.J., Dixon,R., Skillen,I., Jameson,R. & Fernley,J. 1990, MNRAS, 247, 684.
Longmore, A.J. 1993, these proceedings, pp. 19-27.
Mateo,M., Harris,H., Nemec,J., Olszewski,E. & Schombert,J. 1988, BAAS, 20, 717.
Mateo, M., Harris, H.C., Nemec, J.M. & Olszewski, E. 1990, AJ, 100, 469.
Nemec, J. M. In *The Use of Pulsating Stars in Fundamental Problems of Astronomy*, IAU Coll 111, (ed. E.G.Schmidt) pp. 215-245 (Cambridge University Press, 1989).
Nemec, J. M. & Mateo, M. 1990, ASP Conf.Ser. 11, 64.
Nemec, J.M., Wehlau, A. & Oliveira, C. 1988, AJ, 96, 528.
Petersen, J.O. 1993, A&A, preprint.
Pike, C.D. & Meston, C.J. 1977, MNRAS, 180, 613.
Sandage, A.R. 1970, ApJ, 162, 841.
Sandage, A.R.& Katem, B. 1977, ApJ, 215, 62.
Sandage, A.R. 1990, ApJ, 350, 603.
Sandage, A.R., Katem, B. & Sandage, M. 1981, ApJS, 46, 41.
Sawyer Hogg, H. 1973, PDDO, 3, 1.
Seber, G. A. F. 1977, *Linear Regression Analysis*, (Wiley, New York).
Simon, N. & Lee, A.S. 1981, ApJ, 248, 291.

Simon, N. & Clement, C.M. 1993, these proceedings, pp. 304-312.
Smith, H.A., Silbermann, N.A., Baird, S.R. & Graham, J.A. 1992, AJ, 104, 1430.
Walker, A. & Mack, P. 1988, AJ 96, 1362.
Welch, D., Mateo, M., Olszewski, E., Fischer, P. & Takamiya, M. 1993, AJ, 105, 146.
Zinn, R. 1985a, Mem.S.A.It., 56, 223.
Zinn, R. 1985b, ApJ, 293, 424.
Zinn, R. & King, C.R. 1982, ApJ, 262, 700.
Zinn, R. & West, M.J. 1984, ApJS, 55, 45.

DISCUSSION

J. MATTHEWS: In your current formulation, the slope of the P-L relation is defined to be only a function of the type of variable star. What happens to your results, particularly for the anomalous Cepheids and Pop.II Cepheids, if the slope is also free to change from cluster to cluster?

J. NEMEC: If the stars of a given kind in a particular cluster are found to have a P-L slope different from that of the same kind of stars in another system, then one would have to think about the possibility that the stars are not of the same kind. Two relevant situations come to mind: (1) the K-band RR Lyrae photometry of L90 possibly suggests that the slope of the P-L relation for the NGC 5466 RR Lyrae stars is steeper than that for the other RR Lyrae stars. Unfortunately the number of RR Lyrae stars in NGC 5466 is small ($\sim$20) and to check this preliminary result will require more accurate photometry of the available stars; (2) the two brightest Carina 'SX Phe stars' (Nemec & Mateo 1989) appear to lie well above the P-L relation for first-overtone SX Phe stars. Possibly these are higher-overtone pulsators, or a new kind of SX Phe star.

J.O. PETERSEN: Please comment on determination of absolute distances. In two to three years the HIPPARCOS satellite will provide an accurate parallax for SX Phe, and this information will be valuable for determining absolute distances from your relative distances. Is that correct?

J. NEMEC: Having tied our distance estimates to the zero-point cluster M15 the absolute distances depend directly on M15's assumed distance, which at present depends on Pop. II subdwarfs and the application of Lutz-Kelker corrections to the M15 main sequence. The HIPPARCOS results for SX Phe will certainly provide a much improved zero-point for all the P-L-[Fe/H] relations discussed above, and they are eagerly awaited.

N. SIMON: The LNA models (but with old opacities) for short period (<3 days) Type II Cepheids show that these stars lie in a region of the HR-diagram where the F and H blue edges have already crossed. This probably means that there ought not to be any overtone pulsators among these objects.

A. COX: I suggest that your Pop. II Cepheid dichotomy is due to a factor of two in mass (coalescence) rather than pulsation mode. This would give $\sqrt{2}$ in your period ratio and would give your two sequences.

J. NEMEC: I believe Prof. Böhm-Vitense has something to say about your comment and about Prof. Simon's comment. Also, Dr. Moskalik has shown me earlier today a preprint of his that argues for the existence of Pop. II Cepheids pulsating in the first-overtone mode.

E. BÖHM-VITENSE: I think you can have overtone pulsators in regions of the HR-diagram where fundamental mode pulsations are also unstable. I do not know whether there are mass differences in the variables that Jim has studied. Since the lower luminosities occur for the longer periods you would need systematically lower masses for the longer period ones as compared to the shorter period ones. If the anomalous Cepheids are *all* merged stars then all will have the higher mass and you will not have two mass sequences.

A. SANDAGE: The fit of all horizontal branch levels to be the same, independent of [Fe/H], poses a problem doesn't it? What is the range of [Fe/H] of the clusters you used? If $\Delta M_V = 0.3\Delta$[Fe/H], what is the error in the placement of the Type II Cepheids? Can this wipe out your two fundamental and first-overtone lines in the composite P-L relation?

J. NEMEC: Our first models (e.g., Fig. 2) assumed no dependence of HB-luminosity on metallicity, but the P-L relations were still surprisingly tight because metal abundance acts like reddening – an incorrect value translates into an incorrect distance but does not introduce scatter into the diagram. Our subsequent models all assumed metallicity dependences (at $P=0.5$ day) of $M_V(\text{RR}ab, -0.3) = 0.32$[Fe/H]$+1.19$ (L90) and $M_B(\text{RR}ab, -0.3) = 0.35$[Fe/H]$+ 1.53$, and dealt with stellar evolution according to the recent models of Y.-W. Lee. The metallicity range of the stellar systems that contain RR Lyrae stars discussed here is from –2.5 (NGC 5053) to –0.99 (M107); hence, metallicity differences and the effects of stellar evolution are generally rather small.

D. WELCH: The real acid test for your fundamental/first-overtone sequences will be the K-photometry or the use of a colour to remove the temperature-induced scatter.

J. NEMEC: Since March 1992 we have had a program underway at the Apache Point Observatory in New Mexico to obtain B, V, K photometry of the Cepheids in several galactic globular clusters.

Cepheids in IC 4182, Calibration of SNIa 1937C and the Hubble constant

A. Saha[1], A. Sandage[2], N. Panagia[1], G. A. Tammann[3], L. Labhardt[3], H. Schwengeler[3], F. D. Macchetto[1]

[1]*Space Telescope Science Institute, Baltimore, MD, USA,* [2]*Observatories of the Carnegie Institution of Washington, Pasadena, CA, USA,* [3]*Universitat Basel, Basel, Switzerland*

Abstract

Observations with the Wide Field Camera on HST were examined, and 27 Cepheids were discovered. Photometry and period analysis have produced unambiguous light curves free of alias. The observed P-L relation has a slope consistent with seminal calibration studies of Galactic and LMC Cepheids. An apparent distance modulus (in V) of 28.47 ± 0.10 to IC 4182 is derived. This implies that $M_V(max)$ for SNIa 1937C is −19.92 ± 0.2 mag, independent of everything except differential absorption between the Cepheids and the supernova. Using this to calibrate Hubble diagrams for SNIa by several authors, and allowing for a 1-σ uncertainty in the absolute magnitude from the calibration of only one SNIa, we obtain $H_0 = 45 \pm 14\,\mathrm{kms^{-1}Mpc^{-1}}$.

1. Introduction

IC 4182 is a nearby Sdm galaxy which produced a type Ia supernova (SNIa) 1937C (Baade and Zwicky 1938). Hubble diagrams constructed from data on accurately measured SNIa in more distant galaxies ($v > 1600\,\mathrm{kms^{-1}}$) indicate that the dispersion in the peak brightness of SNIa disperse about a mean value by 0.5 mag (1-σ). Thus they are reliable standard candles once the absolute peak magnitudes of a few such SNIa are measured. Here we describe the work to determine the distance to IC 4182 through the discovery and photometry of Cepheids using the Hubble Space Telescope (HST), and thus the calibration of the absolute peak magnitude of SNIa 1937C.

2. Data and Photometry

A field in IC 4182 was imaged at 20 different epochs by the Wide Field Camera (WFC) on the HST in the F555W bandpass (almost equivalent to V). The total time spanned by these observations was 47 days, and the epochs were judiciously spaced in time by amounts varying from 15 hours to 6 days to eliminate aliassing over the period range 3 to 45 days. At each epoch, two back to back exposures were made, each 2100s long. Each such exposure pair was checked for exact registration. To remove the cosmic rays, anti-coincident pixels (at the 6-σ level) were flagged, and the lower of the two values for a given pixel were retained in both the sub-exposures. These were then co-added to a single image, thus yielding 20 images at the 20 epochs that are cleaned of cosmic rays.

A *modified* version of the DoPHOT (Mateo & Schechter 1989) point spread function (PSF) fitting program was run to perform star identification and photometry. In this version an analytic PSF model is fit to the central cores of the stars. The reported brightness measures the rise of the central spiked core (of the spherically aberrated image) of a star *relative to its own 'wings'*. This procedure preserves the relative magnitude difference from one star to another, without having to ascertain the 'true' background under each object. For faint objects this procedure is tantamount to a maximum S/N extraction. The analytic PSF has fewer procedural problems for fitting to the severely undersampled data. The goodness of fit against the model stellar PSF is used to eliminate non-stellar objects and other image pathologies: in particular low level cosmic rays that are not trapped by the anti-coincidence method described above. Pathological pixels are 'turned off' and not used in any further computation. This procedure is crucial to getting good relative magnitudes. The final result is a list of bona-fide stars with positions and *relative* magnitudes with associated individual error estimates. At this point these magnitudes have arbitrary zero-points that vary from chip to chip and exposure to exposure: they are only self-consistent within any chip on any single epoch, and even then there is variation of order 5 percent (from control experiments on HST Science verification data on NGC 1855 that were offset in position and telescope roll angle) depending on position, since even the PSF of the cores of stellar images varies with position.

A 'template' image was constructed for each chip by stacking several exposures (using only integer pixel shifts to adjust the registration), and the PSF fitting process was applied to it as well, resulting in a more comprehensive list of stars and their *relative* magnitudes. We shall call the relative magnitudes obtained on the *template* image the *instrumental* magnitude. Bear in mind that the instrumental magnitudes thus defined are independent from chip to chip, and one must determine the zero-point to go to true magnitudes separately for each chip. For each of the four chips, the stars (from the DoPHOT generated lists) from each epoch were matched by position to those on the template. By examining the difference in relative magnitudes of the stars in any epoch against the instrumental magnitudes of the same stars on the template, the shift needed to go from the relative to instrumental magnitude can be determined and applied. Since this is based on the ensemble average for hundreds or thousands of objects, the shift is extremely robust and precise, to better than 0.02 mag. Note that the error in transforming from relative to instrumental magnitudes is much less than the uncertainty in magnitudes due to the PSF variation across each chip, since the observations were all made in the same position and orientation. For a given star, the relative *systematic* error in the instrumental magnitudes from one epoch to another is only 0.02 mag, although its magnitude relative to other stars in the chip may be uncertain by 0.05 mag.

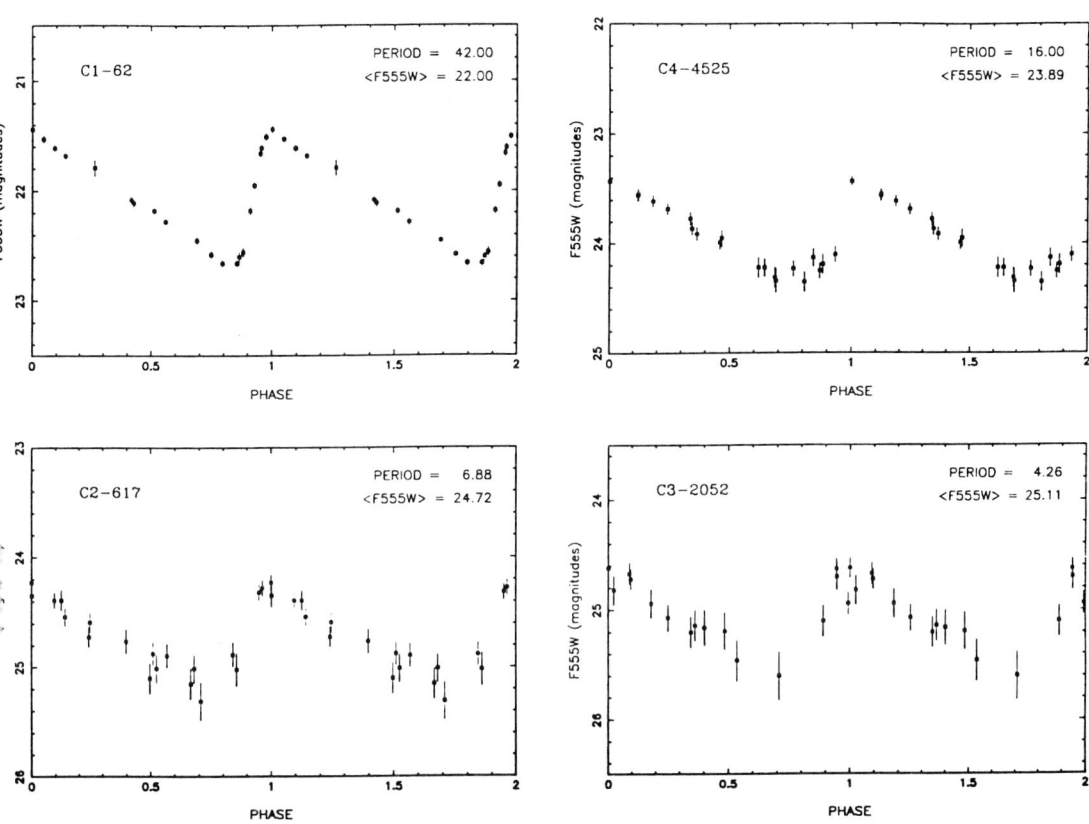

Figure 1. Representative light curves of Cepheids found in IC 4182.

The instrumental magnitudes, as operationally defined above, differ from the 'true' magnitudes by an additive constant, C, which is separate for each chip and must be evaluated independently. To do so, several stars in each chip that are relatively bright and apparently isolated were chosen, and the instrument Data Number (DN) counts within a 30 pixel radius aperture were determined. Using the calibration given for DN to magnitudes in the F555W system in Faber *et al.* 1992 (Table 12.15), the value of C (as above) for each chip was derived. In the more isolated areas the internal consistency (star to star variations within the chip area) in C was found to be within the 0.05 mag expected scatter. In the most crowded areas this concordance is hard to come by, and the uncertainty in the determination of C can be as much as 0.1 mag. Once determined, C can be applied to each object in each epoch to obtain all magnitudes on the standard F555W system.

The search for variables was made using a χ^2 test, followed by a 'periodicity' test based on a variant of the Lafler and Kinman (1965) algorithm. This technique has been demonstrated in Saha and Hoessel (1990). Unambiguous Cepheid light curves were obtained for 27 of the detected variables, ranging in period from 3 to 45 days. The special timing scheme applied in the observations assured the absence of period aliassing in this range. Fig. 1 shows 4 representative light curves, arranged in order of period. Note how as one goes fainter, the error bars increase.

The mean magnitudes in F555W are plotted against $\log P$ in Fig. 2, which is the familiar P-L relation. We consider the Sandage and Tammann (1968) P-L relation which gives:

$$< M_V > = -2.83 \log P - 1.37 \qquad (1)$$

Comparison with the Feast and Walker (1987) and Madore and Freedman (1991) P-L realtions for M_V shows that all three agree to within 0.1 mag over the period range 3 to 40 days. The continuous line in Fig. 2 shows the fit of the Sandage and Tammann (1968) slope to the data for IC 4182, and the dash-dot lines on either side mark the intrinsic scatter in the P-L relation due to the finite width of the instability strip in temperature. Note that the fainter Cepheids appear to be systematically too bright compared to the fitted mean ridge-line. We attribute this to the bias arising from the incompleteness of discovering Cepheids at the very faint limit. The fit is done to all the data irrespective of the bias, which has the effect of deriving a slightly smaller distance modulus. The fitted equation is:

$$< F555W > = -2.83 \log P + 27.16 \qquad (2)$$

The formal difference between $F555W$ and V magnitudes is documented in Harris *et al.* (1991), and the difference is about 0.03 mag. For the present we make no distinction between the two, implying an apparent distance modulus $(m - M)_{app} = 28.53$. Following the same procedure, but using the Feast and Walker (1987) and the Freedman and Madore (1991) calibrations yields $(m - M)_{app} = 28.43$ and 28.45

respectively. We adopt $(m - M)_{app} = 28.47 \pm 0.10$.

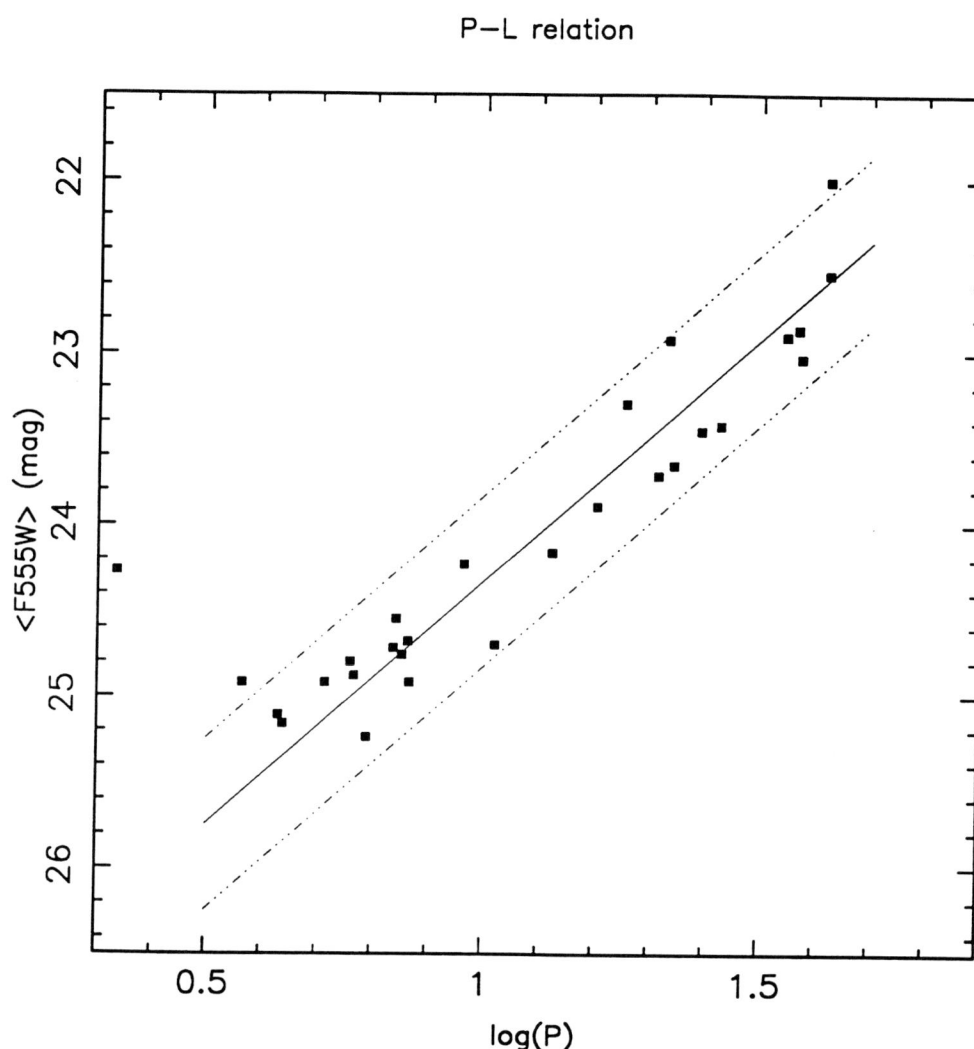

Figure 2 The P-L relation for Cepheids in IC 4182.

3. The Peak brightness of SN 1937C and the value of H_0.

The photographic m_{pg} light curve of SN 1937C were well determined by Baade and Zwicky (1938). The V light curve was determined by Beyer (1939). Standard template light curve fitting (Leibundgut et al. 1991) gives $V_{max} = 8.55$ and $M_{pg}(max)$

= 8.50 for this type Ia supernova. Since our apparent modulus to IC 4182 is determined in V, $M_V(max) = 19.92 \pm 0.2$ mag, as long as there is no *differential* extinction between the Cepheids and the location of SN 1937C.

We see already in Fig. 2 that the Cepheids essentially all lie within the range of the intrinsic scatter, indicating that the differential absorption is less than 0.4 mag in our field, which *includes* the location of SN 1937C. Further, from photometry on the coadded images, it is seen that resolution of the brightest 'normal' giants occurs at $F555W = 25.8$ mag. IC 4182 lies in the direction of the North Galactic Pole, and is thus free of extinction from our own galaxy. Since it is face on, the brightest 'normal' giants seen should be the ones on the closer side of the 'disk', and so free from internal extinction within IC 4182. The giant branch termination point has $M_V \approx -2.5$, thus implying a distance modulus of 28.3, which, as argued above, is free of the extinctions within IC 4182. This value agrees very well with the apparent modulus of 28.47 derived from the Cepheids, thus showing that there is no significant internal extinction to be accounted for, let alone differential extinction (the latter is the only factor that affects our determination of M_V for SN 1937C).

We apply the result obtained above to calibrate the SNIa Hubble diagram. Consider first the one given by Tammann and Leibundgut (1990), for all types of galaxies, where they obtain:

$$M_B(max) = -19.63(\pm 0.09) + 5\log(H_0/50) \tag{3}$$

and a mean value of the color at maximum of $(B-V)_0 = -0.27$, which can be used to transform the above relation to $M_V(max)$:

$$M_V(max) = -19.36 \pm 0.09 + 5\log(H_0/50) \tag{4}$$

Similarly, from Della Valle and Panagia (1992) we obtain (using the appropriate $(B-V)_0 = -0.16$ from their particular sample) we obtain:

$$M_V(max) = -19.96 \pm 0.18 + 5\log(H_0/50) \tag{5}$$

Again, from Sandage and Tammann (1982), who had a sample of SNIa in E and S0 galaxies only (i.e. no internal extinction), using again $(B-V)_0 = -.27$, we obtain:

$$M_V(max) = -19.74(\pm 0.24) + 5\log(H_0/50) \tag{6}$$

The above values bracket other mean calibrations in the literature.

If SN 1937C is a mean type Ia supernova, the above calibrations can be used to give a range for H_0:

$$39 \pm 5 \leq H_0 \leq 51 \pm 7 \ kms^{-1}Mpc^{-1} \tag{7}$$

The dominant uncertainty in this result arises from the fact that SN 1937C may not itself lie on the mean line of the Hubble diagram. The Tammann and Leibungut (1990) and Della Valle and Panagia (1992) Hubble diagrams show an intrinsic scatter in the B magnitudes of 0.50 mag, 1-σ. This implies a 24% uncertainty in the determination of H_0. The conservative estimate from the determination of the absolute magnitude of this one supernova is thus:

$$H_0 = 45 \pm 14 \; kms^{-1}Mpc-1. \tag{8}$$

The calibration of other SNIa in the above manner will significantly reduce the uncertainty.

References:

Baade W., Zwicky, F., 1938, Astrophys. J., **88**, 418.
Beyer. M. 1939, Astron. Nachr., **268**, 341.
Della Valle, M., Panagia, N. 1992, Astron. J. **104**, 696.
Faber, S. *et al.* 1992, Wide Field/Planetary Camera Final Orbital Science Verification Report, STScI.
Feast, M., Walker, A. R., 1987, ARA&A, **25**, 345.
Harris, H. C., Baum, W. A., Hunter, D. A., Kreidl, T. J. 1991, Astron. J. **101**, 677.
Lafler, J., Kinman, T. D., 1965, Astrophys. J. Supp., **11**, 216.
Leibundgut, B., Tammann, G. A., Cadonau, R., Cerrito, D., 1989, Astron. & Astrophys. **89**, 537.
Madore, B., Freedman, W. L., Pub. Astron. Soc. Pac., 1991, **103**, 933.
Mateo, M., Schechter, P. 1989, in *The First ESO/ST-ECF Workshop on Data Analysis*, ed. P. J. Grosbol, F. Murtagh, and R.H. Warmels (ESO, Munich), p. 69.
Saha, A., Hoessel, J. G., 1990, Astron. J., **99**, 97.
Sandage, A., Tammann, G. A., 1968, Astrophys. J., **151**, 531.
Sandage, A., Tammann, G. A., 1982, Astrophys. J., **256**, 339.
Tammann, G. A., Leibundgut, B., 1990, Astron. & Astroph., **236**, 9.

DISCUSSION

E. BÖHM-VITENSE: I would like to suggest that the systematic deviations that you find for your period-magnitude relation are due to a change from fundamental mode pulsation to first overtone pulsation for periods of about 10 days as is expected theoretically.

A. SAHA: We had ascribed the deviations to incompleteness, but it would be worthwhile to test if applying a period shift to account for change to first overtone would reduce the deviations.

W. FREEDMAN: A comment on your statement that the data for IC 4182 are confined to within the ridge lines of the PL relation and therefore suffer low differential extinction. It is also the case that the PL relations for the Cepheids in the inner regions of M31 do not exhibit a large dispersion and yet suffer up to 1 magnitude B absorption in the mean. In any case you do have I data that will allow you to determine the extinction of the IC 4182 Cepheids directly.

A. SAHA: Yes, but the calibration of the brightness of SN1937C depends only on the differential extinction, although the true distance modulus of IC 4182 is extinction dependent. We will have better limits on the value of the absolute extinction, when we've analysed the I data, as you've said.

M. PIERCE: I'm a bit concerned that there are internal inconsistencies in your determination of $M_B(max)$ for SN1937C. If you take the values of $m_V(max)$ and $M_{pg}(max)$ you gave in your talk and the canonical correction for pg to B mags along with the intrinsic color of $(B-V)_{max} = -0.25$ you get $E(B-V) \approx 0.43$ and $A_B \approx 1.4$ mag. The result is that $H_0 \approx 18 km s^{-1} Mpc^{-1}$, a ridiculously small value. Could you comment on the importance of reddening and extinction.

A. SAHA: Your argument would be valid if one could assert that the observed magnitudes and colors of SN1937C as well as the intrinsic color of SNIa's have negligible associated errors and/or uncertainties. When you take your inferred $E(B-V)$ and derive A_B, you have multiplied any measurement errors also by a factor of 4. Thus an entirely plausible error in measurement of the color of SN1937C of 0.2 mag say, plus an uncertainty in our knowledge of the intrinsic color at maximum of SNIa's by a like amount translates to an error that is 4 times as large, i.e. larger than 1 mag. Therefore I think it is misleading to follow your line of argument, since it can lead to fictitious results based solely on amplifying the errors and uncertainties by a factor of 4. The procedure we have chosen to follow, transforms the Hubble diagram obtained in B to one in V using the intrinsic colors and extinctions derived self consistently from within that data-set. This Hubble diagram in V is then compared against the $V(max)$ value of 1937C obtained from Beyer's data. This procedure avoids the problem of a 4× artificial magnification of errors.

Recent improvements to the Cepheid distance scale

Wendy L. Freedman[1] and Barry F. Madore[2]

[1] Carnegie Observatories
813 Santa Barbara Street, Pasadena, California 91101

[2] NASA/IPAC Extragalactic Database
IPAC/JPL/Caltech MC 100-22, Pasadena, California 91125

Abstract

In the course of the last decade significant advances have been made in the observations of Cepheid variables and in their successful application to the extragalactic distance scale. Much of this progress has come about as a result of new CCD and near-infrared photometry. These recent improvements are discussed, and a comparison is given of Population I Cepheids and Population II distances. The correspondence is good, with the zero points agreeing at a level of better than 15% in distance. At this same level of significance, a systematic difference between these distances scales may exist, in the sense that the RR Lyrae distances appear to be smaller than the Cepheid distances (if it is assumed, as has generally been done for extragalactic studies of RR Lyraes, that $M_V(RR) = 0.77$ mag, independent of [Fe/H]). However, several recently-published calibrations of $M_V(RR)$ significantly reduce this discrepancy. Finally, new Cepheid data for the nearby galaxy M81 are presented based on recent Hubble Space Telescope observations.

1. Introduction

Recent refinements to the extragalactic distance scale pale in comparison to the actual discovery of the statistical relation between period and luminosity for Cepheid variables. In the 1920's, refining the Cepheid PL zero point was important, but it was a minor detail in comparison to defining what the basic scale size of the Universe was.

A significant advance came when Baade discovered the (factor of two) distinction between the brightness of Classical Cepheids (which were being observed in external galaxies) and their fainter Population II counterparts, the W Virginis stars (which were acting as the calibrators in our Galaxy). In the 1950's, Arp, Sandage, Irwin, Kraft and others (see Fernie 1969 for a comprehensive historical review) began investigating Cepheids in Galactic clusters as a means of independently setting the Cepheid zero point, this time using Population I main sequence fitting. As photometric errors were decreased, the observed statistical relations among the apparent properties of the Cepheid samples (built up from observations of Cepheids in our Galaxy, the

Large and Small Magellanic Clouds and several other more distant members of the Local Group) were found to be systematically different in many of their properties; for example, (1) The number distributions of amplitude over period were found to be dramatically different. (2) The observed colors at fixed period were observed to differ from galaxy to galaxy. (3) The period-luminosity relations were all found to have a measurably large and what was shown to be intrinsic scatter. And, (4) some extragalactic samples even suggested that the assumption of a universal slope to the Cepheid PL relation might be erroneous. Some of these observed properties (1,2,3) have stood the test of time, whereas photometric scale errors were largely responsible for (4). And while (1) and (3) appear to be reflecting intrinsic properties of the Cepheids themselves, differing amounts of *internal* extinction affecting the extragalactic Cepheid samples appear to be responsible for most of property (2). In some cases in fact, the extinction corrections are so large as to rival the factors of two found so long ago by Baade.

A recent review of the Cepheid distance scale has been presented in Madore and Freedman (1991), and the details will not be repeated here. In that review, the basic physics underlying the Cepheid period-luminosity (PL) and period-luminosity-color (PLC) relations was discussed; observational issues such as the effects of reddening, metallicity and photometric errors were presented in the context of the Cepheid distance scale; the distances to individual galaxies were tabulated; and finally a consistent set of data was used to provide PL relations at 7 wavelengths: $BVRIJHK$. Other recent reviews of the Cepheid distance scale include those of Madore (1986), Feast and Walker (1987), Freedman (1988a), and Jacoby et al. (1992).

The most serious systematic error embedded in the Cepheid distance scale concerned the assumption that the Cepheids in external galaxies suffered no absorption internal to the parent galaxies in which the Cepheids were being studied. This assumption was eventually shown to be false, largely as a result of the availability of new detectors over the past decade (CCDs, IR photometers and arrays). Identifying the problem and finding a solution has been one of the many important aspects of extragalactic Cepheid research undertaken in the last decade.

2. The Significance of Recent Changes Due to New Detector Technology

The opinion that advancements in detector technology have significantly improved the precision of the Cepheid distance scale is not universally shared however. Specifically, Tammann (1992) states that *"despite the considerable progress in Cepheid research, concerning for instance the use of infrared magnitudes (which address the problem of intrinsic absorption and metallicity differences) the resulting distances have changed very little between 1974 and 1991 ([his] Table 1). On the whole the galaxies with known Cepheids have witnessed a distance increase by only 0.05 mag (2.5 percent!)"*. In another context, Sandage et al. (1992) reiterate the similarity of the same two calibrations. The first two columns of our Table 1 (below) list the Sandage and Tammann (1974) [hereafter ST 1974] and the Madore and Freedman (1991) [MF 1991] calibrations as tabulated by Tammann (1992).

The agreement between the ST 1974 and the MF 1991 calibrations is certainly remarkable; and we remark upon it below, because it is almost entirely fortuitous, and as such, it is extremely misleading. Furthermore, while the question repeatedly arises concerning the differences between the Sandage and Tammann calibrations and the more modern determinations, the answers are often not known to those other than distance scale afficionados.

First, as extensively documented in Freedman (1988b), Freedman, Wilson and Madore (1990), Christian and Schommer (1987), Capaccioli et al. (1992) there are enormous scale errors in the photographic photometry on which the ST 1974 calibration was based. At $B = 22$ mag such errors *typically* reach about 0.5 mag.

Table 1. Distances to Local Calibrators

Galaxy	μ_{ST1974}	$\mu_{0,MF1991}$	μ_{ST1984}
LMC	18.59	18.50	18.95
SMC	19.27	18.87	-
NGC 6822	23.95	23.59	-
IC 1613	24.43	24.42	-
M31	24.12	24.44	24.19
M33	24.56	24.63	**25.23**
NGC 2403	27.56	27.51	27.66
M81	27.56	27.59	**28.73**
M101	29.3	29.38	-

Second, the ST 1974 calibration is based almost exclusively on apparent B band photometry (which is all that was readily available before the advent of red-sensitive CCDs and IR detectors). Consequently the ST 1974 calibration provides *apparent* (not true or internal-reddening-corrected) moduli; they have been corrected only for modest amounts of foreground absorption. As clearly shown in numerous studies over the past decade (e.g., Freedman 1986, 1988a,b; Freedman, Wilson, and Madore 1991; Madore 1985; Freedman and Madore 1991) the apparent distance moduli, when measured at different wavelengths systematically decrease in a manner that is well fit by a standard interstellar extinction law. It is necessary to correct any apparent modulus for the effects of total (foreground plus internal) extinction; if the extinction is appreciable (as it is for many late-type spirals), then neglecting this term always leads to systematic overestimates of the distance.

A few examples will show the significance of these effects: (a) In Baade and Swope's Field III in M31 the apparent (CCD) B modulus is 25.36 mag (Freedman and Madore 1991). The true distance modulus, based on $BVRI$ photometry is 24.44 mag, as entered in Table 1, Column 2, in such apparently good agreement with ST 1974. The $BVRI$ CCD apparent moduli are shown in Figure 1 as a function of inverse wavelength. The B band apparent modulus is the relevant modulus to compare directly with the ST 1974 calibration which is also based on B (photographic) photometry. The apparent B moduli differ by 1.24 mag (!). (b) The apparent B band CCD modulus of M33 is 25.04 mag (Freedman, Wilson and Madore 1991). The two B moduli for M33 differ by 0.48 mag.

Third, the distances presented by MF 1991 are all computed relative to an LMC modulus of 18.50 magnitudes. On the other hand, the LMC modulus in the ST 1974 calibration is 18.59 mag. Comparing the two Cepheid calibrations at least requires the zero points to be consistent, while neglecting it introduces a small (but again systematic) difference in zero point of 0.14 mag (as opposed to the 0.05 mag offset quoted by Tammann 1991 above).

In later studies Sandage (1983) and then Sandage and Carlson (1983a) based their zero point for their distance modulus to M33 on an apparent blue LMC modulus of 18.95 mag. This LMC modulus leads to an apparent blue modulus for M33 of 25.35 mag. [That is, Sandage and Carlson (1983 - their Figure 3) found a difference in apparent B moduli between M33 and the LMC of 6.4 mag. The value of 25.35 (= 18.95 + 6.4) thus includes a component due to the known LMC foreground reddening.] However, incorporating the increased distances, Sandage and Tammann (1984) tabulated their then current values for the local calibrator distances useful for the Tully-Fisher relation. These distances are given in Column 4 of Table 1. In that compilation the adopted zero point again corresponds to the apparent B modulus to the LMC of 18.95 mag. Yet these apparent moduli are equated with the true moduli in the application of the Tully-Fisher calibration. And they are increased by almost half a magnitude due purely to LMC foreground extinction, and ignore any contribution from extinction *internal* to either the LMC or M33.

M81 presents a unique case. The ST 1974 calibration lists NGC 2403 and M81 as having similar distance moduli of 27.6 mag. Given that only 2 Cepheids in M81 were known to Sandage, he did not attempt to derive a direct Cepheid distance to M81; rather the galaxies were assumed to be at the same distance on the basis of group membership as defined by Holmberg. Then, much later, Sandage (1984 ; *i.e.*, in the interval between 1974 and 1992) noted that the brightest long-period Cepheids in M81 (for which he did have periods) are fainter than the brightest long-period Cepheids in NGC 2403; on the basis of this comparison he placed M81 at a distance modulus of 28.8 mag, over a full *magnitude* more distant than NGC 2403. (Given the existence of the PL relation however, the higher luminosities of the longest-period Cepheids in NGC 2403 are of course to be expected since they have periods of 67 days, more than a factor of two larger than those in M81 which have periods of only 30 days). Subsequent I-band photometry of the two known (30 day) Cepheids in M81 showed them to have magnitudes very similar to their 30-day counterparts in NGC 2403, and thereby ruled out the larger distance for M81.

The truly significant changes that did occur in the distances to many important calibrating galaxies in the period between 1974 and 1992 led at least some authors at that time (*e.g.* Aaronson 1985) to conclude that any value of the Hubble constant between 50 and 100 km/sec/Mpc could be obtained depending on the specific choice of calibrators. Fortunately, the recent convergence of the distances to nearby galaxies based on new Cepheid, RR Lyrae and other indicators, (discussed below) has largely removed this freedom of choice.

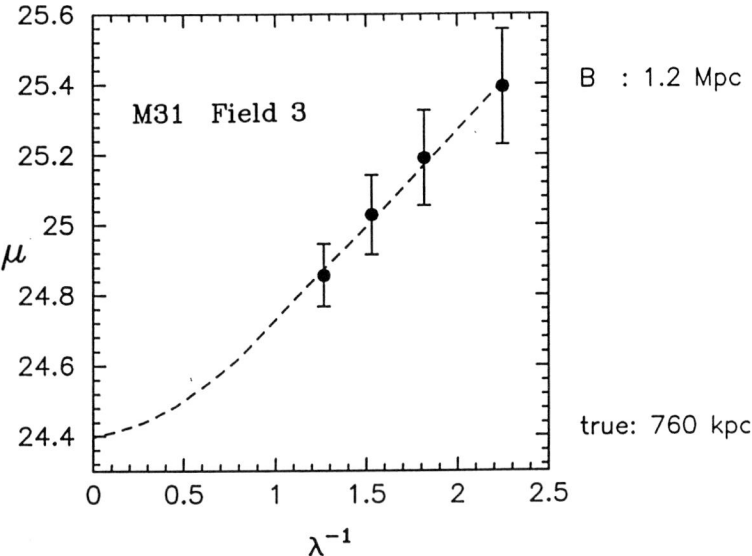

Figure 1. Apparent $BVRI$ distance moduli versus inverse wavelength for Cepheids in M31, Baade and Swope's Field III.

To summarize, the apparent excellent agreement between the Sandage and Tammann (1974) and the Madore and Freedman (1991) calibrations is misleading. The ST 1974 calibration is based on apparent B band distance moduli, uncorrected for the effects of internal interstellar extinction; it is based on photographic photometry demonstrated as having significant scale errors, and finally, the MF 1991 calibration has a different zero point. The remarkable aspect of Tammann's Table 1 may not be the good agreement of those two particular calibrations, but rather that despite the fact that the distances to some calibrators individually changed by almost a factor of two in the interval *between* 1974 and 1992, the value of the Hubble constant derived by him and his collaborators showed very little change within the same period of time.

3. A Comparison of Cepheid Distances with Other Methods

Table 2 presents a comparison of recent distance modulus determinations to the galaxies M31, M33, the LMC and IC 1613. Included are distances obtained using new CCD or infrared photometry for Cepheids, RR Lyraes, and the tip of the red giant branch (TRGB). And finally the Type II supernova expansion parallax measurement of the distance to the LMC is shown for comparison. The RR Lyrae distances listed in Table 2 are based on assuming $M_V(RR) = +0.77$ mag.

Table 2. Recent Distance Determinations to M31, M33, LMC and IC 1613

Method	M31	M33	LMC	IC 1613
Cepheids	24.44 [1]	24.63 [2]	18.50 [3]	24.42 [4]
			18.52 [5]	
RR Lyraes	24.33 [6]	24.45 [7]	18.22 [8]	24.10 [9]
Giant Branch	24.40 [10]	24.8 [11]		24.21 [12]
				24.27 [13]
SN II			18.50 [14]	

References to Table 2: 1. Freedman and Madore (1990); 2. Freedman, Wilson and Madore (1990); 3. Welch et al. 1987; 4. Freedman (1988a); 5. Feast (1988); 6. Pritchet and van den Bergh (1987); 7. Pritchet (1988); 8. Walker (1992); 9. Saha et al. (1992); 10. Mould and Kristian (1986); 11. Mould and Kristian (1986); 12. Freedman (1988b); 13. Lee, Freedman and Madore (1992) 14. Panagia et al. (1991)

The first point to note about Table 2 is that the distances to individual galaxies based on both Population I and II indicators have converged to (*full range*) differences of less than 0.3 mag (or 15% in distance). The Pop I and II distance scales have been completely independently calibrated, the former based on main sequence fitting of Galactic clusters, and the latter based on statistical parallax and Baade-Wesselink analyses of field RR Lyraes. Furthermore, as discussed below, the calibration of the RR Lyraes is still a matter of some controversy, and other published calibrations in fact result in an improvement of the agreement between the Cepheid and RR Lyrae distance scales. Nevertheless, despite remaining subtle calibration uncertainties, the good agreement of the distances to nearby galaxies indicates that the remaining controversy over the Hubble constant is due almost entirely to the uncertainties in the secondary distance methods.

Furthermore, as illustrated graphically in Figure 2, the good agreement amongst the various methods also suggests that there are no large systematic errors in the distance scale still lurking which might be due to the effects of metallicity (say). In Figure 2, differences in distance moduli are plotted versus estimates of [Fe/H] determined for the halo populations in each of these galaxies. Shown are the differences in the Cepheid minus RR Lyrae, Cepheid minus tip of the red giant branch (TRGB), and RR Lyrae minus TRGB distances. Open circles indicate distances based on the assumption that $M_V(RR) = 0.77$ mag, whereas the dots represent distances based on a recent calibration by da Costa and Armandroff (1990) where $M_V(RR) = 0.17$ [Fe/H] + 0.82.

The second point to note about Table 2 is that although the Cepheid and RR Lyrae distance estimates agree to within their stated errors, the differences are systematic. The quoted uncertainties in the RR Lyrae moduli *include* a component to reflect the current uncertainty in the RR Lyrae zero point. In both the LMC and IC 1613 the zero point difference amounts to +0.3 mag (in the sense that the RR Lyrae distances are smaller than the Cepheid distances). Unfortunately the RR Lyrae distance for M33 is a preliminary estimate only, and may not provide additional information concerning the reality of such a systematic difference. Finally, in M31 the discrepancy is small (but the sense of the discrepancy is the same). Further work is clearly needed to establish the magnitude and/or reality of this effect.

As discussed recently by many authors (*e.g.* Sandage and Cacciari 1990; Carney *et al.* 1992; Walker 1992), the dependence of $M_V(RR)$ on metal abundance is not well-established. Extragalactic determinations of distances based on RR Lyraes have adopted the standard $M_V(RR) = 0.77$ mag, a value based on statistical parallax measurements of a sample of Galactic RR Lyraes, in which there are few metal-poor stars. Unfortunately, most of the extragalactic RR Lyrae stars known have been found in the metal-poor halos of the galaxies.

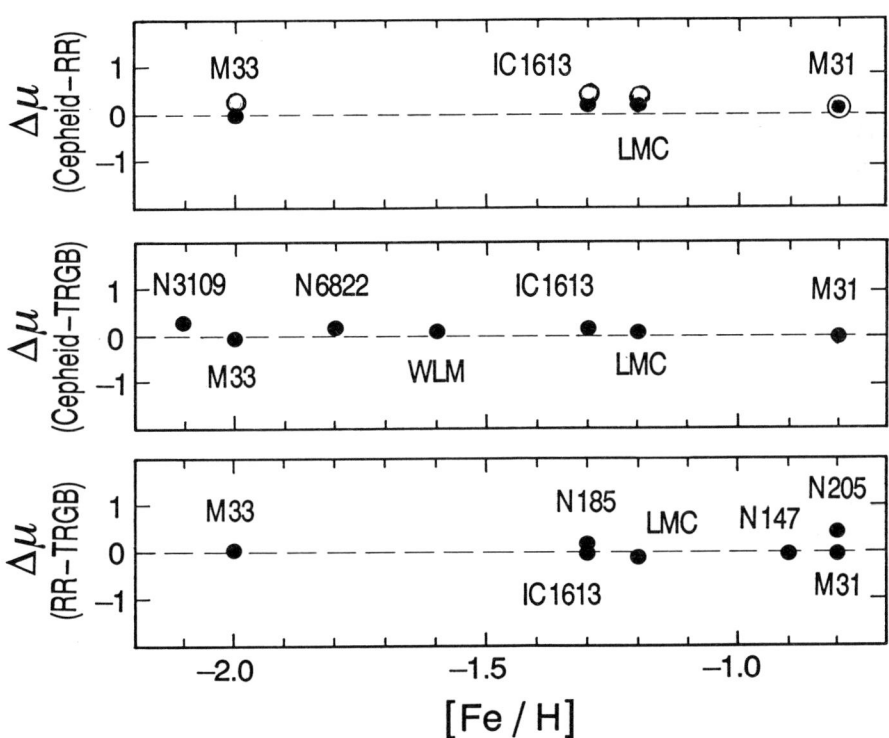

Figure 2. Differences in distance moduli as a function of halo [Fe/H]. (1) (Cepheid - RR Lyrae); (2) (Cepheid - TRGB); (3) (RR Lyrae - TRGB).

As seen in Figure 2, agreement between the Cepheid and RR Lyrae distance scales is improved if, for example, the recent calibration by Da Costa and Armandroff (1990) is adopted. The *slope* of the latter calibration agrees well with recent Baade-Wesselink studies of Galactic RR Lyrae stars (Carney *et al.*), but the *zero point* differs by 0.2 mag, in better agreement with the Cepheid distances. The disadvantage of the Da Costa and Armandroff calibration (which is based on the theoretical models of Lee, Demarque and Zinn 1990), is that it carries with it associated uncertainties in modelling red giant branch evolution; for example mass loss on the red giant branch is not well-understood; the calibration is also dependent on horizontal branch evolution, and is very sensitive to the adopted helium abundance.

Alternatively, is the Cepheid distance scale in error? For example, is the Cepheid distance modulus to the LMC correct? Or are there systematic effects in the Cepheid distance scale due to metallicity? The excellent agreement between the Cepheid and SN 1987A expansion parallax estimates listed in Table 2 suggests that the LMC Cepheid distance is not seriously in error. As discussed by Saha *et al.* (1992), a systematic error in the Cepheid moduli due to metallicity appears unlikely since a theoretically-predicted metallicity-dependence of the Cepheid distance scale would actually increase, rather than resolve, the observed discrepancy.

It is worth noting however that the implications of such a discrepancy, if confirmed, may be profound. In particular, if the Cepheid distances are correct, (the point of view adopted by Walker 1992) it would imply that the absolute magnitudes of RR Lyraes are brighter than currently believed. And an adjustment to the RR Lyrae distance scale zero point of 0.3 mag would result in a decrease in the ages of globular clusters by about 30%. It is interesting to note that while the distances to nearby galaxies now appear to have converged to a level where they have a small impact on uncertainty in the Hubble constant, subtle differences of only a few tenths of a magnitude may still have an impact on cosmology, through the ages determined from stellar evolution. Although the ages of globular clusters are widely regarded as theoretically-determined quantities, in the process of determining ages, it is still necessary to interface theory with observation and transform the apparent magnitudes of globular cluster stars to bolometric luminosities (via an accurate distance scale). It is also worth recalling in this context that this dependence of ages on distance is true for *all* methods of determining absolute (rather than simply relative) globular cluster ages (*e.g.*, Renzini 1991).

4. Using HST for the Discovery of Cepheids

The discovery of Cepheids from space offers a powerful advantage over ground-based searches, even beyond the vital improvement offered in potential resolution. That is, from space, one has the unprecedented luxury of specifying the timing of the observations in order to to minimize aliasing problems in the period determination and maximize phase coverage in the time-averaged magnitude determinations; furthermore, the spacing is not altered by weather conditions.

In December 1991, the first sequence of HST observations of two fields in the relatively nearby galaxy M81 were obtained as part of the Key Project on the Extragalactic Distance Scale.[1] Our only constraint was that the roll angle of the spacecraft

---

[1] Based on observations with the NASA/ESA Hubble Space Telescope, obtained at the Space Telescope Science Institute, which is operated by AURA under contract NAS5-26555.

had to be kept within a narrow range, (consistent with properly illuminating the solar panels yet still allowing us to point back to the same position unrotated throughout the observing sequence). Consequently we were restricted to a 50-day initial observing interval. A power-law spacing algorithm (Madore and Freedman 1993) was developed to space the observations in such a way as to provide almost uniform phase coverage for Cepheids with periods in the range 5 to 50 days. To date 12 epochs of data at V and 4 epochs at I have been obtained for the two fields. Six additional observations have been scheduled to allow us to improve the periods (particularly important for the longest-period Cepheids, for which few, if any, complete cycles have been observed within the 50-day window).

To date the results of the observing strategy and variable star search techniques have been very encouraging. In total, about two dozen candidate Cepheids have been discovered thus far in the two fields, increasing by an order of magnitude the number of Cepheids with known periods in this galaxy. The previously known Cepheid V30 was recovered and its 30-day period confirmed. The data were reduced using a new version of DAOPHOT (named ALLFRAME) developed by another member of the Key Project Team, Peter Stetson. ALLFRAME was designed explicitly for the purpose of reducing a series of frames all covering the same field, and incorporates an empirically-generated PSF that varies with position on the chip. Figure 3 presents a light curve of one of the new Cepheid candidates. The light curve is of excellent quality, indicating that, with care, HST can produce photometry of very high precision. Moreover, with the success of our M81 effort and the similar work on IC 4182 (reported by Saha in this volume) HST is now accomplishing one of the tasks it was originally designed for and intended to do (namely the discovery of Cepheids in external galaxies, and the determination of the Hubble constant).

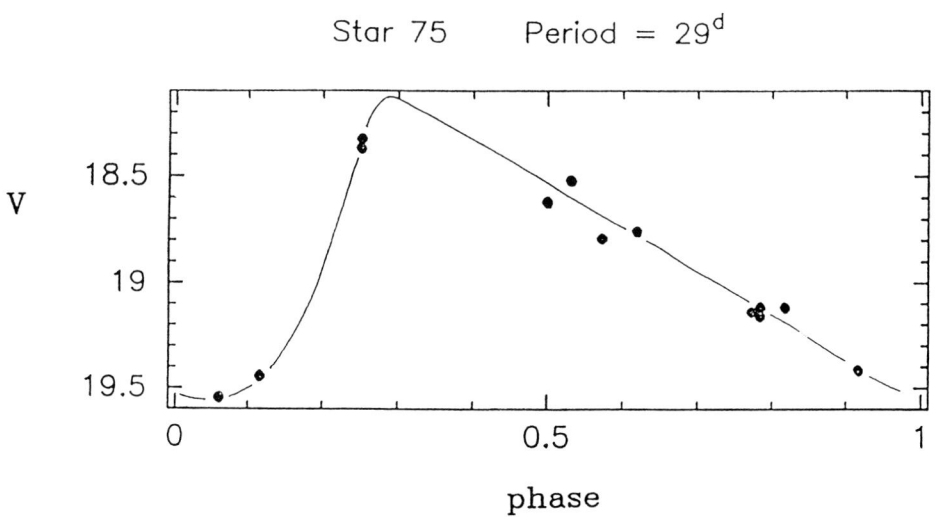

Figure 3. Light curve for one of the new HST Cepheid candidates in M81.

Obviously, with the HST discovery of Cepheids for a single galaxy in our program, it is premature to derive a value of the Hubble constant (however, see Sandage *et al.* 1992).

Encouraged by the quality of the photometry for the M81 Cepheids the HST Key Project Team is intending to look for Cepheids in 2 fields in M101 during upcoming cycles. An inner and an outer field in M101 will provide us with an opportunity to undertake a further test for the metallicity effects in the Cepheid distance scale, as done previously for Cepheids in M31 (Freedman and Madore 1991). When the HST optics have been corrected we intend to then pursue our long-term program of measuring the distances to a statistically significant sample of nearby spirals.

It is a pleasure to acknowledge the efforts of the additional Key Project Team members: Sandy Faber, Holland Ford, Jim Gunn, John Hoessel, John Huchra, Rob Kennicutt, Garth Illingworth, Jeremy Mould, and Peter Stetson, and especially Shaun Hughes and Myung Gyoon Lee. The work on the red giant branch distance scale has been done in collaboration with Myung Gyoon Lee. WLF acknowledges continual stimulating discussions with Allan Sandage, particularly in the Santa Barbara St. library. We have profitted greatly from the use of the NASA/IPAC Extragalactic Database (NED). Support for this work was provided in part by NSF grant AST-91-16496 and by NASA through grant numbers GO-227.04-87A. BFM is supported in part by JPL Caltech, under the sponsorship of NASA's Office of Space Science and Applications.

References:

Carney, B., Storm, J. & Jones, R. V., 1992, ApJ, 386, 663

Da Costa, G.S., Armandroff, T.E., 1990, AJ 103, 1151

Feast, M. 1988, in *"The Extragalactic Distance Scale"*, ASP Conference Series Vol. 4, eds. S. van den Bergh & C. Pritchet (Provo, Brigham Young University Press), p. 9

Feast, M., & Walker, A. R. 1987 ARA&A, 25, 345

Fernie, J. D. 1969, PASP, 81, 707

Freedman, W. L. 1986, in *Galaxy Distances and Deviations from Universal Expansion*, eds. B. F. Madore & R. B. Tully, (Dordrecht: Reidel), p. 21

Freedman, W. L. 1988a, in *"The Extragalactic Distance Scale"*, ASP Conference Series Vol. 4, eds. S. van den Bergh & C. Pritchet (Provo, Brigham Young University Press), p. 24

Freedman, W. L. 1988b, ApJ, 326, 691

Freedman, W. L., & Madore, B. F. 1990, ApJ, 365, 186

Freedman, W. L., Wilson, C., & Madore, B. F. 1991, ApJ, 372, 455

Jacoby, G. H. *et al.*, 1992, PASP, 104, 599

Lee, Y.-W., Demarque, P., and Zinn, R. 1990, ApJ, 350, 155

Lee, M. G., Freedman, W. L., & Madore, B. F. 1993, in preparation

Madore, B. F. 1986, in *"Galaxy Distances and Deviations from Universal Expansion"*, eds. B. F. Madore & R. B. Tully (Dordrecht, Reidel), p. 29

Madore, B. F. & Freedman, W. L. 1991, PASP, 103, 933

Madore, B. F. & Freedman, W. L. 1993, in preparation

Mould, J. & Kristian, J. 1986, AJ, 305, 591

Panagia, N., Gilmozzi, R. Macchetto, F., Adorf, H.-M., & Kirshner, R. P. 1991, ApJ, 380, L23

Pritchet, C. J. 1988, in *"The Extragalactic Distance Scale"*, ASP Conference Series Vol. 4, eds. S. van den Bergh & C. Pritchet (Provo, Brigham Young University Press), p. 59

Renzini, A. 1992, in *"Observational Tests of Inflation"*, eds. T. Banday & T. Shanks, (Dordrecht: Reidel), in press

Saha, A., Freedman, W. L., Hoessel, J G., & Mossman, A. E., AJ, 104, 1072

Sandage, A. R. 1983, AJ, 88, 1108

Sandage, A. R. 1984, AJ, 89, 621

Sandage, A. R. 1988, ApJ, 331, 605

Sandage, A. R., & Cacciari, C. 1990, ApJ, 350, 645

Sandage, A. R., & Carlson, G. 1983a, ApJ, 267, L25

Sandage, A. R., & Carlson, G. 1983b, ApJ, 258, 439

Sandage, A. R., Saha, A., Tammann, G. A., Panagia, N., & Macchetto, F. 1992, ApJ, in press

Sandage, A. R., & Tammann, G. A. 1974, ApJ, 190, 525

Tammann, G. A. 1992, *The Cosmic Expansion and Deviations from It*, to appear in *Physica Scripta*

Tammann, G. A., & Sandage, A. R. 1968, ApJ, 151, 825

Walker, A. R. 1992, ApJ, 390, L81

Welch, D. L., McLaren, R., Madore, B. F., & McAlary, C. W. 1987, ApJ, 321, 162

Calibration of the Cepheid Distance Scale

Wolfgang P. Gieren[1] and Pascal Fouqué[1,2]

[1]*Astrophysics Group, Facultad de Física*
Pontificia Universidad Católica de Chile, Santiago, Chile
[2]*Observatoire de Meudon, Paris, France*

Abstract

The absolute calibration of the Cepheid period-luminosity (PL) relation with galactic Cepheids is discussed. Various methods, most importantly the cluster ZAMS-fitting scale and the Baade-Wesselink scale are found to yield PL zero points which agree within $\sim \pm 0.1$ mag. The present Cepheid calibration sets the Large Magellanic Cloud at μ_o (LMC) = 18.6 $\pm$ 0.1 mag, in good agreement with the distance derived from SN 1987A and other methods except RR Lyrae stars which seem to give a shorter distance scale.

1. Introduction

Cepheid variables continue to be the most important calibrators of the extragalactic distance scale. This is so because they are intrinsically bright, easy to detect and obey a period-luminosity (PL) relation for which we have reasons to believe that it is universal.

In order to derive *absolute* distances to galaxies one has to calibrate the *local* distance scale using Cepheids in our Galaxy. This is a complicated problem and has occupied researchers for some 80 years now without full confidence in the results being attained even today.

Recent reviews dealing with Cepheids as standard candles have been given, among others, by Madore and Freedman (1991), Walker (1988), and Feast and Walker (1987). In the past few years, progress has been made in several respects. Improved Cepheid distances to a number of nearby galaxies have been measured using multicolor CCD photometry (reviewed elsewhere in this volume), and there has been important work on improving our knowledge of the absolute magnitudes of galactic Cepheids using a variety of methods, which will be reviewed here.

It is clear that Cepheid observations in infrared passbands have attained increasing importance in recent years for distance determinations of extragalactic Cepheids, mainly because of less problems with reddening corrections and reduced light amplitudes in the infrared. Still, work on the galactic calibrators has mainly involved observations in the B and V passbands, and for this reason this review will concentrate on the calibration of the PL(V) relation.

Infrared Cepheid PL relations have been discussed elsewhere (e.g. Madore and Freedman 1991; Feast and Walker 1987; Laney and Stobie 1986).

2. The metallicity dependence of the PL and PLC relations

Theory predicts (e.g. Stothers 1988) that the slope of the PL(V) relation is practically metallicity independent. Observations of Cepheids in the Galaxy and the Magellanic Clouds, as well as in some Local Group galaxies seem to confirm this, yielding PL slopes close to -2.9. Different slopes which have occasionally been found can probably be attributed to selection effects due to the observations of only the *brightest* Cepheids in a galaxy which produces a bias of the Malmquist type (Sandage 1988), or to calibration errors in photographic photometry near the plate limit.

Theoretical work also shows that the zero point of the *bolometric* PL relation is metallicity independent (Stothers 1988). The bolometric correction BC is a weak function of metallicity (Laney and Stobie 1986) but since BC is very small in the V band no appreciable metallicity effect on the PL(V) zero point is expected. This has recently been confirmed by the work of Freedman and Madore (1990) on M31 Cepheids which has shown that there are no significant zero point shifts of the PL relations obeyed by Cepheids in fields of widely different radial distances from the center of M31 having metallicities ranging over a factor of ~ 5.

The basic problem in the use of a PLC relation for distance determinations is its strongly increased sensitivity to metallicity differences in the Cepheids (e.g. Stothers 1988). Caldwell and Coulson (1986) have found a PLC zero point offset of 0.51 mag between our Galaxy (at $z = 0.02$) and the SMC (at $z = 0.005$). For galaxies for whose Cepheids accurate metallicities and reddenings are available the use of a PLC relation may be as good or even preferable to the use of a PL relation, but in practice this information is only available in the Magellanic Clouds. Another problem which is still with us is the true size of the color coefficient in the PLC relation which is difficult to determine due to the necessity for disentangling the effects of differential reddening. This is briefly discussed in Section 7 of this review.

3. Calibration on the galactic Cepheid PL(V) relation

There are at least four observational methods to find Cepheid absolute magnitudes. These are:

a) ZAMS-fitting to the color-magnitude diagrams of clusters containing Cepheids

b) Baade-Wesselink methods

c) Statistical parallaxes

d) Binary systems containing Cepheids

The cluster method has traditionally been considered as the most reliable means to obtain Cepheid luminosities. However, there are a number of severe difficulties with this method. There are few rich and nearby clusters containing Cepheids; most clusters are distant and sparse, and Cepheid membership is sometimes quite controversial due to the lack of additional membership information such as radial velocities of the cluster stars. Most clusters have high and variable reddening due to their low galactic latitudes, and have high field star densities contaminating the observed CMD's. The ZAMS is sensitive to the cluster metallicity (e.g. VandenBerg and Bridges 1984) and this is often not known, and there is the problem of the true distance of the standard cluster. The Pleiades are now generally used for this purpose rather than the Hyades cluster because of less problems with evolved stars on the upper main sequence due to its younger age, and of its solar metallicity. Since van Leeuwen's (1983) trigonometric parallax study a value of μ_o(Pleiades) = 5.57 mag has been favored but more recent evidence (e.g. Gatewood et al. 1990; Feast 1991) suggests a slightly larger value of $\sim$ 5.70 which carries directly over into the ZAMS-fitting Cepheid distance scale.

Baade-Wesselink (BW) methods have the advantage of yielding Cepheid luminosities which are *independent* of any of the other methods. BW distances can be found to any Cepheid having the necessary data (photometry and radial velocities) available allowing the PL relation to be calibrated with a large number of stars. Problems with uneven filling of the Cepheid instability strip and a very limited range in period and color of the calibrating Cepheids, as inherent in the cluster method, can be largely avoided in BW studies, but there are possibly severe sources of systematic error, mainly regarding the color indices appropriate to those studies.

The statistical parallax method has most recently been used by Wilson et al. (1991) to derive Cepheid luminosities. The resulting Cepheid magnitudes have relatively large errors but serve as a valuable check on the other methods.

Analyses of Cepheid companions detected in the ultraviolet by the IUE satellite have yielded absolute magnitudes of binary Cepheids in several cases (Evans 1991; Böhm-Vitense 1985) and a typical error in the best studied cases seems to be ± 0.3 mag (see reviews by Evans and Böhm-Vitense in this volume). The binary method is valuable as an additional means to obtain luminosity information on Cepheids, but it is still quite limited regarding the number of studied cases and the accuracy of the results.

4. The ZAMS-fitting Cepheid distance scale

Recent calibrations of the galactic Cepheid PL relation using the ZAMS-fitting method have been given by Gieren and Fouqué (1992)(=GF), Turner (1992), Fernie (1992), Walker (1988), and Feast and Walker (1987). The coefficients of least-squares solutions to the respective data sets used by these authors are given in Table 1. Since the slopes are all close to -2.90 and their mean value is -2.903 ± 0.019, GF argue that -2.90 ± 0.02 should be *adopted* as the appropriate slope of the PL(V) relation. Redetermining the PL zero points of the different data sets with the slopes forced to -2.90 then yields the following results:

FW 87:	-1.251 ± 0.038	$\sigma = 0.203$	N = 28 stars
Wa 88:	-1.228 ± 0.036	$\sigma = 0.191$	N = 28
Fe 92:	-1.203 ± 0.029	$\sigma = 0.26$	N = 28 (+28 δ Sct)
Tu 92:	-1.192 ± 0.022	$\sigma = 0.068$	N = 10
GF 92:	-1.292 ± 0.049	$\sigma = 0.275$	N = 32

There seems to be a strong selection effect in the small Turner sample (10 stars) which further exhibits an unrealistically low dispersion. GF find a zero point of -1.204 ± 0.077 from *their* absolute magnitudes if they restrict their sample to the 10 Cepheids studied by Turner, in agreement with his result, but with a $\sigma = 0.243$ comparable to other studies.

The GF study yields a PL zero point $\sim$ 0.07 mag *brighter* than the other ZAMS-fitting studies. This is essentially due to the fact that they use Cepheid color excesses as given in Fernie (1990) and determined from the Cepheids themselves, rather than calculating the Cepheid color excesses from the cluster OB star reddenings, as done in the other studies. Another improvement in the work of GF, besides of including the most recent data on true cluster distance moduli, is a homogeneous treatment of R, the ratio of total to selective absorption. GF use the values of R for the clusters as given in the original references *if these were measured*; in other cases they standardize the cluster R values to

$$R = 3.06 + 0.25(<B>_o - <V>_o) + 0.05E(OB) \qquad (1)$$

using $<B>_o - <V>_o = -0.2$. The coefficients in (1) are from Olson (1975) and the zero point is chosen so that the value of R agrees with the mean value found for 51 galactic clusters by Turner (1976). All μ_o (cluster) values given in the original references were adjusted to the corrected R (cluster) values and to μ_o(Pleiades) = 5.57. The values of R appropriate to the Cepheids were also calculated from (1) but using the Cepheid intrinsic colors and the Fernie (1990) E(B-V) values. The GF calibrating data are reproduced in Table 2.

5. The Baade-Wesselink distance scale

The most recent and most complete BW study which seems to be representative of other BW work over the past few years using appropriate color indices and techniques (e.g. Coulson, Caldwell and Gieren 1986; Caccin et al. 1981) is that of Gieren, Barnes and Moffett (1992) (=GBM). GBM use the surface brightness version of the BW technique which employs (V-R)$_J$ as the Cepheid surface brightness indicator. Using improved values for the slope (from Thompson's (1975) method) and zero point (from the Cepheid model atmosphere calculations of Hindsley and Bell (1989), and from Cepheid effective temperatures of Pel (1978) and the bolometric correction scale of VandenBerg and Bell (1985)) in the surface brightness relation

$$F_V \equiv \log T_e + 0.1 BC = b + m (V - R)_o \qquad (2)$$

GBM find $< M_V > = -1.32 - 3.06 \log P$ ($\sigma = 0.29$ mag) from 100 galactic Cepheids, and $< M_V > = -1.37 - 3.01 \log P$ ($\sigma = 0.27$ mag) from 79 Cepheids with distance determinations more accurate than 10 percent.

Table 1. Recent Calibrations of the galactic Cepheid PL(V) relation

$$< M_V > = a + b \log P$$

Source	Method	N	a	b	Disp.	$< M_V >(0.8)$
Feast & Walker 87	ZAMS-fitting	28	-1.224	-2.927	0.205	-3.57
Walker 88	ZAMS-fitting	28	-1.222	-2.906	0.193	-3.55
Fernie 92	ZAMS-fitting	28	-1.203	-2.902	0.260	-3.52
Turner 92	ZAMS-fitting	10	-1.153	-2.939	0.070	-3.50
Gieren & Fouqué 92	ZAMS-fitting	32	-1.281	-2.911	0.280	-3.61
GBM 92	BW (surface-brightness)	100	-1.37	-3.01	0.27	-3.78
HB 92	BW (model atmosph.)	23	-1.33	-3.11	0.25	-3.82
BS 84	Hβ	2[1]				-3.59
Schmidt 91	Hβ					-3.50
Wilson et al. 91	Dynamical Parallaxes					-3.46 ± 0.34

[1] CS Vel and TW Nor

6. Comparison of PL results from different methods

The relevant information for a comparison of results is given in Table 1. In the last column the $< M_V >$ at $\log P = 0.8$ as resulting from the different methods are shown. The main conclusions are:

a) The BW method yields a PL zero point ~ 0.15 mag *brighter* than the cluster scale. However, the present cluster scale is tied to μ_o (Pleiades) = 5.57 which is likely to be underestimated, as discussed before. The GBM BW zero point would further be ~ 0.05 mag *fainter* if they had used the same reddening treatment as GF in their ZAMS-fitting study, making the BW scale only ~ 0.10 mag brighter than the cluster scale. This is clearly within the present uncertainties of both methods and *strongly suggests that neither of the two methods contains systematic errors in excess of $\sim \pm 0.10$ mag.*

Table 2. Cluster Cepheid Data used in the calibration of Gieren & Fouqué (1992)

Cepheid	Cluster or association	Q	log P	μ_o	R(OB)	E(OB)	R(cep)	E(cep)	$<M_V>$
SU Cas		C	0.2899	7.07	3.1	0.29	3.178	0.287	-2.01
EV Sct	NGC 6664	B	0.4901	10.88	ass	0.60	3.112	0.679	-2.86
SZ Tau	NGC 1647	A	0.4982	8.68	3.09	0.31	3.212	0.294	-3.09
QZ Nor	NGC 6067	A	0.5782	11.13	ass	0.35	3.232	0.276	-3.16
alpha UMi		C	0.5990	5.19			3.210	0.000	-3.22
CEb Cas	NGC 7790	A	0.6512	12.68	ass	0.54	3.220	0.597	-3.61
CF Cas	NGC 7790	A	0.6880	12.68	ass	0.54	3.241	0.566	-3.38
CEa Cas	NGC 7790	A	0.7111	12.68	ass	0.54	3.240	0.597	-3.70
UY Per	King 4 or Czerny 8	B	0.7296	11.70	3.1	0.99	3.246	0.919	-3.34
CV Mon	Anon van den Bergh	A	0.7307	11.22	3.09	0.77	3.241	0.714	-3.23
V Cen	NGC 5662	C	0.7399	9.11	3.1	0.31	3.218	0.289	-3.22
CS Vel	Ruprecht 79	B	0.7712	12.55	ass	0.794	3.22	0.847	-3.58
V367 Sct	NGC 6649	A	0.7989	11.27	ass	1.35	3.251	1.284	-3.89
BB Sgr	Collinder 394	C	0.8220	9.05	3.1	0.25	3.246	0.284	-3.05
U Sgr	M25	A	0.8290	8.95		0.48	3.260	0.403	-3.57
DL Cas	NGC 129	A	0.9031	11.12	3.20	0.51	3.242	0.533	-3.88
S Nor	NGC 6087	A	0.9892	9.79	3.1	0.19	3.259	0.189	-3.99
Zeta Gem		C	1.0065	7.75			3.256	0.018	-3.89
TW Nor	Lynga 6	A	1.0328	11.41	ass	1.34	3.292	1.338	-4.14
V340 Nor	NGC 6067	A	1.0526	11.13	ass	0.35	3.288	0.315	-3.80
VY Car	Ass Car OB2	B	1.2767	11.42	3.05	0.28	3.302	0.243	-4.76
RU Sct	Trumpler 35	B	1.2945	11.61	3.0	1.03	3.276	0.957	-5.26
RZ Vel	Ass Vel OB1	B	1.3096	11.24	2.88	0.35	3.275	0.335	-5.25
WZ Sgr	C1814-191a	A	1.3394	11.27	3.0	0.61	3.318	0.467	-4.80
SW Vel	Ass Vel OB5	B	1.3700	12.00	3.20	0.38	3.278	0.349	-5.02
T Mon	Ass Mon OB2	B	1.4317	11.02	3.2	0.20	3.310	0.209	-5.59
KQ Sco	Ass Sco OB anon	B	1.4578	12.32	3.04	1.00	3.365	0.896	-5.52
U Car	Ass Car OB2	B	1.5889	11.42	3.05	0.34	3.298	0.283	-6.07
RS Pup		C	1.6172	11.28			3.329	0.446	-5.75
SV Vul	Ass Vul OB1	B	1.6532	11.82	3.0	0.49	3.309	0.570	-6.46
GY Sge	Ass OB anon	B	1.7081	12.66	3.0	1.30	3.390	1.140	-6.29
S Vul	Ass Vul OB2	C	1.8299	13.23	3.0	0.88	3.362	0.827	-7.05

Note: Q = quality. A: cluster membership well established. B: cluster membership more uncertain. C: Cepheid not in cluster or association or membership very uncertain.

b) The results from Hβ photometry of cluster B stars agree with the ZAMS-fitting scale if the Balona and Shobbrook (1984) calibration is used and the TW Nor and CS Vel clusters are rejected because of their very low number of B stars. In this case, $< \mu_o(H\beta) - \mu_o(ZAMS) > $ = -0.03 $\pm$ 0.09 mag from six clusters. Including TW Nor and CS Vel, the difference becomes -0.20 $\pm$ 0.13 mag.

c) The model atmosphere results of Hindsley and Bell (1989), as corrected by Hindsley (1992; private communication) agree with the BW result of GBM.

d) Cepheid absolute magnitudes obtained from statistical parallaxes and binary companions agree, within their large uncertainties, with the other results.

The Cepheid distance scale according to the galactic calibration discussed here sets the Large Magellanic Cloud at μ_o (LMC) = 18.6 $\pm$0.1 mag. This is in excellent agreement with the LMC distance modulus found from the expansion parallax of supernova 1987A of 18.50 $\pm$ 0.13 (Panagia et al. 1991) and seems to be supported by all other distance indicators except RR Lyrae stars (Walker 1992; see review in this volume).

7. A note on the color term in the PLC relation

The PLC relation can be written as

$$< M_V > = \alpha + \beta \, log \, P + \gamma \, (< B >_o - < V >_o) \quad (3)$$

To avoid correlated errors in the solution of (3), we write

$$< V > - \mu_o = \alpha + \beta \, log \, P + \gamma \, (< B > - < V >) + \delta E(B-V) \quad (4)$$

where $\mu_o = < V >_o - < M_V >$ and $\gamma + \delta$ = R (ratio of total to selective absorption).

We have solved for (4) using the GAUSSFIT routine of W.H. Jefferys of Austin and the 100 Cepheid sample of GBM, assuming σ ($log \, P$) = 0, σ $(B-V) = \sigma(V) = 0.02$ for all stars, taking individual σ_i $(E(B-V))$ from Fernie (1990) and σ_i (μ_o) from GBM, and taking R = 3.3. This yields

$$< M_V > = -1.45 - 2.96 \, log \, P + 0.04 \, (< B >_o - < V >_o) \quad (5)$$

which is not different from the GBM PL relation $< M_V > = -1.37 - 3.01 \, log \, P$. This result shows that no significant color term can be extracted from the *galactic* Cepheid data, probably due to the relatively large errors in the individual Cepheid distances.

Walker (1988) gives for the LMC PLC relation, corrected to solar metallicity

$$< M_V > = -2.11 - 3.53 \, log \, P + 2.13 \, (< B >_o - < V >_o) \quad (6)$$

Using a purely statistical approach and assuming that the extra scatter observed in the LMC period-color relation, as compared to the galactic PC relation ($\sim$ 0.11

mag in the LMC (e.g. Martin, Warren and Feast 1979) as compared to ~ 0.07 mag in the Galaxy (GBM) is solely due to larger uncertainties in the determination of color excesses for LMC Cepheids than for their galactic counterparts, Fouqué and Gieren (1992) have shown that the *true* color coefficient in (6) adopts a value of ~ 0.7. This is an interesting finding but it seems to be difficult to reconcile it with the spread in Cepheid magnitudes at a given period, as observed in LMC clusters rich in Cepheid variables (Welch 1992; see review in this volume). It shows, however, that we can still not be completely confident in the true value of the color coefficient of the PLC relation.

We are grateful to those astronomers who have sent data, comments and preprints during the preparation of this work. A grant of the Canadian Institute of Theoretical Astrophysics to WPG is gratefully acknowledged. We thank Dr. W.H. Jefferys for sending us his Gaussfit program and Gisela Hertling for doing part of the calculations of this work.

References:

Balona, L.A. and Shobbrook, R.R., 1984, *M.N.R.A.S.*, **211**, 375.

Böhm-Vitense, E., 1985, *Ap.J.*, **296**, 169.

Caccin, B., Onnembo, A., Russo, G. and Sollazzo, C., 1981, *Astr. Ap.*, **97**, 104.

Caldwell, Z.A.R. and Coulson, I.M., 1986, *M.N.R.A.S.*, **218**, 223.

Coulson, I.M., Caldwell, Z.A.R. and Gieren, W.P., 1986, *Ap.J.*, **303**, 273.

Evans, N.R., 1991, *Ap.J.*, **372**, 597.

Feast, M.W., 1991, *SAAO Preprint* 709. v

Feast, M.W. and Walker, A.R., 1987, *Ann. Rev. Astr. Ap.*, **25**, 345.

Fernie, J.D., 1990, *Ap.J. Suppl.*, **72**, 153.

Fernie, J.D., 1992, *A.J.*, **103**, 1647.

Fouqué, P. and Gieren, W.P., 1992, in preparation.

Freedman, W.L. and Madore, B.F., 1990, *Ap.J.*, **365**, 186.

Gatewood, G., Castelaz, M., Han, I., Persinger, T., Stein, J., Stephenson, B. and Tangreu, W., 1990, *Ap.J.*, **364**, 114.

Gieren, W.P. and Fouqué, P., 1992, *A.J.*, submitted.

Gieren, W.P., Barnes, T.G. and Moffett, R.J., 1992, *Ap.J.*, submitted.

Hindsley, R.B. and Bell, R.A., 1989, *Ap.J.*, **341**, 1004.

Laney, C.D. and Stobie, R.S., 1986 *M.N.R.A.S.*, **222**, 449.

Madore, B.F. and Freedman, W.L., 1991, *Pub. A.S.P.*, **103**, 933.

Martin, W.L., Warren, P.R. and Feast, M.W., 1979, *M.N.R.A.S.*, **188**, 139.

Olson, B.I., 1975, *Pub. A.S.P.*, **87**, 349.

Panagia, N., Gilmozzi, R., Macchetto, F., Adorf, H.M. and Kirshner, R.P., 1991, *Ap.J.*, **380**, L23.

Pel, J.W., 1978, *Astr. Ap.*, **62**, 75.
Sandage, A., 1988, *Pub. A.S.P.*, **100**, 935.
Stothers, R.B., 1988, *Ap.J.*, **329**, 712.
Thompson, R.J., 1975, *M.N.R.A.S.*, **172**, 455.
Turner, D.G., 1992, private communication.
Turner, D.G., 1976, *A.J.*, **81**, 1125.
VandenBerg, D.A. and Bridges, R.B., 1984, *Ap.J.*, **278**, 679.
VandenBerg, D.A. and Bell, R.A., 1985, *Ap.J. Suppl.*, **58**, 561.
van Leeuwen, F., 1983, Ph.D. Thesis, Leiden University, Netherlands.
Walker, A.R., 1988, in The Extragalactic Distance Scale, ed. S. van den Bergh & C.J. Pritchet ASP Confer. Series, Vol. 4, p. 89.
Wilson, T.D., Barnes, T.G., Hawley, S.L. and Jefferys, W.H., 1991, *Ap.J.*, **378**, 708.

Discussion

A. SANDAGE: I do not understand why there is any question about the value of the color term in the PLC relation. Its value is one of the best determined numbers in the relation because it depends only on the slope of the lines of constant period in the HR diagram. The only way the color term can be zero is if the lines of constant period are *flat* in the HR diagram which is impossible. I have always believed one must fix the color coefficient from the pulsation equation rather than attempting to determine it empirically.

W. P. GIEREN: I fully agree that the slope of the lines of constant period in the HR diagram cannot be zero, so there *must* be a color term in the PLC relation. However, there are uncertainties in the theoretical and semiempirical steps needed to calculate the color term which seem larger to me than usually assumed, and so there *is* a question if the true value is not significantly different from ~2.5. One empirical way to better clarify the question is to look at LMC clusters which contain lots of Cepheids of very similar period.

D. TURNER: I have a comment regarding the PL relation for cluster and association Cepheids, which seems to exhibit rather large scatter at the low period end. It is my belief that at least four of these Cepheids are overtone pulsators, so it makes a difference which period they are plotted for.

W. P. GIEREN: It is certainly possible that there are overtone pulsators in the sample, but the existing evidence is not convincing to me except for SU Cas. Anyway, even if one adopts all four shortest period Cepheids as overtone pulsators this does not make a significant difference in the PL coefficients although the scatter maybe somewhat lower.

D. WELCH: I have three comments. First, it is worth pointing out that the dominant error in most extragalactic distances is still the galactic calibration. Second, if most of the errors are with reddening, work at longer wavelengths. Third, the angular diameter situation should improve dramatically with SUSI coming into existence.

Cepheid and Long – Period Variables in Virgo Cluster Galaxies

M.J. Pierce[1], R.D. McClure[2], D.L. Welch[3], R. Racine[4] S. van den Bergh[2]

[1]*Kitt Peak National Observatory, U.S.A.*, [2]*Dominion Astrophysical Observatory, National Research Council, Canada*, [3]*McMaster University, Canada*, [4]*Université de Montreal, Canada*

Abstract

We are currently undertaking a *ground-based* imaging survey which attempts to discover and determine periods for variable stars in Virgo Cluster galaxies. Such a survey is now feasible thanks to the high resolution imaging (FWHM $\leq$ 0.50 arcsec) routinely obtained with the High Resolution Camera on the Canada-France-Hawaii Telescope. The Virgo Cluster has long been considered a crucial "stepping-stone" in the extragalactic distance scale problem given that the cluster is at a "cosmologically interesting" distance and that there is little controversy in the *relative* distance between Virgo and more distant clusters, such as Coma. Consequently, much of the controversy regarding the extragalactic distance scale and the Hubble Constant can be eliminated with a determination of the Virgo Cluster distance. Some preliminary results and the prospects for establishing the distance to the Virgo Cluster using Cepheids and LPVs are discussed.

1. The Extragalactic Distance Scale Problem

It is generally believed that the historical discrepancy in estimates of the Hubble Constant prevents a definitive test of cosmological models. Much of the problem lies with the fact that the best understood distance indicators (e.g., Cepheid and RR Lyrae variables) are typically applicable over only modest distances, while the various secondary techniques (e.g., $D_n - \sigma$, luminosity fluctuations, novae, planetary nebulae and globular cluster luminosity functions, supernovae, and the Tully-Fisher relations) are more uncertain and must rely on the primary methods for calibration. As a result, the uncertainties in the estimated distances to galaxies grow beyond the Local Group. However, the situation has recently improved now that quantitative prescriptions and cross-checks of the various secondary techniques have been developed. A comparison of these methods yields remarkably good agreement for the distance to the Virgo Cluster (e.g., Jacoby et al. 1992). Nevertheless, a more direct distance to the Virgo Cluster based on Cepheids would add considerable confidence to the current situation.

The Virgo Cluster is a key "stepping stone" in the extragalactic distance scale problem. Not only is it rich in galaxies over the full range of morphological types, but it contains enough systems that such relatively rare events, as supernovae, can be studied. Thus, Virgo appears to be an ideal environment for comparing and ultimately calibrating the secondary techniques. In addition, several of these methods can be used to estimate the relative distance between the Virgo and Coma clusters and hence sample the Hubble flow out to ~ 7000 km sec^{-1}. The various estimates of the relative distance between these two clusters is also in very good agreement, so we can effectively "by-pass" any local peculiar velocities and estimate the Hubble Constant, provided that we can establish the distance to Virgo. It is generally believed that the detection and photometry of Cepheid variables in members of the Virgo Cluster will provide the most definitive estimate of the distance. Unfortunately, the current status of HST prevents this problem from being addressed for a few more years. In the meantime, it appears that the recent technical strides in high resolution imaging make a *ground-based* attempt at this problem feasible. In addition, we believe that Long-Period Variables (LPVs) may offer an interesting compliment/alternative to the Cepheids. An illustration of the promise of LPVs can be found in Pierce & Crabtree (this volume). Here we report on some preliminary results of our attempts at discovering variables in Virgo Cluster spirals using HRCam at the CFHT.

2. The Promise of High Resolution Imaging

High resolution imaging produces significant gains in the problem of stellar photometry in external galaxies. Not only are the effects of crowding reduced with decreasing FWHM, but signal-to-noise increases linearly with decreasing FWHM in the background limited regime. Finally, "seeing" is a logarithmic quantity. That is, a 0.1 arcsec improvement in the FWHM at 1.2 arcsec is not nearly as significant as a similar improvement at a FWHM of 0.5 arcsec. Consequently, what might appear to be relatively modest gains in image quality at an already good site like Mauna Kea, are in fact quite significant. We also note that Cook, Aaronson, and Illingworth (1986) found Cepheids in M101 with the KPNO 4-m with FWHM ~ 1.2 arcsec. Given that Virgo Cluster is at most a factor of 3 more distant, we can expect a similar chance of success if we can attain a FWHM ~ 0.4 arcsec. Such a goal is now within reach thanks to recent progress in ground-based imaging; in particular the success of the High Resolution Camera (HRCam) on the CFHT (see below). A similar argument for the potential gains from high resolution imaging of Virgo Cluster galaxies and some preliminary results can be found in Tanvir et al. (1991).

HRCam is essentially a "fast guider" which operates at frequencies up to 200 Hz. It is capable of correcting for the "tip-tilt" component of atmospheric seeing by monitoring a nearby guide star. The characteristics of HRCam have been described by Racine & McClure (1989) and McClure et al. (1989). This instrument has been demonstrated to produce significant gains in FWHM, as is evident from Figure 1.

The distribution of image FWHM with HRCam is seen to peak at FWHM = 0.55 arcsec and appears to have a "wall" at ~ 0.40 arcsec. This lower limit to the FWHM is probably imposed by the optical aberrations of the CFHT primary and the optics within HRCam (e.g., Racine et al. 1991). Efforts are ongoing to push this limit down to ~ 0.3 arcsec in the future. In any event, we will demonstrate that the current performance of HRCam is sufficient to address the problem of obtaining photometry for the brighter stellar populations in Virgo Cluster galaxies.

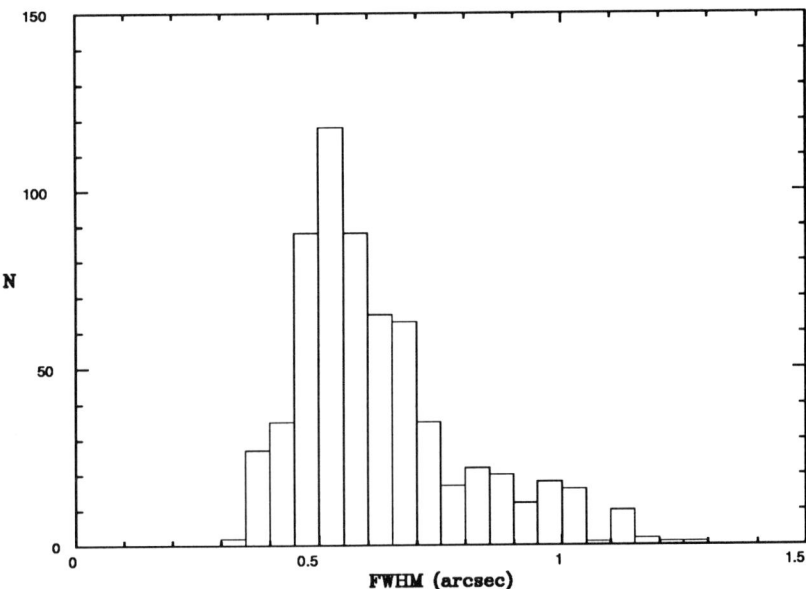

Figure 1 Histogram of measured FWHM with the CFHT with HRCam. The data are for every image (Jan. – Nov., 1991) with an integration longer than 10 sec.

The Virgo Cluster contains a considerable population of spiral galaxies from which we could choose our sample. Unfortunately, there is accumulating evidence for significant line-of-sight structure in the direction of the Cluster (e.g., Pierce and Tully 1988; Pierce 1991; Jacoby et al. 1992). This circumstance makes associating any small sample of spirals with the ellipticals, which might be presumed to lie in the inner core of the cluster, a somewhat dangerous approach. Nevertheless, it is the only way in which we can hope to establish the distance to the Virgo Cluster with Cepheids. To guard against this problem we have chosen two galaxies which are among the most likely to be in the inner core of the cluster. These are NGC 4571 and NGC 4321. The first system is a relatively face-on spiral, of moderate luminosity, which has a very low radial velocity (348 km sec^{-1}). This implies it is within the core of the cluster, as any reasonable estimate of the Hubble Constant would place a foreground system so close that it would have been resolved into individual stars long ago. In fact, Virgo is the only region outside the Local Goup where galaxies have negative radial velocities, due to the large velocity dispersion in the cluster. NGC 4571 also shows strong evidence for being stripped of its outer envelope of H I (e.g., van der Hulst et al. 1987). These two facts almost certainly place NGC 4571 in the core of the cluster. The second galaxy, NGC 4321, is a relatively rare, luminous

Sc I spiral, and has been argued to be a member of the cluster core on this basis (e.g., Sandage 1992). In addition, this system has produced a well observed type II supernova which allows a distance estimate via the "Baade-Wesselink", or expanding photosphere method (e.g., Schmidt, et al. 1992), and we can thus provide a direct test of this method.

3. A Search for Variables in the Virgo Cluster with HRCam on the CFHT

For our search program, we typically take 1 – 1.5 hours of integration in the R-band, breaking this into individual 15-min exposures to allow for cosmic ray discrimination and to give us some flexibility during conditions of variable seeing. The images are registered and combined with a clipping algorithum to produce a single image. To date, we have acquired 6 epochs with a FWHM $\leq$ 0.50 arcsec for NGC 4571 and an additional epoch with FHWM = 0.55 arcsec. We have only acquired 2 epochs for NGC 4321 (FWHM $\sim$ 0.55 arcsec), so we will discuss only NGC 4571 here. An example of our "plate material" is shown in Figure 2.

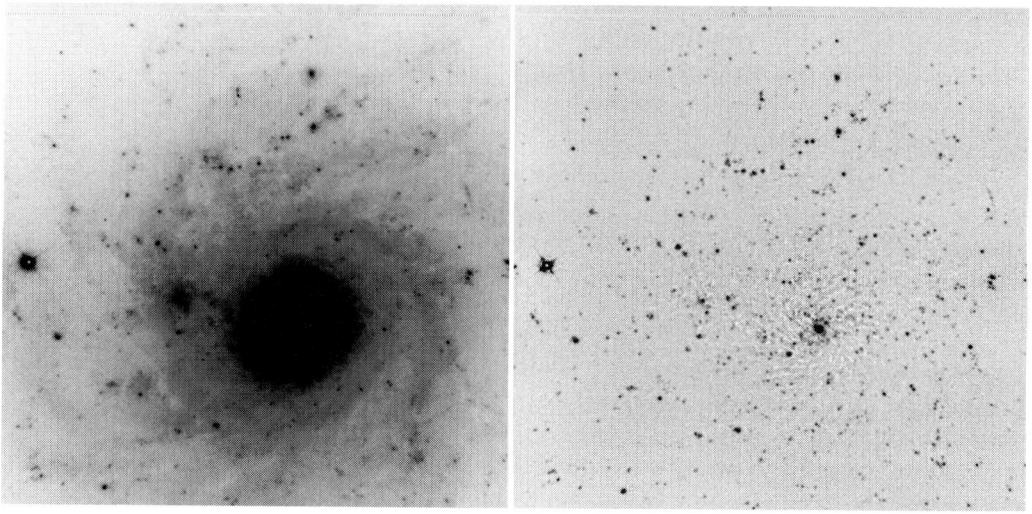

Figure 2 a) R-band image of NGC 4571 (1.5 hours, FWHM = 0.42 arcsec). b) The same image following the removal of the background light of the galaxy (see text). The individual stars are clearly resolved and range from R $\sim$ 21 for the brightest to R $\sim$ 26 for the faintest.

In the first panel we show one of our images (FWHM = 0.42 arcsec). The resolution of the galaxy into stars is obvious, as is the rapidly varying background upon which they are found. In order to suppress the background and obtain stellar photometry we go through a few iterations of DAOPHOT, followed by smoothing. That is, we: 1) construct a PSF from the brightest isolated stars, 2) identify stars and subtract them, 3) smooth the subtracted image to suppress the residuals, and 4) subtract this "model" of the background from the original image. The process is repeated a few times, progressively decreasing the size of the smoothing kernel. The second panel illustrates the result of this procedure. The background light of the galaxy is suppressed, revealing numerous individual stars. In addition, the procedure reveals the presence of considerable dust complexes associated with the spiral structure of the galaxy. This image can be compared with that shown in Sandage and Bedke (1988) to give some idea of the improvement made possible with HRCam and the CFHT.

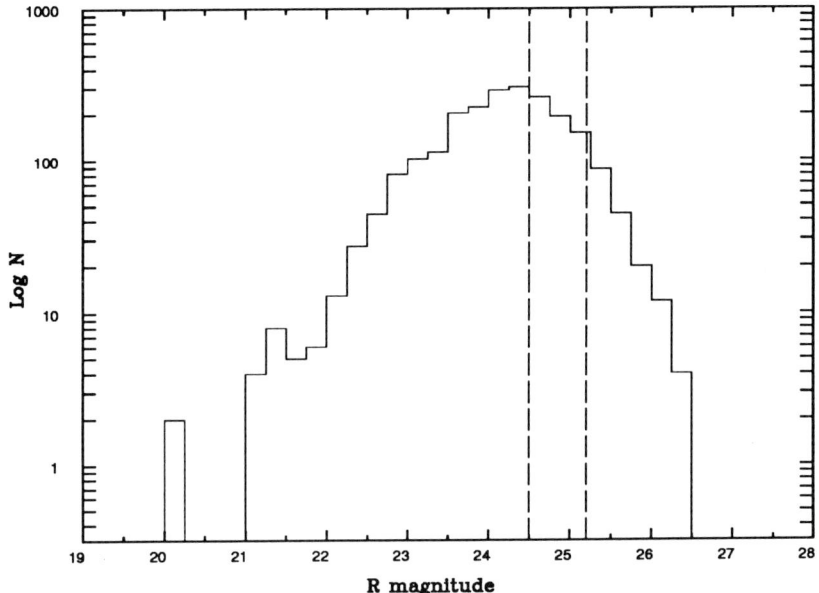

Figure 3 R-band luminosity function for stars in NGC 4571 from DAOPHOT. Note that incompleteness begins below $m_R \sim 24.0$ and that our limiting magnitude is about $m_R \sim 26$. The dashed lines indicate the expected R mag. at mean light for 50 day Cepheids, assuming distances of 15 and 22 Mpc (see text).

Some preliminary results of our photometry of the resolved stellar population in NGC 4571, and a more extensive discussion of the feasibility of detecting of Cepheids, can be found in Pierce, McClure, & Racine (1992). Figure 3 shows the R-band luminosity function of NGC 4571 taken from that paper. The two vertical lines indicate the expected R magnitude for 50 day Cepheids at mean light assuming Virgo distances of 15 and 22 Mpc. Since the luminosity function is empirical, we can estimate the completeness as $\sim 50\%$ and 10% for the two distances by assuming a simple power law form. The brightest Cepheids ($P \geq 50$ days) are relatively rare and tend to have lower amplitudes, while those with $P \leq 30$ days suffer from aliasing due to the lunar cycle. The shortest periods ($P \leq 20$ days) are simply too faint for us to detect. Consequently, we should be sensitive to only a narrow range in period and any aliasing problems should be minimal. With the 6 epochs acquired to date ~ 40 variable star candidates have been identified in NGC 4571. We have used both blinking and the photometry to assess the reality of the variations. A few of the variables we have found are shown in Figure 4. The data span approximately 600 days. The first two epochs are from the 1991 season, while the remainder were obtained in 1992. The variation is clearly seen by comparing the apparent brightness of each of the identified stars with their neighbors. The variables we have identified have $24 \leq m_R \leq 25$ and amplitudes of ~ 1 mag. The photometric errors at this level are ~ 0.15 mag. per epoch and this should provide adequate period determination with about 10 epochs.

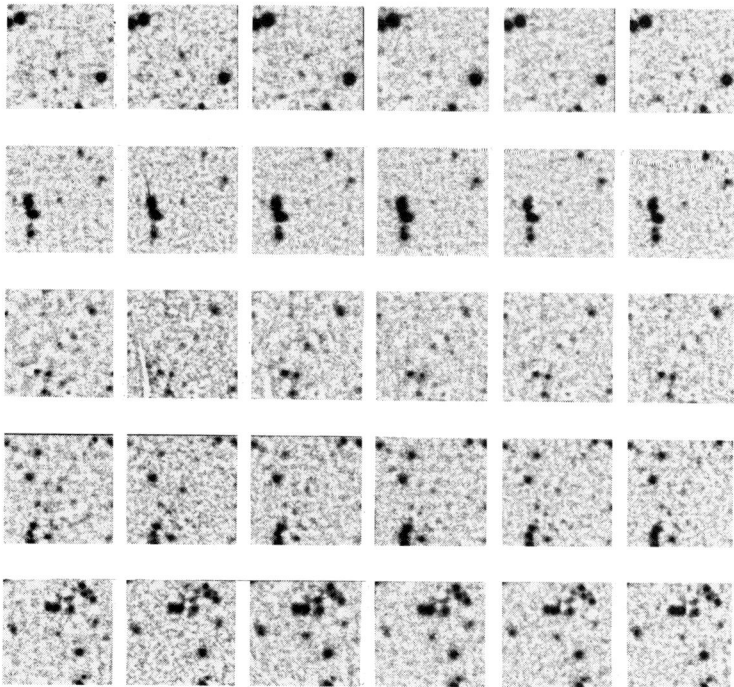

Figure 4 Subrasters centered around selected variables in NGC 4571 illustrating the six epochs acquired to date. The variations are made fairly obvious by the presence of nearby stars of comparable brightness. The background fluctuations due to the underlying population of K giants is also apparent. The non-uniformities in the background due to this incipient resolution and dust complexes are the limiting factor in our photometry.

Scaling the number of variables in the LMC by the relative luminosity of the galaxies, and accounting for our incompleteness, we should detect $\sim 3 - 8$ Cepheids in NGC 4571, depending on the actual distance. The crowding of stellar images should have no effect on the periods but will bias the mean light of the variables to brighter levels. On the other hand, crowding will result in decreased amplitudes and lower the probability of detecting the most contaminated variables. We plan to assess the result of these competing effects through simulations.

It is interesting to note that the variables in NGC 4571 are 5 magnitudes fainter than similar variables in members of the Local Group. This clearly demonstrates that significant gains are possible with high resolution imaging. With the small number of epochs in hand, we cannot yet determine periods. Nevertheless, we can still get some idea of the nature of these variables by looking at colors. Figure 5 shows a color-magnitude diagram for a selection of stars in NGC 4571 for which we have adequate V-band photometry.

The "plume" of blue supergiants near $V - R \sim -0.2$ is evident and suggests that the brightest blue stars suffer relatively little extinction. We might expect the same for any Cepheids we find since B main sequence stars are the progenitors. The brightest red supergiants are apparent near $R \sim 22$ with $V - R \sim 1.0$. They imply a distance modulus of ~ 31 mag., or ~ 15 Mpc (Pierce, McClure, & Racine 1992). The majority of the variables ($\sim 80\%$) have $V - R$ colors of ~ 1.0, but a few have $V - R \sim 0.3$. The former are almost certainly LPVs, while we suspect the latter are Cepheids.

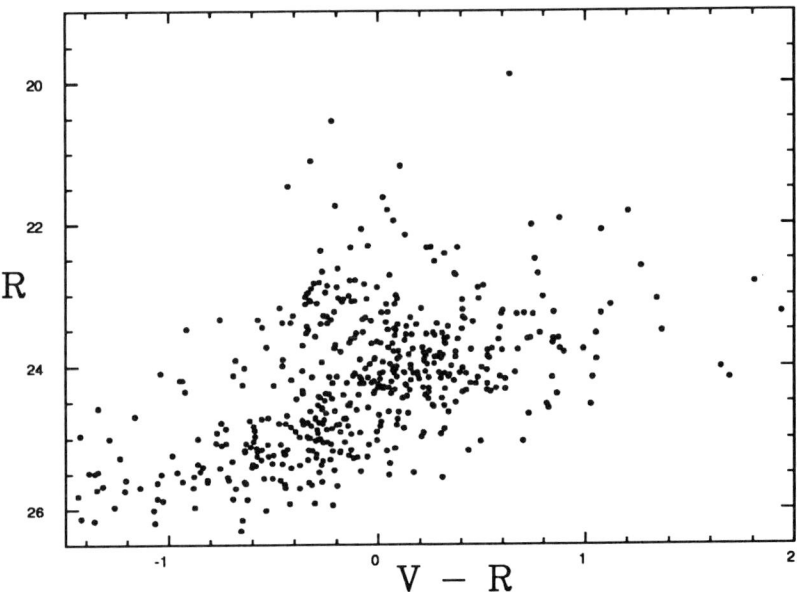

Figure 5 R vs. V − R color magnitude diagram for the resolved stars in NGC 4571. The "blue plume" of supergiants is obvious at $V - R \sim -0.2$, as well as the red supergiants at $R \sim 22$ and $V - R \sim 1.0$. The distribution of the fainter stars is due primarily to limitations in our V-band photometry.

LPVs offer an additional tool for estimating the distance to Virgo Cluster galaxies. A period-luminosity relation has been found for LPVs in the K-band with a dispersion comparable to that of the Cepheids (e.g., Wood, Bessell, & Paltoglou 1985).

However, the LPVs in Virgo are expected near K $\sim$ 21 and photometry will be extremely difficult. An alternative is to apply the optical P $-$ L relation found for these variables (Pierce & Crabtree, this volume). By using a narrow band filter to isolate the continuum between the strong TiO found in these stars a random-phase P $-$ L relation with a dispersion of $\sim$ 0.25 mag. has been found. There are no signs of any obvious systematics due to differences in metallicity. Given that LPVs are $\sim$ 5 times more common than Cepheids of similar brightness means that we can expect to detect considerably more LPVs in distant galaxies than Cepheids. The same arguement also applies to the problem of the Galactic calibration as well. In fact, 10 LPVs are known in the Per OB1 association alone. So, it would appear that LPVs may turn out to be a useful alternative to Cepheids in estimating extragalactic distances.

References:

Cook, K., Aaonson, M., & Illingworth, G. 1986, ApJ, **301**, L45.
Jacoby, G., et al. 1992, PASP, **104**, 598.
McClure, R. et al. 1989, PASP, **101**, 1156.
Pierce, M. 1991, in *Observational Tests of Cosmological Inflation*, ed. T. Shanks, A.J. Banday, R.S. Ellis, C.S. Frenk, & A. W. Wolfendale, NATO ASI Series C **348**, p. 173.
Pierce, M., McClure, R., & Racine, R. 1992, ApJ, **393**, 523.
Racine, R., & McClure, R. 1989, PASP, **101**, 731.
Racine, R., Salmon, D., Cowley, D., & Sovka, J., 1991, PASP, **103**, 1020.
Sandage, A., & Bedke, J. 1988, Atlas of Galaxies Useful for Measuring the Cosmological Distance Scale, NASA Scientific and Technical Information Division, (Washington: NASA)
Sandage, A. 1992, preprint
Schmidt, B., et al. 1992, preprint
Tanvir, N., et al. 1991, MNRAS, 253, 21P
van der Hulst, J., Skillman, E., Kennicutt, R., Bothun, G. 1987, AA, **177**, 63.
Wood, P., Bessell, M., & Paltoglou, R. 1985, ApJ, **290**, 477.

Questions

J. Percy: In view of the fact that supergiant LPVs can be rather irregular, are you sure that you can use them reliably for distance determination? Will they be comparable in galaxies with perhaps different metallicities?

M. Pierce: The nice feature of the P $-$ L relation for LPVs is that it is so flat that in some sense we don't even need accurate periods. We don't see any evidence of metallicity effects between the Galaxy and the LMC but it is something we are continuing to look into.

J. Matthews: Do the monthly gaps induced by the spacing of "dark time" really *improve* the aliasing situation? There could be Cepheids with periods as short as, say, 20 d for which you could not reliably identify a period!

M. Pierce: Well, "improve" is certainly a poor choice of words. My point is only that the Cepheids as faint as, say, 20 d will be too faint to confuse with those at 40 d, so we are likely to get periods with a minimal number of epochs.

N. Simon: What are the prospects for observing Cepheids with periods of about 15 days?

M. Pierce: I can't speak for HST, but the prospects using *ground-based* high resolution techniques appears slim. With FWHM ~ 0.4 arcsec they will be impossible to detect and more exotic adaptive optical systems will likely have such small fields as to make such a survey prohibitively expensive in terms of telescope time.

The Calibration of Colours and Luminosities for Classical Cepheid Variables

D. G. Turner

Saint Mary's University, Halifax, Nova Scotia, Canada

Abstract

Current progress is described for an ongoing program to secure new observational data (star counts, photoelectric, photographic & CCD photometry, and spectroscopy) for stars in the large sample of potential Cepheid calibrating clusters and associations.

New Baade-Wesselink radii have been obtained for 11 Cepheids using a variant (Turner 1988) which uses KHG spectrophotometric data to isolate phases of equal effective temperature in Cepheid light curves. The resulting best fit to the data is described by $\log R/R_{\rm sun} = 1.07 + 0.75 \log P_0$, with a scatter of only ~2-4%, versus 5-8% with more traditional methods. Space reddenings for cluster Cepheids, combined with KHG-based reddenings (Turner *et al.* 1987) and similar reddenings (all < 0.5) tied to this system, are a useful tool for testing the photometric reddening systems in the literature. The Fernie (1990) compilation, for example, suggests that there is a large, period-dependent scatter in Cepheid intrinsic colours, but this is not supported by the space reddening data, which suggest a smaller colour width for the strip and no striking change in strip width with the pulsational period.

Recent studies of main sequence stars in clusters of comparable age to classical Cepheids indicate that stellar rotation is likely to be an important parameter affecting distances derived from ZAMS-fitting or Hβ photometry, as well as Cepheid space reddenings. It may also produce the main sequence colour spread observed in reddening-free colour-magnitude diagrams for Cepheid clusters, and have important ramifications for Cepheid luminosities derived from UV observations of their hot companions. A newly discovered circumstellar reddening effect tied to rapid rotation of cluster B stars may be an additional concern for cluster ZAMS fitting (Turner 1993, in press). A *P-L* calibration has been made using the 10 program clusters which have been observed to date. The resulting calibration for SU Cas, SZ Tau (an overtone pulsator), CV Mon, V367 Sct, DL Cas, S Nor, ζ Gem, WZ Sgr, SW Vel and SV Vul gives $<M_V> = -1.15 - 2.94 \log P_0$, with a scatter much reduced from earlier calibrations. The strip filling is not yet complete enough to derive the colour term unambiguously, although initial studies indicate it may lie in the range 2.5 to 3.5. The cluster Cepheid BB Sgr is very important in this regard, since it appears to lie on the red edge of the instability strip.

References:

Fernie, J.D. 1990, Astrophys.J. 354, 295
Turner, D. G. 1988, Astron.J. 96, 1565
Turner, D.G., Leonard, P. J. T & English, D.A. 1987, Astron.J. 93, 368

I-band Cepheid Distance to NGC 6822

Myung Gyoon Lee [1], Wendy L. Freedman[1], Barry F. Madore[2]

[1] *Carnegie Observatories, Pasadena, USA,* [2] *IPAC/JPL/Caltech, Pasadena, USA*

NGC 6822 was the first galaxy outside the Galaxy where the Cepheid period-luminosity relation was applied as a distance indicator (Hubble 1925, ApJ, 62, 409). Later, thirteen Cepheids in this galaxy were studied by Kayser (1967, AJ, 72, 134) using BV photographic photometry. We have obtained $BVRI$ CCD photometry of stars in two $2.2' \times 3.5'$ fields of NGC 6822. Field 3) using Figure 1 shows an $I - (V - I)$ diagram of $\sim 6,400$ measured stars in NGC 6822 including seven known Cepheids. The reddening is estimated to $E(B - V) = 0.28 \pm 0.03$ from the $(B - V)$-$(V - I)$ diagram of the stars with $V < 20$ mag. The distance to NGC 6822 has been estimated using the random-phase I band photometry of seven Cepheids (V7(I=16.31), V5(I=18.97), V6(I=18.44), V3(I=17.88), V1(I=18.05), V28(I=17.61), and V2(I=17.29)) combined with their periods given by Kayser. We have obtained $\Delta(m - M)_I(\mathrm{NGC\,6822} - \mathrm{LMC}) = 5.48 \pm 0.17$ and $(m - M)_0 = 23.62 \pm 0.20$, adopting $(m - M)_0(\mathrm{LMC}) = 18.5$ and $A_I(\mathrm{NGC6822}) = 0.58$. The distance to NGC 6822 has also been estimated independently using the brightness of the tip of the red giant branch (RGB) of low mass stars following the method described in Da Costa & Armandroff (1990, AJ, 100, 162). The tip of the RGB is detected at $I = 20.05 \pm 0.10$. This estimate yields a distance modulus of $(m - M)_0 = 23.46 \pm 0.20$, which is in good agreement with the Cepheid distance modulus. Taking the average of the Cepheid and TRGB distances, we obtain f WLM, $(m - M)_0 = 23.54 \pm 0.11$ corresponding to a distance of 510 ± 30 kpc.

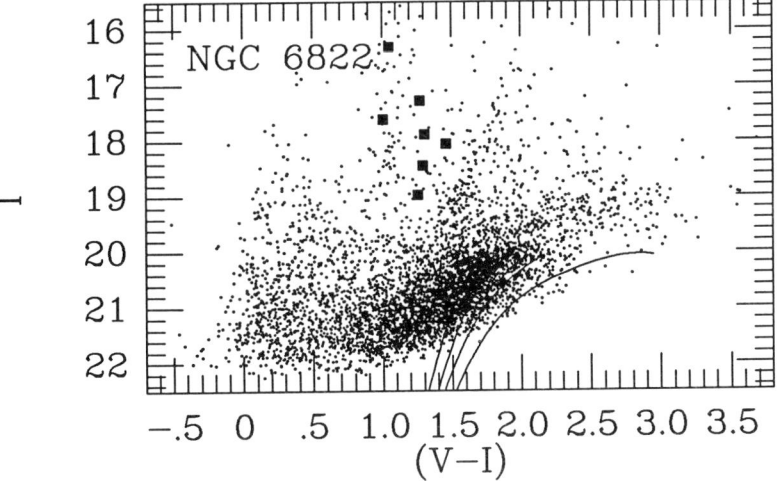

Fig. 1: $I - (V - I)$ diagram of $\sim 6,400$ measured stars in NGC 6822. The filled squares represent the Cepheids. The solid lines show the loci for the RGB of Galactic globular clusters, M15, M2, NGC 1851, and 47 Tuc (from Da Costa & Armandroff).

I-band Cepheid Distance to WLM

Myung Gyoon Lee [1], Wendy L. Freedman[1], Barry F. Madore[2]

[1]*Carnegie Observatories, Pasadena, USA*, [2]*IPAC/JPL/Caltech, Pasadena, USA*

WLM (DDO221) is a highly resolved dwarf irregular galaxy in the Local Group. Fifteen Cepheids in this galaxy have been studied by Sandage & Carlson (1985, AJ, 90, 1464) using photographic photometry. We have obtained $BVRI$ CCD photometry of stars in the central area ($2.2' \times 3.5'$) of WLM. Figure 1 shows an $I-(V-I)$ diagram of $\sim 2,600$ measured stars including five known Cepheids. The distance to WLM has been estimated using the random-phase I band photometry of five Cepheids (V7(I=20.52), V29(I=20.77), V48(I=20.43), V66(I=21.19), V67(I=21.14)) combined with their periods given by Sandage & Carlson. We have obtained $\Delta(m-M)_I(\text{WLM} - \text{LMC}) = 6.25 \pm 0.11$ and $(m-M)_0 = 24.92 \pm 0.21$, adopting $(m-M)_0(\text{LMC}) = 18.5$ and $A_I(\text{WLM}) = 0.04$. The distance to WLM has also been estimated independently using the brightness of the tip of the red giant branch (RGB) of low mass stars, following the method described in Da Costa & Armandroff (1990, AJ, 100, 162). The tip of the RGB is detected at $I = 20.85 \pm 0.1$ mag. This estimate yields a distance modulus of $(m-M)_0 = 24.81 \pm 0.15$, which is in excellent agreement with the Cepheid distance modulus. Taking the average of Cepheid and TRGB distances, we obtain M, $(m-M)_0 = 24.87 \pm 0.08$ (940 ± 30 kpc).

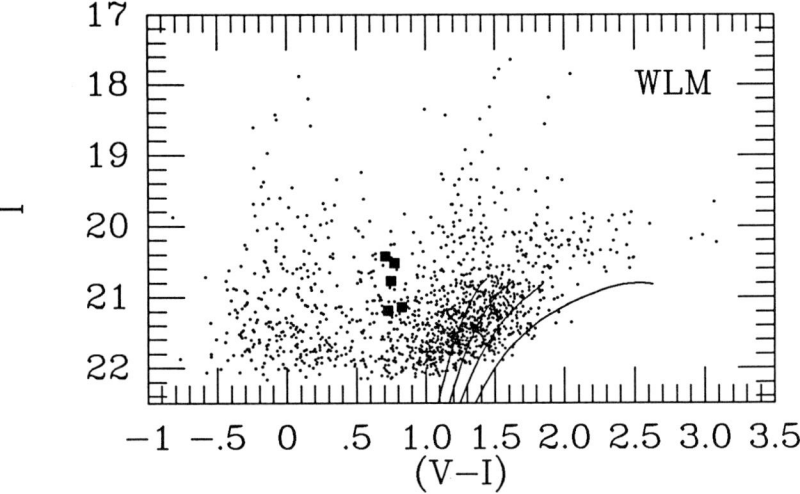

Fig. 1: $I-(V-I)$ diagram of $\sim 2,600$ measured stars in WLM. The filled squares represent the Cepheids. Note that there is a clearly distinguishable concentration of faint red stars with $I < 20.85$ mag, which represents the red giant branch (RGB) of low mass stars. The solid lines show the loci for the RGB of Galactic globular clusters, M15, M2, NGC 1851, and 47 Tuc (from Da Costa & Armandroff 1990).

A Preliminary Distance to the Small Magellanic Cloud by the Surface Brightness Technique

Thomas G. Barnes III
McDonald Observatory
The University of Texas at Austin

Thomas J. Moffett
Department of Physics
Purdue University

and

Wolfgang P. Gieren
Astrophysics Group
Physics Faculty
Pontificia Universidad Católica de Chile

Abstract

We present a new distance determination to the Small Magellanic Cloud from the surface brightness technique applied to the Cepheid variable HV 829. Although this is a preliminary distance based on only one star, it illustrates the power of the surface brightness technique to extragalactic Cepheid distances, it develops the technique which we will apply to additional SMC and LMC Cepheids, and the distance is of intrinsic interest because of the current controversy concerning distances for the Magellanic Clouds.

For HV 829 itself we obtain a distance modulus of 18.91 ± 0.20 mag. From other evidence we infer that HV 829 is slightly in front of the SMC centroid distance. Correcting to the SMC centroid yields a distance to the SMC of 19.05 ± 0.20 mag. We stress that this distance modulus is fully independent of any other distance modulus for the SMC, including those based upon Cepheids. Even so, our result agrees more closely with other, independent Cepheid distances than with RR Lyrae distances and main sequence fitting distances.

Full details are being published in the Astrophysical Journal.

An Adjustment to Cepheid Distances Calculated Using Model Atmospheres

Robert Hindsley[1] and R.A. Bell[2],

[1] *U.S. Naval Observatory, Washington, D.C.* [2] *University of Maryland*

Hindsley and Bell (1989, Ap. J., 341, 1004) determined distances for 23 Cepheids using the surface brightness method. The color $V - I_C$ on the Cousins system or $V - R_J$ on the Johnson system was used to measure visual surface brightness. Combined with the observed visual magnitudes, angular diameters were then calculated. Comparison of angular diameters with radius differences in linear units, obtained by integrating the radial velocity curve, yielded each Cepheid's distance.

Hindsley and Bell calibrated the relationship between color and surface brightness using model atmospheres. The zero point of the relationship was found using observations of Vega and an appropriate model. More recently the spectral scans of Gunn and Strycker (1983, Ap. J. Suppl., 52, 121) and the Cousins filter response curves of Hindsley and Bell were used to calculate colors and these were compared to the corresponding observed colors. It was found that the colors produced from the Gunn and Strycker data were redder than the observed colors. A transformation equation of the form

$$C_{OBS} = constant \times C_{SYN} \qquad (1)$$

with C_{OBS} being the observed colour and C_{SYN} that calculated from the Gunn and Strycker data, was found to be sufficient. For the Cousins colour $V - R_C$, the constant was 0.977, while for the colour $V - I_C$ the constant was 0.966. Note that no offset is needed. When the synthetic colours $V - I_C$ and $V - R_J$ were thus corrected and the analysis repeated for twelve stars, the resulting period-luminosity relationship was:

$$< M_V > = (-3.24 \pm 0.43)(log P - 0.8) - (3.82 \pm 0.44) \qquad (2)$$

This zero point is 0.21 magnitudes fainter than the previous result, but more in agreement with the value of -3.75 found by Barnes, Gieren, and Moffett 1990 (in Confrontation Between Stellar Pulsation and Evolution, ed. C. Cacciari and G. Clementini, Astr. Soc. Pac. Conf. Series, 11, 221). Absolute magnitudes derived using the Johnson system colour $V - R_J$ in the analysis are about 0.25–0.5 mag brighter than absolute magnitudes obtained using the colour $V - I_C$ on the Cousins system. This is due to the well-known difficulty with the Johnson R filter response curve. Further work is needed to resolve this problem. It is expected that the final result will be a zero point very close to that obtained using only $V - I_C$ observations, -3.60.

Galactic Cepheid Kinematics as a Probe of Large Scale Non-Axisymmetry of the Galaxy

J. A. R. Caldwell[1], C. Koen[1], I. M. Avruch[2], M. R. Metzger[2], P. L. Schechter[2], M. J. Keane[2]

[1]*South African Astronomical Obs., P.O.Box 9, Observatory 7935, S. Africa*
[2]*Massachusetts Inst. of Technology, Cambridge MA 02139 U.S.A.*
[3]*University of California, Santa Cruz CA 95064 U.S.A.*

Abstract

We investigate whether the currently available Galactic Cepheid kinematic data can put interesting constraints on large scale low amplitude non-axisymmetry of the Galactic plane rotation pattern. In this connection we address the experimental design problem of where in the Galactic plane additional Cepheids would prove the most useful for the axisymmetric and the non-axisymmetric modeling of the kinematics.

1. Optimal Design Selection of Galactic Cepheids to Constrain the Axisymmetric Rotation Model

Caldwell and Coulson (1987) (hereafter CC87) analyzed the available photometry, reddenings, and radial velocities of Galactic Cepheids in terms of an axisymmetric Galactic rotation model, yielding R_0 relative to the Cepheid luminosity scale. Caldwell et al. (1992) (hereafter CAMSK92) reviewed the further observational progress over the intervening years, specifically the additional photometry and velocities, the improved reddenings, and the rapid relative increase of knowledge about distant Cepheids across the Galaxy. They redetermined the Cepheid rotation model with the new information, and derived $R_0 = 8.5 \pm 0.5$ kpc, similarly to before, but with smaller uncertainty due to the accrued data. Two issues raised in that paper are the subject of further consideration here: the optimal choice of further Galactic Cepheids to search for and measure in future, and the possibility of large scale non-axisymmetry of the rotation pattern influencing the results.

The axisymmetric kinematic model fits the observed radial velocities of Galactic Cepheids by a 5-parameter function of the input variable, namely the Cepheid position in the Galactic plane. Four of the parameters appear linearly in the model (cf. CC87), namely the velocity shear $2AR_0$, the solar motions u_0 and v_0, and the additive velocity zeropoint δv_r. The fifth parameter m_0 enters nonlinearly as a relative distance modulus scale. The problem of choosing where in the Galactic plane further Cepheids are to be detected and measured, in order to yield the most efficacious information about the parameters, comes under the statistics topic of optimal

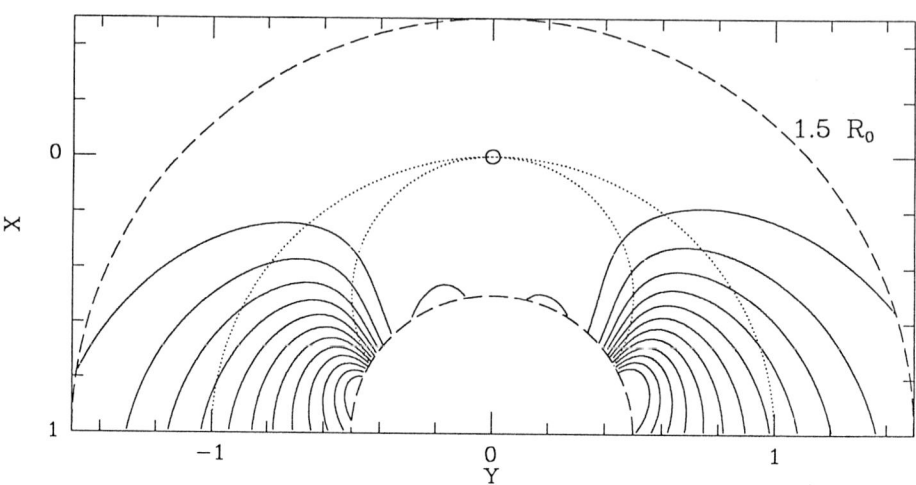

Fig. 1. The design parameter of the Cepheid axisymmetric kinematic problem. Increasing contours specify the desirability of obtaining new Cepheid data in an optimal design sense.

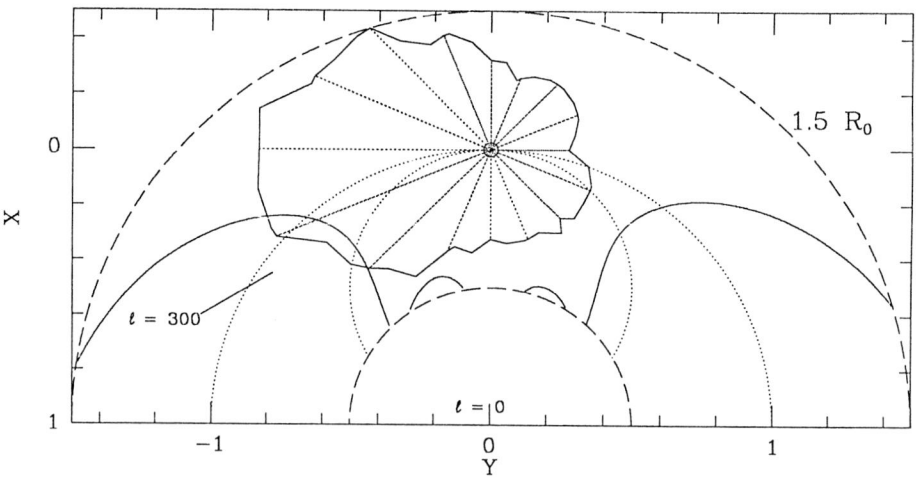

Fig. 2. The irregular outline encloses the region of the Galactic plane amenable to an analysis for Cepheid streaming velocities, with current data.

design. Khuri and Cornell (1987, chap. 8) present a method for the optimal design of the values of the variables used to constrain a nonlinear model. In essence the design criterion is to maximize the design parameter, which is namely the $N \times N$ determinant of the variance-covariance matrix formed from the existing data plus one (or more) free design points, where N is the number of model parameters. The maximization is carried out with respect to the design point(s).

Fig. 1 shows as solid lines the increasing contours of the design parameter, as a function of location on the Galactic plane. The design parameter, as remarked, indicates the relative efficacy of a new data point in maximizing the knowledge of the model parameters, as a function of where that new point is obtained. The origin at (X,Y) = (0,0) is the position of the sun, and the solar circle and tangent point circle are shown as dotted lines. Fig. 1 confirms that new Cepheids brought to bear on axisymmetric kinematic problem are most useful to the extent that they breach the tangent point distance at longitudes from 30-70° and 290-330°. An interesting detail of the figure is the north/south asymmetry of the design parameter, resulting from the imbalance in the currently available data (CAMSK92). Distant *northern* Cepheid data are now perceptibly more useful in constraining the model parameters than equivalently distant southern data. While at face value Fig. 1 gives high priority to new Cepheid data from deep chords across the Galaxy, one must bear in mind that the usefulness of such Cepheids depends critically on the assumption that the rotation curve is linear over the Galactocentric distance range concerned.

2. Search for Large Scale Streaming Motions

CAMSK92 presented the results of a kinematic model using a PLC relation (cf. CC87) to infer the Cepheid distances. We examine here whether the radial velocity residuals from the predictions of that model show any hint of "streaming" motions that could be the signature of large scale low amplitude non-axisymmetric distortion of the rotation pattern. Recent work especially by Blitz and Spergel (1991) and by Kuijken and Tremaine (1992) give the impetus to test the utility of Cepheids as tracers of such elliptical kinematic perturbations. We calculated a projected streaming velocity and projected distance for all Cepheid residual velocities from the axisymmetric solution prediction. This was done for a number of longitudes from 180° to 337.5° in steps of 22.5°, including for each direction only those Cepheids with a projection factor in absolute value within 10% of unity. The individual Cepheid velocities generally scatter with ~11 km s^{-1} around the solution. To be more sensitive to possible streaming at the 5 km s^{-1} level, the points have been binned by 4. **Fig. 2** illustrates the direction vectors and heliocentric distance extent that the present binned data allow us to say anything about. Also shown is the lowest contour of design parameter discussed above. Clearly, as a handle on large scale non-axisymmetry at the few km s^{-1} level, the existing Cepheid data can offer only very limited information.

Fig. 3 shows the projected velocities in the specified longitude directions as a function of the projected distances in those directions. The only remarkable finding is a net contraction trend in $\ell = 180°$ direction and some possibly discrepant velocities belonging to distant Cepheids towards $\ell = 112\text{-}135°$. The longitude 180° trend resembles the differential velocity effect of an ellipsoidal distortion with the long axis pointing from the Galactic center into the first quadrant, and with radially decreasing amplitude. Otherwise there appears to be no possibly significant "streaming" trend in the data.

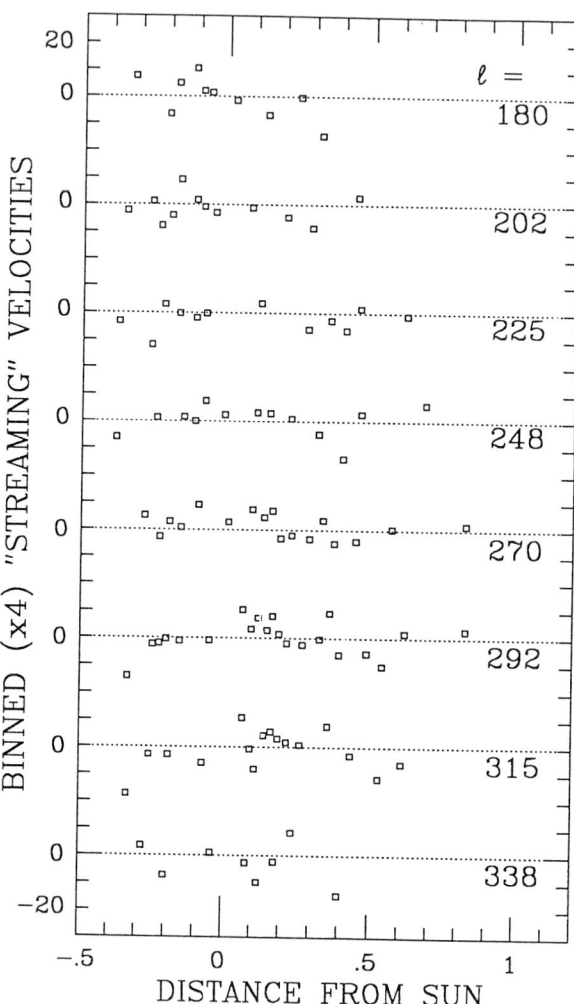

Fig. 3. Inspection of the velocity residuals from the axisymmetric kinematic solution, to look for "streaming" trends.

3. Non-Axisymmetric Simulation

The program code used by CAMSK92 was expanded by the addition of a three-parameter ellipsoidal velocity perturbation. The ellipses are modeled by an axis-ratio-squared, κ, a tilt angle with respect to the anticenter vector, θ, and a velocity normalization at the solar circle, v_p. The perturbation velocity magnitude is taken to behave in keplerian fashion with Galactocentric radius, and its direction is taken as following the elliptical streamlines. To keep the velocity perturbation more nearly differential, the purely circular velocity pattern corresponding to the same v_p is subtracted. The perturbed motion of the LSR is taken into account in computing perturbed radial velocities.

Fig. 4 illustrates the perturbation velocity pattern resulting from the model just described, with κ and θ chosen to be 2° and –45° respectively, that is with the long axis of the ellipses extending from the Galactic center into the first quadrant. This orientation is the same as adopted by Blitz & Spergel (1991) and Whitelock (1992). The velocity contours indicate the relative perturbation radial velocity magnitude, and the signs are positive towards $\ell = 90°$ and 270°, and negative towards the center and anticenter. The hashed zones denote nearly zero net perturbation radial velocity. The expansion trend implied along the solar circle is if anything contradictory to the trend of the points in Fig. 3.

Fig. 5 shows the partial derivative of the radial velocity with respect to the principal other ellipsoidal parameter that one would ideally hope to be able to constrain by possible future Cepheid observations, θ. Here the sign is again positive at the large peak in the first quadrant. It is apparent that one would need Cepheid velocities much farther across the Galaxy in this direction than are now available (cf. Fig. 2), to have good sensitivity to this parameter.

A number of nonlinear least squares solutions were attempted with the 8-parameter model (five from the axisymmetric case, plus v_p, κ, and θ). The conclusions are that the present data do not constrain κ and θ. Further, when fixing κ and θ at a variety of trial values, the radial velocities were best fit with *no* ellipsoidal perturbation added: v_p converged toward zero in all attempts. Although this negative result was disappointing, but not surprising, plots such as Figs. 4 and 5 hold out the hope that the discovery of new Cepheids half way to the solar circle at $\ell = 45$-$60°$ will be directly useful in testing for large scale non-axisymmetry in future.

References:

Blitz, L., and Spergel, D. N. 1991, ApJ, 370, 205.
Caldwell, J. A. R., and Coulson, I. M. 1987, AJ, 93, 1090.
Caldwell, J. A. R., Avruch, I. M., Metzger, M. R., Schechter, P. L., and Keane, M. J. 1992, in: *Variable Stars and Galaxies*; symposium in honor of M. W. Feast, ASP Conference Series (in press).
Khuri, A. I., and Cornell, J. A. 1987, *Response Surfaces: Designs and Analyses*, Statistics: textbooks and monographs volume 81.
Kuijken, K., and Tremaine, S. 1991, in: *Dynamics of Disk Galaxies*, 71.
Whitelock, P. A. 1992, in: *Variable Stars and Galaxies*; symposium in honor of M. W. Feast, ASP Conference Series (in press).

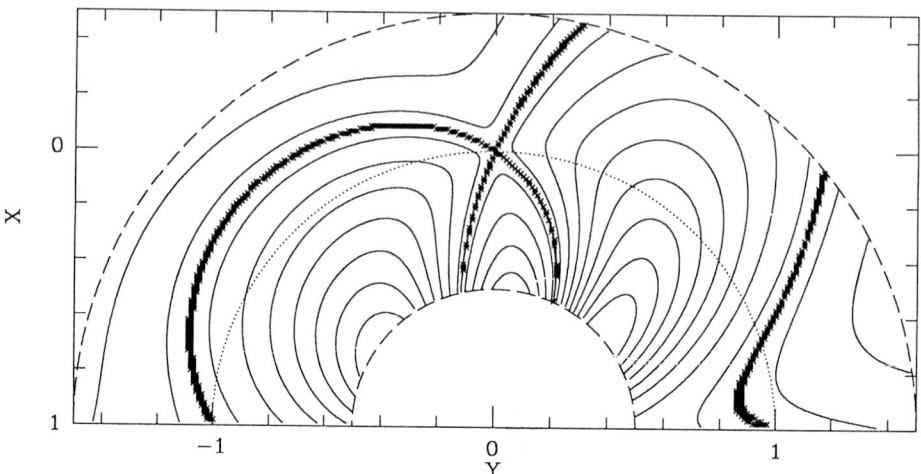

Fig. 4. Contours of the partial derivative of the observed Cepheid radial velocity with respect to the ellipsoidal perturbation velocity coefficient, v_p, as a function of Cepheid location in the Galactic plane, for a model discussed in the text.

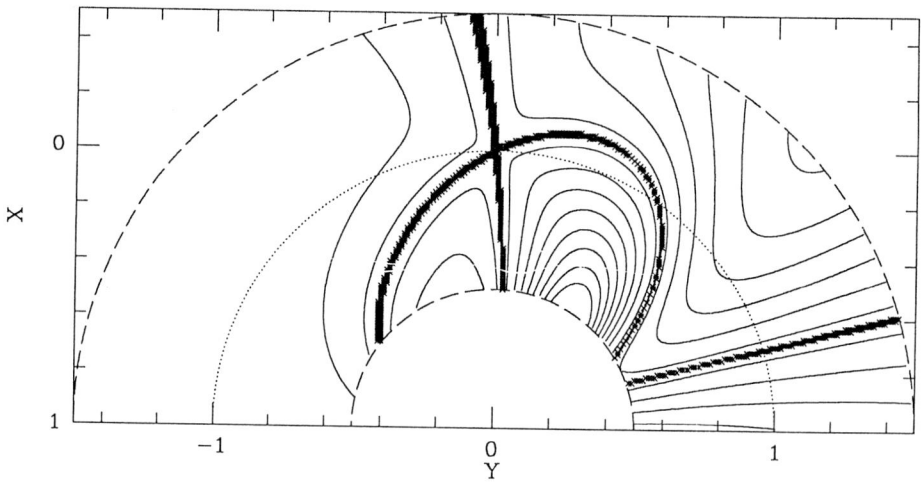

Fig. 5 Contours of the partial derivative of the observed Cepheid radial velocity with respect to the ellipsoidal perturbation tilt angle, θ, as a function of Cepheid location in the Galactic plane, for the same model.

DISCUSSION

D. TURNER: Is the solar motion relative to the LSR solved for in your analysis, and how does a spiral density wave affect your solution?

J. CALDWELL: Yes, the u_0 and v_0 velocities of the Sun are free parameters, that come out typically at 7 and 14 km s^{-1}, respectively. I haven't yet looked into the effect of a spiral wave pattern. I suspect that there would be only a very minor effect because of the scaller in the Cepheid velocities and positions.

D. WELCH: I'm kind of surprised that you haven't mentioned outer disk Cepheids. Also, there is an asymmetry in the outer disk Cepheid velocities.

J. CALDWELL: The outer disk Cepheids will be extremely useful probes of the metallicity dependences of, *e.g.*, the Cepheid *PL* and *PC* laws. Those towards $l \sim 135°$ and $225°$ strongly constrain $2AR_0$. For the present though, I feel that additional *distant* solar circle or inner disk Cepheids would add the most to our understanding, for reasons I have presented previously. I will be very interested to know more about the asymmetry of the outer disk Cepheid velocities.

An Optical P − L Relation for LPVs

M.J. Pierce[1], D.R. Crabtree[2],

[1]*Kitt Peak National Obs., U.S.A.,* [2]*Dominion Astrophysical Obs., Canada*

1. An Optical P-L Relation for LPVs

Supergiant Long-Period Variables (LPVs) are both more luminous and more common than Cepheids. The fact that a P − L relation exists at K (e.g. Wood et al. 1983 ApJ, **272**, 99; Mould et al. 1990 ApJ, **349**, 503) and not at visible wavelengths is usually attributed to the strong temperature sensitivity of TiO, which dominates the spectra of these stars in the visible. This has led us to apply a narrow bandpass ($\lambda_0 = 8250 \text{Å}$, $BW = 350 \text{Å}$), avoiding TiO, in the hopes that an optical P − L relation can be found.

2. Estimated Distances for the LMC and M 33

The data for Per OB1, the LMC, and M 33 are shown in figure 1. The small dispersion ($\sigma_{LMC} = 0.22$ mag.) and flat slope are evident. Combining the data we find:

$$M_I = -0.00267 P(days) - 6.31, \qquad (1)$$

with the absolute calibration from the ten LPVs in Per OB1, assuming $m - M = 11.8 \pm 0.2$ (Garmany & Stencel 1992 AAS, **94**, 211). This results in estimated distance moduli of 18.33 ± 0.05, and 24.79 ± 0.08 for the LMC and M 33, vs. 18.47 ± 0.15, and 26.64 ± 0.09 using Cepheids (e.g. Feast & Walker 1987 ARAA, **25**, 345; Freedman et al. 1991 ApJ, **372**, 455). Evidently, any systematics from metallicity differences must be small. We conclude that this P − L relation for LPVs is a powerful tool for estimating extragalactic distances and offers a useful compliment/alternative to Cepheids. LPVs are being detected in Virgo Cluster spirals (Pierce et al. this volume).

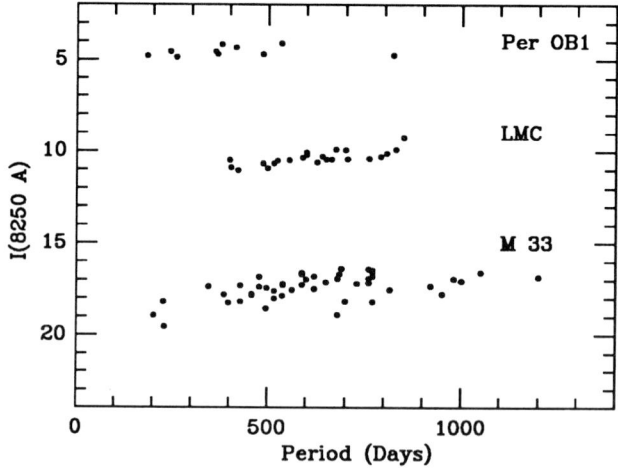

Implication of a *P-L* Relation of Mira Variables

Hiromoto Shibahashi

Department of Astronomy, University of Tokyo, Japan

Feast et al. (1989) obtained observationally a period-luminosity (*P-L*) relation of Mira variables in LMC. Basically, fundamental periods of stellar pulsation should be determined by two quantities —a mass and a radius of the star. The existence of a *P-L* relation implies a special condition that reduces a function of two variables into a function of only a single variable. Though the AGB appears as a thin line in the HR diagram like the Cepheid instability strip, it is an "asymptotically" merged line of evolutionary tracks of stars in a range of about 0.5-3 solar masses. This fact means that the masses of stars on a point of the AGB are not unique though the radii and the effective temperatures are unique, respectively. Therefore we cannot expect, in the case of AGB stars, that a period is reduced to a function of a single variable. The periods of an AGB star should be a function of two variables —the luminosity and the mass of the star. The observationally obtained *P-L* relation should be interpreted as such a period-luminosity-mass relation, and scattering of the relation must be recognized as reflection of the mass range of Mira variables.

The luminosity of an AGB star is strongly dependent on the core mass and almost independent of the total mass of the star. However, the pulsation period is dependent on the total mass of a star, and becomes shorter with increasing mass for a given luminosity (and hence a given effective temperature). This fact means that we would be able to distinguish masses of AGB stars from evolutionary tracks drawn on a (P, L)-diagram while we cannot do so from evolutionary tracks in the HR diagram. Preliminary calculations show that an evolutionary track of a star on a (P, L)-diagram is less steeply inclined than the observationally fitted period-luminosity relation of Mira variables. If we plot the observed period and luminosity of each Mira variable and draw evolutionary tracks of various mass stars on the same (P, L)-diagram, we can determine the mass of each star. If we can deduce the mass and the radius of each Mira variable in this way, we can determine the position of the star in the HR diagram by mapping again a theoretical (mass, radius)-diagram onto the HR diagram. The distribution of Mira variables on a (P, L)-diagram has a little bit steeper inclination than an evolutionary track of a star. This feature is conserved by mapping the (P, L)-diagram onto the HR diagram. Hence, the distribution of Mira variables on the HR diagram must be inclined with respect to the asymptotic giant branch from the left above to the right below. This means that Mira variables concentrate in a more specific region of the HR diagram, and implies the existence of an instability region for Mira variables.

References:

Feast, M.W., Glass, I.S., Whitelock, P.A., and Catchpole, R.M. 1989, MNRAS, **241**, 375.

Studies of Large-Amplitude δ Scuti Variables

W. J. F. Wilson, E. F. Milone and D. J. I. Fry

Rothney Astrophysical Observatory (RAO)
Department of Physics and Astronomy, The University of Calgary

The determination of Baade–Wesselink radii and luminosities for pulsating stars are long-standing and highly desired goals since they provide the promise of being standard candles. In a modest contribution towards these goals, we have undertaken a programme to determine the radii and luminosities of the large-amplitude δ Scuti stars DY Herculis, EH Librae and DY Pegasi from optical and infrared photometry and cross-correlated radial velocity data. We use Fourier representations for V, I and J light curves and for the radial velocity curves in Baade–Wesselink analyses to derive minimum radii over the pulsation cycles. These radii and their errors and the mean bolometric luminosities and absolute magnitudes will be discussed here and in papers to follow. As a check, we also apply our method to the data and results of other groups.

Each V light curve was divided into 200 equally spaced phase points, and a value was calculated for R_{min} at each phase point in the descending portion of the colour index curve using phase pairs of equal colour index. The error δR_{min} was calculated at each point from the computed errors in the Fourier coefficients. R_{min} and δR_{min} were then plotted against phase and the mean values $<R_{min}>$ and $<\delta R_{min}>$ were calculated over the phase range where δR_{min} was smallest.

We obtain minimum radii from the phase ranges shown of

EH Lib ($0.1 \leq \phi \leq 0.5$): $<R_{min}> \pm <\delta R_{min}> = 2.54 \pm 0.30\ R_\odot$
DY Her ($-0.03 \leq \phi \leq 0.33$): $<R_{min}> \pm <\delta R_{min}> = 3.26 \pm 0.30\ R_\odot$

For EH Lib, we combine our values of R_{min} + computed displacement with the values of T_{eff} from Joner (1986) to obtain mean values of the bolometric luminosity and absolute bolometric magnitude of

$$<L_{bol}> = 23.1\ (+5.6, -5.1)\ L_\odot$$
$$<M_{bol}> = 1.34\ (-0.24, +0.27).$$

Fuller descriptions will be published elsewhere. This work has been supported by the Department of Physics and Astronomy, and NSERC grants to EFM.

References:

Joner, M.D., 1986, PASP 98, 651

Stellar Seismology

White Dwarf and Pre-White Dwarf Oscillations

Arthur N. Cox

Los Alamos Astrophysics

Abstract

Compact stars that result from extreme mass loss on the asymptotic giant branch and planetary nebula formation are observed to pulsate in a very large surface effective temperature range as they cool to become the classical white dwarfs. The hottest and most luminous of these display periods in excess of 1000 seconds because they are large, but when the stars arrive on the cooling line on the Hertzsprung-Russell diagram, their periods become generally less than 1000 seconds. Then the stars have masses near 0.6 $M_\odot$ and radii near 10^9 cm. Their luminosity depends then almost entirely on the surface effective temperature as the entire star with its legacy of complicated internal luminosity peaks cools to the classical simple electron degenerate structure. Very thin surface layers of hydrogen and helium cover the bulk of the carbon- and oxygen-rich mass that results from hydrogen and helium burning in earlier intermediate mass stellar evolution. The cause of the nonradial pulsations of low angular degree, but rather high radial order, for the most luminous of these stars is the cyclical ionization of carbon and oxygen in layers not too deep that their effectiveness is limited by a long luminosity time scale. Thus the surface hydrogen and helium must be thin, probably thinner than the current period spacings interpretation for PG 1159-035 suggests. For the classical DBV and DAV pulsators, it appears that neither the hydrogen ionization κ and γ effects or convection blocking at the bottom of a hydrogen convection zone can destabilize the observed pulsations when the overriding short time scale effects of time-dependent convection are included. It appears, however, that a thin CO convection shell can produce pulsations by its time-dependent effects, but again only very thin H and He surface layers are allowed. This new pulsation mechanism can alleviate the serious problem that DAV variables are observed hotter than the hydrogen κ and γ effects and convection blocking can predict. The appearance of non-pulsators in the DAV and DBV instability strips can be explained by a too-thick hydrogen and helium surface layer that interferes with (poisons) the CO ionization convection zone. Finally time-dependent convection predicts that only a few of the many possible modes exist due to their internal amplitude structure that can result in both strong driving and strong damping. Thus actually observed pulsating modes can assist in mapping individual internal white dwarf composition structures, not only by their periods but also the fact that they pulsate.

1. Pre-white dwarf evolution

After asymptotic giant branch evolution for intermediate mass stars with efficient mass loss, the CO core resulting from He burning is almost exposed with only a thin layer of primordial H and of He produced from earlier stages of shell H burning. Some of the envelope evaporates to create a very compact pre-white dwarf with a planetary nebula surrounding it. When this nebula blows away, the star is small, with its deep He still creating nuclear energy. Its surface temperature can be over 150,000 K. Classical evolution (Figure 1) suggests that layers of H 10^{-4} and He 10^{-2} of the mass thick comprise the envelope. More mass loss is being investigated now. The high gravity may not be as inhibiting as earlier thought. Pulsation theory, now including time-dependent convection effects, shows that such thick layers will not give linear theory instability for the observed nonradial modes. For the hottest pre-white dwarfs, this H and He layer cannot be any thicker than 10^{-9} of the stellar mass to pulsate, but the many non-pulsators, with H and He poisoning in the CO pulsation driving layers, can have these evolution suggested thick layers.

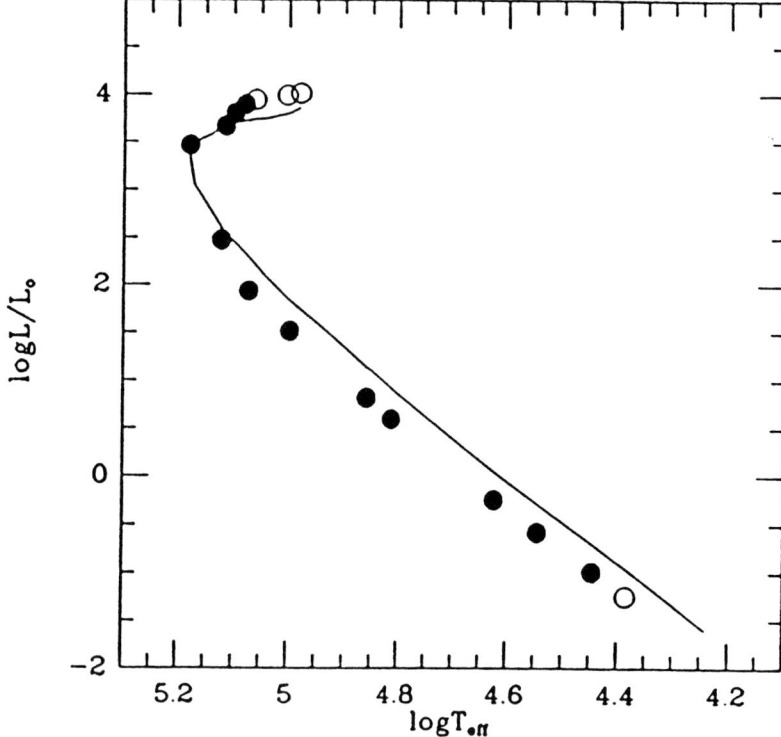

Figure 1 The evolution track for Iben-Hollowell 0.6 $M_\odot$ models with standard surface H and He. Solid dots are for pure CO models that are pulsationally unstable, while the open circles indicate stable models.

2. The three classes of white dwarfs

Various considerations indicate that all these stars have a mass near 0.6 $M_\odot$. The most luminous pre-white dwarfs are named the DOV stars. These multiperiodic variable stars are observed with periods of from 500 to over 1500 seconds at surface temperatures over 100,000 K. They have significant surface CO with rather little H and He. Due to rapid diffusive settling of CO and floating of H and He, the surface compositions probably do not reflect the composition in mass layers 10^{-9} deep where the cyclical ionizations of CO produce pulsational instability by the classical κ and γ effects. Stars on the evolution track across the top of the Hertzsprung-Russell diagram are larger than those on the cooling track and thus have larger periods for their high radial order nonradial modes. Pure CO models pulsate over most of this post-planetary nebula state, but the presence of He in the pulsation driving region dilutes (poisons) the κ and γ effects driving, and pulsations then occur only in a smaller temperature range. The blue edge for any pulsation is at about 150,000 K. Red edges are determined when the pulsation driving effects layers retreat to deep, adiabatic regions. Recent theoretical work is reported by Stanghellini et al. (1991).

DBV white dwarfs with essentially pure He at their surfaces and in the pulsation driving region pulsate in a narrow temperature range of about 20,000 K to 25,000 K. Periods are from about 200 to about 1000 seconds. The driving mechanism for these pulsations is related to the deep convection zone in the He. A mechanism called convection blocking periodically valves the luminosity at the bottom of the convection zone at about 10^{-13} of the mass deep in a similar way that the classical κ and γ effects do to convert the emergent luminosity to pulsation motions. Time-dependent convection destroys much of this surface convection driving, but other contributions by the classical κ and γ effects caused by He ionization 10^{-14} of the mass deep and also time-dependent convection in a deep convecting CO shell about 10^{-6} of the mass deep seem to add enough additional driving to produce the observed instability effective temperature band. It has been predicted by two independent groups that a very thin non-convecting H layer 10^{-15} thick can exist on top of these stars to disguise their true DBV character and make an apparent hot DA star pulsate. The instability strip red edge is due to deep convection that mixes He into the CO, forcing the pure CO core surface so deep that no convection and its time-dependent effects occur in that material.

The classical DA pulsating white dwarfs have periods between about 200 and 1200 seconds. They have surface effective temperatures between about 11,500 K and 13,500 K. Recent stellar atmosphere studies including convection in the atmosphere may reduce these temperatures somewhat. Just as for the DBV variables, the DAV pulsators are driven at the bottom of a convection zone about 10^{-13} of the mass deep. Very efficient convection is needed to get the convection zone deep enough, and no nonadiabatic calculation has yet predicted pulsations at the observed blue edge of the instability strip. Time-dependent convection destroys much of the convection blocking driving, but small amounts of classical H ionization κ and γ effects 10^{-14} of the mass deep and time-dependent convection driving in CO 10^{-6} deep can possibly make the blue edge DA stars pulsate. The instability strip red edge is caused by deep convection that mixes just a little He into the CO, but an insignificant amount of CO to the surface, forcing the pure CO core surface so deep and hot that no convection and its pulsation driving effects occur in that material. Some recent nonadiabatic stability calculations are given by Cox et al. (1987).

3. Pulsation mechanisms for pre-white dwarfs

The κ effect depends on a large positive logarithmic derivative of the opacity with respect to temperature as shown in Figure 2 for a DOV model.

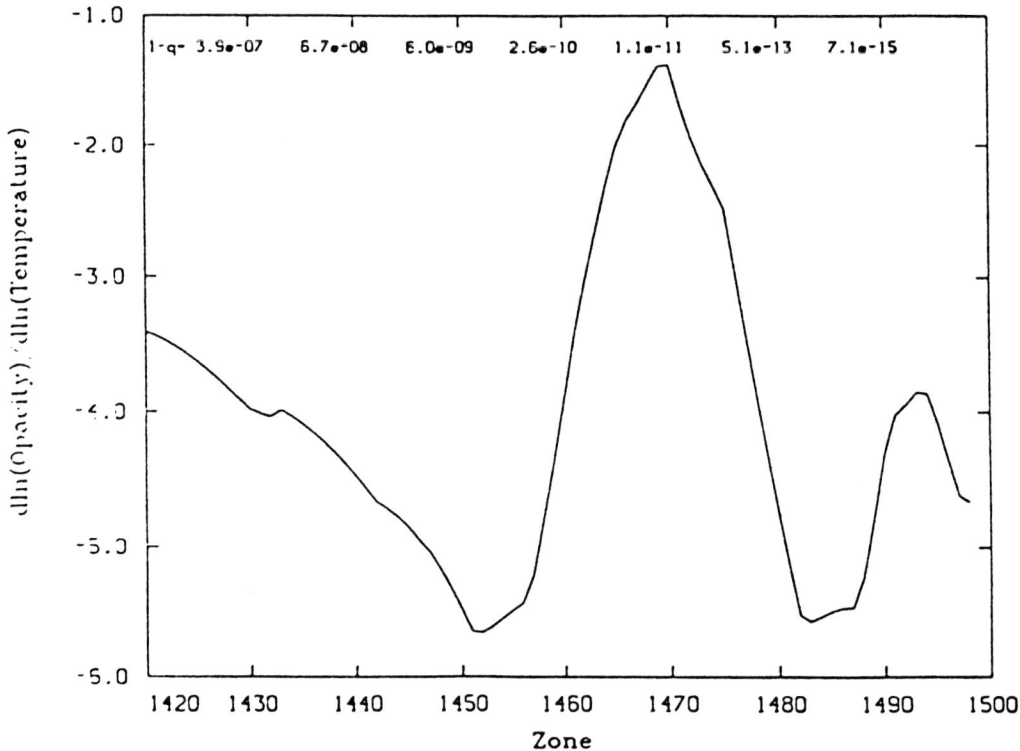

Figure 2 The logarithmic derivative of the opacity with respect to the temperature versus zone number for a pre-white dwarf model at 64,780 K with a surface and pulsation region composition of half He and half C.

The γ effect can hide luminosity during compression stages when it is small as shown in Figure 3. Figure 4 then shows the work to drive or damp pulsations, and the important region is 10^{-9} of the mass deep.

There are two problems with such a thin H and He layer on top of the pulsation driving CO core. PG 1159-035 is observed to have a period decrease for its strongest mode of about 2.5×10^{-11} seconds per second, whereas existing evolution calculations (with considerable surface H and He) predict a density deconcentration during this early evolution, and that gives a period increase. Current efforts try to trap modes in rather thick layers (10^{-6} of the mass) to predict this observation. Possibly just shrinking of this very hot star now with very little surface H and He can more easily allow a correct period decrease prediction.

With only 10^{-9} of the mass in the surface layer of H and He to prevent pulsation poisoning, mode trapping in these layers that can give unequal period spacings may not occur at all. Yet recent observations (Winget et al., 1991) suggest strong mode trapping. I have reviewed the Whole Earth Telescope fourier intensity spectrum kindly supplied by Paul Bradley, and I propose that the largest period spacing fluctuations are incorrect due to wrong mode identifications.

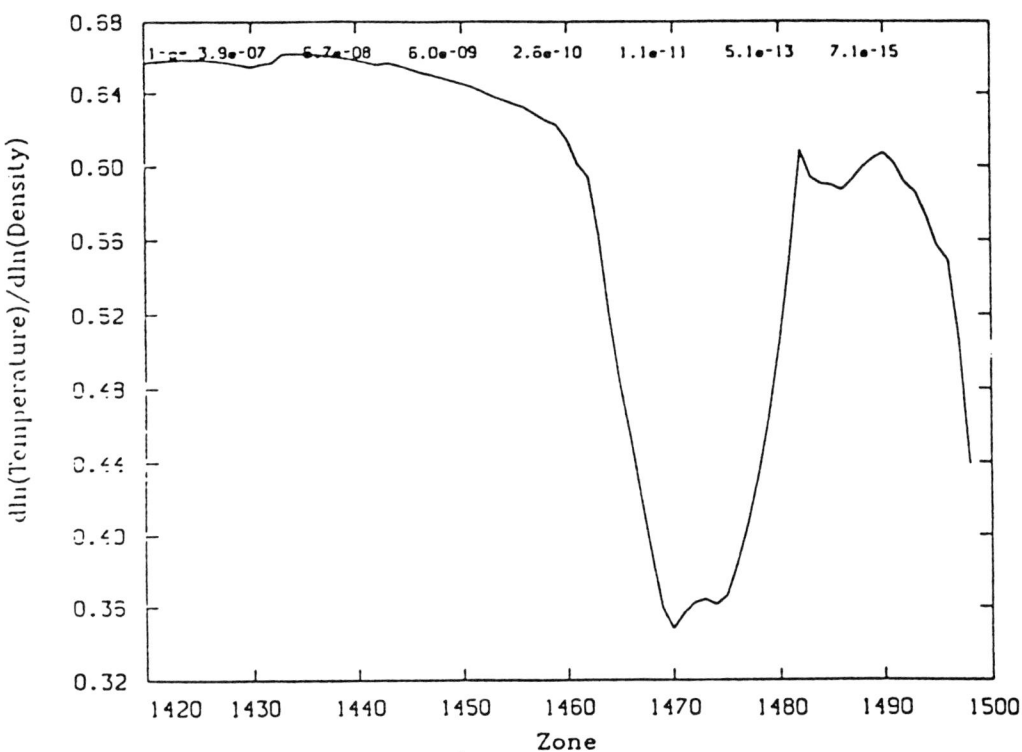

Figure 3 $\Gamma_3 - 1$ versus zone number for a pre-white dwarf model at 64,780 K with a surface and pulsation region composition of half He and half C.

4. Pulsation mechanisms for DBV white dwarfs

The pulsation driving region for a 25,000 K DBV model is shown in Figure 5. for the conventional frozen-in convection case. For this model near the DBV instability strip blue edge, driving is mostly the κ and γ effects in helium, because the surface convection is weak with my ratio of mixing length to pressure scale height of 2.5. There is a slight convection blocking contribution at zone 500 just barely visible. The radiation damping noise between zones 300 and 400 is caused by my three point Lagrangian interpolation, which assures that opacities are accurately interpolated, but there are frequent density and temperature derivative discontinuities.

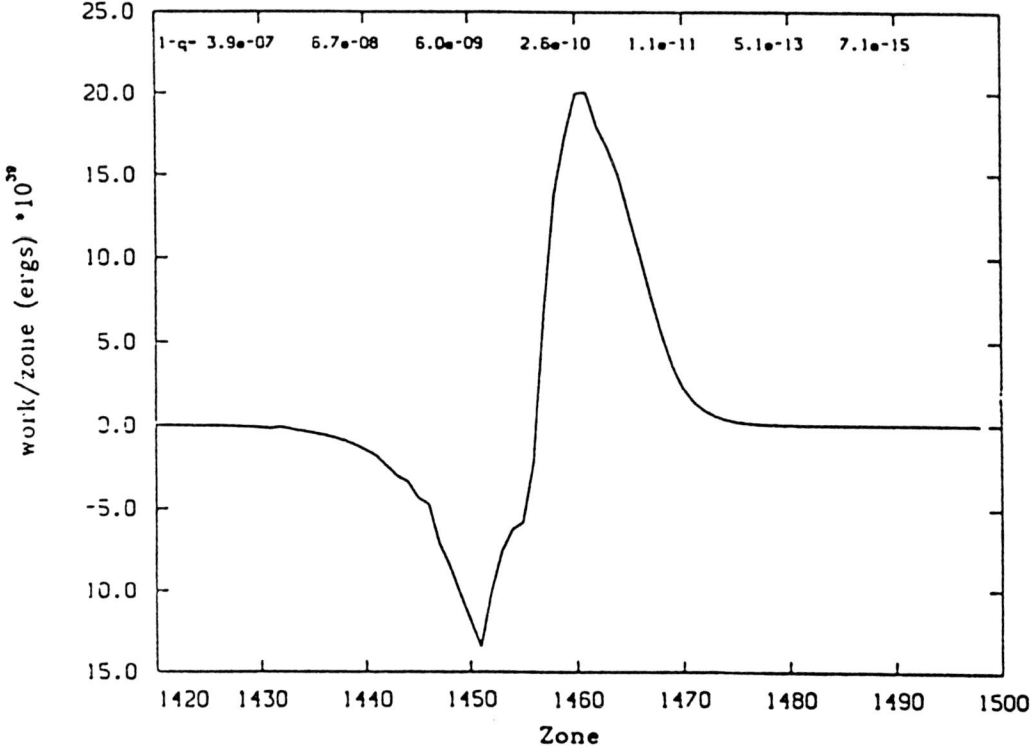

Figure 4 Work versus zone number for a pre-white dwarf model at 64,780 K with a surface and pulsation region composition of half He and half C. This is for the g_{14}, $\ell = 1$ mode at 461 seconds period.

Time-dependent convection using the philosophy of Cox et al. (1966) has been developed for my linear nonadiabatic code. It uses numerical derivatives of the local convection luminosity with respect to the density and temperature on both sides of Lagrangian interfaces and the lagging formulas discussed by Cox and Giuli (1968). When the quickly adapting convection in the surface layers only is considered, most of the radiation blocking is destroyed by convection luminosity, and driving no longer is larger than the deeper damping. This is shown in Figure 6 for the same 25,000 K model. Figure 7 shows the case when the lagging convection in the deep CO convecting layer, carrying up to 75 % of the luminosity is allowed. Now net pulsation driving obtains. This lagging results in an amplitude for the convection luminosity variations of only about 0.4 of the instantaneous luminosity, calculated using no lagging and the instantaneous configuration. The strong nonadiabatic effects greatly influence the eigensolution.

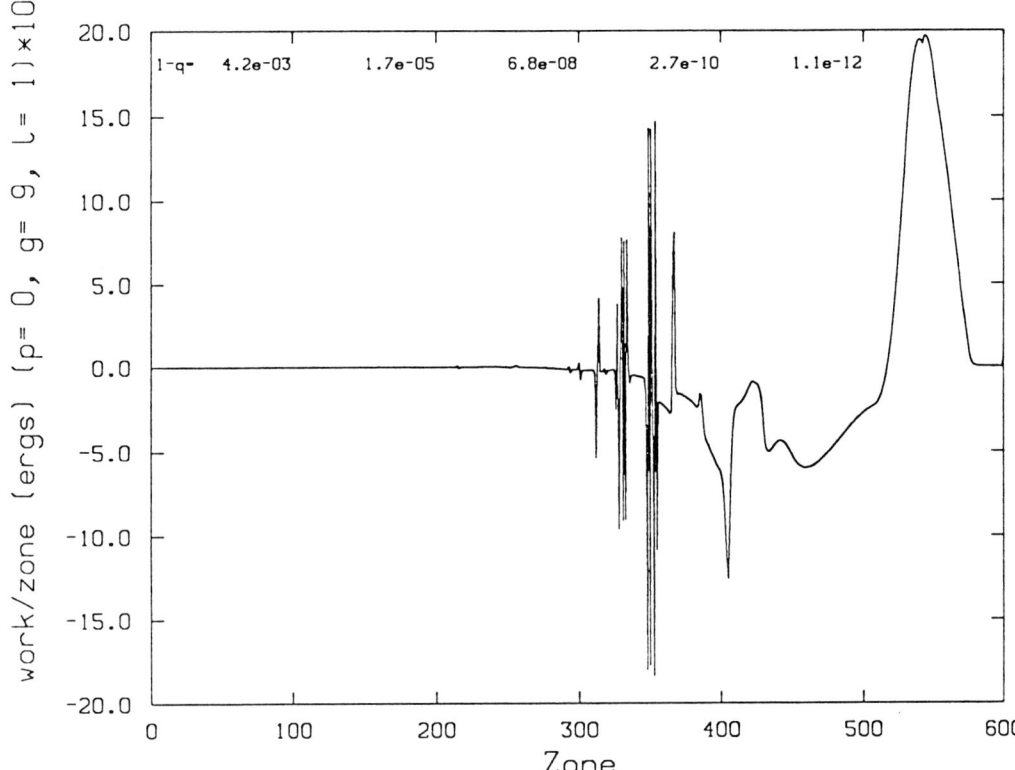

Figure 5 Work versus zone number for a pure He surface and pulsation driving region composition DBV model at 25,000 K. The noise between zones 300 and 400 comes from my three point Lagrangian interpolation in the He opacity table. Zone numbers are used so that eye integration can be made to see if the driving is larger than the damping. This is for the g_9, $\ell = 1$ mode at 523 seconds period.

Note that it is necessary that there be deep convection to give this time-dependent convection driving and pulsational instability. If the surface He layer is too thick, the transition from surface helium to the CO core will have a diffusion composition shape instead of a steeper convection overshooting μ gradient shape. The fact that this DB model is unstable requires that the surface He have a thickness of less than about 10^{-6} of the mass. Fortunately, period spacings for $\ell=1$ modes derived by Bradley (private communication) for the DBV variable GD 358 confirm that this He layer thickness is reasonable. When the He thickness gets over 3×10^{-6} of the mass, this lower opacity material occupies the region where CO material would be convecting.

Then there is no deep convection and probably no pulsations.

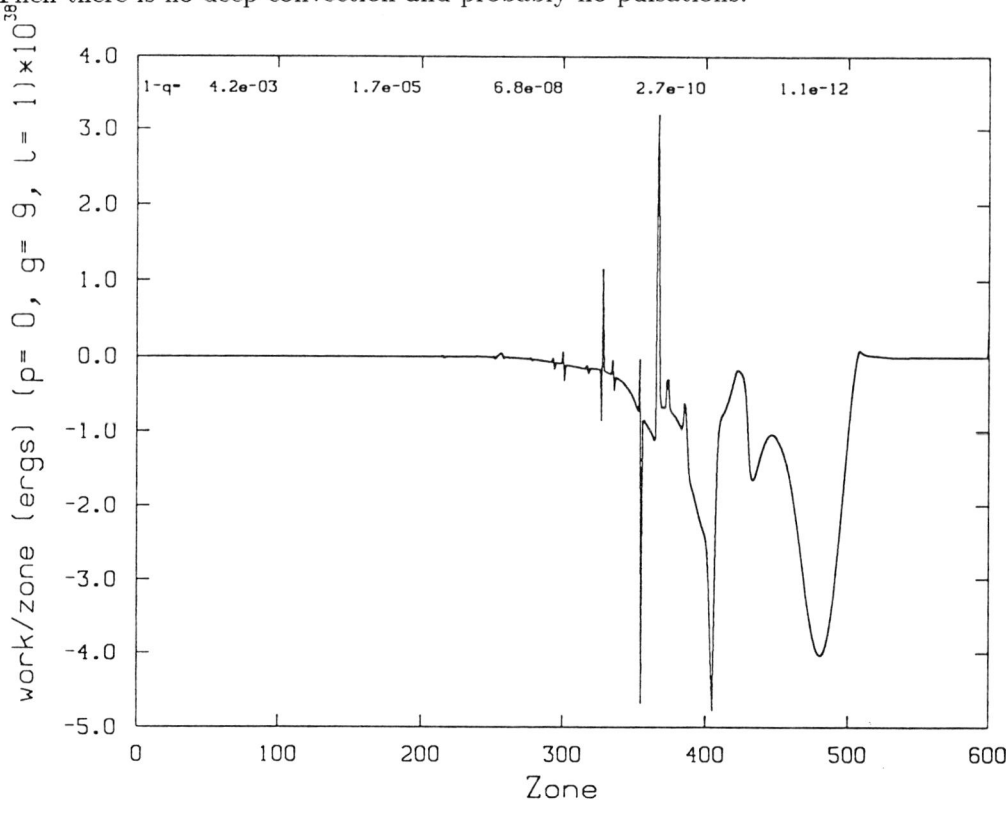

Figure 6 Work versus zone number for a pure He surface and pulsation driving region composition DBV model at 25,000 K considering time-dependent convection only in the surface convection zone. This is for the g_9, $\ell = 1$ mode at 520 seconds period.

5. DAV white dwarfs and discussion

The situation is similar for the DAV variables, where frozen-in convection, nonadiabatic calculations do not give instability as hot as 13,500 K. This is true even though Fontaine et al. (1984) inferred instability by only looking at the depth of the convection rather than its strength and ability to block luminosity during compression stages. Again time-dependent convection destroys almost all surface driving (unless inhibited by strong magnetic fields), but driving in a CO convecting shell only 3×10^{-6} deep on top of the CO core gives unstable modes for models as hot as 13,500 K.

Time-dependent convection damps pulsations when the density and temperature variations increase toward the surface, as in the H and He ionization layers. However, deeper, below the second node, these variations decrease toward the surface, and they together with the natural convection lagging driving and residual surface driving give adequate pulsation driving for DBV and DAV stars. There could be driving also just below the first node and exterior to the density and temperature variations maximum, but are no luminosity blocking mechanisms in this few hundreds of thousand kelvin region.

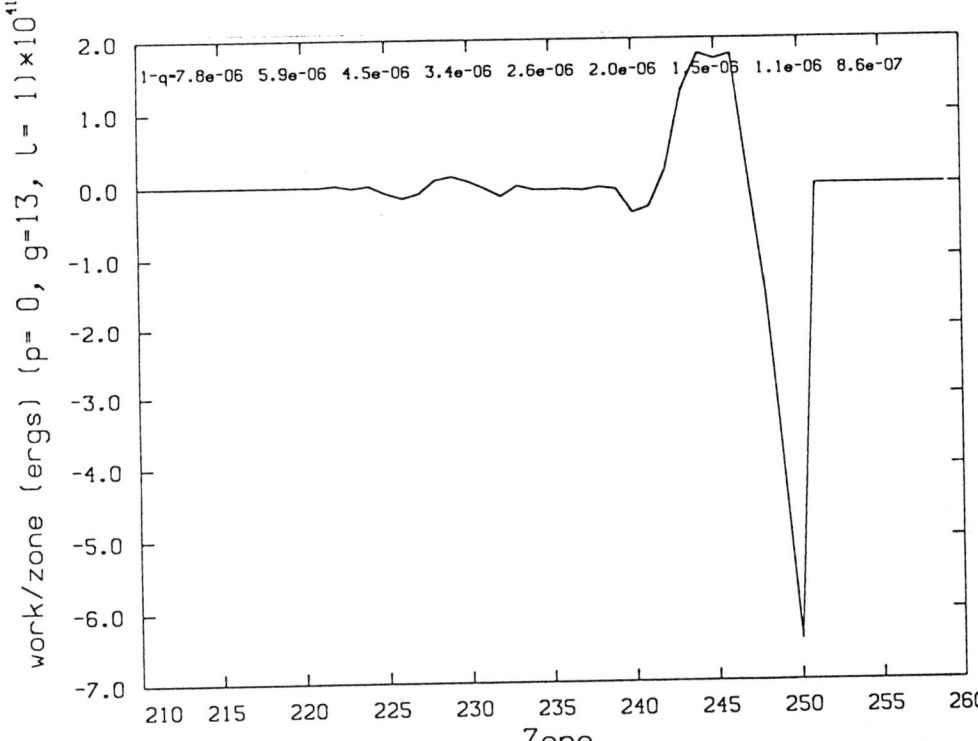

Figure 7 Work versus zone number for a pure He surface composition DBV model at 25,000 K considering time-dependent convection in the CO core convection shell that lies between zones 234 and 250 and between 5 and $7 \times 10^6 K$. This is for the g_{13}, $\ell = 1$ mode at 758 seconds period. Not all modes are unstable. Stability depends on the convection derivatives, the convection lagging, and how these influence the eigensolution.

Time-dependent convection gives different drivings from mode to mode. Thus matching observed periods to theoretical ones for various He-CO interface structures allows detailed probing of white dwarfs.

References:

Cox, A. N., Brownlee, R. R. and Eilers, D. D., 1966, Astrophys. J., **144**, 1024.
Cox, A. N., Starrfield, S. G., Kidman, R. B., and Pesnell, W. D., 1987, Astrophys. J., **317**, 303.
Cox, J. P. and Giuli, R. T., 1968, "Principles of Stellar Structure" (New York: Gordon and Breach), p. 1045.
Fontaine, G., Tassoul, M., and Wesemael, F., 1984 in, "Theoretical Problems in Stellar Stability and Oscillations," Universite de Liege. p. 328.
Stanghellini, L., Cox, A. N., and Starrfield, S. G., 1991, Astrophys. J., **383**, 766.
Winget, D. E., et al., 1991, Astrophys. J., **378**, 326.

An Example Demonstrating the Potential for Asteroseismology of DB White Dwarf Stars

P.A. Bradley[1] and M.A. Wood[2]

[1] *McDonald Observatory and Department of Astronomy, University of Texas*
[2] *Department of Physics and Space Sciences, Florida Institute of Technology*

Abstract

We present the results of a parametric survey of evolutionary models of compositionally stratified white dwarfs with helium surface layers (DB white dwarfs). Because white dwarfs are the most common final end state of stellar evolution, determining their internal structure will offer us many clues about stellar evolution, the physics of matter under extreme conditions, plus the history of star formation and age of the local Galactic disk. As a first step towards determining the internal structure of DB white dwarf stars, we provide a comprehensive set of theoretical g-mode pulsation periods for comparison to observations.

Because DB white dwarfs have a layered structure consisting of a helium layer overlying the carbon/oxygen core, some modes will have the same wavelength as the thickness of the helium layer, allowing a resonance to form. This resonance is called mode trapping (see Brassard et al. 1992 and references therein) and has directly observable consequences, because modes at or near the resonance have eigenfunctions and pulsation periods that are similar to each other. This results in much smaller period spacings between consecutive overtone modes of the same spherical harmonic index ℓ than the uniform period spacings seen between non-trapped modes. We demonstrate with an example how one can use the distribution of pulsation periods to determine the total stellar mass, the mass of the helium surface layer, and the extent of the helium/carbon and carbon/oxygen transition zones. With these tools, we have the prospect of being able to determine the structure of the observed DBV white dwarfs, once the requisite observations become available.

We are grateful to C.J. Hansen, S.D. Kawaler, R.E. Nather, and D.E. Winget for their encouragement and many discussions. This research was supported by the National Science Foundation under grants 85-52457 and 90-14655 through the University of Texas and McDonald Observatory.

References:

Brassard, P., Fontaine, G., Wesemael, F. & Hansen, C.J. 1992, ApJS, 80, 369

The Internal Structure of White Dwarf Stars Using the Whole Earth Telescope

P.A. Bradley

McDonald Observatory and Department of Astronomy, University of Texas

Abstract

White dwarfs are the final end state for the majority of stars, and hold clues to help solve many current pressing astrophysical problems. We can perform asteroseismology on the pulsating white dwarfs to better understand their internal structure and input physics, paving the way to a better understanding of astrophysics, stellar evolution, and the history of our Galaxy. I describe briefly the potential of asteroseismology by using it to infer the internal structure of PG1159–035.

1. What Are White Dwarfs and Why Are They Interesting?

White dwarfs have attracted considerable interest from astronomers because they offer insight into many problems of physics and astrophysics. In the following discussion, I will concentrate on the pre-white dwarfs known as PG1159 stars. Like the white dwarfs they will become, the PG1159 stars are the product of stellar evolution with main sequence masses up to $\sim 8 M_\odot$, and have an average mass of $0.5 - 0.6 M_\odot$ (see Bergeron, Saffer, & Liebert 1992; Weidemann 1990). We believe the PG1159 stars have carbon/oxygen cores comprising more than 99 % of the mass, and spectroscopy (Werner, Heber, & Hunger 1991) reveals ionized carbon, helium, oxygen, and sometimes nitrogen in an atmosphere with a temperature of 100,000 to 150,000 K. The PG1159 stars will eventually become either helium atmosphere (DB) white dwarfs or hydrogen atmosphere (DA) white dwarfs. Nature is kind enough to supply us with pulsating PG1159 stars, DB and DA white dwarfs for seismology so that can determine the evolutionary fate of the PG1159 stars. (See the conference proceedings by Vauclair & Sion 1991; Wegner 1989; and Philip, Hayes, & Liebert 1987 for articles concerning the structure, evolution, and pulsations of white dwarfs.)

The intense gravitational fields ($\log g \sim 7 - 8$) of the PG1159 stars and white dwarfs implies that they should have a layered structure, with the heavier elements (carbon + oxygen) sinking to become the core, while the lighter helium and hydrogen float to the surface. The compositional stratification expected of white dwarf stars means that a g-mode can become trapped within a given region of the star when the wavelength of the g-mode matches the thickness of the trapping layer, allowing a resonance to form. (See Brassard et al. 1992 and references therein for more details on mode trapping in white dwarfs.) Observationally, mode trapping reveals itself through minima in the period spacing between consecutive overtone modes, the result of modes in or near the mode trapping resonance having very similar eigenfunctions and pulsation periods. Models with thicker surface layers require lower overtone modes for the first mode trapping resonance to form, and the number of nontrapped modes in between the trapped modes is smaller. Thus, if the mass of the surface layer is close to that predicted by stellar evolution calculations (see Iben 1991; Mazzitelli & D'Antona 1990; and references therein) the interval between period spacing minima will be shorter than for the much thinner surface layer masses predicted by spectral evolution theory (Fontaine & Wesemael 1991, 1987).

2. Sample Results

New observations from the Whole Earth Telescope ($\equiv$ WET; Nather et al. 1990) have resolved the light curves of several pre-white dwarf and white dwarf stars with unprecedented detail. The analysis of WET data on the pre-white dwarf PG1159–035 (Winget et al. 1991) demonstrates that asteroseismology can yield much new information about these stars with existing asymptotic theory and models. However, these WET observations, along with those of other white dwarfs, demonstrate that we need a new generation of models for asteroseismological determinations of their structure.

As an example of the potential of the new models (Wood 1990), we show plots of the period spacing between consecutive overtone modes versus the pulsation period for PG1159–035 (a DOV star) against the predicted periods from some of the new models. For PG1159–035, the interval between period spacing minima–trapped modes–is best matched by models with a transition region at a few $\times 10^{-3} M_*$ (see Figure 1). At long periods, the period spacing curve for the model is much flatter than the observations, suggesting that we don't yet have the composition transition region accurately modelled. We can add this to what we already know about PG1159–035 from the previous efforts of Winget et al. (1991), who determined the total stellar mass ($0.586 M_\odot$), the rotation rate (1.38 d), set limits on any magnetic field that may be present (< 6000 G), and demonstrated the star must already be stratified. More work on the model matching should allow us to arrive at a better match to the observations and a better knowledge of PG1159–035's internal structure, although we can already say that the surface layer mass is several times less than the value predicted by stellar evolution theory.

Clearly, the prospect of asteroseismology for stellar structure determination is quite promising. With time, we expect to resolve and interpret the observed period structure of many pulsating white dwarfs to better understand their structure and place in the universe.

I am grateful to C.J. Hansen, S.D. Kawaler, R.E. Nather, D.E. Winget, and M.A. Wood for their help, encouragement, and many discussions. This research was supported by the National Science Foundation under grants 85-52457 and 90-14655 through the University of Texas and McDonald Observatory.

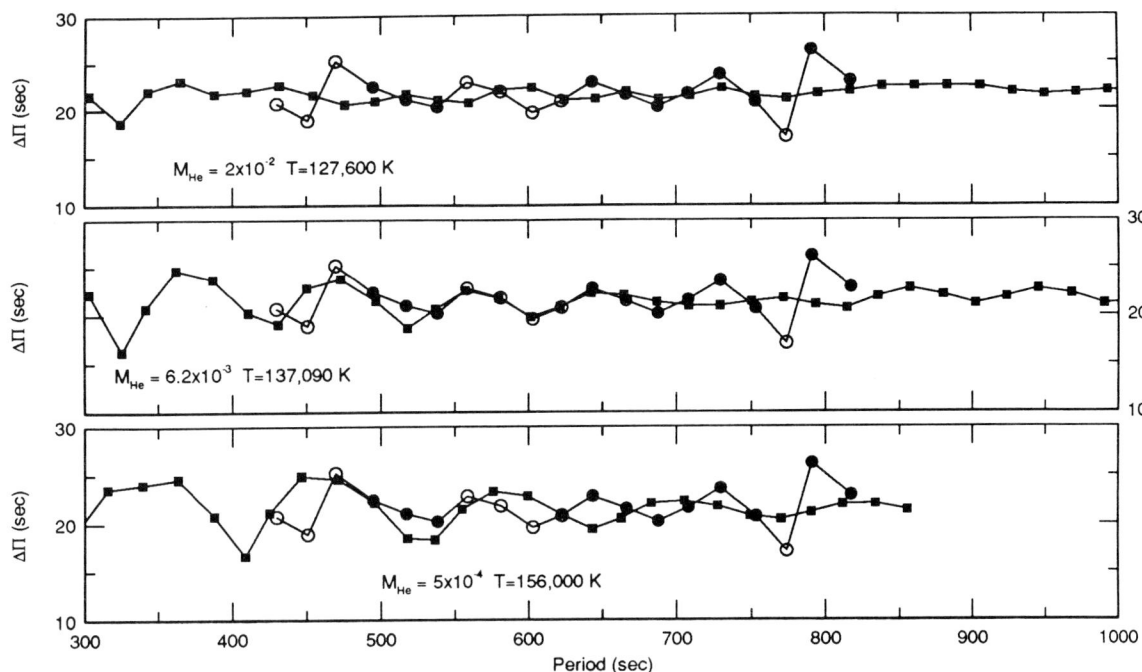

Figure 1: Sample fits of the $\ell = 1$ period spacings for PG1159-035 (dots) versus period spacings from 0.59M⊙ evolutionary models (squares).
The open dots represent less certain period determinations than the solid dots.

References:

Bergeron, P., Saffer, R.A., & Liebert, J. 1992, ApJ, 394, 228
Brassard, P., Fontaine, G., Wesemael, F. & Hansen, C.J. 1992, ApJS, 80, 369
D'Antona, F., & Mazzitelli, I., 1990, ARA&A, 28, 139
Fontaine, G., & Wesemael, F. 1991, in IAU Symp. 145, The Photospheric Abundance Connection, ed. G. Michaud & A. Tutukov, (Dordrecht: Reidel), 421
Iben, I.,Jr. 1991, ApJS, 76, 55
Nather, R.E., Winget, D.E., Clemens, J.C., Hansen, C.J., & Hine, B.P. 1990, ApJ, 361, 309
Philip, A.G.D., Hayes, D.S., & Liebert, J. 1987, IAU Colloq. 95, The Second Conference on Faint Blue Stars, (Schenectady: Davis)
Starrfield, S.G., Cox, A.N., Kidman, R.B., & Pesnell, W.D. 1984, ApJ, 281, 800
Vauclair, G., & Sion, E.M. 1991, in the Seventh European Workshop on White Dwarfs, White Dwarfs, (NATO ASI Ser.), (Dordrecht: Kluwer)
Wegner, G. 1989, IAU Colloq. 114, White Dwarfs, (Berlin: Springer)
Werner, K., Heber, U., & Hunger, K. 1991, A&A, 244, 437
Weidemann, V. 1990, ARA&A, 28, 103
Winget, D.E. 1988, in IAU Symposium 123, Advances in Helio- and Asteroseismology, ed. J. Christensen-Dalsgaard & S. Frandsen, (Dordrecht: Reidel), 305
Winget, D.E., et al. 1991, ApJ, 378, 326
Wood, M.A. 1990, Ph.D. thesis, University of Texas at Austin

White Dwarf Seismology at the Université de Montréal

G. Fontaine, P. Brassard, P. Bergeron, F. Wesemael

Département de Physique, Université de Montréal

Over the last several years, we have developed a comprehensive program aimed at better understanding the properties of pulsating DA white dwarfs (or ZZ Ceti stars). These stars are nonradial pulsators of the g-type, and their study can lead to inferences about their internal structure. For instance, the period spectrum of a white dwarf is most sensitive to its vertical chemical stratification, and one of the major goals of white dwarf seismology is to determine the thickness of the hydrogen layer that sits on top of a star. This can be done, in principle, by comparing in detail theoretical period spectra with the periods of the observed excited modes. Likewise, because the cooling rate of a white dwarf is very sensitive to the specific heat of its core material (and hence to its composition), it is possible to infer the core composition through measurements and interpretations of rates of period change in a pulsator.

In order to provide the needed theoretical framework within which to analyze observed period spectra and measurements of rates of period change, we have recently completed a thorough adiabatic survey of evolutionary models of white dwarf stars (Ap.J. Suppl., **72**, 335; Ap. J. Suppl. **81**, 747). In the process, we have developed new analytical and numerical tools, including an analytical model for the phenomenon of trapping of gravity-modes at a composition discontinuity in a star (Ap. J. Suppl. **80**, 369), as well as a high-performance pulsation code based on finite-element techniques (Ap. J. Suppl. **80**, 725). We believe that these results will be instrumental in reaping the most important benefits from white dwarf seismology. We have also developed a new technique (based on numerical simulations of the emergent energy distribution of a pulsating white dwarf) which allows us to exploit the information contained in the relative amplitudes of a pulsation mode and its harmonics in a Fourier spectrum. This is particularly important for interpreting the high signal-to-noise ratio observations which we have started gathering at the Canada-France-Hawaii Telescope with the help of "LAPOUNE", the Montréal 3-channel photometer. This observational effort constitutes an integral part of our white dwarf seismological program, and has produced, so far, data of unprecedented quality. In the immediate future, we expect that our observations coupled to our adiabatic survey data will allow us to infer the amount of hydrogen floating on top of several DA white dwarfs.

The Effect of a Vertical Magnetic Field on the Periods of Trapped g-modes in White Dwarfs

Bradley W. Carroll

Dept. of Physics, Weber State University, Ogden, UT 84408-2508

Abstract

A white dwarf model with cylindrical symmetry is used to investigate the effect of a uniform vertical magnetic field ($\vec{B} = B_o \hat{z}$) on the periods of g-modes trapped in the compositionally stratified surface layers. For those modes for which trapping is most effective, the periods wander by approximately 10% about their zero-field values as B_o increases. As the field strength increases further, the period may abruptly increase by some 25% to a new, more stable value; see Figure 1. The periods of less efficiently trapped modes do not show this sensitivity for field strengths up to 1 MG, the upper limit examined.

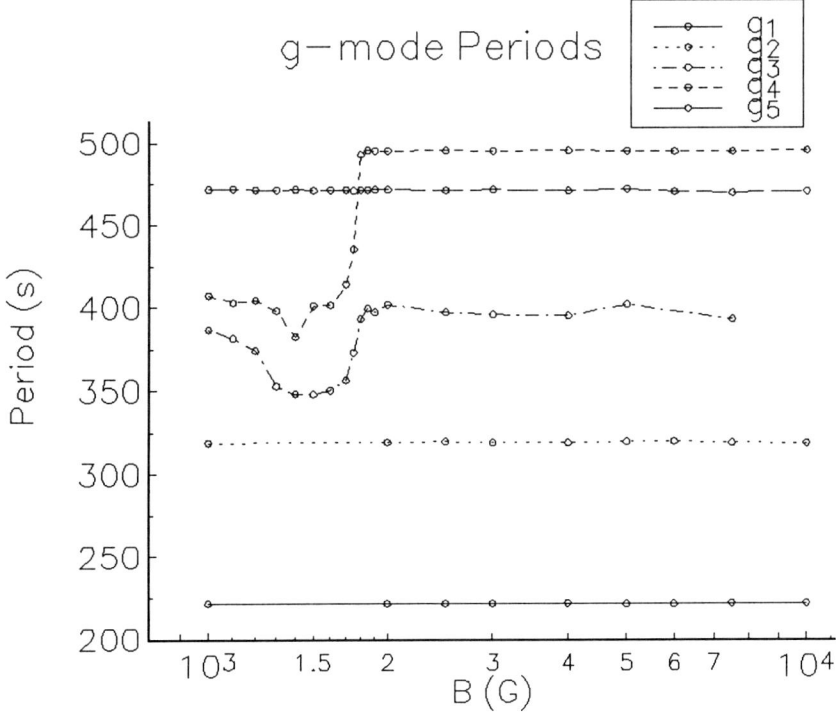

Figure 1 Pulsation periods of $g_1 - g_5$ modes with $\ell = 2$ for a compositionally stratified 0.6 M$_\odot$ white dwarf model ($T_e = 9960$ K, $M_{He}/M_* = 10^{-5}$, $M_H/M_* = 10^{-10}$)

Seismology of Pulsating Ap Stars:
Results From The Past Decade, Prospects For The Next

Jaymie M. Matthews

Département de physique, Université de Montréal

Abstract

Since the discovery of the first rapidly oscillating Ap (roAp) star in 1978 by Kurtz, this class of magnetic chemically-peculiar pulsators has grown to over two dozen. The eigenfrequency spectra of roAp stars (with periods of $\sim 6-15$ min) are consistent with nonradial p-modes of low degree ℓ and high overtone n, not unlike the Sun's five-minute oscillations seen in integrated light. However, unlike the Sun, the strong global dipole fields of roAp stars significantly affect the pulsations.

Although much of the effort in the last decade has been towards detecting new roAp candidates and refining the frequencies of known variables, initial "seismic" analyses have already yielded important results. Measurements of fundamental frequency spacings (ν_0) constrain the luminosities and radii of some roAp stars. In addition, mode splitting provides: (1) an independent determination of rotation period, even in the absence of longer-term light variations; (2) limits on the rotational inclination i and magnetic obliquity β; and (3) an indication of the relative *internal* field strengths of certain roAp stars. Very recently, the temperature – optical depth structure of the atmosphere of HR 3831 was inferred from optical and IR photometry of its oscillations.

Judging from current developments, the next decade promises exciting results on both observational and theoretical fronts. Several roAp stars have now been monitored for over a decade, allowing us to investigate long-term period changes due to evolution, binarity, etc. Eigenfrequency models for stars in the mass and radius range appropriate for Ap stars are becoming available, as well as explicit treatments of the perturbations due to magnetic fields. Armed with these, we may be able to place some roAp stars on a theoretical $\nu_0 - \delta\nu_{21}$ (or "asteroseismological H–R") diagram to derive independently their masses and main-sequence ages.

1. Introduction

Imagine a star with a magnetic field thousands of times stronger than that of the Sun; a star whose spectrum is dominated by lines of holmium; a star which – although it shouldn't be unstable to pulsation at all – is in fact vibrating persistently in something like the 25th overtone of its fundamental resonance period. Now imagine that such a bizarre object may actually reveal something useful about 'normal' main-sequence stars.

What you've pictured is Pryzbylski's Star (HD 101065). Even though this star is possibly the most peculiar example of a class already considered "peculiar" relative to other A–F stars, its pulsations are not unique. HD 101065 was the first star to join the ranks of the *rapidly oscillating Ap* or *roAp* stars. These variables were all identified by their photometric oscillations with short periods ($\sim 4 \leq P \leq \sim 20$ min) – compared to expected fundamental periods of several hours – and low amplitudes ($\Delta B \leq 0.015$ mag).

The observed characteristics of roAp stars have been well summarised in various reviews, including Kurtz (1990), Matthews (1991), and Weiss (1986). Shibahashi (1987) has reviewed some of the theoretical aspects of these stars. Therefore, I'd like to concentrate here on the applications of roAp stars as probes of stellar astrophysics, through the techniques of asteroseismology.

2. The *p*-mode pulsation spectrum

Many of the roAp stars are multiperiodic; two of them – HR 1217 (HD 24712) and HD 60435 (see Figure 1) – have very rich eigenspectra in which it is clear that the frequencies are nearly equally spaced from each other. The regular frequency spacing seen in both the roAp and solar oscillations is the signature of nonradial *p*-mode pulsations where the modes are of high overtone; i.e. $n \gg \ell$. According to the asymptotic pulsation theory of Tassoul (1980), the frequencies of such modes can be expanded in a fit of the form (Christensen-Dalsgaard 1986):

$$\nu_{n,\ell} \simeq \nu_0(n + \frac{\ell}{2}) - D_0[\ell(\ell+1)] + \text{higher order terms} \quad (1)$$

where

$$\nu_0 = [2\int_0^R \frac{dr}{c(r)}]^{-1} = (\text{sound travel time across the star})^{-1} \quad (2)$$

and

$$D_0 \propto [\frac{c(R)}{R} - \int_0^R \frac{dc}{dr}\frac{1}{r}dr]^{-1}. \quad (3)$$

$c(r)$ is the local sound speed at radius r. The higher-order terms in equation (1) and the proportionality constant in equation (3) depend on the detailed structure of the star. Due to the properties of the expansion, the term D_0 can be expressed as $\frac{1}{6}(\nu_{n,\ell} - \nu_{n-1,\ell+2})$. The term in brackets is often designated $\delta\nu_{02}$: the fine-splitting between the same overtone of an $\ell = 0$ and and $\ell = 2$ mode.

The higher-order terms on the right side of equation (1) are small, so to first order, one expects a comb of equally spaced frequencies. If a set of consecutive overtones n is present for modes of *only* odd or even degree ℓ, then the spacing will be ν_0. If modes of *both* odd and even degree are present, the predicted spacing is $\nu_0/2$. The second-order term introduce slight deviations from this equal spacing. In Figure 1, ν_0 could be either 26 or 52 μHz (2.246 or 4.492 d^{-1}). Tentative mode identifications and other arguments suggest $\nu_0 \simeq 52$ μHz for HD 60435.

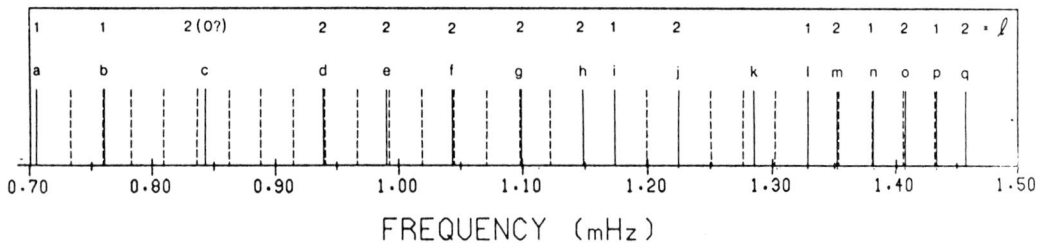

Figure 1. Oscillation frequencies identified in HD 60435, compared to the expected values for exactly equal spacing (from Matthews *et al.* 1987).

From equation (2), it is clear that the observed frequency spacing ν_0 depends on the average sound speed across the diameter of the star, which is proportional to its mean density. On the other hand, the fine structure of the frequency spectrum, due to the $\frac{1}{r}$ dependence in equation (3), is sensitive to conditions near the core of the star.

3. Measuring luminosities

A basic parameter like luminosity is extremely difficult to determine for an Ap star. The normal photometric indicators of log g do not apply to the heavily line-blanketed flux distributions of such stars.

However, we can exploit the dependence of ν_0 in equations (1) and (2) to place certain roAp stars on a theoretical H-R diagram. Gabriel *et al.* (1985) have shown that, for A-G V stars, $\nu_0 \sim 0.20 \left(\frac{GM}{R^3}\right)^{\frac{1}{2}}$. Hence, lines of constant ν_0 are straight lines in a plot of log L vs. log T_{eff}; in fact, they are lines of nearly constant radius. By measuring ν_0 in a multiperiodic roAp star, and finding an independent determination of its T_{eff}, one can in principle derive its luminosity and radius.

Of course, there are problems with this approach. Many of the multiperiodic roAp stars have only two or three observed frequencies, so it is unclear if the observed frequency spacings are in fact directly related to ν_0. Even in the cases of HD 60435 and HR 1217, whose p-mode eigenspectra are rich enough to make the pattern more obvious, there is an ambiguity. Is the observed spacing ν_0 (indicating modes of only even *or* odd ℓ) or $\frac{\nu_0}{2}$ (modes of even *and* odd ℓ)? Estimates of the evolutionary state of the star through long-term period changes may resolve the ambiguity (see the discussion in §6) but such measurements are themselves confused by the multiperiodicity of the star.

There may be another way to decide upon the actual value of ν_0. I had noticed (Matthews 1988, 1991) that the ratio $\frac{<\nu>}{\nu_0}$ observed in roAp stars tends to fall near two preferential values. Even the solar value of this ratio fits into this pattern. This tendency has endured as new stars have been added to the original sample. The ratios are shown in Table 1, in which values of mean frequency and possible spacings have been drawn from a variety of published sources. Be warned that, while some of these values are well established, others are still uncertain.

To first order, from equation (1), the ratio $\frac{<\nu>}{\nu_0}$ is roughly equal to the average overtone $<n>$ of the modes, as long as $n \gg \ell$. In other words, it appears at first glance that both roAp stars and the Sun choose to pulsate in two restricted ranges of overtone, near $n \sim 25$ and 40. If confirmed by increasing numbers of roAp variables, or if a simple theoretical mechanism can explain it, then we may be able to use this rule to decide between two possible values of ν_0.

To demonstrate the diagnostic potential of ν_0, Kurtz (1992) has taken published estimates of ν_0 for roAp stars – wary of their ambiguities and varying quality – to try and locate these stars on the H–R diagram. He has adopted effective temperatures based on the H_β calibration by Moon & Dworetsky (1985), which should be relatively insensitive to metallicity. Kurtz's results are shown in Figure 2.

Table 1.

Star	$<\nu>$ (μHz)	ν_0 (μHz)	$\frac{<\nu>}{\nu_0}$
Sun	3300	135	24
33 Lib	2105	80	25
HD 101065	1373	58	24
γ Equ	1370	58	24
HD 60435	1380	52	26
10 Aql	1385	51	27
HD 203932	2800	72	39
HR 1217	2710	68	40
HD 218495	2240	61	37
HD 119027	1930	52	37
HD 166473	1870	50	37

Note that all the roAp stars appear to lie within the boundaries of the lower instability strip defined by the δ Scuti pulsators, arguing for a common κ driving mechanism. Also, the stars seem to be relatively evolved, lying well above the ZAMS. However, some of the values Kurtz takes as ν_0 would appear to be $\frac{\nu_0}{2}$ according to the pattern of Table 1. This would bring at least two of the "most highly evolved" stars much closer to the ZAMS, and one would fall outside the classical instability strip.

Fortunately, the question may soon be settled when accurate parallaxes become available for some of these stars, courtesy of the Hipparcos satellite. Then we can invert the process: specifying the luminosities and using ν_0 to determine T_{eff} to high accuracy. This will have the added benefit of testing the validity of the H_β temperature calibration for these peculiar stars.

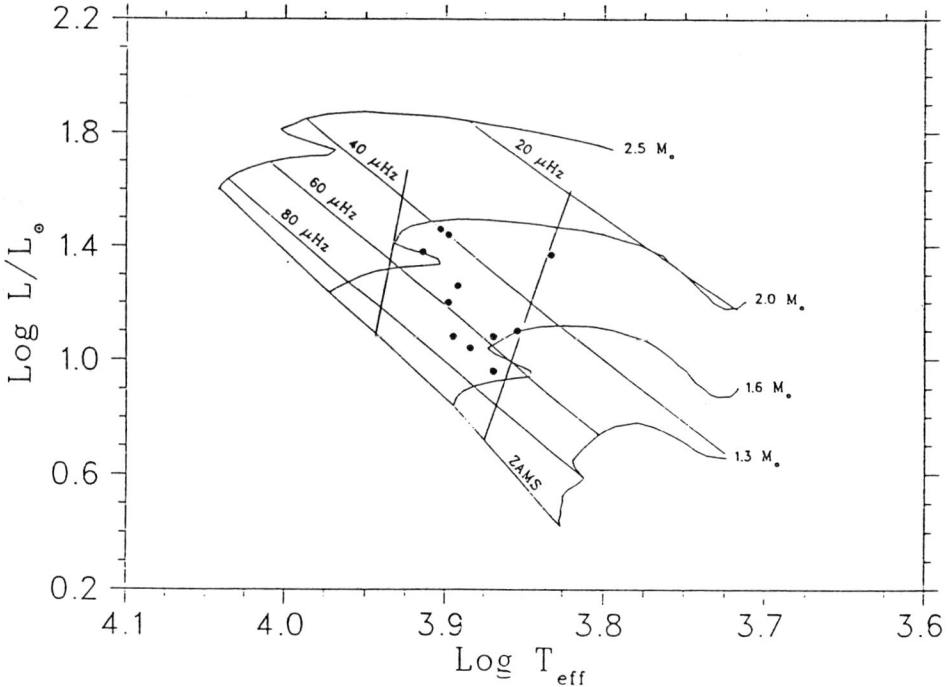

Figure 2. A theoretical H–R diagram (taken from Kurtz 1992) which shows tentative locations for roAp stars for which ν_0 has been estimated. The parallel diagonal lines are contours of constant ν_0; the borders of the δ Scuti instability strip are also shown.

4. Measuring internal magnetic fields?

The oscillations of several roAp stars exhibit periodic amplitude modulation and π^{rad} phase shifts synchronised with their magnetic variations. This behaviour can be explained by the *Oblique Pulsator Model* (Kurtz 1982, 1990): the nonradial pulsation pattern is tilted towards the magnetic dipole axis, rather than being aligned with the rotation axis. (The observations are consistent with zonal pulsation patterns; in particular, the $\ell = 1$, $m = 0$ mode). In the Fourier domain, a single pulsation frequency $\nu_{n,\ell}$ would be split into a multiplet of $2\ell + 1$ frequencies, centred about the rest-frame frequency and spaced by the rotation frequency Ω_{rot}. If the magnetic field were to completely dominate over rotation, then the magnetic and pulsation poles would be identical. In this case, the Fourier multiplet would be symmetric in amplitude.

However, the pulsations will also feel Coriolis forces due to the star's rotation which would tend to disrupt this alignment and bring the pattern back towards the rotational pole. When neither the Lorentz forces ($\propto B^2$) nor the Coriolis forces ($\propto \Omega_{rot}$) totally dominate, the pulsation and magnetic poles will be at the same stellar longitude, but not exactly aligned. (In other words, the pulsation pole will be at a point directly between the magnetic and rotational poles.) In this instance, the Fourier multiplet is asymmetric in amplitude. The amplitude differences of the sidelobes contain information about the orientation of the magnetic field (i.e., inclination i and obliquity β) and its average strength through the pulsating *interior* of the star.

Analytical treatments of the magnetic perturbations on the spherical harmonics of the pulsation pattern have been developed by Dziembowski & Goode (1985), Kurtz, Shibahashi & Goode (1990), and Shibahashi & Takata (these proceedings), among others. Several roAp oscillations show triplet structure in the Fourier domain, consistent with $\ell = 1$. Estimates of the *relative* internal field strengths based on the triplet amplitudes are available for four stars (see, e.g., Matthews 1991): HD 6532, HR 1217, HD 60435 and HR 3831. The results suggest that HD 60435 has the weakest global field of the four[1] and HR 3831, the strongest.

5. Measuring atmospheric structure

(a) Critical frequencies

By considering the pulsations of an roAp star to be standing waves in a potential well (defined by the stellar structure), one can treat them as a Sturm-Liouville-type problem:

$$\frac{d^2v}{d^2r} + \frac{1}{c^2}\left[\sigma^2 - \phi(r)\right] v = 0 \qquad (4)$$

valid for

$$\sigma^2 >> \ell(\ell+1)\frac{c^2}{r^2} \qquad (5)$$

where σ = angular frequency, $\phi(r)$ = potential, $v = \rho^{\frac{1}{2}} cr \xi_r$, and ξ_r = radial displacement.

In this picture, there must be an acoustic cutoff frequency for p-modes above which the waves will not be reflected by the density falloff in the upper atmosphere, and the mode becomes evanescent. This *critical* frequency

$$\nu_{crit} \sim \frac{c}{2H_p} \qquad (6)$$

(where $\nu = \sigma/2\pi$ and H_p is the pressure scale height) is a sensitive function of the atmospheric structure.

Shibahashi & Saio (1985) first recognised that this was a potentially useful diagnostic of the atmosphere of an roAp star. They found that standard Kurucz model atmospheres for $2M_\odot$ stars produced values of ν_{crit} of about 50–75% of the highest

---

[1] Attempts to measure the longitudinal field of HD 60435 by Landstreet and coworkers (*cf.* Matthews 1987; Matthews *et al.* 1987) have produced only low upper limits.

frequencies observed in HR 1217 and HD 60435. Since those frequencies were long-lived (recurring in observations spaced by years), they could not be evanescent modes. Hence, the true critical frequencies of these atmospheres must be higher than inferred from standard models (or those modes must experience incredibly strong driving!). Shibahashi & Saio noted that $\nu_{\rm crit}$ could be raised by *steepening* the $T-\tau$ gradient of the atmosphere; i.e, making the surface temperature $T(\tau=0)$ cooler than one would infer from $T_{\rm eff}$.

(b) Limb darkening

Support for this interpretation came from an unlikely source: rapid multicolour photometry of the roAp star HR 3831 (HD 83368). Observers had already recognised that the oscillation amplitude of an roAp star dropped rapidly with increasing wavelength – much more so than for other known pulsators. Based on their ESO K-band photometry of HR 3831, Matthews *et al.* (1990) argued that this was a result of limb darkening.

The pulsations of HR 3831 appear to be dominated by an $\ell = 1$, $m = 0$ mode. Simulations show that limb darkening acts as a filter to enhance the integrated amplitude of that mode, no matter how much the pulsation pole is inclined to the observer (except for 90°, where the amplitude is zero). Stronger limb darkening at shorter wavelengths leads to higher apparent amplitudes at those wavelengths. By measuring the amplitude at various wavelengths, one can estimate the limb darkening coefficients and infer the general features of the $T-\tau$ structure of the star's atmosphere.

Matthews *et al* (1992a, 1992b) have applied this approach to simultaneous rapid photometry of HR 3831 in $vbyRI$ and JHK bandpasses. They find that the stellar atmosphere must have a steeper $T-\tau$ gradient than the solar atmosphere in order to account for the data, just as Shibahashi & Saio predicted. They also find evidence for a temperature inversion near $\log \tau_{5000} \sim -0.7$, suggesting an additional source of opacity at that optical depth.

6. The immediate future

(a) Period changes

As the baseline for observations of known roAp stars increases, we can begin to investigate their long-term behaviour. Heller & Kawaler (1988) have predicted the rates of period change for models of stars in the roAp mass range, based on evolution in and beyond the main-sequence band. Values of $|dP/dt| \sim 10^{-12} - 10^{-13}$ are expected.

However, an $O-C$ diagram spanning roughly 400,000 cycles of the oscillation of HD 101065 (Martinez & Kurtz 1990) implies a rate of period change about $10-100\times$ higher than expected through evolution (see Figure 3). If due to light-time-delay effects in a binary system, the unseen companion would have a mass $M \leq 0.1 M_{\rm Earth}$! Perhaps the $O-C$ diagram in Figure 3 doesn't represent a continuous smooth change in period, but abrupt shifts due to mode switching, which may be occurring in other roAp stars (e.g., HD 217522; Kreidl *et al.* 1991). Clearly, our unusual friend HD 101065 holds more surprises in store for us.

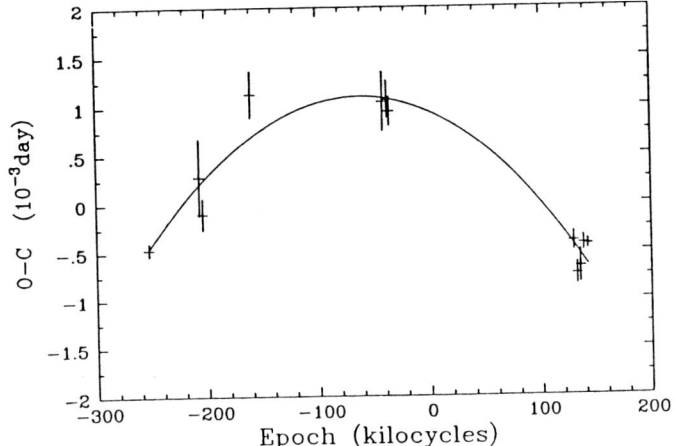

Figure 3. $O - C$ diagram for the oscillations of HD 101065 (from Martinez & Kurtz 1990).

HR 3831 also exhibits anomalous phase shifts over long intervals of time (Kurtz 1992), which are as yet unexplained. One must be cautious, though; cycle-count ambiguities are always a hazard when dealing with periods of a few minutes and gaps in the data sets of months or years. An apparent phase shift of f cycles could easily be $N + f$ cycles, where $N = \pm 1, 2, 3...$, depending on the accuracy of the period. Thus, the shape of the $O - C$ curve may be difficult to specify.

(b) Estimating main-sequence ages

Recall equations (2) – (3), which describe the frequencies of the high-overtone pulsations of roAp stars and the Sun. The fundamental frequency spacing ν_0 is a function of the star's mean density; hence, mass and radius, which change very slowly during the mean-sequence lifetime of a star. The deviations from that spacing, represented by D_0 (equation 3), are most sensitive to the sound-speed gradient $\frac{dc}{dr}$ near the core of the star. The convective core of an Ap star should be isothermal; the only changes in sound speed are due to the changing composition of the gas during core H-burning. Thus, the second-order spacing in the eigenfrequency spectrum should be sensitive to the main-sequence age of the star. Ulrych (1986) and Christensen-Dalsgaard (1986) suggested that the $\nu_0 - D_0$ plane would be a mass – age diagram.

Such "seismological H-R diagrams" have been generated for models in a narrow mass range around $1 M_\odot$, but to date, only one solar-type p-mode oscillator has been detected with certainty: the Sun. On the other hand, observations of several roAp pulsators supply the data for such a diagram, but the theoretical calculations for masses near $2 M_\odot$ were until recently unavailable.

That situation is changing. Audard & Provost (1992) have produced eigenspectra for their models of 1.5 and 2.0 $M_\odot$ stars. They find values of ν_0 and D_0 which are quite consistent with the available roAp observations. Pedersen & Vandenberg (work in progress) are pursuing similar lines. The prospects are bright that we may be able to place roAp stars on seismological H-R diagrams to test our developing notions of their ages and other global properties.

References:

Audard, N., Provost, J., 1992, to appear in "Inside The Stars" (IAU Colloquium 137), ed. W.W. Weiss & A. Baglin, A.S.P. Conference Series

Christensen-Dalsgaard, J., 1986, in "Advances in Helio- and Astero-seismology" (IAU Symposium 123), ed. J. Christensen-Dalsgaard & S. Frandsen (Dordrecht, Reidel), p. 295

Dziembowski, W., Goode, P.R., 1985, ApJ, 296, L27

Gabriel, M., Noels, A., Scuflaire, R., Mathys, G., 1985, A&A, 143, 206

Heller, C.H., Kawaler, S.D., 1988, ApJ, 329, L43

Kreidl, T.J., Kurtz, D.W., Kuschnig, R., et al., 1991, MNRAS, 250, 477

Kurtz, D.W., 1982, MNRAS, 200, 807

Kurtz, D.W., 1990, ARA&A, 28, 607

Kurtz, D.W., 1992, to appear in "Peculiar and Normal Phenomena in the A-type and Related Stars" (IAU Colloquium 138), ed. M. Hack, A.S.P. Conference Series

Kurtz, D.W., Shibahashi, H., Goode, P.R., 1990, MNRAS

Martinez, P., Kurtz, D.W., 1990, MNRAS, 242, 636

Matthews, J.M., 1987, Ph.D. thesis, University of Western Ontario

Matthews, J.M., 1988, in "Seismology of the Sun and Sun-like Stars", ESA SP–286, p. 547

Matthews, J.M., 1991, PASP, 103, 5

Matthews, J.M., Kurtz, D.W., Wehlau, W.H., 1987, ApJ, 313, 782

Matthews, J.M., Wehlau, W.H., Walker, G.A.H., 1990, ApJ, 365, L81

Matthews, J.M., Wehlau, W.H., Rice, J., Walker, G.A.H., 1992a, to appear in "Inside The Stars" (IAU Colloquium 137), ed. W.W. Weiss & A. Baglin, A.S.P. Conference Series

Matthews, J.M., Wehlau, W.H., Rice, J., Walker, G.A.H., 1992b, to appear in "Peculiar and Normal Phenomena in the A-type and Related Stars" (IAU Colloquium 138), ed. M. Hack, A.S.P. Conference Series

Moon, T.T., Dworetsky, M., 1985, MNRAS, 217, 305

Shibahashi, H., 1987, in Lecture Notes in Physics, 274, "Stellar Pulsation", ed. A.N. Cox et al. (Berlin, Springer-Verlag), p. 112

Shibahashi, H., Saio, H., 1985, PASJ, 37, 245

Tassoul, M., 1980, ApJS, 43, 469

Ulrych, R.K., 1986, ApJ, 162, 993

Weiss, W.W., 1986, 1986, in IAU Colloquium 90, "Upper Main Sequence Stars with Anomalous Abundances", ed. C.R. Cowley et al. (Dordrecht, Reidel), p. 219

Discussion

T. J. KREIDL: HD 134214 also has $\nu > \nu_{crit}$ for its pulsation mode. It appears to be both unstable as far as its pulsation frequency *and* its phase coherence are concerned. Might the fact that $\nu > \nu_{crit}$ have an influence on the pulsational stability of roAp stars?

J. M. MATTHEWS: It's possible that some observed modes, particularly in roAp stars with rich eigenspectra, might be evanescent. Certainly a star like HD 60435 seems to show mode growth and decay on timescales of a day or less. If such a mode dies out and is re-excited, there is no reason for it to maintain its original phase. However, it should return with the same frequency. A pronounced shift in frequency would suggest that a *different* mode has appeared to replace it.

B. CARROLL: You mentioned that $\ell = 1$, $m = 0$ is the dominant mode for these stars. Is it that the roAp stars actually prefer this mode, or that you could not detect them in they were pulsating in another mode?

J. M. MATTHEWS: The evidence for this mode lies in the amplitude modulation and phase shifts which are observed. For maximum oscillation amplitudes to coincide with magnetic extrema and π^{rad} phase jumps to take place at zero cross-overs of the B_{eff} field, one must be seeing something very much like a zonal pulsation mode. A different pulsation pattern would produce quite different modulation and phase behaviour during the star's rotation cycle, as well as different fine-splitting in the Fourier domain.

C. AERTS: (1) The Coriolis force induces more than 1 spherical harmonic for one mode. Do you take this into account? (2) What are the rotation periods of these stars?

J. M. MATTHEWS: (1) We do allow for the additional spherical harmonics. In fact, recent observations of HR 3831 by Kurtz et al. (1992) require a decomposition of its oscillation amplitude spectrum into a series of 7 spherical harmonics. A treatment of the effects of rotation and magnetic field by Shibahashi & Takata (these proceedings) predicts this level of complexity. (2) Typical rotation periods of Ap stars range between a few days to a few weeks, although periods as long as years have been inferred for some stars. Generally, Ap stars are slower rotators than their "normal" counterparts.

Magnetic fields of rapidly oscillating Ap stars

G. Mathys

European Southern Observatory, Chile

Magnetic field appears to play a major rôle in the pulsations of rapidly oscillating Ap (roAp) stars. Understanding of the behaviour of these objects thus requires knowledge of their magnetic field. Such knowledge is in particular essential to interpret the modulation of the amplitude of the photometric variations (with a frequency very close to the rotation frequency of the star) and to understand the driving mechanism of the pulsation. Therefore, a systematic programme of study of the magnetic field of roAp stars has been started, of which preliminary (and still very partial) results are presented here.

Magnetic fields of Ap stars can be diagnosed from the Zeeman effect that they induced in spectral lines either from the observation of line-splitting in high-resolution unpolarized spectra (which only occurs in favourable circumstances) or from the observation of circular polarization of the lines in medium- to high-resolution spectra.

The best studied roAp star, both from the point of view of the oscillations as from that of the magnetic field, is HD 83368. From circular polarization observations, the magnetic field of this star does not seem to have cylindrical symmetry about an axis passing through the centre of the star. Additional observations are being obtained to establish this more definitely and allow detailed modelling of the field. It can already be noticed, however, that this lack of symmetry, if confirmed, might possibly account for the presence of spherical harmonics with even values of ℓ in the deconvolution of the observed pulsations (Kurtz 1992).

In high-resolution ($\lambda/\Delta\lambda = 10^5$) unpolarized spectra, HD 134214, HD 137949, and HD 201601 are observed to have lines resolved into several magnetically split components, while HD 24712 and HD 176232 are found to have sharp unresolved lines. The lines of HD 128898 are obviously significantly broadened by rotational Doppler effect. This is puzzling as the variations of the longitudinal field of this star (measured from circularly polarized spectra), presumably due to its rotation, seem to be inconsistently slow. HD 24712 and HD 201601 have also been observed in circular polarization. In the former, the maximum of the mean longitudinal magnetic field occurs at the same phase as the pulsation amplitude maximum. The latter appears to have a rotation period of the order of a century.

More details about these magnetic observations are to be found in Mathys (1991; 1992, to be submitted to Astron. Astrophys.), and Mathys et al. (1992).

References:

Kurtz, D., 1992, Monthly Not. r. Astron. Soc., in press.
Mathys, G., 1991, Astron. Astrophys. Suppl., **89**, 121.
Mathys, G., Landstreet, J.D., Lanz, T., 1992, IAU Coll. 138, in press.

Chaos in Pulsating Variable Stars: Preliminary Analysis of Photometric Photometry and Observational Constraints of Detection

T.J. Kreidl

Lowell Observatory, Flagstaff, AZ 86001 U.S.A.

Abstract

Chaos theory has been applied to a variety of variable stars, but few convincing candidates for chaos have been identified. Here, well-established analysis methods have been applied to some very extensive data sets of rapidly oscillating Ap (roAp) stars and one white dwarf. It it shown that in spite of the amount of data, the signal-to-noise ratio makes positive detection of chaos extremely difficult, especially due to scintillation noise. A new form of dimension computation is presented and discussed. Simple models were constructed to show what noise levels can be tolerated before the detection of chaos is no longer possible and comparisons are drawn with data that could be obtained in the future from space. The lack of phase and amplitude stability in HD 134214 and mode switching in HD217522 and HD 137949 are pointed out as the possible results of chaos, making frequent monitoring of roAp stars desirable.

1. Present and future hopes of detecting chaos

One of the best-established means of calculating a fractal dimension is via the method of Grasberger and Procaccia (1983). I propose here an alternative method of dimension calculation, which makes use of only the existing data, even with gaps. As remarked earlier by Theiler (1987), it is unnecessary to compute the correlation distance from every point to every other point to adequately estimate the correlation dimension. It has been possible to make use of this to get a good dimension estimate.

Experiments with artificial data show that for detecting chaos, S/N levels of 10^5 or 10^6 will be needed. Long, uninterrupted data sets are necessary and hence satellite data, e.g. PRISMA, offer the best promise.

The evidence of irregular periodicity in pulsating variables implies the possibility that chaos may be responsible. In particular, the roAp stars HD 137949 and HD 217522 have undergone mode switching. HD 134214 seems to be unstable both in period and phase over many years.

References:

Grasberger, P. and Procaccia, I., 1983, Phys. Rev. A, **28**, 2591.
Theiler, J., 1987, Phys. Rev. A, **36**, 4456.

Pulsation of Rotating Magnetic Stars

H. Shibahashi, M. Takata

Department of Astronomy, University of Tokyo, Japan

Recently, one of the rapidly oscillating Ap stars, HR 3831, has been found to have an equally split frequency septuplet, though its oscillation seems to be essentially an axisymmetric dipole mode with respect to the magnetic axis which is oblique to the rotation axis (Kurtz et al. 1992; Kurtz 1992). In order to explain this fine structure, we investigate oscillations of obliquely rotating magnetic stars by taking account of the perturbations due to the magnetic fields and the rotation. We suppose that the star is rigidly rotating and that the magnetic field is a dipole field and its axis is oblique to the rotation axis. We treat the effects of the rotation and of the magnetic field as perturbations. In doing so, we suppose that the rotation of the star is slow enough so that the effect of the rotation on oscillations is smaller than that of the magnetic field. Under these assumptions, we can show that the oscillation eigenfunction represented in the zero order by the spherical degree l has mainly perturbed components represented by the degree $l \pm 2$. In the case of $l = 1$ like the roAp stars, those first-order components are octapole, and it is these octapole components that are responsible for the observed septuplet fine structure of the amplitude spectrum of HR 3831 (Shibahashi and Takata 1992). The spherical harmonic expressed in terms of the spherical coordinates with respect to the magnetic axis is written in terms of $(2l + 1)$ spherical harmonics of the same degree l with respect to the line-of-sight. Hence the pulsation of the spherical degree l in the zero-order of obliquely rotating magnetic stars is observed as a $(2l + 5)$-fine structure of the frequency spectrum $\{\omega - (l+2)\Omega, ..., \omega, ..., \omega + (l+2)\Omega\}$. The fine structure is asymmetric with respect to its central frequency because of the perturbation due to the Coriolis force. These features are qualitatively consistent with Kurtz et al.'s (1992) recent observational results, and provide theoretical justification of Kurtz's (1992) decomposition of the amplitude spectrum of HR 3831 by a series of spherical harmonics. The degree of asymmetry of the fine structure is independent of the geometrical configuration but depends on the ratio of the Coriolis force and the Lorentz force. Then it allows us to get information about the magnetic fields, the rotation, and the geometrical configuration of the star from the oscillation data.

References:

Kurtz, D. W. 1992, MNRAS, in press.
Kurtz, D. W., Kanaan, A., and Martinez, P. 1992, MNRAS, in press.
Shibahashi, H., and Takata, M. 1992, in *Inside the Stars, Proc IAU Colloq. No. 137*, ed. W.W. Weiss & A. Baglin, Astron. Soc. of Pacific Conf. Series, in press.

Nonradial Pulsation among δ Scuti Stars

Michel Breger

Institute of Astronomy, University of Vienna, Austria

Abstract

The review examines recent observational and theoretical results on the nonradial pulsation of δ Scuti stars. The dominant pulsation modes are rotationally split p modes with $\ell = 1$ and 2, and radial orders 1 to 4. Line-profile analyses also reveal the existence of additional high-degree modes. Nonradial pulsation is usually accompanied by slow amplitude and period variability. The small-amplitude single-mode variables are shown to be typical nonradial pulsators.

The interesting cases of 1 Mon and θ^2 Tau are examined in more detail. The present status of asteroseismolgy of the δ Scuti stars is reviewed.

The periods of the four evolved δ Scuti variables studied are found to be decreasing, although stellar evolution predicts period increases due to larger radii. New work on the multiple periods of AI Vel shows that, apart from stellar evolution, an additional mechanism must exist to cause temporary period changes. It is argued that for a larger sample of stars the average period change must nevertheless reflect evolutionary changes.

1. Introduction

The δ Scuti stars are situated in the classical cepheid instability strip. A few are radial pulsators, while the majority pulsate with a large number of nonradial p modes simultaneously. They represent a transition between the cepheid-like large-amplitude radial pulsation of the classical instability strip and the ocean of nonradial pulsation occurring in the hot half of the H-R Diagram. Many excited modes show photometric amplitudes in excess of 0.001 mag, which makes it possible to study these stars photometrically. The position of δ Scuti stars on and slightly above the main sequence permits the asteroseismological comparison between oscillation data and stellar models in a region where the basic stellar structure is regarded as relatively well known.

The relationship between the δ Scuti stars and other pulsators in the hot part of the H-R Diagram is illustrated in Figure 1. We notice that, according to the recent studies, the δ Scuti stars are only one of many types of nonradial pulsators. The difference between the pulsation properties of these different types of stars must be a reflection of the different stellar structure and evolution.

Stars with dominant radial pulsation are found among Pop. II as well as slowly rotating Pop. I δ Scuti stars. The known radial pulsators usually have only one or two known periods as well as very large amplitudes with typical amplitudes of 0.5 mag. This is a factor 10 or 100 larger than the amplitudes of the typical nonradial pulsators.

The nonradial pulsations of δ Scuti stars found photometrically are low-order ($\ell \leq 3$) and low-degree (n = 0 to 4) p modes. A typical example is the star 4 CVn, for which the observed frequencies could be matched by rotationally split p_1, p_2 and p_3 modes with $\ell = 2$ (Breger et al. 1990, Breger 1990a).

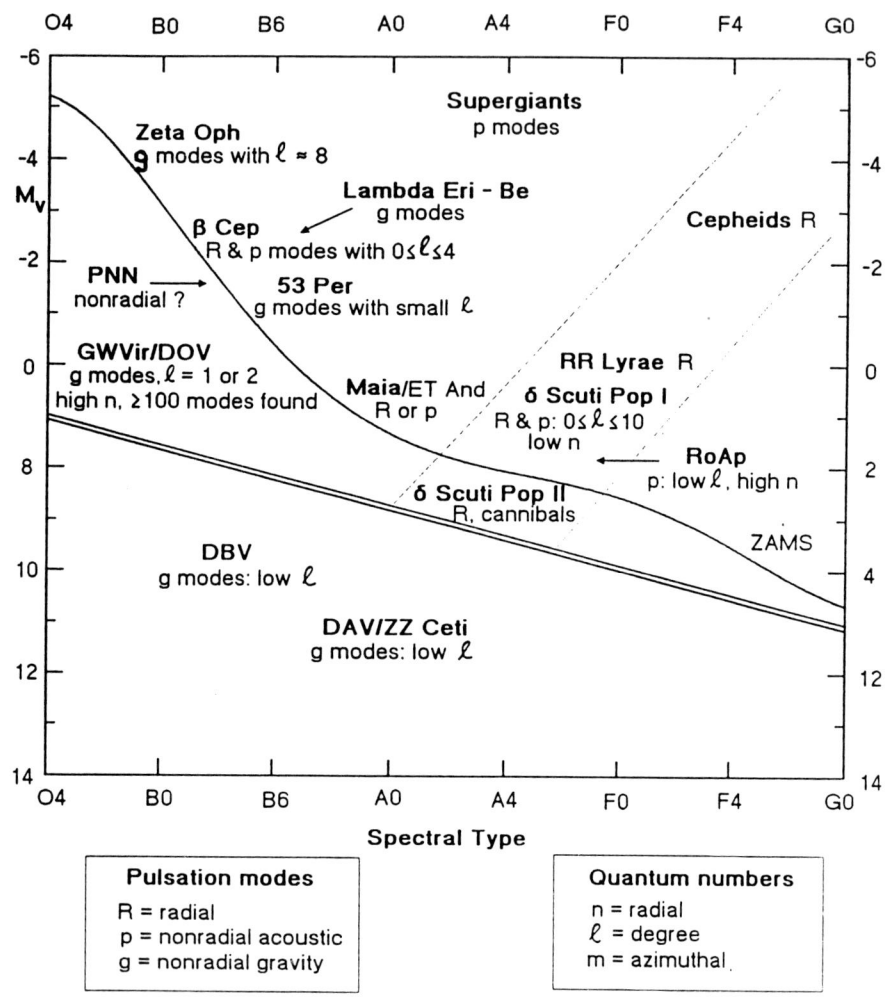

Figure 1 The position of the δ Scuti variables in relation to other types of pulsators in the hot part of the HRD. Some of the mode identifications and other recent information may be speculative. There is no universal agreement that the light variations of the λ Eri stars are indeed caused by pulsation.

While these nonradial low-order p modes are probably the most common modes found among δ Scuti stars, it must be noted that these modes are also the easiest to detect photometrically. On the other hand, the study of line-profile variations favors the detection of high-degree sectorial modes with $\ell = |m|$. These high-degree modes are being studied by the group at the University of British Columbia (e. g. $|m| \sim 10$ for γ Bootis, Kennelly et al. 1992). For κ^2 Bootis, the available data can be matched by a low-degree mode ($\ell = 0$ to 2) and a high-degree $\ell = |m| \sim 12$ mode (Kennelly et al. 1991).

The multiperiodicity and the small amplitudes make the observational determination of the multiple periods quite difficult. In order to limit the effects of aliasing, photometric multisite campaigns around the globe have given the most promising answers. Only about a dozen nonradially pulsating δ Scuti stars can be regarded as being observed well enough for the multiperiod structure to be regarded as 'known'. Future work should probably concentrate on some very large campaigns utilizing telescopes on several continents in order to detect more of the large number of excited pulsation modes.

2. Stars with only one or two periods

A few δ Scuti stars appear to show only a single period of pulsation with a small amplitude of pulsation near 0.01 mag. Could these stars be radial pulsators such as the single or double mode stars with large amplitudes? Or are they nonradial pulsators such as the small-amplitude multiperiodic variables?

During the last few years three photometrically monoperiodic stars were examined in more detail in order to identify the pulsation mode.

The star τ Peg shows a single photometrically dominant frequency of 18.4052 cycles d^{-1} with an amplitude changing slowly from 0.005 mag (or less) to 0.012 mag over 13 years. The amplitude ratios and phase differences between observed light and color variations rule out radial pulsation. A nonradial p_3 or p_4 mode with $\ell = 2$ fits the observed data.

The results obtained for τ Peg were dramatically confirmed by the simultaneous $uvby$ photometry of 28 And and β Cas (Rodriguez et al. 1992a, 1992b). In 28 And only the dominant 14.429 cycles d^{-1} frequency could be detected photometrically. Its amplitude varied from season to season, while the phase shifts and amplitude ratios showed that 28 And also is a nonradial pulsator. The mode could be identified as p_3, $\ell = 2$. Furthermore, for the star β Cas the pulsation mode could be identified as a p_2 or p_3 mode with $\ell = 1$.

We can conclude that the photometrically single-period pulsators with small amplitudes are not analogs of the large-amplitude variables such as AI Vel, which pulsate mainly with radial modes. These measurements of the three stars demonstrate that the low-amplitude single-mode pulsators resemble the nonradial multifrequency δ Scuti stars. Even the values of the radial overtone, $n \sim 2$, and $\ell = 1$ or 2 are typical for nonradially pulsating δ Scuti stars.

Table 1: Observed and computed pulsation modes for 1 Mon

Frequency	Value (cycles d^{-1})	Identification
Observed		
f_1	7.217139	$\ell = 1$, m = +1, probably p_1
f_2	7.346146	Radial, probably 1H
f_3	7.475268	$\ell = 1$, m = -1, probably p_1
Computed		
f_0	7.3555	$\ell = 1$, m = 0

3. The nonradial pulsation modes of 1 Mon

One of the simplest (?) observed examples of a nonradial δ Scuti pulsator is the star 1 Mon, which has been extensively studied and analysed by Balona and Stobie (1980). They could represent the observed variations of 1 Mon by a set of three almost equidistantly spaced frequencies and various harmonics and combinations of these frequencies. The relatively large amplitudes together with the excellent data and analyses suggest that the frequency determinations should be reliable. Furthermore, the mode identifications by Balona and Stobie on the basis of color phase and amplitude relations were confirmed through line-profile analyses by Smith (1982).

It can be seen in Table 1 that the three frequencies found by Balona and Stobie almost form an almost exact frequency split with a deviation of only $\Delta f = (f_3 - f_2) - (f_2 - f_1) = 0.000115$ cycles d^{-1}. This value corresponds to the inertial (observer's) frame.

The fact that an almost exact frequency split is observed can hardly be accidental. Due to the second-oder effects, for rotating stars rotational mode splitting does not lead to equidistantly spaced frequencies. On the basis of the identification of f_1 and f_3 with $\ell = 1$, we can calculate the expected value of the central m = 0 frequency through the following equation

$$\sigma_m = \sigma_0 + (C_L - 1)\, m\Omega + D_L m^2 \Omega^2 / \sigma_0,$$

where σ represents the frequency of pulsation of mode with azimuthal quantum number m, Ω is the rotational frequency, while C_L, D_L are coefficients depending on the model and pulsation mode. For the present calculation we determine the rotational frequency, Ω, from the observed frequency separation of the m = +1 and m = -1 modes for a p_1 mode and find $\Omega = 0.131$ revolutions per day. This value is reasonable and with a radius of 2.2 solar radii (given by Balona and Stobie) predicts the rotational velocity of 14.5 km s^{-1} (observed $vsini \sim 15$ km s^{-1}).

Three models were chosen for the C_L and D_L coefficients: the polytrope model by Saio (1981), the ZAMS model as well as the evolved model, both computed by Dziembowski and Goode (1992). It was found that for the p_1, $\ell = 1$ modes all three models give similar answers, so that only the results for the ZAMS model are shown in Table 1.

The calculations show that even the low rotational velocity of ~ 15 km/s leads to nonequal frequency splitting. In fact, the observed spacing is more than a factor of 100 times more equidistant than the predictions for the three $\ell = 1$ modes! If the central mode is indeed radial (and not the central $\ell = 1$ mode), then the observed equidistant splitting may not be a problem. Since in the stellar co-rotating frame the three frequencies have essentially the same value, the nonradial modes might be in resonance with the radial mode.

The picture presented above is internally consistent, but leaves two questions:

(i) Why do the frequency values of the $\ell = 0$ (radial) and the two $\ell = 1$ modes agree to better than one part in 10^4 in the co-rotating frame?

(ii) Why are the central $\ell = 1$, $m = 0$ mode and its beating with the radial mode (beat period of about 100 days) not seen photometrically?

4. Multiple nonradial pulsation modes and asteroseismology

The resolution of the discrepancies between observations and theory concerning the *radial* period ratio, P_0/P_1 (Petersen 1992) can be regarded as an asteroseismological success. The observed value of 0.773 for Pop. I δ Scuti stars can now be matched by the new opacity tables with higher opacity values.

Among the *nonradial* oscillators, the simultaneous excitation of a large number of pulsation modes makes most δ Scuti stars very difficult to study observationally. On the other hand, the amount of astrophysical information that can be derived from the study of these pulsation modes depends directly on the number of modes that can be successfully identified in a star. The procedure can be divided into three stages:

(i) the observational determination of the excited multiple frequencies. This usually requires lengthy multisite photometric campaigns in order to decrease the 1 cycle d^{-1} aliasing problem,

(ii) the identification of the multiple frequencies with specific pulsation modes. The methods involve pattern recognition in frequency space, consideration of Q values (see Breger 1989), amplitude and phase information between light and color variations (Balona and Stobie 1979, Watson 1988) as well as matching line-profile variations with models (Smith 1982),

(iii) the comparison of the frequencies with those predicted by a variety of models in order to obtain additional information on stars.

Dziembowski and Goode (1992) point out that patterns seen in the frequency spectrum could be misidentified since different orders can overlap and since we only see the amplitudes above a certain observational limit. For evolved stars, this difficulty is amplified by the possible development of g-mode type behavior in the interior for low-order nonradial p modes (see Dziembowski and Krolikowska 1990, Dziembowski and Goode 1992). On the other hand, the trapped modes can provide an asteroseismological test for convective overshooting theories (Dziembowski and Pamyatnykh 1991).

The challenge could be met by two improvements to the present technique:

(i) for each star examined observationally specific stellar models should be computed,

(ii) more (true) modes need to be identified for each star studied observationally. For many δ Scuti stars only the modes with larger amplitudes can be extracted from the photometric data due to the presence of noise and observational errors. Better statistical methods to extract multiple periods will probably only lead to a small improvement. The major improvement will probably be achieved through a further increase in the quantity and quality of the photometric data. In practice this might mean that more attention should be given to actually achieving the $\pm$ 1 millimag precision presently possible with standard photoelectric photometers.

The second point concerning additional frequencies can be illustrated with an example. The star θ^2 Tau was examined in two large multisite campaigns (Breger et al. 1987, 1989). The five frequencies found in the first campaign were confirmed by the second campaign, which also showed the constancy of the amplitudes. However, the data spanning six years contains more frequencies of pulsation. Their small amplitudes of pulsation less than 0.001 mag makes the identification of specific peaks in the power spectrum somewhat speculative; for this reason their values were not published.

Figure 2 shows the power spectrum of θ^2 Tau after prewhitening of the five published frequencies in the 13.23 to 14.61 cycles d^{-1} range. The remaining power is concentrated in two regions. The high-frequency power occurs around the value of twice the frequencies already identified. We can speculate that, if real, this corresponds to the sum of two coupled frequencies. The lower frequency band is in the frequency range of the five frequencies which had already been identified and prewhitened. This means that more frequencies are probably excited. If one accepts the mode identifications derived from color and amplitude data ($\ell = 0$, 1 and 2), the values of the two highest peaks (near 12.2 and 12.8 cycles d^{-1}, respectively) fit the expected pattern! However, until more observations are available, these arguments must regarded as speculative.

Apart from illustrating the improvements suggested above, the example demonstrates that the star θ^2 Tau shows considerable asteroseismological potential. A model for a star similar to θ^2 Tau has been calculated by Dziembowski and Goode. If the new frequencies are indeed confirmed, then this star might provide the opportunity to compare the rotational splitting of $\ell = 1$ and $\ell = 2$ together with a 'normalization' of the frequencies relative to the radial mode at 13.23 c/d.

We expect that future observations and models will lead to direct comparisons of the rotational mode splitting coefficients and permit the probing of the depth dependence of Ω. Furthermore, the study of pulsational amplitudes and their selection rules provides additional means to test stellar models.

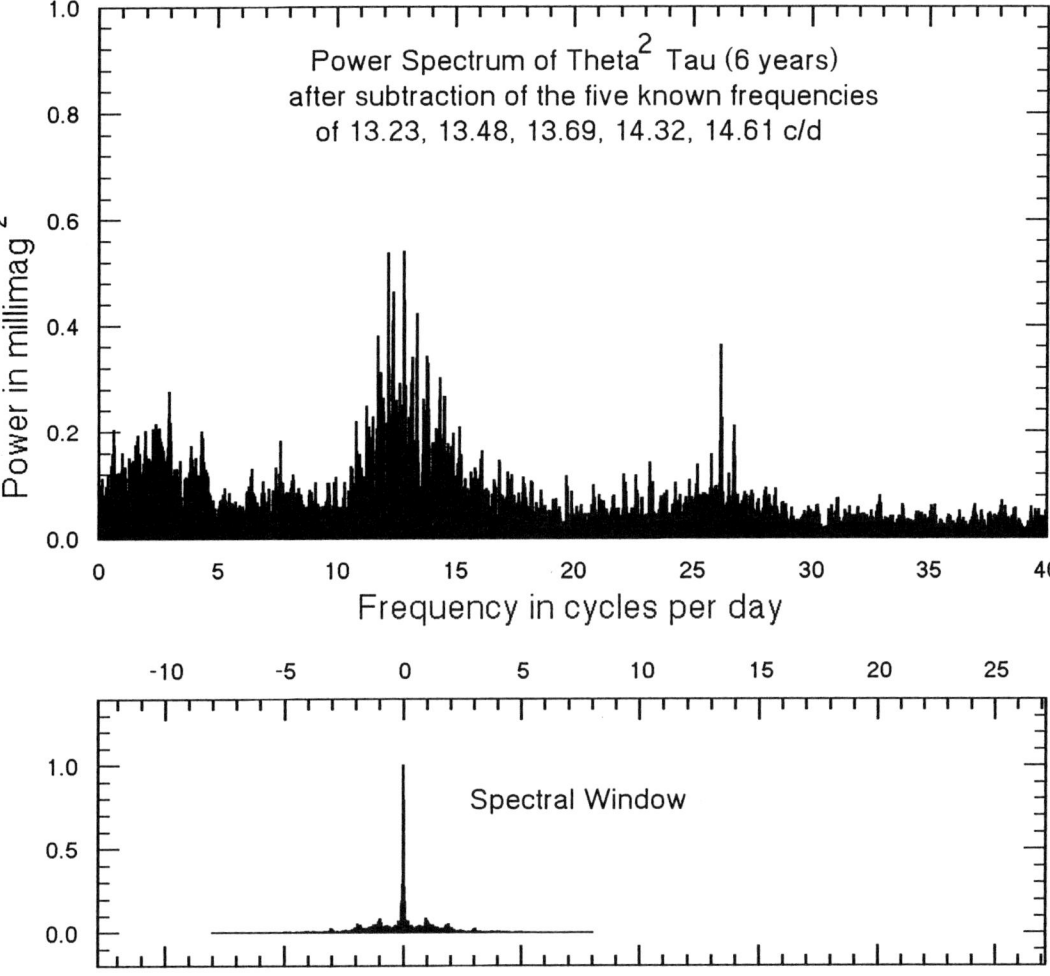

Figure 2 Power spectrum of θ^2 Tau after prewhitening of the five published frequencies. The diagram shows that several additional frequencies with small amplitudes are also excited in the star.

5. Period changes and evolution

The period changes caused by the stellar evolution in and across the Lower Instability Strip permit an observational test of stellar evolution theory, provided that other physical reasons for period changes can be excluded. The period-density relation predicts

$$(1/P)\,dP/dt = -0.69\,dM_{\mathrm{bol}}/dt - (3/T_{\mathrm{eff}})T_{\mathrm{eff}}/dt + (1/Q)\,dQ/dt.$$

The predicted values of $(1/P)dP/dt$ from 10^{-10} to 10^{-7} per year should be observable.

At the last meeting of this pulsation series in Bologna, the observed period changes were compared with values calculated from theoretical evolutionary tracks (Breger 1990b). For the four evolved δ Scuti stars the observed period decreases were in contradiction to the radius increases predicted from stellar evolution. Could there be other physical explanations for the observed period decreases?

Guzik and Cox (1991) have examined possible explanations for the period decreases seen in evolved δ Scuti stars. While their models indeed predicted smaller period increases, period decreases were not found.

The recent results for the star AI Vel (Walraven, Walraven and Balona 1992) may indicate a solution to the dilemna. They have found that only one of the two radial frequencies shows a period increase, while the second frequency was essentially constant. The example of AI Vel shows that the (slow) stellar evolution is not the only mechanism to generate changes in the periods of pulsation. This means that for an individual star the conversion of observed period changes into stellar evolution rates (e.g. radius changes) has to be applied with caution. On the other hand, the nonevolutionary period changes should cancel out for a larger group of stars. This can be seen from the fact that the δ Scuti stars on the whole obey a period-luminosity-color relation.

The fact that all four evolved δ Scuti stars show period decreases may conceivably be a reflection of an additional period changing mechanism (leading to random changes) coupled with small-number statistics. A doubling of the number to eight stars should provide the required statistical significance.

Part of this investigation has been supported by the Austrian Fonds zur Förderung der wissenschaftlichen Forschung.

References:

Balona, L. A. and Stobie, R. S. 1979, MNRAS, 189, 649
Balona, L. A. and Stobie, R. S. 1980, MNRAS, 190, 931
Breger, M. 1989, Comm. Asteroseismology, 7, 1, Vienna: Austrian Acad. Sciences
Breger, M. 1990a, A&A, 240, 308
Breger, M. 1990b, ASP Conf. Ser., 11, 263
Breger, M., Garrido, R., Huang, L., Jiang, S.-Y., Guo, Z.-h., Frueh, M., Paparo, M. 1989, A&A, 214, 209
Breger, M., Huang, L., Jiang, S.-y., Guo, Z.-h., Antonello, E., Mantegazza, L. 1987, A&A, 175, 117
Breger, M., McNamara, B. J., Kerschbaum, F., Huang, L., Jiang, S.-y., Guo, Z.-h., Poretti, E. 1990, A&A 231, 56
Dziembowski, W. A., and Goode, P. R. 1992, ApJ, in press
Dziembowski, W. A. and Krolikowska, M. 1990, Acta Astron. 40, 19
Dziembowski, W. A. and Pamyatnykh, A. A. 1991, A&A, 248, L11
Fitch, W. S. 1981, ApJ, 249, 218
Guzik, J. A. and Cox, A. N. 1991, in Delta Scuti Newsletter, 3, 6, Vienna: Austrian Acad. Sciences
Kennelly, E. J., Walker, G. A. H. and Hubeny, I. 1991, PASP, 103, 1250

Kennelly, E. J., Yang, S., Walker, G. A. H. and Hubeny, I. 1992, PASP, 104, 15
Petersen, J. O. 1992, A&A, in press
Rodriguez, E., Rolland, A., Lopez de Coca, P., Garrido, R. and Gonzalez-Bedolla, S. F. 1992a, A&AS, in press
Rodriguez, E., Rolland, A., Lopez de Coca, P., Garrido, R. and Mendoza, E. E. 1992b, A&A, in press
Saio, H. 1981, ApJ, 244, 299
Smith, M. A. 1982, ApJ, 254, 242
Walraven, Th., Walraven, J. and Balona, L. A. 1992, MNRAS, 254, 59
Watson, R. D. 1988, A&SS, 140, 255

DISCUSSION

BALONA: *The reason why we do not see a beating between the $\ell = 0$ and the $\ell = 1$, $m = 0$ modes in 1 Mon may be due to the inclination of the axis of rotation, i. For $i \sim 90$ deg, the observed amplitude of $\ell = 1$, $m = 0$ is close to zero whereas $\ell = 1$, $m = \pm 1$ is close to the maximum value.*

BREGER: Such an inclination would explain the observations. It is regrettable that the small value of $vsini \sim 15$ km s^{-1} makes it difficult to derive the inclination by comparing this value with the predicted value of the rotational velocity, v.

MATTHEWS: *What limit does the frequency spacing of 1 Mon set upon the second-order rotational splitting term D_L and what could this mean for the star's structure?*

BREGER: If the central frequency of the observed triplet were the missing $\ell = 1$, m = 0 mode, no model I have seen can predict such a small D_L value. After all, waves travelling in the same direction as stellar rotation have a weaker effective gravity (and therefore a different frequency) than waves travelling in the opposite direction.

The Pulsation Characteristics of HD 93044

Liu Zong-li,   Li Zhi-ping

Beijing Astronomical Observatory, Chinese Academy of Sciences
100080 Beijing, China

Abstract

The amplitude and period of the first frequency increased, but the amplitude and period of the second frequency decreased. Two new frequencies (10.5011 and 2.0484 c/d) were found. The frequency (2.0484 c/d) might be an interaction term from which resonances can occur and then lead amplitude and period variations.

1. Observations and period analysis

The star was observed from February 7 to March 5, 1992 with the 60-cm telescope at Beijing Astronomical Observatory. The period analysis was completed using a combination of single-frequency Fourier transforms and multifrequency least squares of brightness residuals (Breger 1990, Hao Jin-xin 1991). The 5 frequencies were: 11.9085, 16.801, 10.498, 2.052 and 22.085 cycles per day. The power spectra of the 5 frequency solution were shown in Figure 1. The data were combined together with the data obtained by Liu Zong-li in spring of 1991 as a new set of data, and the period analysis was made for it using the same methods. The 5 frequencies were: 11.90803, 16.8017, 10.5011, 2.0484 and 22.0833 cycles per day.

2. Results

Comparing with the result obtained by Liu Zong-li (1992) the amplitude and period of the first frequency increased. And the amplitude and period of the second frequency decreased. Two new frequencies (10.5011 and 2.0484 cycles per day) were found. They are very obvious in the power spectra of HD93044 (see Figure 1). The frequency (2.0484 cycles per day) might be an interaction term from which resonances can occur and then lead to amplitude and period variations. The resonance hypothesis as an explanation for long-term amplitude and period variations is supported.

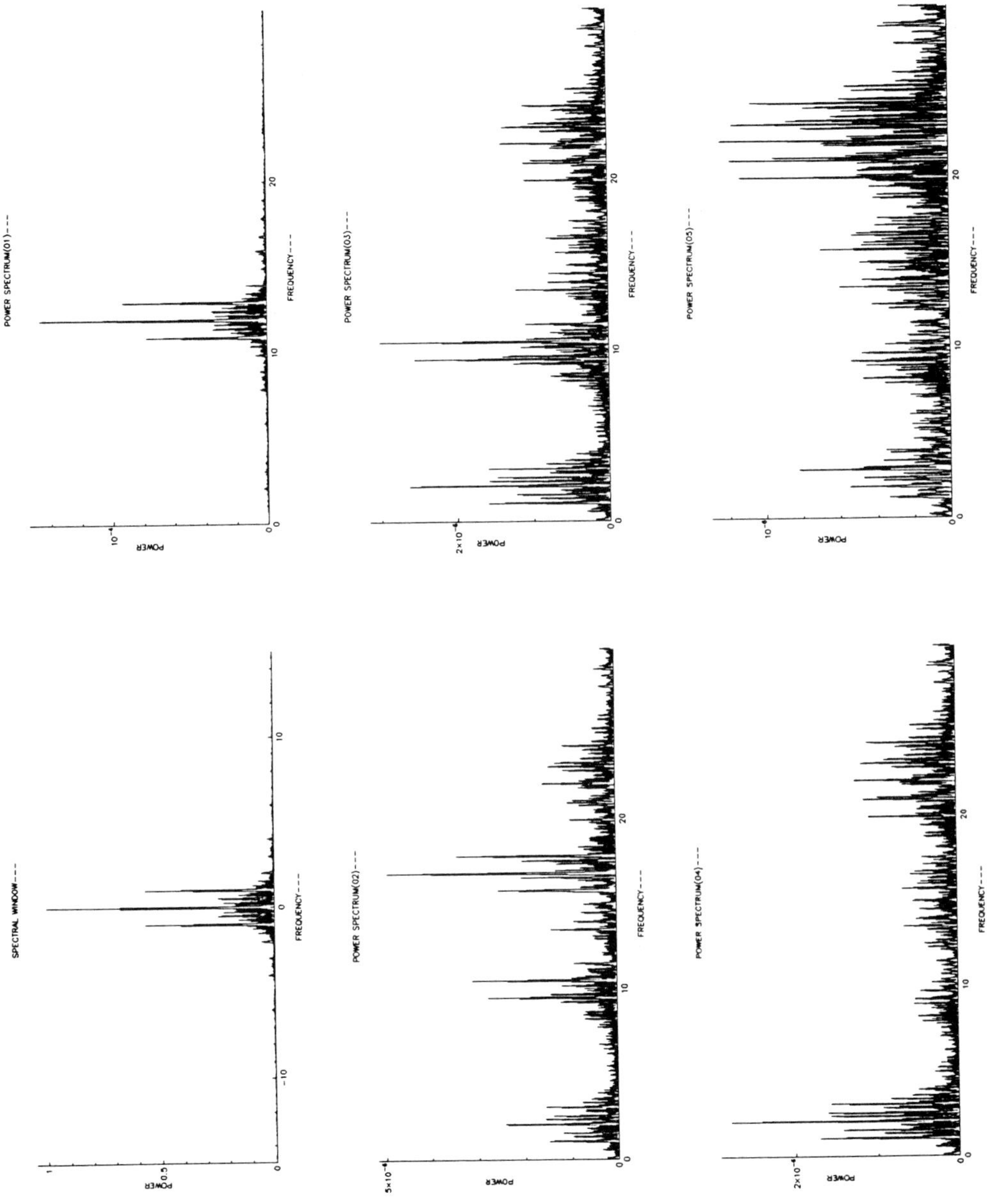

Figure 1. The power spectra of five frequency solution of HD93044

References:

Breger, M. 1990, Comm. Asteroseismology 20,Vienna:Austrian Acad. Science
Hao, J.-x. 1991, Publications of Beijing Astronomical Observatory No.**18**
Liu,Z.-l. 1992, Astrophys. and Space Science, in press

Discussion:

Tobias Kreidl: I would like to comment that I am very pleased to see how small telescopes can be used to obtain good data on δ Sct stars and provide a valuable contribution to the studies of these satrs, since the errors are mainly limited by transparency and photon counting statistics.

Li Zhi-ping: Thank you.

Michel Breger: The freqency of 2 c/d is very interesting. Have you eliminated the possibility of observational errors causing the 2 c/d peak? Is the frequency present in different subsets of your extensive data?

Li Zhi-ping: Before doing period analysis all the measurements of every night were averaged to obtain the average brightness of every night. The average brightness was subtracted from the measurements of each night to make the average brightness of each night zero to improve the zero-point drifts between the different nights. Therefore, the possibility of observational errors causing the 2 c/d peak was eliminated. The frequency was present in different subsets of our extensive data.

Jaymie Matthews: Does your frequency solution include the cross-coupling terms among the principal frequencies you identify?

Li Zhi-ping: No. you can see that the spectral window is very sharp, and the principal frequencies we identify are independent each other.

Mode Determination by Fourier Analysis of Line-Profile Variations: Application to τ Peg

E.J. Kennelly[1], G.A.H. Walker[1], W.J. Merryfield[2], J.M. Matthews[3]

[1] *University of British Columbia,* [2] *University of Victoria,* [3] *Université de Montréal*

Abstract

The identification of modes of oscillation is an important first step towards the seismology of stars. Low- and high-degree ($0 \leq \ell \leq 16^+$) nonradial modes of oscillation may appear as variations in the line profiles of rapidly rotating δ Scuti stars. We present a technique whereby complex patterns in the line profiles are decomposed into Fourier components in both time and "Doppler space". The technique is applied to the 7.3-hour time series of high-resolution data obtained from CFHT for the δ Scuti star τ Peg. In addition to the low-degree mode which has been identified in photometric studies (Breger 1991), we find evidence for at least three high-degree modes near $\ell = |m| = 7$, 11 and 15. Correcting for the rotation of the star, most of these modes appear to oscillate with frequencies near 17 cycles day^{-1}. Our results are found to be in good agreement with the theoretical limits imposed on the frequencies of oscillation by the models of Dziembowski (1990).

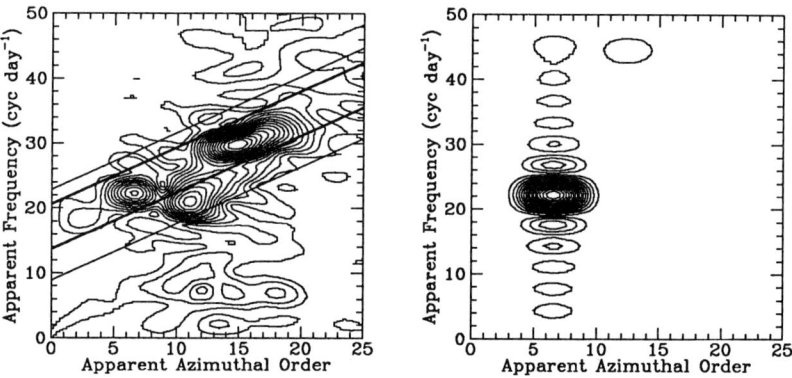

Figure Caption. The two-dimensional Fourier spectrum of the line-profile variations of τ Peg (left) and a window function generated from a time series of synthetic line profiles resulting from an $\ell = |m| = 6$ mode (right). Theoretical limits to the pulsation frequency are plotted for Dziembowski's standard (thick lines) and He-rich models (thin lines).

References:

Breger, M., 1991, Astron. Astrophys. **250**, 107.
Dziembowski, W., 1990, in: *Progress of Seismology of the Sun and Stars*, eds. Y. Osaki and H. Shibahashi, Lecture Notes in Physics **367**, p 359.

Fourier Analysis of Line-Profile Variations: Toward Stellar $m - \nu$ Diagrams?

W.J. Merryfield[1], E.J. Kennelly[2]

[1]*University of Victoria,* [2]*University of British Columbia*

Abstract

Numerous rapidly rotating δ Scuti stars exhibit variable line profiles containing traveling subfeatures [1]-[2], which may be signatures of nonradial pulsations having relatively high azimuthal order $|m|$. We describe a procedure whereby a time series of spectral line profiles is Fourier analysed both in time and in a wavelength variable that is presumed to correspond to azimuthal position ϕ on the star. What such an analysis can tell us is examined by analysing artificially-generated data. For an ideal example in which $\sin i = 1$ and a single mode having $\ell = -m = 10$ is present, the two-dimensional Fourier transform yields a power spectrum in frequency ν and an apparent azimuthal order $\hat{m}$ that provides a good indication of the actual ν and m. Such a straightforward interpretation is also possible when $\sin i < 1$, and when multiple sectoral modes ($\ell = |m|$) are present. For tesseral modes ($\ell > |m|$), $\hat{m}$ may correspond more closely to ℓ than to m.

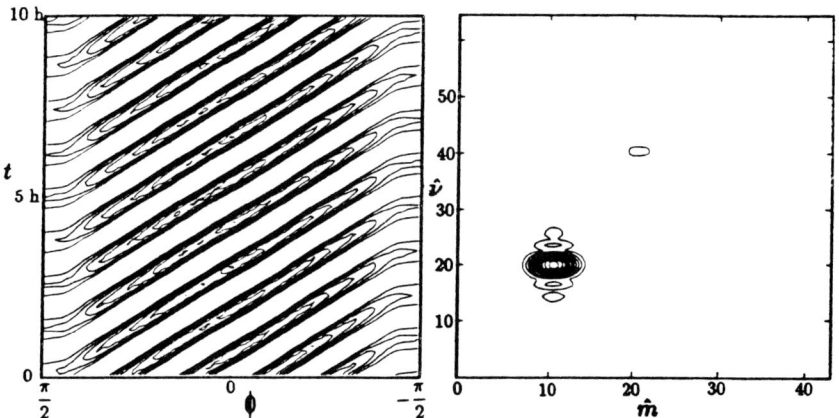

Figure Caption. (a) Residual line-profile variations for a ten-hour observation when a single mode is present having $\ell = -m = 10$, $\nu = 20$ cycles d^{-1}, and radial velocity amplitude $V_r = 2.5$ km s^{-1}; (b) Two-dimensional Fourier transform of (a). The position of the primary peak corresponds closely to the input values of m and ν.

References:

Walker, G.A.H., Yang, S. & Fahlman, G.G. 1987, *Astrophys. J.* **320**, L139.
Kennelly, E.J., Yang, S., Walker, G.A.H. & Hubeny, I. 1992, *P.A.S.P.* **104**, 15.

Frequency analysis of multiperiodic δ Scuti stars

L. Mantegazza[1], E. Poretti[2], F.M. Zerbi[1]

[1] *Dipartimento di Fisica Nucleare e Teorica, Università di Pavia, Italy*
[2] *Osservatorio Astronomico di Brera, Merate, Italy*

We report some of the recent results of our studies on δ Scuti star pulsation which are based on observations made at Merate and La Silla Observatories, sometimes in two site campaigns. Our recent experiences on X Caeli and 44 Tauri have shown that quite complicate light curves can be resolved even with observations obtained from one site only, if compact and accurate datasets are collected on sufficiently long time baselines (Mantegazza and Poretti, 1992; Poretti et al., 1992). Here we report the preliminary results of the light curve analysis of three more stars surveyed for more than 120 hours each: BI CMi, HD 18878 and HD 224639.

BI CMi has been observed in a two-site campaign in January–February 1991, the observations being collected during 17 nights. The frequency analysis of the light curve allowed us to identify unambiguously 4 pulsation modes at 8.247, 8.863, 8.514 and 7.424 c/d with semi-amplitudes of 22, 19, 5 and 5 mmag respectively. Even if these components explain almost 90% of the data variance, there is evidence of the presence of other components of smaller amplitude (<2.5 mmag), whose correct identification is however more ambiguous. The period ratios between the 4 frequencies evidences that at least some of them are ascribable to non–radial pulsations.

HD 18878 was observed from November 1991 to January 1992 at Merate Observatory, our data revealing clearly a multiperiodic behaviour. A preliminar frequency analysis yields four predominant frequencies (amplitudes ranging from 23 to 5 mmag): 6.86, 11.22 , 11.18 and 10.01 c/d. The fit with these four frequencies leaves a rms residual of 7.3 mmag and explains the 87% of the data variance. Further analysis is necessary to unambiguously evidence the other periodic terms with very small amplitudes present in the light curves.

HD 224639 has been previously observed by us in 1989 (Mantegazza and Poretti, 1990) in a short two-site campaign. Those data were insufficient to allow a light curve solution, and therefore new observations were collected by us at La Silla Observatory in September and October 1991 during 20 nights. The data frequency analysis is now in progress. As a preliminary result we can report that the star has a very rich pulsational spectrum with 8 components with semi-amplitudes between 20 and 5 mmag, and some more with smaller ones.

References:

Mantegazza, L. and Poretti, E., 1990, in: *Confrontation between stellar pulsation and evolution*, Astr.Soc. of Pacific Conf.Series **11**, p 324.
Mantegazza, L., Poretti, E., 1992, Astron.Astrophys. **255**, 153.
Poretti, E., Mantegazza, L., Riboni, E., 1992, Astron.Astrophys. **256**, 113.

Search for a Secondary Frequency in the Large-Amplitude δ Scuti Star CY Aqr

D. Coates[1], J. Fernley[2], K. Sekiguchi[3],
T. G. Barnes[4], M. Frueh[4]

[1] Department of Physics, Monash University, Clayton, Victoria, Australia
[2] IUE-Vilspa, P.O. Box 50727, 28080 Madrid, Spain
[3] South African Astronomical Observatory, P.O. Box 9, Cape, South Africa
[4] McDonald Observatory, University of Texas at Austin, Austin, Texas

Abstract

The large-amplitude δ Scuti star CY Aqr was observed from sites in the U.S.A., South Africa and Australia during August 1988. Coates et al. (1991) published 48 new times of maximum light derived from these observations and assembled, from the literature, previous times of maximum light. It is clear that the period of the star is changing with the balance of evidence favouring discrete changes in 1951 and 1966, rather than a continuous change.

It has been suggested by Fitch (1973) and Else (1972), from an analysis of the observations of Zissell (1968), that there is a secondary frequency present in CY Aqr. Coates et al. (1992) have analysed both the 1988 observations and those of Zissell. After subtracting the primary frequency and its harmonics, they find no stable secondary frequency above the noise level of two millimagnitudes.

References:

Coates, D.W., Barnes, T.G., Fernley, J.A., Frueh, M.L., Sekiguchi, K., 1991, Delta Scuti Star Newsletter 4, 10

Coates, D.W., Fernley, J.A., Sekiguchi, K., Barnes, T.G., Frueh, M.L., 1992, MNRAS, submitted

Elst, E.W., 1972, A&A 17, 148

Fitch, W.S., 1973, A&A 27, 161

Zissell, R., 1968, AJ 73, 696

Interpretations of Solar Oscillations

Arthur N. Cox

Los Alamos Astrophysics

Abstract

The current theoretical status of understanding solar oscillations is reviewed. Interpretation of the thousands of well-determined frequencies refines our knowledge of the composition and convection structure of the Sun, since its mass, radius, luminosity, and age are better known from other sources. Recent issues that have been discussed are the solar center structure, bearing on the missing solar neutrino problem, the convection zone helium content, validating helium settling by diffusion, the variations of the oscillation frequencies over the solar cycle, indicating cyclical structure changes in the very outer magnetic layers, and the fine structure splittings of mode frequencies, revealing the internal rotation. Our ability to match observed frequencies to now within only a few microhertz has been enhanced by the recently improved MHD equation of state and the new Livermore OPAL opacities. Thus solar oscillations not only reveal solar structure data, but also they guide improvements for stellar astrophysics material properties. A new discussion of current investigations of the convection zone helium abundance and its depth is presented.

1. Current issues in solar oscillation research

Solar oscillation frequencies allow us to probe the Sun to determine its composition, convection, and rotation structure. Other parameters that are usually sought in stellar astrophysics, such as mass, radius, luminosity, and age are all well known for the solar case. After a short introduction, this review considers only issues about the solar convection zone in any detail. The references and other recent papers cover the many other solar oscillation topics of current interest.

A very important problem has been the case of the missing neutrinos. Now with data from the ^{37}Cl, electronic, and ^{71}Ga detectors from the Homestake mine, the Kamiokande detector and from the SAGE and Gallex experiments, it seems that indeed the ^8B, ^7Be, and $p-p$ reaction neutrinos are deficient relative to predictions for so-called standard solar models. Though even today there are statements that our understanding of solar structure could be defective and the cause of the solar neutrino deficiency, that position is really out of the question. It has been so for over 25 years. The Mikeyev, Smirnov, and Wolfenstein effect, where electron neutrinos undergo a transformation by interactions with the electrons in the solar material to change their electron flavor to muon or tauon flavors, seems to be the correct reason why less than the full neutrino flux is detected here on earth.

A related question has been whether there might be weakly interacting massive particles (WIMPs) orbiting within the inner 10% of the solar mass and radius that can more efficiently transport energy than radiation processes. Then the solar center would be significantly cooled. Two recent papers, including one by Cox, Guzik, and Raby (1990), have shown that this is not the case, even though at least three earlier ones concluded these WIMPs could effect the desired cooling. With such a cool and almost isothermal central temperature, predicted low angular degree solar oscillation frequencies just do not agree with those observed.

The lifting of the frequency degeneracy for modes with different ℓ values, caused by modes traveling with both prograde and retrograde motions, can allow mapping of the internal solar rotation structure. Interesting details about rotation near the surface and within the convection zone have been discovered in the last 10 years, but probing the expected rapid central rotation has not yet been accurately done. The problem is mostly due to the fact that only the lowest angular degree modes have any amplitude in the inner 20% of the solar radius, and they have only a few $(2\ell+1)$ separate modes. Separations between the frequencies of these distinct modes need to be measured to small fractions of a microhertz out of about 3000 μHz, and that has not been achieved yet.

Attempts have been made to discover solar-like oscillations in other stars. The best case is for Procyon, but this bright F5 IV star is still not bright enough to acquire adequate statistical accuracy to identify individual modes. Probing stars like we do for the Sun will allow similar results and will increase our knowledge of stellar structures and evolution.

Predictions of solar oscillation frequencies require very accurate material properties for the model construction. Fortunately equations of state (Däppen et al., 1988) and opacities (Rogers and Iglesias, 1992) are now available, and solar oscillations have inspired further refinements in these data. Guzik and Cox (1991) discuss this matter.

I do not discuss here the recently confirmed discovery that very high angular degree oscillation mode frequencies vary with the solar cycle. This correlation helps to map magnetic field and density structure variations with time in the solar surface layers. Also I leave out a discussion of the highest observed frequencies that are not like the usual trapped modes in stars.

2. Standard solar models

Table 1 lists only a few of the recent standard solar models. The latest models are by Guenther et al. (1992). Some of these listed have been constructed only to discuss the solar neutrino problem, but most have been used to calculate solar oscillation frequencies. Note that the most recent models, which use the most modern material properties, derive a helium mass fraction in the primordial composition of close to Y=0.28. They all predict about 8 solar neutrino units (SNU or captures in 10^{36} atoms per second). The widely quoted, and very unreliable Turck-Chieze et al. results, give only 5.8 SNU. Our 11.4 SNU was found to be caused by a poor approximation for the equation of state, and now our best result is 8.5 SNU.

The solar radius is obtained by models with adjustment of the mixing length in the mixing length theory of convection. Recent models, using the best material properties, need a mixing length of about 2.0, but not much theoretical significance should be attached to this number, since it merely is an adjustable parameter. In section 4, I show how arbitrarily changing this value changes the convection zone depth and significantly changes predicted solar oscillation frequencies.

Table 1

Some Recent Standard Solar Models

Authors	Y	SNU	ℓ/H_p
Bahcall, Ulrich, 1988	0.271	7.9	–
Lebreton, Däppen, 1988	0.278	7.6	2.16
Guenther, Sarajedini, 1988	0.240	–	1.35
Lebreton, Berthomieu, Provost, Schatzman, 1988	0.287	8.0	2.18
Lebreton, Berthomieu, Provost, Schatzman, 1988	0.291	8.4	2.11
Turck-Chieze, Cahen, Casse, Doom, 1988	0.276	5.8	1.55
Christensen-Dalsgaard, Däppen, Lebreton, 1988	0.237	–	–
Korzennik, Ulrich, 1989	0.271	8.2	–
Guenther, 1989	0.282	–	1.25
Guenther, Jaffe, Demarque, 1989	0.28	–	1.24
Cox, Guzik, Kidman, 1989	0.291	11.4	1.89
Cox, Guzik, Raby, 1990	0.28	8.0	1.89
Sackman, Boothroyd, Fowler, 1990	0.278	7.7	2.1
Sienkiewicz, Bahcall, Paczynski, 1990	0.280	7.7	1.62
Christensen-Dalsgaard, Gough, Thompson, 1991	0.273	–	2.184
Guzik, Cox, 1991	0.270	8.5	2.291
Guenther, Demarque, Kim, Pinsonneault, 1992	0.289	–	1.894

3. The convection zone helium content

Figure 1 shows the variations of the p-, f-, and g-mode frequencies versus angular degree (ℓ, or the number of node lines on the stellar surface) according to Christensen-Dalsgaard (1988). All the observed p- and f-modes can now be predicted to within a few microhertz with current precision solar models. The detection of g-modes is controversial, because they are so weak at the solar surface. If they do occur in the Sun, they must tunnel through the solar convection zone where they are evanescent, and their surface manifestation can barely be observed. Predictions of g-mode frequencies do agree closely with observations, and even rotation mode splittings seem reasonable. I do not discuss these modes any more here, even though a theoretical review of these g-modes is overdue.

Table 2 (Guzik and Cox, 1992) gives our predicted frequencies for modes that probe the part of the solar convection zone that is sensitive to the helium abundance. The helium content in the solar mixture varies throughout the Sun, since in the central regions the primordial value is enhanced by hydrogen burning, and in the convection zone, helium settles to deeper non-convecting layers. Recent studies of this diffusive settling are given by Cox, Guzik, and Kidman (1989) and by Proffitt and Michaud (1991). Kosovichev et al. (1992) present the most recent study for the convection zone helium abundance using many oscillation frequencies.

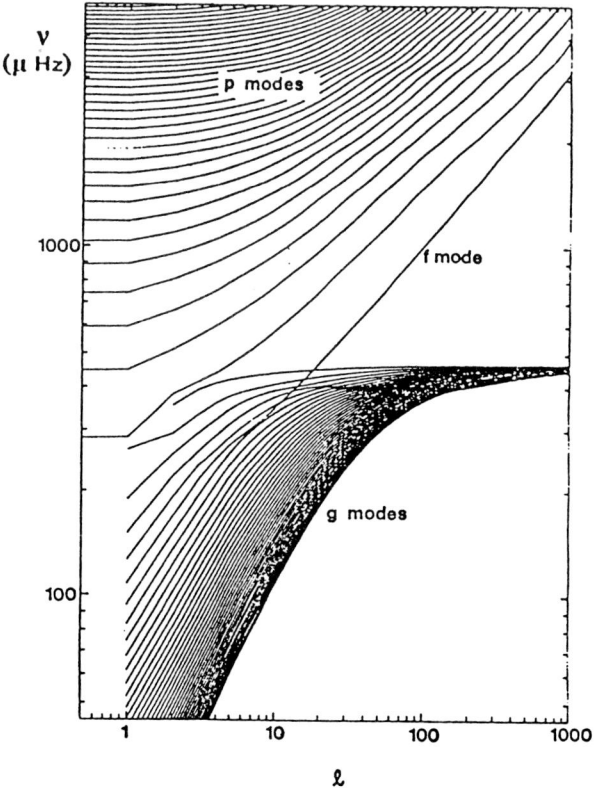

Figure 1 The p- f- and g-mode oscillation mode frequencies are plotted against the angular degree ℓ according to Christensen-Dalsgaard (1988). Individual mode points are connected for each ℓ value.

Modes with ℓ values between 300 and 600 seem to have sensitivity to the convection zone helium content. Those with ℓ smaller have their weight for period determination considerably deeper than around layers with a temperature near 40,000 K, where helium undergoes its second ionization. Higher ℓ modes are too shallow. Our best fit to the observed frequencies is for Y=0.24, and that is 0.03 in Y less than needed for the deeper layers just below the convection zone that reflect the primordial composition. This decrease is almost exactly what diffusion calculations give.

Table 2

FREQUENCY SENSITIVITY TO HELIUM ABUNDANCE

p-Mode		Frequency (μHz) $Y_{ss} = 0.27$	$(O-C)_L$	$(O-C)_K$	Frequency (μHz) $Y_{ss} = 0.24$	$(O-C)_L$	$(O-C)_K$	Difference (μHz) $\nu(0.27) - \nu(0.24)$	Observational Error (μHz)	
l	n								L	K
(1)	(2)	(3)	(4)	(5)	(6)	(7)	(8)	(9)	(10)	(11)
200	1	1968.0	7.1	−1.0	1965.0	10.1	2.0	3.0	7.0	0.2
	2	2392.4	3.1	−0.0	2391.1	4.5	1.3	1.3	1.0	0.3
	3	2765.8	0.2	0.3	2765.1	0.9	1.0	0.7	0.4	0.2
	4	3134.4	−4.2	−1.3	3133.5	−3.3	−0.4	0.9	0.3	0.4
	5	3476.2	−4.1	−2.2	3475.9	−3.8	−1.9	0.3	0.4	0.4
	6	3803.2	−3.0	0.2	3802.9	−2.7	0.5	0.3	0.5	0.4
	7	4112.6	3.8	5.8	4111.8	4.6	6.6	0.8 (−2.5)	0.9	0.4
	8	4404.6	20.4	17.3	4401.9	23.1	20.0	2.7 (−4.7)	0.9	0.4
300	0	1740.9	2.7	...	1740.9	2.7	...	0.0	5.5	...
	1	2291.0	0.2	−2.3	2285.8	5.4	2.9	5.2	1.5	0.3
	2	2785.7	−5.4	−4.7	2779.6	0.7	1.4	6.1	0.4	0.2
	3	3239.0	−5.3	−2.8	3236.7	−3.0	−0.4	2.3	0.3	0.2
	4	3652.7	−2.6	−0.2	3652.6	−2.4	−0.1	0.1	0.4	0.3
	5	4055.3	6.3	6.9	4053.6	8.0	11.6	1.7 (−2.8)	0.7	0.3
	6	4429.3	29.5	25.6	4425.3	33.5	29.6	4.0 (−6.1)	1.0	0.6
395	1	2549.8	−4.6	...	2544.1	1.1	...	5.7	0.6	...
	2	3083.6	−9.2	...	3076.6	−2.2	...	7.0	0.3	...
	3	3606.2	−10.7	...	3598.7	−3.2	...	7.5	0.4	...
	4	4079.2	3.8	...	4074.9	8.0	...	4.3 (−3.6)	0.7	...
	5	4500.6	37.1	...	4494.8	42.9	...	5.8 (−7.9)	1.7	...
400	0	2006.7	−8.7	6.6	2006.7	−8.7	6.6	0.0	5.5	0.6
	1	2562.7	−16.0	−5.4	2556.9	−10.2	0.4	5.8	7.0	0.3
	2	3098.5	−8.9	−8.8	3091.3	−1.7	−1.6	7.2	12	0.2
	3	3623.9	−17.4	−9.3	3615.9	−9.4	−1.3	8.0	10	0.2
	4	4099.7	−7.6	6.2	4095.0	−2.9	10.9	4.7 (−3.7)	9.2	0.3
	5	4522.0	20.8	38.3	4515.7	27.4	43.4	6.3 (−8.4)	9.2	1.2
600	0	2452.6	−14.6	...	2452.5	−14.5	...	0.1	4	...
	1	3042.5	−17.5	−11.7	3037.8	−12.8	−7.0	4.7	4	0.8
	2	3645.5	−23.5	−12.4	3638.5	−16.5	−5.4	7.0	5	0.3
	3	4230.0	−5.0	9.18	4219.0	6.0	20.18	11.0 (−4.3)	5	0.5
800	0	2828.8	−7.8	...	2828.7	−7.7	...	0.1	4	...
	1	3479.7	−28.7	...	3477.1	−26.1	...	2.6	4	...
	2	4136.9	−16.9	...	4129.2	−9.2	...	7.7 (−3.4)	7	...
1000	0	3160.5	−20.5	...	3160.4	−20.4	...	0.1	3	...
	1	3890.2	−16.2	...	3889.4	−15.4	...	0.8	5	...
	2	4575.1	25.9	...	4563.9	37.1	...	11.2 (−8.8)	14	...

4. The depth of the solar convection zone

The depth of the solar convection zone can be determined to great accuracy by matching observed frequencies for high ℓ modes with predictions for various cases. Figure 2 shows how the weight for frequency determination is all concentrated near the bottom of the model convection zone. With only about 10 modes each for about 5 ℓ values that display great sensitivity to the structure at the convection zone bottom, its depth can be measured to about 0.002 in radius.

Figures 3, 4 and 5 show how the observed minus calculated (O-C) p-mode frequency differences vary with the mode frequency for various ℓ values and for various convection zone depths. The observed frequencies are from Libbrecht, Woodard, and Kaufman (1990) and Korzennik (1990).

Points for individual modes for each ℓ are connected by a line, but one can see the mode frequencies at the kinks. Modes for $\ell = 20$ and above seem to be too shallow for this depth measurement, but they do reveal model structure problems higher in the convection zone.

Inspection of these three figures shows that the best fit is for the intermediate case with the convection zone radius near 0.711 of the solar photospheric radius.

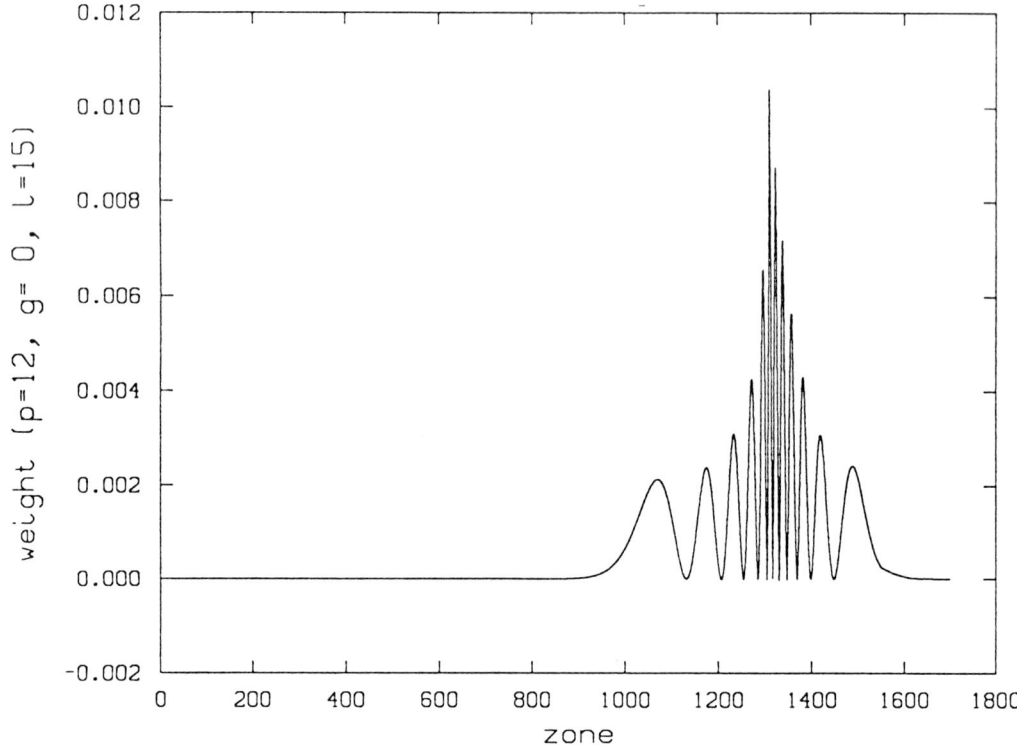

Figure 2 The weight for the p_{12} mode with $\ell=15$ versus mass zone number, showing that there is considerable weight at the bottom of the convection zone at zone 1234.

5. Discussion

Many results of probing the solar structure by comparing observational and theoretical oscillation frequencies are now available. Our new detailed results on the convection helium content and zone depth are now being followed by investigations of the shape of the helium composition structure in this layer. These results are reported by Guzik in another paper at this conference and in a paper being prepared for publication. For this work, we need nonadiabatic eigensolutions that include the effects of radiation diffusion in the upper convection zone to model the phenomena for the real Sun.

A long-standing question is how much overshooting occurs at the convection zone bottom. Another is the diffusion composition shape that results from the differing Y value in the convection zone and the deeper primordial value below. Some combination of these mechanisms, rotation induced turbulence, and the μ gradient, that stabilizes convection can now be mapped with good precision.

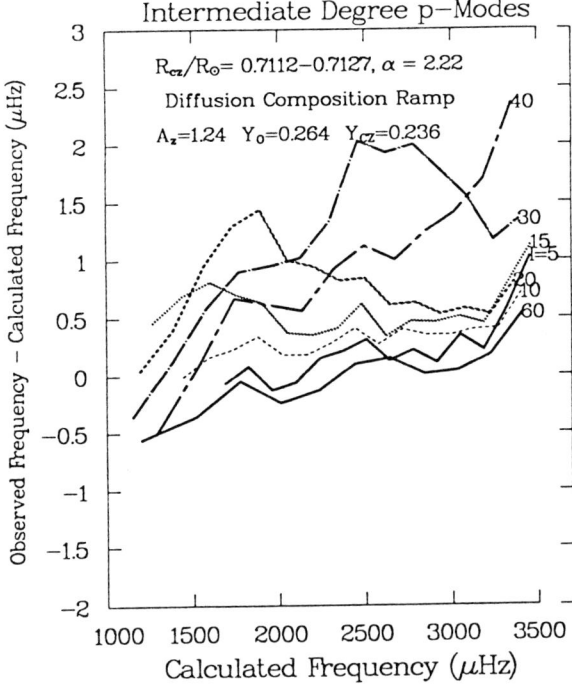

Figure 3 The O-C for intermediate ℓ p-mode frequencies versus mode frequency for the mixing length that produces a convection zone bottom at about 0.712 of the solar radius.

6. Questions

Hiromoto Shibahashi: How many free parameters do you have in your treatment of diffusion? I suppose that you adjust the radius of your solar model to the present solar value by changing the mixing length. Is the mixing length determined after fixing the parameters for diffusion?

The several parameters for the diffusion part of the solar evolution calculation are uncertain, but they are considered as given, just like the material properties. Indeed the mixing length required for the model to have the solar radius is somewhat influenced by these diffusion parameters, because the helium abundance in the convection zone is much smaller with diffusion than without. We now need an even larger mixing length.

Geza Kovacs: What happens with the pulsation spectrum if you completely neglect convection? Are there any calculations done with the Iglesias-Rogers (1992) opacities?

Solar models with no convection do not come anywhere near to matching the accurately known solar radius. Thus none have ever been calculated even 35 years ago! Such models will be much too large, and frequencies will be much too small. Surface convection cannot be ignored in the Sun, even though it often is for yellow giant pulsators. Since 1988, all solar models have used the then current Livermore OPAL opacities. As I discussed in my "Inside the Sun" review in Versailles, the only effect of these new opacities is in the few million kelvin region, where opacities 15% higher than the Los Alamos ones are needed to predict the correct intermediate ℓ p-mode frequencies. All the interesting blue and yellow giant variable star opacity effects are hidden in the extensive solar convection zone, where opacities are really very large, but they do not matter at all.

Jayme Matthews: Do the pulsation models you use to constrain the convection zone correspond to sectorial modes? Different $m \neq \ell$ modes will have different latitudinal weightings, so, for example, $m = \ell$ will be very sensitive to properties at the solar equator, while other modes will represent different latitude kernels.

The observers are aware of this behavior of spherical harmonics, and they consider these shapes when analyzing their data. Theoretically, the assumed spherical shape of a non-rotating Sun means that no consideration of the degenerate m parameter is necessary for calculating oscillation frequencies. Our solar oscillation frequencies depend on our spherical model approximations such as frozen-in convection and only on the radial order and the angular degree ℓ.

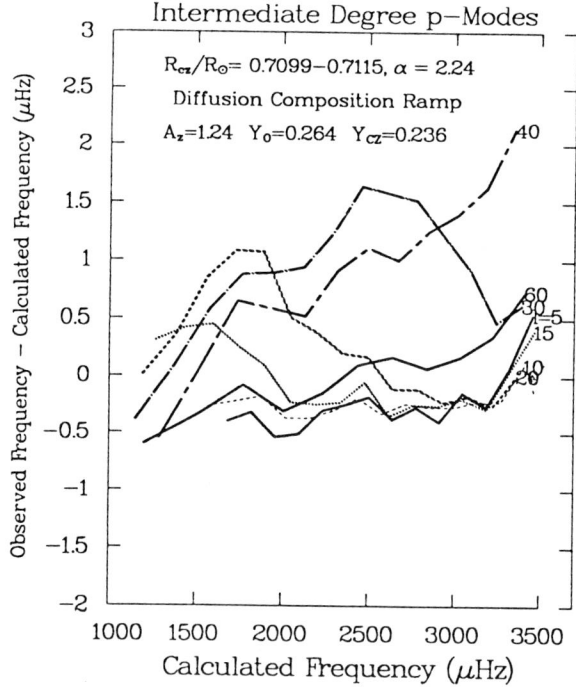

Figure 4 The (O-C) for intermediate ℓ p-mode frequencies versus mode frequency for the mixing length that produces a convection zone bottom at about 0.711 of the solar radius.

The high ℓ modes that display departures from sphericity are the ones that reveal the surface layer rotation.

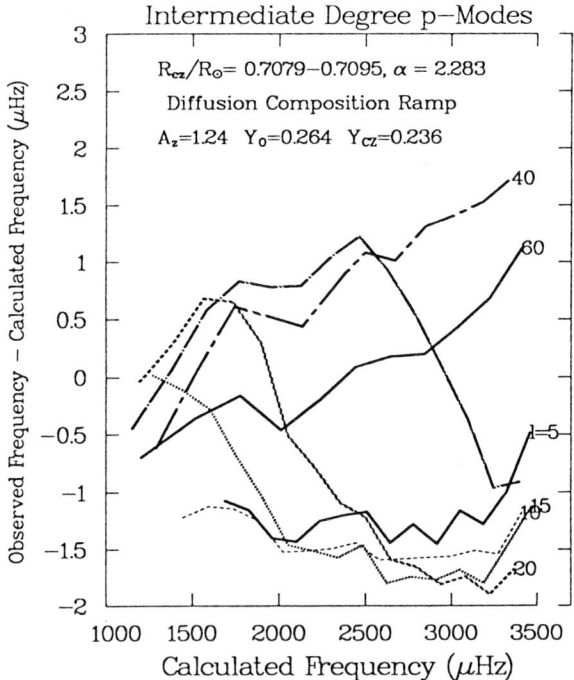

Figure 5 The (O-C) for intermediate ℓ p-mode frequencies versus mode frequency for the mixing length that produces a convection zone bottom at about 0.709 of the solar radius.

References:

Christensen-Dalsgaard, J., 1988, IAU Symposium 123, "Advances in Helio- and Asteroseismology," eds. J. Christensen-Dalsgaard and S. Frandsen (Dordrecht: Reidel), p. 3.
Cox, A. N., Guzik, J. A., and Kidman, R. B., 1989, Astrophys. J., **342**, 1187.
Cox, A. N., Guzik, J. A., and Raby, S., 1990, Astrophys. J., **353**, 698.
Däppen, W., Mihalas, D., Hummer, D. G., and Mihalas, B., 1988, Astrophys. J., **332**, 261.
Guenther, D. B., Demarque, P., Kim, Y.-C. and Pinsonneault, M. H., 1992, Astrophys. J., **387**, 322.
Guzik, J. A. and Cox, A. N., 1991, Astrophys. J., **381**, 333.
Guzik, J. A. and Cox, A. N., 1992, Astrophys. J., **386**, 729.
Kosovichev, A. G., Christensen-Dalsgaard, J., Däppen, W., Dziembowski, W., Gough, D. O., and Thompson, M. J., 1992, MNRAS, submitted.
Korzennik, S. G., 1990, Ph.D. dissertation, UCLA.
Libbrecht, K. G., Woodard, M. F., and Kaufman, J. M., 1990, Astrophys. J. Suppl., **74**, 1129.
Proffitt, C. R. and Michaud, G., 1991, Astrophys. J. **380**, 238.
Rogers, F. J. and Iglesias, C. A., 1992, Astrophys. J. Suppl., **79**, 507.

A Striking Similarity Between the Sun, Binary Systems and RR Lyrae Stars

V. A. Kotov[1,2]

[1]*Crimean Astrophysical Observatory, Crimea, USSR*
[2]*Stanford University, Stanford CA 94305 U.S.A.*

Abstract

The periodicity of $P_0=160.0101$ (± 1) min discovered in the Sun's global oscillations, then in rapid variability of active galactic nuclei (AGNs) (Kotov & Lyuty 1990), might have a cosmological origin. Analysing the distribution of periods of $\sim$4000 close binaries we found that the most commensurate (with an odd-even parity taken into account) period of galactic binaries is equal to 160.0($\pm$1) min ($\sim 4\sigma$). Accordingly, it is argued that the famous 2-3 hr gap of cataclysmic binaries might be closely associated with the "universal" P_0-oscillation. A similar analysis applied to the $\sim$1200 RR Lyrae variables in galactic globular clusters (GCs) and 72 RR Lyraes in the LMC GCs (Graham 1985; Nemec et al. 1985) showed that for all those stars the P_0-period appears to be the most "resonant" one (Fig. 1). This strongly supports the cosmological origin of the P_0-oscillation. Since the P_0-resonance for RR Lyraes appears to be most pronounced when the "odd-even" parity is taken into account, it is hypothesized that variability in the majority of GC RR Lyraes should be related to their binary nature and may be relevant to explanation of Blazhko effect.

References:

Graham, J. 1985, PASP, 97, 676.
Kotov, V.A. & Lyuty, V.M. 1990, Compt. Rend. Acad. Sci. Paris, 310, 743.
Nemec, J.M., Hesser, J.E. & Ugarte, P.P. 1985, ApJS, 57, 287.

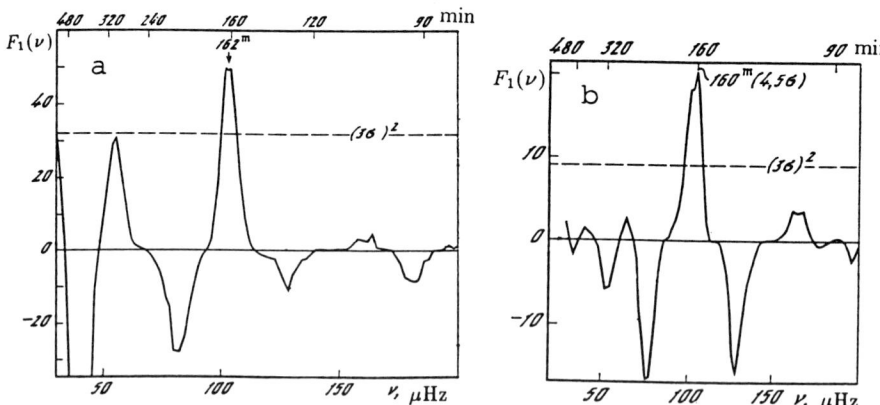

Fig. 1. The "resonance spectra computed for (a) 1211 RR Lyrae stars in galactic globular clusters, and (b) 72 RR Lyrae stars discovered in the LMC globular clusters.

Beyond the Classical Instability Strip

New Opacities and the Beta Cephei Stars

L.A. Balona

South African Astronomical Observatory, P.O.Box 9, Observatory 7935, Cape, South Africa

Abstract

The recent revision of metal opacities has opened up the possibility that the long-sought mechanism for driving pulsations in these stars has at last been found. This hypothesis makes a testable prediction — that no β Cep variables should exist among metal-poor B-type stars. We report on the results of an intensive CCD monitoring campaign to test this prediction. The question of the pulsation mode of β Cep stars and the consequence of a revision of the absolute magnitudes to accommodate the fundamental radial mode among these stars is also discussed. Pulsational instability due to driving by ionization of metals has many other repercussions for the incidence of pulsation among the B-type stars. A classification scheme for other intrinsically variable B-type stars is suggested. It is shown that if, as generally supposed, pulsation is common among B-type stars then at least two different mechanisms must be in operation.

1. Introduction

The β Cep stars were first recognized nearly a century ago by their short-period radial velocity variations and light variations. The periods (generally 3 to 6 hours) are too short to be ascribed to stellar rotation or binary interaction. The best definition of what constitutes a β Cep star will ultimately depend on the physics of stellar pulsation in early-type stars. One definition is that a β Cep star must have at least one radial pulsation mode. While this may yet prove to be a good definition, I am of the opinion that we do not yet know enough of the physics of the pulsation, of what constitutes a radial mode in rapidly-rotating stars and how one might identify modes (especially in faint stars) to make this definition very useful at the present time. The best definition for such a star still appears to be the one in general use for most of this century: an early B- or late O-type star with a radial velocity or photometric period too short to be explained by anything other than pulsation. In terms of pulsation theory, this merely implies a p-mode pulsation of low spherical harmonic order.

With the recent revision of metal opacities (Rozsnayi 1989, Iglesias & Rogers 1991), the possibility now exists of understanding the mechanism which drives pulsations in the β Cep stars (Cox & Morgan 1990, Cox *et al.* 1992, Moskalik & Dziembowski 1992, Kiriakidis *et al.* 1992). The sudden appearance of a tremendous number of iron lines at a temperature of about 150,000 K gives a high sensitivity of opacity to temperature at the very low densities found in these stars. In this paper we discuss the implications of this mechanism for β Cep and other pulsating B stars.

2. Dependence of driving on metal abundance

The clearest prediction, and one in which there is agreement among all investigators, is the high sensitivity of the pulsations to metal abundance. Pulsation is found only for stars with a metal abundance somewhat larger than the solar value, but this depends very sensitively on the opacity and may change as the opacities are improved. Nevertheless, detection of β Cep pulsations in a metal-poor star would pose a major problem.

The detection of a β Cep star, PHL 346, situated at more than 5 kpc from the galactic plane (Waelkens & Rufener 1988) is of particular interest in this regard. Waelkens & Rufener (1988) argue that the star could not have been formed in the galactic plane and belongs to Population II. However, there is no direct evidence that metals are underabundant in this star. More recently, Waelkens, Van den Abeele & Van Winckel (1991) found that the β Cep stars in the direction of the galactic centre seem to be hotter than normal, indicating a possible dependence on metal abundance. Again, the evidence is not very compelling.

3. A search for β Cep stars in LMC clusters

The mechanism should cease to be effective when the metal abundance is sufficiently low. Therefore β Cep stars should not exist in metal-poor systems such as the Magellanic Clouds. This was found to be the case in the cluster NGC 330 in the SMC (Balona 1992a). A large proportion of early B-type stars in NGC 330 are Be stars (Grebel, Richtler & de Boer, 1992); indeed Balona (1992a) found a substantial number of periodic Be stars (λ Eri variables) in this cluster. The β Cep phenomenon is rare among rapid rotators such as Be stars and in any case would be difficult to detect owing to the irregular light variations in these stars. Under the circumstances, the lack of β Cep stars in NGC 330 is not too surprising and is probably not sufficient evidence in favour of the metal opacity mechanism.

With this in mind, Balona & Jerzykiewicz (1992) and Balona (1992b) obtained intensive CCD photometry of two young clusters in the LMC: NGC 2004 and NGC 2100. These are not known to contain significant numbers of Be stars and no confirmed λ Eri variables were detected. Over 150 observations for each star were obtained over a period of 12 nights. A total of 178 stars lying in the β Cep instability strip were examined. The intrinsic colour-magnitude diagrams of the two clusters, together with the extreme magnitude limits of the β Cep instability strip, are shown in Figs. 1 and 2.

Not a single star was found with the characteristic short-period variability,

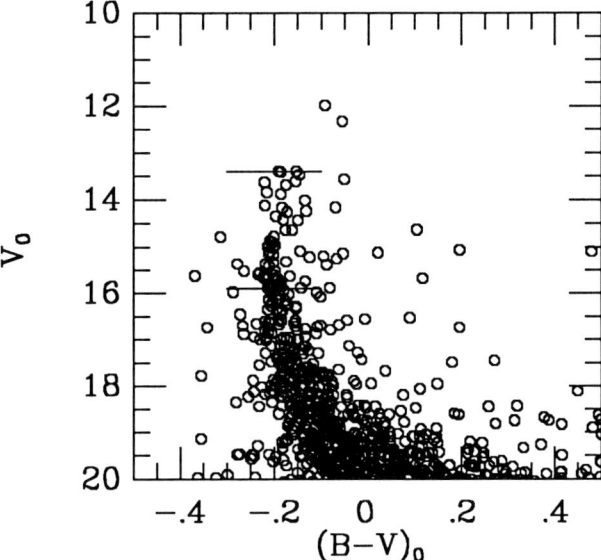

Figure 1 The intrinsic colour-magnitude diagram for the blue main sequence in the LMC cluster NGC 2004. The extreme magnitude limits of the β Cep instability strip is shown by the two lines.

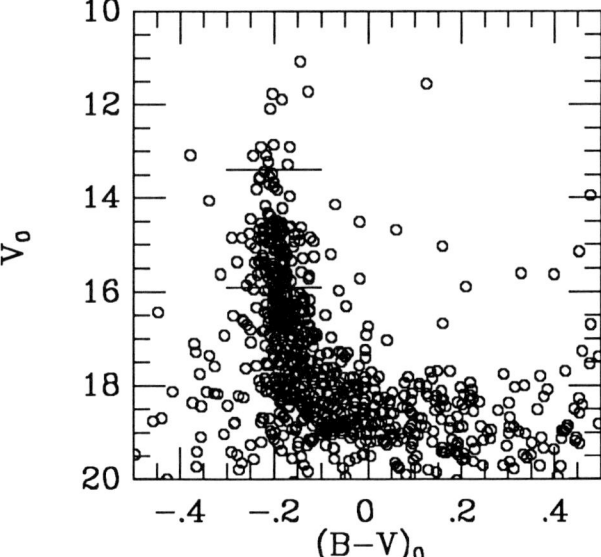

Figure 2 The intrinsic colour-magnitude diagram for the blue main sequence in the LMC cluster NGC 2100. The extreme magnitude limits of the β Cep instability strip is shown by the two lines.

though the observations would have been able to detect such stars with amplitudes as small as 0.01 mag. The approximate proportion of B0-B2 main sequence and giants which are β Cep variables can be estimated from the *Bright Star Catalogue*. There are 404 stars of this kind in the catalogue, of which 31 appear as definite β Cep stars in the catalogue of Jerzykiewicz & Sterken (1992). This gives a value of 7.5 percent for the expected proportion of β Cep stars among galactic B0-B2 dwarfs and giants, but it must be taken as a lower limit because not all stars of this type in the *Bright Star Catalogue* have been surveyed for short-period variability. Certainly the proportion among Galactic open clusters is considerably higher.

On the assumption that the fraction of β Cep stars in the two LMC clusters is the same as in the Galaxy, the chance of not finding a single variable in the 178 stars surveyed in NGC 2004 and NGC 2100 is less than 0.1 percent. This is very strong evidence in favour of a dependence on metallicity.

In spite of this strong support for the current theory, the possibility exists that the age of the clusters are not suitable for producing β Cep pulsation. Although both NGC 2004 and NGC 2100 contain stars which are classified as B0-B2 dwarfs and giants and therefore should be good candidates for β Cep variations, the clusters are actually older than galactic clusters such as NGC 3293 and NGC 4755 which also contain stars with these spectral types (Balona 1992b). The reason for this is the lower metallicity which has the effect of increasing the effective temperature at a given luminosity. It is just possible that the instability strip at low metal abundances may be at a different location from the one with normal metal abundances. It would be important to survey a younger LMC cluster to make sure that this is not the case.

4. The mode of pulsation

Moskalik & Dziembowski (1992) and Kiriakidis *et al* (1992) both find that only fundamental *p*-modes are excited. Cox & Morgan (1990) also find instability in one or two overtones, a result which may not be real (see Moskalik & Dziembowski, 1992). A test for the mode of pulsation depends on an estimate for the pulsation constant, Q, which in turn depends mostly on knowing the absolute magnitude of the star. Most β Cep stars are multiperiodic, so it is certain that non-radial modes are also excited. This complicates the issue because rotational splitting of these modes will give erroneous Q values. Nevertheless, the mean Q-value of a group of stars may be taken to approximate the predominant pulsation mode.

Shobbrook (1983) found that the β Cep stars in NGC 3293 were predominantly pulsating in the fundamental radial mode ($Q = 0.034$), in agreement with current theoretical predictions. However, this work was based on a preliminary calibration of the absolute magnitudes in terms of Strömgren indices which was subsequently revised by Balona and Shobbrook (1984). The revised calibration showed that for most β Cep stars $Q = 0.029$ which corresponds more closely with the first overtone radial mode (Shobbrook 1985). This is consistent with all other attempts at estimating this quantity (Balona & Feast 1975, Shobbrook 1979, Jakate 1980). A detailed investigation of the pulsation modes in the β Cep stars in NGC 3293 (Engelbrecht 1986) shows that most of these stars are overtone pulsators.

The estimation of Q depends sensitively on the absolute magnitudes of the early B-type stars. It is important to investigate whether the current calibrations may not overestimate the luminosity of these stars by about 0.4 mag, which is required for fundamental radial pulsation. There is little doubt that the absolute magnitude calibration in terms of Strömgren indices is susceptible to systematic errors (Balona & Shobbrook 1984). Millward & Walker (1985) attempted to overcome these errors by using high signal-to-noise spectra to measure the $H\gamma$ equivalent width. No direct comparison was made between these absolute magnitudes and the Balona & Shobbrook (1984) calibration, but the distance modulus for the one young star cluster in common, NGC 2244, is the same. This calibration is not likely to make a significant difference in the Q values for β Cep stars. Perhaps the most comprehensive investigation of this problem is the work of Wolff (1990) who used spectroscopic measurements of $H\gamma$ to re-calibrate the luminosities of the B-type stars. This calibration gives absolute magnitudes which are about 0.15 mag fainter than those of Balona & Shobbrook (1984) in the range appropriate to β Cep stars. This is in the right direction, but too small to push the mean Q value towards fundamental radial pulsation. My conclusion is that while there may well be systematic errors in the present absolute magnitude calibration of early B-type stars, they are unlikely to be as large as the 0.4 mag required to give a mean pulsation constant appropriate to the fundamental radial mode. A systematic error of this magnitude would, moreover, almost certainly affect the zero-point of the Cepheid distance scale and destroy the agreement now reached between pulsation and evolution masses for these stars.

5. Some outstanding problems

One of the most puzzling aspects of the β Cep phenomenon is the well-established fact that the pulsations can decay in a time span of a few years. The most famous of these cases are Spica and 16 Lac. In the former case, the low spherical harmonic order pulsations which were observable in the radial velocities and photometry completely disappeared in a period of less than 10 years. However, high-order oscillations are still present (Smith 1985). In 16 Lac the oscillations have not completely disappeared. Both stars are close binaries, which prompted Balona (1985) to suggest that the reason for the decay in amplitude is due to a changing geometrical aspect caused by precession.

An even more tantalizing case where pulsations were seen to grow from zero amplitude in less than two years has recently been reported for the Be star 27 CMa

(Balona & Rozowsky, 1991). This star is unique in two respects, being the only one in which β Cep pulsations have suddenly appeared, and the only well-studied Be star which is also a β Cep variable. An earlier example of a Be/β Cep star is 19 Mon (Balona & Engelbrecht 1979, Balona et al. 1992), but the $H\alpha$ emission in this star is very weak indeed and may possibly be attributed to the pulsation. Another chapter in the intriguing Be/β Cep story was recently revealed by Mathias et al. (1991) who found that β Cep itself had developed strong emission in $H\alpha$. Henrichs et al. (this volume) suggest that the development of emission could be related to a variable magnetic field.

These observations constitute a considerable challenge to our understanding of the driving mechanism. At this stage it is probably to early to address these issues as much work still needs to be done to make sure that the driving mechanism as we now understand it is indeed the correct one.

6. Other B-type variables

It has been know for some time that β Cep stars are by no means the only intrinsically variable B-type stars. The development of high-resolution spectroscopy led to the discovery of the 53 Per stars, a group of sharp-lined stars characterized by low-order line profile variations with a period of a day or longer. Little spectroscopic work has been done on these stars, though some photometrically variable stars have been tentatively classified as members of this group or the group which has recently been termed "slowly pulsating B stars" (Waelkens, 1991).

A group which seems to be physically distinct is characterized by the prototype, the Be star ζ Oph. These are rapidly rotating stars showing high-order line profile variations of short period. These are not confined to Be stars, a counterexample being the non-Be star Spica (Smith 1985). Although the period between consecutive "moving bumps" in the line profile is measured in hours, the period in a frame of reference rotating with the star is very long and could even be infinite (Balona 1990). If this is a pulsation, it cannot be due to p-modes.

Finally, there is a third group which consists only of Be stars and which exhibits strictly periodic photometric variability and low-order line profile variations. These have been called λ Eri variables, after the prototype. A statistical analysis shows that the period is not significantly different from the expected rotational period of the star (Balona 1990) and that the most likely cause of the variability is a nonuniform surface brightness distribution. Very often the light curve is of double-wave form which is phase locked from season to season and is highly suggestive of a magnetic dipole configuration (Balona et al., 1991). This type of variation is somehow connected with the Be phenomenon in a manner which is not yet understood. The discovery of λ Eri stars in the SMC cluster NGC 330 (Balona 1992a) shows that if these stars are pulsating, the mechanism must be different from the one causing β Cep pulsations because the metal abundance in NGC 330 is very low.

Calculations show that g-modes in B stars are not excited with the current opacities, though the extreme sensitivity of the pulsation on small details of the opacity greatly diminishes the predictive power of the models. The most promising explanation for exciting g-modes is the mechanism originally advocated by Osaki (1974) and refined by Lee & Saio (1986). In this mechanism overstable modes in the convective core penetrate to the envelope where they are seen as g-modes. The periods of these modes should be closely related to the rotation period of the core. A prediction of this theory is that the oscillations should not be produced if the convective core is too small. Further progress needs to be made in defining the predicted instability strip before the hypothesis can be put to the test.

7. Conclusions

The hypothesis that β Cep pulsations are driven by the classical κ-mechanism involving the ionization zone of metals is strongly supported by the result that no such stars are found in metal-poor systems such as the two young clusters in the LMC (Balona 1992b). However, the prediction that only modes with periods close to that of the fundamental radial mode should be excited presents a problem which needs a more careful study both observationally and theoretically.

It is clear that while this mechanism may well resolve the problem of the β Cep stars, it cannot explain the λ Eri stars which are found in the metal-poor cluster NGC 330. At present, the models do not seem to be able to drive g-modes found in 53 Per and ζ Oph stars, though the latest work indicates that these may be excited by small changes in the opacities. The most pressing need at the moment are predictions of the expected limits of instability and modes of pulsation for both the κ-mechanism and by the Osaki mechanism so that these can be compared to observations.

References:

Balona, L.A., 1985. *Mon. Not. R. astr. Soc.*, **217**, 17p.
Balona, L.A., 1990. *Mon. Not. R. astr. Soc.*, **245**, 92.
Balona, L.A., 1992a. *Mon. Not. R. astr. Soc.*, in press.
Balona, L.A., 1992b. *Mon. Not. R. astr. Soc.*, submitted.
Balona, L.A., Cuypers, J. & Marang, F., 1992. *Astron. Astrophys. Suppl.*, **92**, 533.
Balona, L.A. & Engelbrecht, C.A., 1979. *Mon. Not. R. astr. Soc.*, **189**, 171.
Balona, L.A. & Jerzykiewicz, M., 1992. *SAAO Circ.*, submitted.
Balona, L.A. & Feast, M.W., 1975. *Mon. Not. R. astr. Soc.*, **172**, 191.
Balona, L.A. & Rozowsky, J., 1991. *Mon. Not. R. astr. Soc.*, **251**, 66p.
Balona, L.A. & Shobbrook, R.R., 1984. *Mon. Not. R. astr. Soc.*, **211**, 375.
Balona, L.A., Sterken, C. & Manfroid, J., 1991. *Mon. Not. R. astr. Soc.*, **252**, 93.
Cox, A.N. & Morgan, S.M., 1990. *Confrontation between Stellar Pulsation and Evolution*, p. 293, eds Cacciari, C. & Clementini, G., Astr. Soc. Pacif., San Francisco.
Cox, A.N., Morgan, S.M., Rogers, F.J. & Iglesias, C.A., 1992. *Astrophys. J.*, in press.
Engelbrecht, C.A., 1986. *Mon. Not. R. astr. Soc.*, **223**, 189.

Grebel, E.K., Richtler, T. & de Boer, K.S., 1992. *Astr. Astrophys.*, **254**, L5.
Iglesias, C. & Rogers, F.J., 1991. *Astrophys. J.*, 371, L73.
Jakate, S.M., 1980. *Astr. Astrophys.*, **84**, 374.
Jerzykiewicz, M. & Sterken, C., 1992. In preparation.
Kiriakidis, M., El Eid, M.F. & Glatzel, W., 1992. *Mon. Not. R. astr. Soc.*, **255**, 1p.
Lee, U. & Saio, H., 1986. *Mon. Not. R. astr. Soc.*, **221**, 365.
Mathias, P., Gillet, D. & Kaper, L., 1991. in Proc. ESO Workshop *Rapid variability of OB-Stars: Nature and Diagnostic Value*, p. 193 ed. D. Baade, ESO, Garching.
Millward, C.G. & Walker, G.A.H., 1985. *Astrophys. J. Suppl.*, **57**, 63.
Moskalik, P. & Dziembowski, W.A., 1992. *Astr. Astrophys.*, **256**, L5.
Osaki, Y., 1974. *Astrophys. J.*, **189**, 469.
Rozsnayi, B.F., 1989. *Astrophys. J.*, **341**, 414.
Shobbrook, R.R., 1979. *Changing Trends in Variable Star Research,* IAU Coll. **46**, p.512 eds. Bateson, F.M., Smak, J. & Urch, I.H., Univ. Waikato, Hamilton, New Zealand.
Shobbrook, R.R., 1983. *Mon. Not. R. astr. Soc.*, **204**, 47p.
Shobbrook, R.R., 1985. *Mon. Not. R. astr. Soc.*, **214**, 33.
Smith, M.A., 1985. *Astrophys. J.*, **297**, 206.
Waelkens, C., 1991. *Astr. Astrophys.*, **246**, 453.
Waelkens, C. & Rufener, F., 1988. *Astr. Astrophys.*, **201**, L5.
Waelkens, C., Van den Abeele, K. & Van Winckel, H., 1991. *Astr. Astrophys.*, **251**, 69.
Wolff, S.C., 1990. *Astr. J.*, **100**, 1994.

Discussion

Moskalik: I understand that W. Dziembowski, using the latest OPAL opacities which include spin-orbit interaction, now finds that the *first overtone* is unstable and that the minimum metal abundance for instability is somewhat lower than previously obtained.

Cox: New opacities from OPAL and OP show that doubling the Z is no longer necessary. β Cep stars can now pulsate with Z = 0.03 (using OPAL but not OP). Unfortunately the Z = 0.03 OP opacities that are a bit smaller than OPAL do not give pulsation. Anyway, large Z is no longer needed, but at Los Alamos we do not get pulsation at Z = 0.02.

Walker: The Galactic β Cep stars have spectral types B0-B3. Do the B-type stars in the three Magellanic Cloud clusters you observed have spectral types this early?

Balona: The few early-type giants for which spectral types are available do indeed fall in this range.

Focal Points in Contemporary β Cephei Star Research

C. Sterken[1]

University of Brussels (VUB), Pleinlaan 2, 1050 Brussels, Belgium

Abstract

We review the latest results on searches for new β Cephei stars, and discuss the most recent findings from high-resolution spectrographic observations. The implications of analyses of archival data and new observations in the light of period changes, binarity, and aspects of the evolutionary state of β Cephei stars are evaluated.

1. Introduction

Attempts to find a pulsation mechanism for β Cephei stars went on since decades, and the problem of the unknown driving mechanism gradually grew to become a real challenge for stellar oscillation theory. The problem may have been solved now: the application to the β Cephei star problem of new opacity tables published by Iglesias and Rogers (1991a,b) (see Cox and Morgan 1990, Kiriakidis et al. 1992, Moskalik and Dziembowski 1992) leads to the understanding that the driving is caused by the usual κ mechanism acting in a zone with temperature near 200,000 K where there is an opacity bump. The pulsations are extremely sensitive to metal abundance. The theoretical instability strip agrees well with the observational one, and extends well above the β Cephei star region in the H-R diagram, which suggests that the same mechanism may be responsible for oscillations observed in Luminous Blue Variables (Moskalik and Dziembowski 1992, see also Sterken 1988).

For many years, the justification to carry out a specific topic of observational β Cephei star research was extracted from the state of affairs that no adequate pulsation mechanism was at hand. Now that a plausible mechanism seems to have been found, one might argue that there are much less reasons to carry on observational studies of β Cephei stars. This opinion, however, is unfounded, as is shown in the present paper.

First of all, the new pulsation mechanism needs to be tested observationally, and several papers of the 1991-1992 period deal with such measurements (see also Balona's contribution to this conference). In addition, high-precision spectrographic and photometric observations (at high time-resolution) of selected "classical" β Cephei stars have revealed hitherto unknown facts and have opened up new horizons for observational research on β Cephei stars: the last two years disclosed an outstanding element

---

[1] Belgian Fund for Scientific Research (NFWO)

in observational studies of β Cephei stars, viz. the large research effort put into the observation of a couple of **bright** β Cephei stars. The stars concerned are BW Vul, σ Sco and β Cephei and were the subject of several independent investigations.

2. Searches for β Cep stars

Monitoring of β Cephei stars is still being done (see, for example, Heynderickx 1991), but few systematic searches for β Cephei stars seem to be going on. Delgado et al. (1992) carry on their systematic observations of young open clusters with the detection of β Cephei stars and related pulsating variables as primary aim. In their last paper ($uvby\beta$ photometry of NGC 1502 and NGC 2169) particular attention is given to the problem of internal calibration of their $uvby$ photometric system, and to possible influence of interstellar reddening in the photometric transformation procedures. They report the discovery of two new β Cephei candidates, viz. 1502-A and 1502-26, with respective pulsation periods of 0.19 and 0.10 days. The latter star has also a secondary pulsation period of about 0.06 days, and the former exhibits long-term variations that may reflect binarity. These new candidates, of course, need confirmation, and the values of their pulsation periods need to be refined.

An interesting search project for β Cephei stars in LMC clusters was initiated by Balona and Jerzykiewicz (Balona 1992a, Balona and Jerzykiewicz 1992, see also Balona 1992b). The underlying idea is that β Cephei stars should not exist in metal-poor systems such as the Magellanic Clouds. Such considerations, among others, prompted Sterken and Jerzykiewicz (1988) to initiate a first search for β Cephei stars among field stars in the LMC. Out of six programme stars they selected from among the LMC stars which fell in the same region in the $Q - M_V$ diagram where are the galactic β Cephei stars, only one star showed β Cephei star-like variations, and was classified as a candidate β Cephei star. Kubiak (1992) searched the young LMC cluster NGC 1712 for new β Cep stars, without success. Balona (1992a), and Balona and Jerzykiewicz (1992) monitored 178 stars lying in the instability strip of the young LMC clusters NGC 2004 and NGC 2100. For each star 150 observations were obtained over a period of 12 nights, but not a single star was found that is variable with a short period and with an amplitude in excess of 0.01 mag. Balona (1992b) concludes that the chance of not finding a variable among the stars surveyed is less than 0.1 per cent. As he points out, one should take into consideration that the studied clusters are older than the galactic clusters that contain β Cephei stars, so that one should await the results of a survey of younger LMC clusters. The NGC 2004 and NGC 2100 results, however, give additional evidence that there is a dependence on metallicity. In this respect, attention must be drawn to the results of Waelkens et al. (1991) who conclude that β Cephei stars in the direction of the galactic center seem to be hotter than normal, so that the blue edge of the β Cephei instability strip moves blueward, what is another point of support for the hypothesis of metallicity dependence. Conclusive evidence on this point may be obtained from similar surveys

3. High-resolution spectrography

Fundamental contributions to the understanding of the atmospheres of the bright β Cephei stars σ Sco and 12 Lac have been published by Mathias et al. (1991, 1992). Hα and Si III spectra of σ Sco reveal, for the first time, the line of the 33-day companion of the star. The observed line-profile changes and radial-velocity changes are consistent with a double-shock model (as for BW Vul, see Crowe and Gillet 1989). 12 Lac resembles σ Sco and BW Vul (the two β Cephei stars which have the largest 2K-amplitudes, respectively 120 and 200 km s^{-1}), which undergo shock waves during one pulsation period and exhibit a stillstand, which is the consequence of the simultaneous combination of inward and upward motions of the atmosphere. 12 Lac, however, has a rather chaotically variable radial-velocity curve, in strong contrast to σ Sco and BW Vul. Mathias et al. (1992) understand the phenomenon as the superposition of radial and non-radial modes. In 12 Lac there is no sign of duplicity.

Kaper et al. (1992) report on new observations of strong Hα emission-activity in β Cephei characterized by a (slow) decline in strength of the emission component, and Henrichs et al. (1992) confirm a periodicity of 12 days in the equivalent widths of the ultraviolet resonance lines of $C\,IV$, $Si\,III$, $Si\,IV$ and $N\,V$. Their measurements of the variable magnetic-field support a model in which the stellar wind from the star is rotationally controlled ($P = 12$ days) by an asymmetric oblique magnetic field. The on-and-off appearance of emission in β Cephei is an interesting case of transition from a "normal" B star to a star of type Be. A most interesting example of reverse transition is the appearance of β Cephei-like pulsations in the Be star 27 CMa (Balona and Rozowsky 1991) with a periodicity of 2^h12^m in addition to the 1.257-day variability. This is the first time the growth of β Cephei pulsations has been witnessed, and 27 CMa is the only Be star known which shows such pulsations.

4. Period changes, pulsation and binarity

During a star's lifetime, the star changes its radius in its evolution off the main sequence, and, as is well-known, the pulsation period, which varies with the mean density, will vary as well. Other processes may cause changes in the pulsation period of a star, such as the presence of binary companions, the occurrence of discontinuous mass loss, or mixing events within the star itself. In Cepheids, for example, parabolic $O - C$ diagrams[2] are observed (see e.g. Szabados 1977, 1980, 1981), though two other kinds of period changes were observed too, viz. cyclic behaviour caused by the light-time effect, and stepwise period changes, where the period returns to its earlier value (so-called phase jumps, which always seem to occur in those Cepheids having a companion star). But, as Hall (1990) points out, most $O - C$ curves which show more than one complete cycle of period variation do not have equal amplitudes, shapes or periods for successive cycles, as would be required by Keplerian motion.

---

[2] i.e. corresponding to continuous period changes as a result of stellar evolution

In addition, several Cepheids with demonstrable binary companions show variable $O-C$ curves, even after the light-travel-time effect is removed. Moreover, sudden jumps can be produced by cycle miscounts, see for example the famous case of Polaris (Fernie 1984).

The large scatter in rates of period changes in β Cephei stars is not explainable by evolutionary effects alone, but if averaged over a sufficiently long time interval, one may expect to obtain a value of $\dot{P}$ that is in agreement with the calculated evolutionary rate of period change. From an analysis of photometric and radial-velocity data, Chapellier (1985) had come to the conclusion that probably no secular variations of pulsation periods do occur in β Cephei stars. In σ Sco, BW Vul and β Cep the amplitudes of the changes are largest, and a better fit to the data is obtained assuming that positive and negative period jumps occur. Hence, the large $\dot{P}$ cannot be reconciled with evolution, and from this he concludes that all β Cephei stars are in the core hydrogen burning state of evolution.

Chapellier (1990) considered that the apparent changes of the pulsation period of β Cephei can be explained by the light-time effect, induced by the orbital motion of the variable in a highly-eccentric double-star system. The high eccentricity is then responsible for the swiftness of the change of period around the time of periastron passage. Pigulski and Boratyn (1992) derived the spectroscopic orbital elements of the system from the $O-C$ diagram *assuming* that the orbital motion is the *only* cause of the observed variation of the pulsation period. The solution is supported by speckle observations that roughly cover one fifth of the orbital period of 91.6 years. The time of periastron passage coincides with the time of the "abrupt" change of period found by Chapellier (1985)[3], and the next periastron passage is to be expected in A.D. 2003. The agreement between the observed and calculated pulsation period is excellent, though discrepancies up to about 0.1 s do remain. The scatter of the observed radial velocity around the synthetic radial-velocity curve remains substantial, and it is not clear whether this difference is due to systematic errors or to a real physical phenomenon. The $O-C$ curve derived from the orbital elements leaves irregular deviations up to about 0.03 days.

In a subsequent paper, Pigulski (1992a) shows that the changes in the dominant pulsation period of σ Sco can be completely[4] explained by a superposition of two effects: an evolutionary increase of the intrinsic pulsation period, and a variation due to the light-time effect in a binary (the $O-C$ diagram reveals time intervals of increasing and decreasing period, which means that an evolutionary effect alone cannot be held responsible, and the presence of a speckle tertiary companion suggests that the light-time effect may contribute to the observed changes). He derived the spectroscopic elements of the system from radial-velocity data, and from elements obtained by Mathias et al. (1991). He finds a satisfactory explanation for the period change: *assuming* an evolutionary period increase of 3.3 s cen^{-1}, together with proper

---

[3] see also Chapellier 1990, p52

[4] the explanation of the changes of the period of σ Sco in terms of the light-time effect by Chapellier (1990) ceases to function for data obtained after 1955

adjustment of the ambiguous cycle-count for the time interval from 1925 to 1947 (a gap of 35000 cycles to be bridged), a resulting $O - C$ diagram is obtained that is commensurable with an orbital period of the order of 100-350 years, as was suggested by Evans et al. (1986). A complete orbit solution, of course, could not be obtained.

In a similar approach, Pigulski (1992b) explains the changes in the period of BW Vul by an evolutionary increase of the period with constant rate equal to 2.34 s cen^{-1}, and a periodic term caused by an *hypothetical* companion (as suggested by Odell 1984, who pointed out that the periodic variations in the $O - C$ residuals of BW Vul could indicate that the star is a binary with period 24.9 ± 6.5 years, see also Jiang 1985, who found P=26.3 years). The derived orbital period is 33.3 years ± 0.3 and the expected time of the next perisatron passage is A.D. 1992.3 ± 1.5. He concludes that the light-time effect, combined with an effect of evolutionary origin, may be responsible for at least some of the "unexpected" or "sudden" period changes. An alternative hypothesis, proposed by Odell (1984), is in terms of simultaneous excitation of two very closely spaced pulsation frequencies causing a beat phenomenon (the periods differ only 0.3 s and have an amplitude ratio of 1/3).

Note that, since β Cep and σ Sco are known to have distant companions, any period analysis *must* incorporate the light-time effect, whereas in BW Vul the changes seen in the $O-C$ diagram are the *only* manifestations of the presence of a companion. It should be stressed that all $O - C$ diagrams used by Pigulski and Boratyn (1992) and by Pigulski (1992a,b) are based both on photometric and on radial-velocity data, an approach that involves the assumption of constant average time difference between the times of maximum of the radial-velocity curve and the light curve (Lloyd and Pike 1984, and Chapellier 1986, demonstrated that the phase difference between light- and radial-velocity maximum is variable).

σ Sco has also been the subject of an investigation by Chapellier and Valtier (1992). From all available spectroscopic data they derive an orbital period of 33.0114 ± 0.0026 day, a value which is in very good agreement with the findings of Pigulski (1992a), Mathias et al. (1991) and Goossens et al. (1984). From their investigation Mathias et al. (1991) also concluded that, despite variations in the principal pulsation period, the value of the beat period remains constant and is strictly equal to a quarter of the orbital period. They are convinced that the secondary pulsation period does not really exist and is only an artifact of period analysis, and they conclude that σ Sco is a monoperiodic nonradial pulsator.

5. The $O - C$ diagram: its power and its deficiencies

The real power of the $O - C$ diagram lies in the fact that quasi-sinusoidal portions of it may indicate binary motion with K-values that are too small to be detected by radial-velocity observations. As was demonstrated above, such studies of $O - C$ diagrams have proven to be very useful in unveiling unknown orbital elements of binary β Cephei stars. The remaining discrepancies, if of proper magnitude, could reflect a long-term modulation in the pulsation itself, or could be explained by orbital-element variations due to mass exchange inside the system or due to stellar wind from the system, or even reflect the evolution of the binary (especially in close binary systems) through degradation of the orbit. But sometimes, the orbital period is the only one physical property known accurately for a binary system, and also then an $O - C$ study would be extremely beneficial.

However, one must also ask what is the role of observational selection in all such studies of period changes. Jerzykiewicz (1986), for example, showed that the correlation between abrupt period change and period (Hoffleit 1976) has arisen as a result of observational selection and from limitations in the quality of data used to derive these quantities. One may not forget that for the evaluation of the period history, archival data are used, some of which were obtained more than half a century ago, and one must ponder whether these data may be combined with modern data as straightforwardly as is currently been done.

Before discussing the interpretation of observed period changes, one should remember that the $O - C$ diagrams have a number of inherent uncertainties and inaccuracies. In particular, *all* such representations are of non-homogeneous character, because

- data points from which the times of minimum (or maximum) light were derived have been obtained with very different time resolution, and the number (and the quality) of these underlying measurements often is insufficient

- the time distribution of the used times of maximum differs in different parts of the $O - C$ diagram

- the fact that, when one goes back in time, the uncertainties on the times of maximum are much larger than they are now (see Fernie 1990 and Sterken 1992). Times of minimum or maximum light could be routinely measured to an accuracy of better than 0.001 day. But, to quote Jerzykiewicz 1986: "No attempt is usually made to take full advantage of the precision of modern observations, the excuse being that the early ones are not so good anyway"

- the fact that, when one goes back in time, there is less and less certainty that the data have been properly corrected for the light-time effect due to the orbital motion of the earth (heliocentric correction)

- the response functions of the equipment (through unequal effective wavelength of filters, and due to the application of cooled and uncooled photomultipliers with differing response curves) deviate significantly, and multiple-periodic stars (like σ Sco, where the ratio of amplitudes varies from about 2 in y to about 5 in u, see Jerzykiewicz and Sterken 1984), will, for measurements at shorter wavelengths, yield smaller differences between time of observed maximum and time of maximum associated with the dominant pulsation period

However, assuming a homogeneous $O - C$ diagram, one should then very carefully consider the following points:

- a few scattered epochs of maximum light are sometimes used to determine a period

- a single large gap in the sequence of times may cause cycle-count errors, which induce large apparent period variations

- an improper type of extremum is sometimes used (like time of light maximum in the case of BW Vul, see Sterken et al. 1987)

- a non-uniform time system (UT instead of ET) has been used

- one should ask whether the available data are accurate enough to furnish reliable evidence on evolutionary period changes, specifically for those $O-C$ diagrams that are constructed under the assumption of constant average phase difference between the radial-velocity curve and the light curve

- $O-C$ diagrams for well-studied β Cephei stars never cover more than one full orbital cycle, and thus give us no indication of occurrence of strict repetition, so that one is not sure whether the cyclical behaviour is truly periodic or only quasi-periodic

- the separation of evolution and light-time effect is sometimes rather arbitrary, and this is even more so for stars with multiple pulsation periods with appreciable amplitudes

- Fernie (1990) makes a very pertinent remark concerning the common acceptance of the expectation that evolutionary changes necessarily yield *linear* changes of the pulsation period of a variable star. After all, changes in luminosity and temperature are nonlinear in time, and the motion along the evolutionary track has a variable rate. Fernie advocates the introduction of quartic or even higher-order polynomials in the $O-C$ diagrams of classical Cepheids, a move that would eliminate most evidence supporting sudden period changes

- dP/dt should, in fact, be computed using two independent techniques, viz. study of the $O-C$ diagram, and direct non-linear least squares fit to the entire dataset of a sine curve incorporating additional terms (linear or quadratic in time)

In BW Vul, for example, some period variability (sudden glitches or smooth changes) does remain after a linear evolutionary period change has been removed and after the effect of binarity has beeen taken into account (Sterken 1992). Thus, a nonlinear evolutionary effect could be present, since the star appears to be in the shell hydrogen-burning phase of evolution (Sterken and Jerzykiewicz 1990), where such nonlinear effects are more likely to become detectable. Such interpretation of residual variations in terms of a nonlinear evolutionary effect, or in terms of secondary aspects of binarity (is the remaining $\dot{P}$ correlated with the period, does mass transfer[5] play a role, and does this give rise to disk-like structures which might even account for some of the intermittent emission phenomena?) intricately depends on the accuracy of the available observational data. As Irwin (1952) puts it: "The problem of the determination of the light-time orbit will occur with increasing frequency as the observational data become more accurate and extend over greater stretches of time..."

6. Conclusions

Despite the discovery of a suitable pulsation mechanism for β Cephei stars, there is a surge of observational data of high quality with strong astrophysical impact. These observations make up direct tests of the proposed pulsation mechanism, and reveal new facts of fundamental importance concerning the binary nature of some of these stars. One must not forget that binary stars are the main sources of fundamental data on stellar masses and radii. Andersen (1991) presented a sample of 45 detached, double-lined eclipsing binary systems with mean errors $\leq 2\%$ in both mass and radius. These are fundamental data, of lasting value, independent of changes in temperature and flux scales, model atmospheres, abundance data and stellar models. The sample not only highlights the paucity of accurate masses and radii for normal giants, the sample also does not contain a single β Cephei star. This fact alone should be a driving force to further motivate continuous monitoring of a suitably defined sample of bright β Cephei stars.

References:

Andersen, J., 1991, The Astron. Astrophys. Review 3, 91

Balona, L.A., 1992a, MNRAS In press

Balona, L.A., 1992b, This Conference

Balona, L.A., Jerzykiewicz, M., 1992, *SAAO Circ.* In press

Balona, L.A., Rozowski, J., 1991, MNRAS 251, 66p

Chapellier E., 1985, A & A 147, 135

Chapellier E., 1986, A & A 163, 329

Chapellier E., 1990, *Aspects de la variabilité des étoiles de type β CMa et 53 Per*, Thèse de Doctorat, Université de Nice et de Sophia-Antipolis

Chapellier, E., Valtier, J.-C., 1992, A & A 257, 587

Cox, A.N., Morgan, S.M., 1990, In *Confrontation between Stellar Pulsation and Evolution*, eds. C. Cacciari & G. Clementini, ASP Conf. Ser. 11, 293

Crowe, R., Gillet, D., 1989, A & A 211, 365

Delgado, A.J., Alfaro, E.J., Garcia-Pelayo, J.M., Garrido, R., 1992, AJ 103, 891

Evans, D.S., McWilliam, A., Sandmann, W.H., Frueh, M., 1986, AJ 92, 1210

---

[5] e.g. in a binary star with a highly-eccentric orbit

Fernie, J.D., 1984, ApJ 231, 841

Fernie, J.D., 1990, PASP 102, 905

Goossens, M., Lampens, P., de Maerschalck, D., Schrooten, M., 1984, A & A 140, 223

Hall, D. S., 1990, in *Active Close Binaries*, ed. C. Ibanoglu, Kluwer Ac. Publ., p95

Henrichs, H.F., Bauer, F., Hill, G.M., Kaper, L., Nichols-Bohlin, J.S., Veen, P.M., 1992, This Conference

Heynderickx, D., 1991, *A Photometric Study of β Cephei Stars*, Ph.D. Thesis, K.U. Leuven

Hoffleit, D., 1976, Inf. Bull. Variable Stars 1131

Iglesias, C.A., Rogers, F.J., 1991a, ApJ 371, 408

Iglesias, C.A., Rogers, F.J., 1991b, ApJ 371, L73

Irwin, J.B., 1952, ApJ 116, 211

Jerzykiewicz, M., 1986, Acta Astron. 36, 147

Jerzykiewicz, M., Sterken, C., 1984, MNRAS 211, 297

Jiang Shi-yang, 1985, Chin. Astron. Astrophys. 9, 191

Kaper, L., Henrichs, H., Mathias, P., 1992, *La Lettre de l'OHP*, 8, 3

Kubiak, M., 1990, Acta Astron. 40, 297

Lloyd, C., Pike, C.D., 1984, Observatory, 104, 9

Mathias, P., Gillet, D., Crowe, R., 1991, A & A 252, 245

Mathias, P., Gillet, D., Crowe, R., 1992, A & A 257, 681

Moskalik, P., Dziembowski, W.A., 1992, A & A 256, L5

Odell A.P., 1984, PASP 96, 657

Pigulski A., 1992a, A & A 261, 203

Pigulski A., 1992b, Atlanta Conf In Press

Pigulski A., Boratyn D., 1992 A & A 253, 178

Sterken, C., 1992 submitted to A & A

Sterken, C., 1988, In *Physics of Luminous Blue Variables*, ed. K. Davidson, A.F.J. Moffat and H.J.G.L.M. Lamers, Kluwer Academic Publishers Dordrecht, 59

Sterken, C., Jerzykiewicz, M., 1988, MNRAS 235, 565

Sterken, C., Jerzykiewicz, M., 1990, In *Confrontation between Stellar Pulsation and Evolution*, eds. C. Cacciari & G. Clementini, ASP Conf. Ser. 11, 236

Sterken C., Young A., Furenlid I., 1987, A & A 177, 150

Szabados, L., 1977, *Konkoly Obs. Comm* 70

Szabados, L., 1980, *Konkoly Obs. Comm* 76

Szabados, L., 1981, *Konkoly Obs. Comm* 77

Waelkens, C., Van den Abeele, K., Van Winckel, H., 1991, A & A 251, 69

Slowly Pulsating B Stars

C. Waelkens

Instituut voor Sterrenkunde KU Leuven, Belgium

Abstract

We discuss two new slowly pulsating B stars for which it can be convincingly shown that they are multiperiodic variables. The group now contains ten confirmed and likely multiperiodic mid-B stars with periods between 0.7 and 4.4 days. The multiperiodicity and the length of the periods point to g-mode pulsations. The very high order of the modes implies that the modes are excited in the core or in the very outer layers of the star. A recently proposed mechanism, in which both the inner and outer layers intervene, is discussed.

1. Introduction

The slowly pulsating B stars were introduced as a group of B-type variable stars by Waelkens and Rufener (1985) and by Waelkens (1992). They are characterized by spectral types in the range B3-B8 (masses between 3 and 6 solar masses), multiple periods of the order of days, and photometric amplitudes which increase with decreasing wavelength. Probably the line-profile variable stars 53 Persei and ι Herculis also belong to the class. However, line-profile variabilty is rather common among B-type stars, and is not always linked to periodic photometric variability; it may thus not be sufficient for the definition of a class of variables. It is for this reason that we prefer to speak about "slowly pulsating B stars" rather than "53 Persei stars".

2. Multiperiodicity of the new slowly pulsating B stars

The periods observed for the slowly pulsating B stars are of the order of twenty times that of the fundamental radial mode. If only one period were observed in these stars, the case for pulsation would appear fairly unconvincing; rotation would be a more promising explanation for the variations. Observations gathered over several years, however, clearly indicate multiple periods in all well observed stars.

Data gathered during a single observing season most often do not offer sufficient proof for multiperiodicity, because only a few beat periods or even a fraction of one can be observed during a single season. We have now found two examples for which the frequency spacing is large enough so as to enable us to distinguish different frequencies in the data of individual years. The clearest case is HD 45284, for which three well resolved frequencies with similar power are found; the periods amount to 0.664 days, 0.807 days, and 0.887 days, respectively. A somewhat more complex star is HD 34798 in which at least four periods in the range between 1.02 and 1.48 days occur. Both stars will be discussed extensively in a forthcoming paper.

3. Discussion of the excitation mechanism

High-order g-modes in B-type stars attain appreciable amplitudes only in the core and in the very outer layers; in the intermediate layers the displacement is very small. The cause of the pulsation of these stars must then reside in the deep interior or the outmost layers, or in both. The outer layers contain only a small amount of mass, so that oscillations that are only excited there would be damped in the core.

Degryse et al. (1992) have carried out a vibrational stability analysis of an evolved five-solar-mass model, and found overstability, due to the combined effects of the ϵ-mechanism in the core and the HeII partial ionization in the outer layers. Some contribution to the excitation was also found in the μ-gradient zone, which may prove spurious if a better treatment of semi-convection is applied. Nevertheless, it is promising that actual destabilization is found in the layers were the oscillations in these stars attain there largest amplitudes. It is also found by Degryse et al. (1992) that the modes are less unstable (though not stable) for smaller radial orders. The instability occurs only for a limited range of models, with masses near five solar mass and near the end of core-hydrogen burning, in reasonable agreement with observations.

Earlier attempts to detect overstability in B-type stars from the ϵ-mechanism (e.g. Osaki 1976) all have failed. We note that these attempts were directed at understanding the β Cephei stars, which imply more massive models and modes of lower radial order. In such hot stars the HeII partial ionization zone contributes less to the excitation of g-modes. Moreover, low-order modes have appreciable amplitudes throughout the star, and so also in the zones where damping is important. It is the particular aspect of high-order modes to be practically confined to a few layers in the star which may allow excitation of these modes in our stars.

The main weakness of the analysis by Degryse et al. may reside in the treatment of the μ-gradient zone and semiconvection. To our knowledge a stability analysis of high-order g-modes in B-type stars has never been performed before. We would welcome an independent analysis, with a more sophisticated code, to check the results by Degryse et al.

References:

Degryse, K., Noels, A., Gabriel, M., Waelkens, C., Smeyers, P., 1992, Astron. Astrophys., in press.
Osaki, Y., 1976, Publ. Astron. Soc. Japan 28, 105.
Waelkens, C., 1992, Astron. Astrophys. 246, 453.
Waelkens, C., Rufener, F., 1985, Astron. Astrophys. 152, 6.

Line Profile Variations of Rotating Pulsating Stars

C. Aerts & M. De Pauw

Instituut voor Sterrenkunde, KU Leuven, Belgium.

Line-profile variations (LPVs) are often seen in early-type stars. They were first detected in the β Cephei stars, where they are due to radial as well as to nonradial pulsations (NRP). It is still a matter of debate whether NRP is also the cause of LPVs in broad-lined B- and Be stars.

The study of LPVs of nonradially pulsating stars is often considered to be especially interesting in the case of rapid rotators, because in such stars the line profiles may offer a Doppler image of the stellar surface. Almost all investigators conclude that only sectorial modes with high ℓ-values can properly account for the observations of LPVs in broad-lined stars; on the other hand, high-degree modes are almost never seen in slowly rotating stars.

We point out here that the expression for the velocity field which is usually adopted, is not valid for rapid rotators. The Coriolis force implies that an individual pulsation mode should not be expressed in terms of a single spherical harmonic. We derived an expression for the components of the Lagrangian displacement field $\vec{\xi}$ which is exact to the first order in Ω/ω, where Ω is the rotation frequency and ω is the pulsation frequency. The approximation is valid in the case of p- or low-order g-modes. Two toroidal correction terms, one with degree $\ell-1$ and one with degree $\ell+1$, and one spheroidal correction term with degree ℓ, all having azimuthal number m, are found for $\vec{\xi}$ (where ℓ and m are the degree and the azimuthal number of the mode in the non-rotating case).

From this result, we determined the velocity of a point on the stellar surface in the direction of the observer and studied the effect of the correction terms due to the Coriolis force on line profile variations. Theoretical profiles for $\ell=2$ and $\ell=4$ modes were calculated in the case of $\Omega/\omega = 1, 20, 50\%$ for an equatorial rotation velocity of $v_\Omega = 200$ km/s ($i = 90°$) and a pulsation amplitude $v_p = 20$ km/s. In the case of axisymmetric modes, the Coriolis force has little effect on the line profiles. However, for tesseral and sectorial modes the line profiles are drastically affected by the correction terms, if Ω/ω is of the order of 20% or larger, *i.e.* in the conditions which are met in broad-lined B stars and Be stars. It appears that "moving bumps" can then also occur for modes with low ℓ-values (Aerts, in preparation).

The correction terms are so huge that they cannot be ignored if the aim is to obtain a reliable mode determination from line-profile fitting. In stars where the pulsation period is about equal to the rotation period it is probable that even the formalism presented here is not sufficient and that higher-order effects are important.

Did Beta Canis Majoris Quit Pulsating?

Andrew P. Odell[1,2,3], Robert D. Watson[2]

[1] *University of Canterbury, Christchurch, New Zealand,*
[2] *University of Tasmania, Hobart, Tasmania, Australia,*
[3] *On sabbatical from Northern Arizona University*

Abstract

The star β CMa, prototype of the β CMa variables, is known to have been a double mode pulsator since the beginning of the century, with amplitudes of 12 and 7 km/sec for the 6.00 and 6.03 hour oscillations (Shobbrook, 1973). Recently, Dziembowski, Moskalik, and Pamyatnykh (1992) have produced models of B stars using OPAL opacities which have many radial and non-radial modes unstable. As only a few modes are observed in these stars, these authors suggest that modes might change amplitude in such a way that modes can appear, disappear, or be replaced by other modes. Based on unsubstantiated rumour that β CMa had, like α Vir, quit pulsating (Pesnell, priv. comm.), we undertook observations with the Mt. John University Observatory (New Zealand) echelle spectrograph and CCD camera on the 1-m McLelland telescope. We found that β CMa *is* still pulsating, but with reduced amplitude and substantial period change compared with earlier data.

In particular, quadratic ephemerides fit to all known times of spectroscopic maxima agree with our times of maxima [P1: HJD 2448611.8436; P2: HJD 2448611.8278] to better than two minutes, but an earlier linear ephemeris by Chappelier (1985) disagrees by over two hours. Our data is best fit with amplitudes of about 7 km/sec for both modes–this gives a nominal beat amplitude of nearly zero at the 49-day beat minimum (at a time when the linear ephemeris would have predicted a maximum amplitude). This may be the cause of the rumour mentioned above. Additional support for our conclusion of continuing pulsation comes from the fact that obvious line profile changes occur on the pulsation time scale.

References:

Chappelier, E., 1985, Astron. Astrophys. **147**, 135.
Dziembowski, W. A., Moskalik, P., and Pamyatnykh, A. A., 1992, in proceedings of IAU Coll 137, *Inside the Stars*, Vienna, Austria, April 1992, in press.
Shobbrook, R. R., 1973, M.N.R.A.S. **161**, 257.

β Cephei Pulsation Anomalies: Potential New Windows into the Instabilities and Evolution of Early B Stars

B.A. Goldberg[1], R.S. Polidan[2], R.A. Crowe[3], G.C.L. Aikman[4], R.J. Bambery[1], J.T. Gathright[3], G.J. Odgers[4]

[1]*Jet Propulsion Laboratory/Caltech, Pasadena, CA, U.S.A.,* [2]*NASA Goddard Spaceflight Center, Greenbelt, MD, U.S.A.,* [3]*University of Hawaii at Hilo, HI, U.S.A.,* [4]*Dominion Astrophysical Observatory, Victoria, BC, Canada*

Abstract

We have obtained *Voyager* Ultraviolet Spectrometer (UVS) measurements of well-known β Cephei stars, which now total more than 1500 hours (> 300 pulsation cycles!) and which constitute the most comprehensive coherent data set that can address fundamental pulsation properties of a significant cross-section of the group. The extended measurement sequences for individual stars, which cover many successive pulsation cycles at wavelengths where pulsation amplitudes reach a maximum, can provide more comprehensive tests of pulsation stability than any ground-based data. During 1990-91, we acquired more than 100 hours of ground-based high-resolution spectroscopic observations and UBV photometric observations, simultaneous and near-simultaneous with the UVS data set. Analysis has been initiated at NASA's Goddard Spaceflight Center (NASA/GSFC), the University of Hawaii at Hilo (UH Hilo), and the Dominion Astrophysical Observatory (DAO).

1. Description of Voyager UVS and Ground-Based Data

The two *Voyager* spacecraft have objective grating spectrometers with wavelength coverage of 500-1700 Å. Spectral resolutions of approximately 18 Å for point sources and 30 Å for diffuse sources are achieved. Instrumental sensitivity is optimized for the 800-1200 Å region. Typical limiting fluxes at 1050 Å in the far-UV are 1.0×10^{-12} ergs cm^{-2} sec^{-1} Å^{-1} for *Voyager 1* (5.0×10^{-13} for *Voyager 2*). In-flight performance of the UV spectrometers has been reviewed by Broadfoot *et al.* (1981, *Journal of Geophysical Research*, **86**, p. 8259). During 1990-91, we obtained UVS data on the stars BW Vulpeculae, β Cephei, ν Eridani, δ Ceti and 12 Lacertae.

Measurements at the DAO were made using CCD sensors on both the Cassegrain spectrograph of the 1.8-meter telescope and the coudé spectrograph of the 1.2-meter telescope. Spectral resolutions were approximately 0.3 Å and 0.1 Å respectively; time resolution was typically in the range of 2-4 minutes with a S/N exceeding 50:1. The lines of Si III, He I, Mg II, O II and C III in the wavelength range 4450-4600 Å were among those observed. For observations with the UH 2.2-meter telescope on Mauna Kea, a CCD sensor was used on the coudé spectrograph. The spectral resolution was 0.2 Å and the time resolution was 10-15 minutes with a S/N in the range 100-150. In addition, simultaneous UBV photometric observations were obtained on the UH 0.6-meter "Air Force" telescope during May 1991. The ground-based program has been an outstanding success, having produced almost 100 hours of simultaneous data on the stars BW Vul, β Cep, ν Eri, δ Cet and 12 Lac.

2. Scientific Objectives and Analysis

The scientific objectives which can be addressed directly with the available data are as follows: **(1)** determine the origin and significance of an "instability anomaly" present in the far-UV light curves of β Cep and ν Eri; **(2)** determine cycle-to-cycle pulsational stability from *Voyager* UVS data; **(3)** determine the role of atmospheric shock waves; investigate the relationship of shock waves to proposed pulsation mechanisms and to the range in behaviour within the group; **(4)** determine whether there exist unique relationships between pulsation amplitude and line profile variations, light and velocity curve shapes, etc.; **(5)** institute an analysis to discriminate radial from non-radial pulsation modes using temperatures derived from the *Voyager* UVS data; **(6)** continue the analysis of the long-term (decades) pulsational stability of stars such as BW Vul and assess the evolutionary significance. The more general program goals are as follows: **(a)** further define the evolutionary status of β Cephei stars; **(b)** explain the wide range in pulsation amplitude within the β Cephei group; **(c)** identify, if possible, the pulsation mechanism; **(d)** determine the significance of β Cephei instability within the context of the variability of early B stars.

Data analysis is currently in progress at NASA/GSFC, UH Hilo, and DAO. Examples of *Voyager* UVS light curves are given below. Observations have been folded into one cycle. The "instability anomaly", shown for β Cep in Fig. 2, is manifested in the form of unexplained fluctuations in intensity near the phase of maximum UV flux. In Fig. 1, UVS data for β Cep obtained during May 1991 show that the fluctuations near maximum are still present a decade later. Moreover, the magnitude of the fluctuations during 1991 appears to be comparable to that found in the 1981 data.

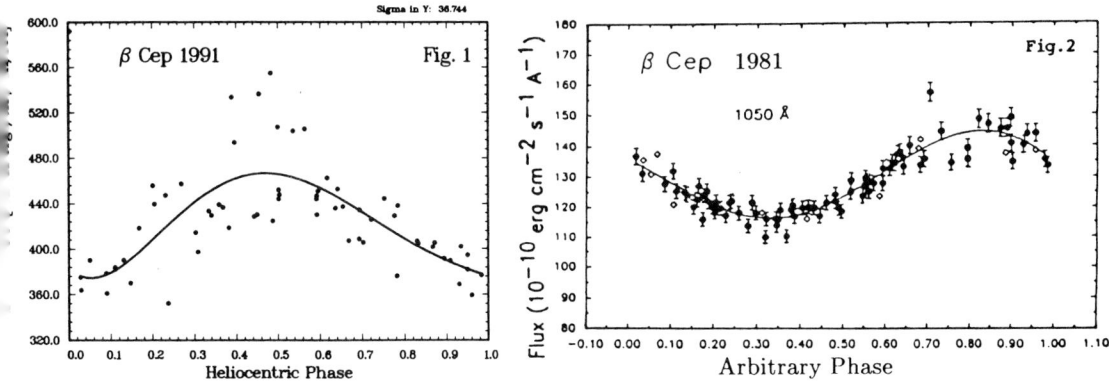

Fig. 1: *Voyager* UVS observations of β Cep (P = 4h 34m) taken over a 5-day observing period in 1991. **Fig. 2:** *Voyager* UVS observations of β Cep taken over a 26-hour observing period in 1981 (Rautenkrantz and Polidan, in prep.); note the "instability anomaly" near maximum light.

β Cep: a Magnetic Be Star?

H.F. Henrichs[1], F. Bauer[2], G.M. Hill[3], L. Kaper[1], J.S. Nichols-Bohlin[4], P.M. Veen[1]

[1]*Astronomical Institute Univ. of Amsterdam*, [2]*Universitäts-Sternwarte, München*,
[3]*Dep. of Astronomy, Univ. of Western Ontario*, [4]*CSC/NASA, Greenbelt*

Abstract

Besides the well-known pulsation period of 4^h34^m, the B1 IV star β Cep has a very significant period of 6 or possibly 12 days in the equivalent width of the ultraviolet resonance lines. This was discovered by Fishel and Sparks (1972) with the OAO-2 satellite, and later confirmed with IUE data. Until now, no explanation has been put forward for this period.

We propose that the UV periodicity arises from a 12 day rotational period of the star and that the stellar wind is modulated by an oblique dipolar magnetic field at the surface.

Support for this hypothesis is given by the striking similarity between the UV-line behavior of β Cep and of known rotating magnetic B stars, for example the B2 V helium-strong star HD 184927 (Barker et al. 1982), and by the measured magnetic field strength of B $\pm$ σ = (810 $\pm$ 170) G for β Cep itself by Rudy and Kemp (1978). A rotational period of 12 days corresponds well with an adopted radius between 6 and 10 $R_\odot$, given the reported values of 20 – 43 km/s for $v \sin i$.

To verify our hypothesis we carried out new magnetic field measurements simultaneously with UV spectroscopy. We confirm the 12 day UV period in the equivalent width of the stellar wind lines of C IV, Si III, Si IV and N V (see Figs. 1 and 2), but find a lower and likely variable field strength (Fig. 3), which is consistent with a 12 day period, but not conclusive.

It remains puzzling why our new magnetic field measurements show a lower field than in 1978. It is interesting to recall the recent discovery (Mathias et al. 1991) that β Cep has entered a new Be phase (July 1990), when Hα turned into emission (see Fig. 4). This opens the suggestion that the lower magnetic field is related to the emission phase. Because the UV period is still the same, the field must still be strong enough to modulate the wind. A possibly higher equivalent width of the 1991 UV data with respect to the 1979 data might also be related to this transition, but this needs to be confirmed.

The star β Cep appears to be the only star in its class which shows this wind variability and in this respect β Cep is an exceptional β Cep star.

References:

Barker, P., Brown D., Bolton, C. and Landstreet, J., 1982, in *Advances in UV Astronomy*, NASA CP-2238, Y. Kondo, J.M. Mead and R.D. Chapman (Eds.), p. 589

Fishel, D., Sparks, W.M., 1972, in *The Scientific Results From the Orbiting Astronomical Observatory (OAO-2)*, NASA SP-310, p. 475

Kaper, L., Henrichs, H.F. and Mathias, Ph., 1992, OHP Newsletter, Feb.

Mathias, P., Gillet, D. and Kaper, L., 1991, in *Proc. ESO Workshop "Nature and Diagnostics of OB star Variability"*, D. Baade (Ed.)

Rudy, R.J. and Kemp, J.C., 1978, MNRAS **183**, 595

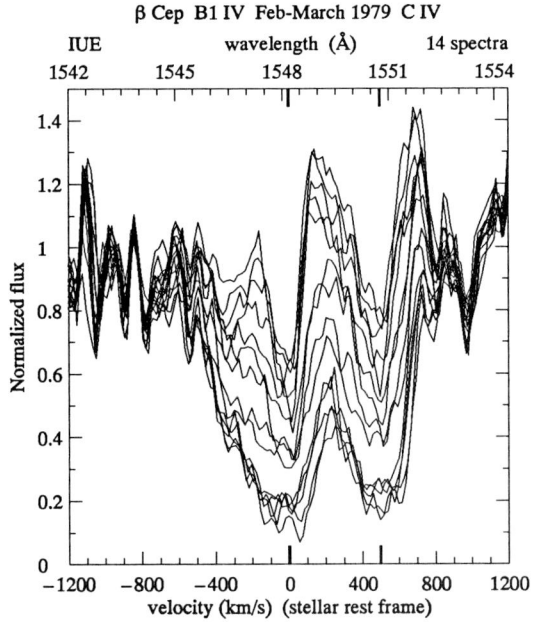

Figure 1 Significant variations in sample spectra of the ultraviolet C IV stellar wind line in β Cep in 1979 (IUE archival data). New data taken in 1991 are very similar.

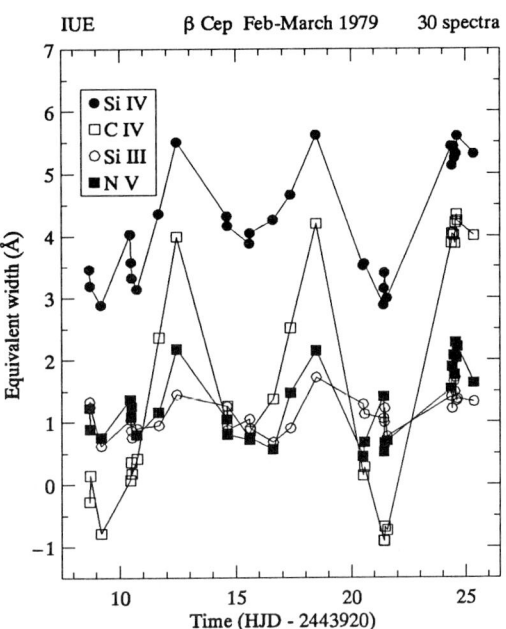

Figure 2 Equivalent widths of C IV, Si III, Si IV and N V of the 1979 data, showing a 12, rather than 6 day period. The new 1991 data might have a larger equivalent width.

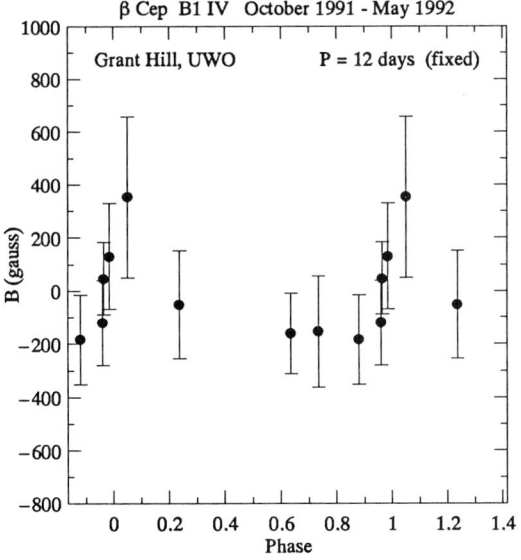

Figure 3 Longitudinal magnetic field measurements over 7 months. Each datapoint is an average of 8–16 measurements. The data are folded with a 12 day period.

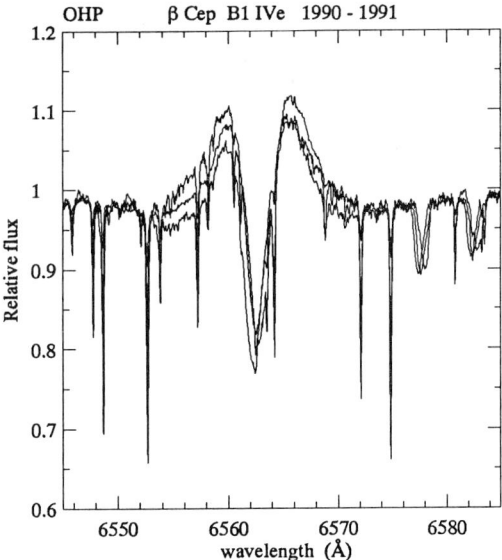

Figure 4 (From Kaper et al. 1992) Hα spectra of β Cep on 1990 July 18, 1991 Feb. 5 and Oct. 28 (left peak decreasing, respectively). Hα is usually observed in absorption.

An Extraordinary Early-Type Eclipsing Binary*

L.A. Balona[1], J. Cuypers[2]

[1]*South African Astronomical Observatory, South Africa,*
[2]*Koninklijke Sterrenwacht van België, Belgium*

HR 2680 (B5V) was used as a comparison star in a multi-site (ESO & SAAO) campaign organised in 1988 to observe Be stars. We found that the star is an eclipsing binary with a period of 8.1 days. The eclipse is partial with a depth of 0.18 mag. Radial velocity observations confirmed the period.

A light variation with an amplitude of as much as 0.03 mag was seen outside the eclipse (Fig. 1). This variation can be interpreted as two oscillations with approximate periods of 1.19 and 1.28 days. Further photometric observations were obtained in 1989, 1990 and 1991 at SAAO. The multiperiodicity was confirmed, but the periods were not constant from season to season.

We suspect that the star is a pulsator of the 53 Per class of line-profile variables. Being an eclipsing binary, this unique system is of potentially great importance as a test bed for stellar dynamics and nonradial pulsations.

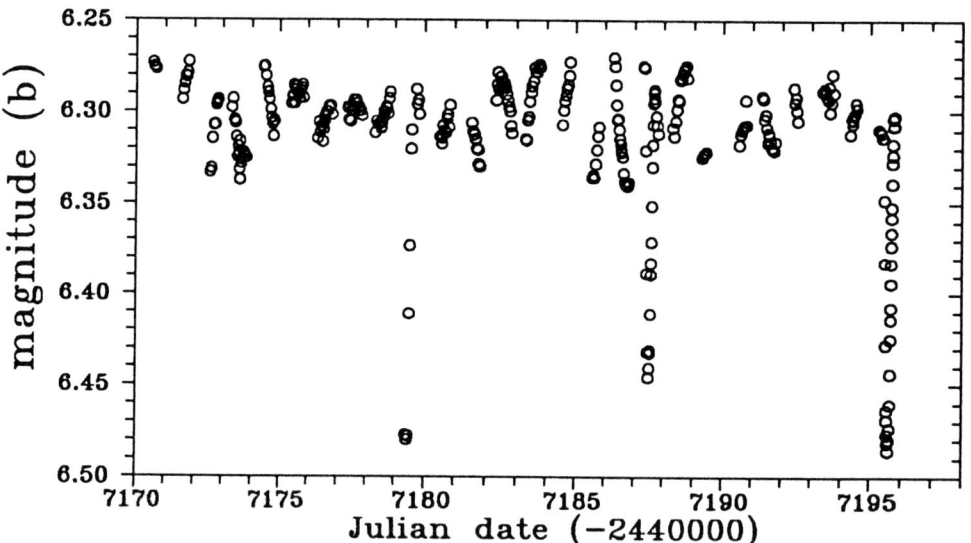

Figure 1. Photometric variability of HR 2680 during January 1988 showing the eclipses and variation between the eclipses.

* Based on observations obtained at the European Southern Observatory and the South African Astronomical Observatory.
The full text of this paper has been submitted to Monthly Notices of the Royal Astronomical Society.

HR 8762: Low-amplitude Photometric Variation in a Pre-shell Phase

S. González–Bedolla[1], J. P. Sareyan[2],
J. Chauville[3], P. J. Morel[2], M. Alvarez[1]

[1] *Instituto de Astronomiá, UNAM, Apdo. Postal 70-264, México, D.F. México*
[2] *Observatoire de la Côte d'Azure, B.P. 139, F-06003 Nice Cedex, France*
[3] *Observatoire de Meudon, F-92195, Meudon Cedex, France*

Longitude-coordinated high-precision photometry has been obtained a few weeks before the beginning of a strong Be and shell phase (1988) in HR 8762 (o And). The star showed variations of a few millimagnitudes in amplitude; i.e., just over the detection threshold. The classical 1.57-day double-wave period is still detected, showing that it probably never fades out completely, whatever the phase of the star. These variations can be interpreted as normal photospheric activity in a regular rotating B star. Although the variations of HR 8762 during our campaign were quite small, we could detect their amplitudes at a level of a few mmag.

We still have to check whether they are still "in phase" with previous photometric observations: if this is the case, it means that in the "spot" hypothesis, these (superficial?) features would remain in the same position on the photosphere, changing only in surface area and/or brightness with the star's activity. In this pre-shell phase, these variations were much smaller than their equivalent, observed two shell phases before by Guerrero and Mantegazza (1979), and they are probably the counterpart of "normal" photospheric activity in the B star HR 8762.

We still do not have any idea of the time constants involved in the beginning of the shell phases; i.e., how the 1.6-day period amplitude increases. Is it a question of a few days, hours, or even minutes? Is it progressive, or does it start abruptly? Due to the historical lack of observations around the beginning of shell phases in o And, we can set only an upper limit of 2-3 weeks, although in κ Dra (Sareyan et al., in preparation), we think the "oscillation" could start in a few days. If we accept the "spot" and rotation model, this means that the inhomogeneities appear in a few rotations, last for several months, and disappear (progressively? see λ Eri) in a few weeks.

Of course, such behaviour must be tested in each case, as probably all Be variables do not behave in the same manner around the start and end of their shell phases. Continuous monitoring is needed at these phases, in order to determine the time constants involved in the increase and decrease of the photometric amplitude.

References:

Guerrero, G., Mantegazza, L., 1979, A&A 36, 471

Time-Series Spectroscopy of ζ Ophiuchi

A.H.N. Reid[1], C.T. Bolton, R.A. Crowe,
M.S. Fieldus, A.W. Fullerton, D.R.Gies, I.D. Howarth,
D. McDavid, R.K. Prinja, & K.C. Smith.

[1]*Dept. of Physics & Astronomy, University College London, Gower St., London WC1E 6BT, England, U. K.*

Abstract

We have undertaken a multi-site, multi-wavelength observing campaign on the archetypal O stars ζ Puppis (O4 I(n)f) and ζ Ophiuchi (O9.5 V). Both stars are well known for the strength of their line profile variations (lpv's), and represent extremes of O spectral type and luminosity class. UV time-series spectroscopy of ζ Pup and ζ Oph is described by Prinja *et al.* (Ap.J. 1992, **390**, 266), and Howarth *et al.* (Ap.J. 1992, *submitted*) respectively. The optical spectroscopic results of ζ Oph are reported by Reid *et al.* (1992, ApJ *submitted*), of which some of the principal results are given here.

During late April, and early May, 1989, we obtained high-resolution, high signal-to-noise optical spectra of the late O-type, rapid rotator ζ Oph. Time-series analysis, using the CLEAN algorithm, has shown that the characteristic lpv seen in He I λ4471Å, Si III $\lambda\lambda$4552, 4567, 4575Å, and Mg II λ4481Å can be satisfactorily represented as a set of 4 sinusoids. No substantial variation is observed in He II λ4541, or N III λ4517Å. We attribute this behaviour to a combination of equatorial gravity-darkening and a latitudinally-confined origin for the lpv.

The phase changes over the line profiles indicate repetitive patterns of axial symmetry, rotating prograde in the co-rotating frame of the star. The periods are 3.339 hours ($-m = 4$), 2.435 hours ($-m = 5$ or $-m = 6$), 1.859 hours ($-m = 9 \pm 1$), and either 1.366 hours or 1.292 hours ($-m = 11 \pm 1$); $-m$ represents the spatial frequency around the stellar equator. The first three periods confirm those found at earlier epochs, and we conclude that some lpv characteristics are reproduced over at least a 2-year interval.

Since no commensurate superperiod ($|m|$P) exists, and since the super-periods are less than our estimated minimum rotation period for ζ Oph (> 18 hours), we reject a rotational modulation origin for the lpv and conclude that the star is undergoing multi-mode, sectorial, non-radial pulsations.

Pulsation and Mass Loss in LPVs

George H. Bowen

*Department of Physics & Astronomy,
Iowa State University, Ames, Iowa 50011 U.S.A.*

Abstract

The large-amplitude pulsation of long-period variables, together with a number of other interacting processes and phenomena, causes a rich variety of effects on the structure and behavior of the stars. Outflowing winds result, causing extensive mass loss, with profound consequences for stellar evolution. The present status of modeling calculations for LPVs will be discussed first, with various examples. Emphasis will be given to the great importance of complex, nonlinear, time-dependent interactions between things such as the waves and atmospheric shocks that result from pulsation; non-LTE radiative transfer; non-equilibrium chemistry; the growth, changing optical properties, and dynamics of grains; and radiation pressure on both grains and molecules. I will then survey the developing implications and insights from new results and from work now in progress. Some of these concern the structure and the behavior of individual stars (e.g. determination of the pulsation mode and limiting amplitude; properties of more massive stars); some relate to the evolution of individual stars (e.g. evolution of the wind and the mass loss rate; the wind and circumstellar region during helium shell flashes; effects of the star's metallicity); and some relate to the evolution of populations of stars (e.g. the white dwarf mass distribution). All of these, and many more, offer new perspectives and new understanding concerning the character of LPVs and their role in stellar evolution.

Long Period Variables in the Magellanic Clouds and the Galaxy

S.M.G. Hughes

Palomar Observatory 105-24, California Institute of Technology, Pasadena CA 91125, USA

Abstract

New results (∼last two years) on mainly observational properties of Long Period Variables (LPVs) in the Magellanic Clouds and the Galaxy are reviewed. These properties include the effects of metallicity variations on their mass loss rates, the use of AGB LPVs to map the stellar distributions of the Galactic disk and bulge, and using detailed observations of nearby Miras to investigate their structure and to obtain new parallax distances, with implications for the pulsation mode of Miras.

1. Introduction

It is a tradition to remark on the supposed confusion about what to call these things. The term LPV is simply a general classification of any red star with a long period of pulsation ($\gtrsim 80$ days). The color of the star isn't critical as a classification criterion, but since $P \propto R^2/M$ (for the fundamental mode), only red stars are cool enough (and hence large enough) to pulsate at such long periods. In terms of evolutionary status, LPVs have been mainly identified to either be members of the asymptotic giant branch (AGB, which are shell-burning (helium and/or hydrogen) stars, around a mostly degenerate CO core), or more massive red supergiants (RSG, which are massive enough to burn helium, carbon, oxygen etc in their cores). In the past, most studies of LPVs have been of the Miras, which are LPVs with large ($\Delta V > 2.5$ mag) amplitudes of pulsation, and hence are generally the easiest to detect (they also have a well-established PL relation). LPVs with $\Delta V < 2.5$ mag are termed semi-regular variables (SRVs), which can be a misnomer as their periods are often just as stable as the Miras' (which, it should be pointed out, are known to vary). Because Miras are almost certainly all members of the upper AGB (eg. Hughes & Wood 1990), they have tended to become the most commonly studied AGB stars. But our knowledge of the AGB is now being extended through studies of non-Mira LPVs both at the end of the AGB (variable OH/IR stars; eg Wood et al 1992) and near the beginning (SRVs; eg Willems & de Jong 1988 and Kerschbaum & Hron 1992; Jura & Kleinmann 1992b). Since a large proportion of the AGB variables are likely to be non-Miras — Hughes & Wood (1990) found that up to half the likely AGB LPVs in the LMC are the so-called SRa type, while in the Galaxy, Jura & Kleinmann (1992b) found that ∼40% of the Galactic AGB LPVs are SRVs (with ∼30% in higher overtones of pulsation) — and most, or even all, red giants are likely to be variable (eg Eggen 1992), an attempt will be made to refer to all LPVs by their evolutionary status (ie RSG LPVs, AGB LPVs, and one possible RGB LPV).

This review is mainly of results that have appeared since the last variable star conference (Wood 1990b; Whitelock 1990), with a personal bias towards the observational aspect of LPVs in the Galaxy and the Magellanic Clouds. The PL relations of both the AGB and RSG LPVs continue to be of great interest, and are being used to probe the distributions of AGB LPVs (and the stellar populations they represent) in both the Galaxy's disk (thick and thin), and bulge, and also to potentially derive distances to galaxies at least as far as Virgo. Closer to home, high resolution studies of the very nearest LPVs are establishing new absolute distances and revealing detailed morphology of their atmospheres, and providing possible evidence to suggest that Mira may not be in the fundamental mode of pulsation.

2. The Magellanic Clouds

Wood, Moore & Hughes (1991) have reviewed work done on LPVs in the Magellanic Clouds prior to July 1990. Since then, the only new LPVs to have been reported are a sample of variable OH/IR stars in the LMC (Wood et al 1992), which are the most difficult to find, as they are in their final stages of high mass loss and so are hidden (optically) behind thick circumstellar envelopes. In terms of AGB evolution, the OH/IR stars are important as they are the link between the upper AGB and the planetary nebulae. Wood et al (1992) searched for OH maser and strong IR emission in IRAS-selected candidates in both the LMC and SMC, finding 6 point sources with OH, and 19 with strong IR, all in the LMC. Many were found to be variable in the IR, with periods ranging from 930 to 1390 days. The stellar wind velocities were only 0.6 of those found for comparable Galactic OH/IR stars, which is consistent with the LMC's lower metallicity. Mass loss rates of $\sim 10^{-5}$ $M_\odot$ yr^{-1} were derived, based on the wind velocities and IRAS and OH maser fluxes. Evidence for a possible metallicity-mass loss relation (for O-rich AGB LPVs) was found in the existence of two optically visible LPVs in the SMC with $P > 1000$ days and $M_{bol} < -6$, which show no sign of large mass loss (ie no IRAS flux), in contrast to AGB LPVs of similar periods in the LMC and the Galaxy. Such a relation would imply that low metallicity stars would end up with higher white dwarf masses, and more will explode as supernovae. However, there is at least one SMC LPV, with a period of 800 days, that does have a dusty shell and therefore a high mass loss rate, but its luminosity is consistent with it being a carbon star. Therefore, it may be that low metallicity AGB stars can in fact have high mass loss rates if they turn into C stars. The Wood et al IRAS sources, when plotted on the same (M_{bol}, $\log P$) plane as the other AGB LPVs in the LMC (Figure 1a), clearly shows all but one of them to be an extension of the optical AGB LPV sequence.

The non-nuclear-burning core of the AGB stars means that there exists a direct relation between their luminosities and their core masses, so that there is a natural upper luminosity limit, at $M_{bol} \sim -7.1$ (corresponding to the Chandrasekhar limiting mass for degenerate matter of $M \sim 1.4 M_\odot$). The inexactness of this luminosity limit is mainly due to uncertainties about mass loss rates, and while new models by Boothroyd & Sackmann (1992), which include more realistic molecular opacities and the effects of any hot bottom burning, indicate the core mass-luminosity relation may not be applicable at high mass (which may explain the group of higher luminosity M stars at $\log P \sim 2.8$), their model luminosities still do not go beyond the classic AGB luminosity limit. In contrast, the models by Blöcker & Schönberner (1991), which also incorporate hot bottom burning, predict AGB luminosities well above this limit. However, this is not supported by the observations of Wood et al (Figure 1a) who find that the only OH/IR star above the AGB luminosity limit is a probable member of the LMC cluster NGC 1984, which would imply it has a mass of 15-20 $M_\odot$, and is therefore a supergiant (which is also consistent with its K amplitude, which is very low for an OH/IR variable, at just 0.2 mag).

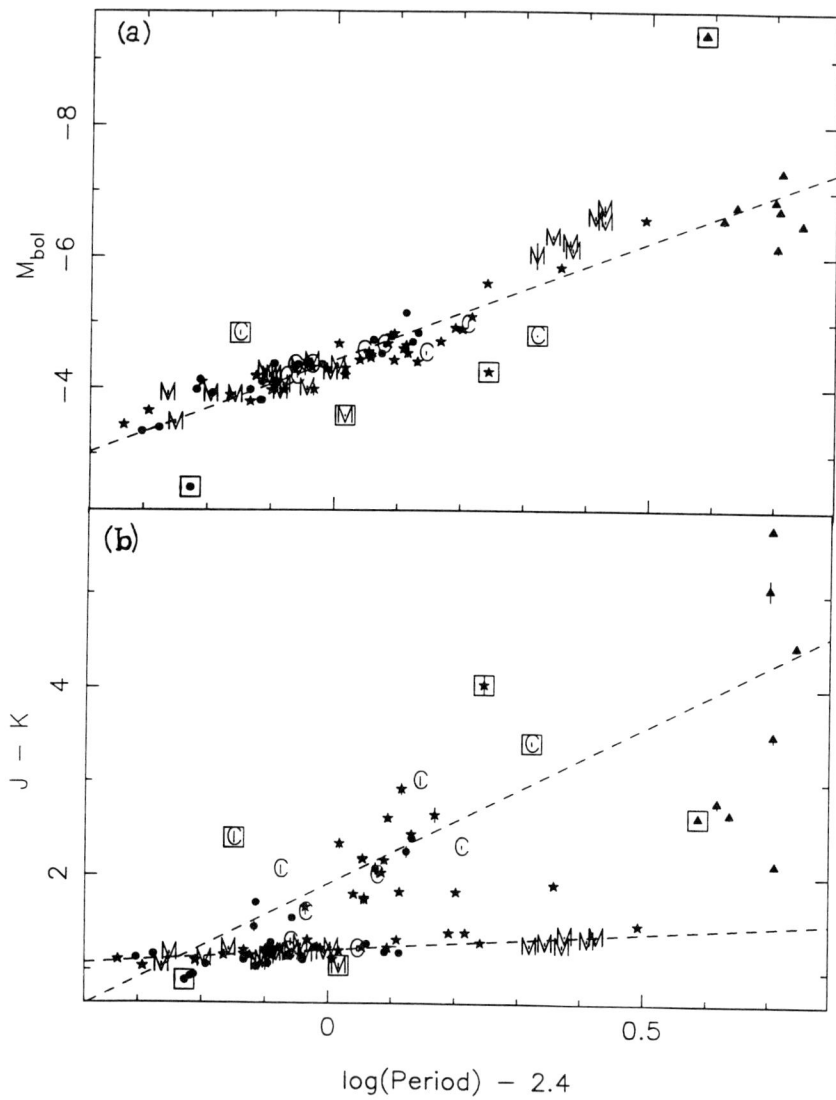

Figure 1 M_{bol} and $J - K$ vs log$P$ diagrams for 102 AGB LPVs (and one RSG LPV) in the LMC. The symbols have the following meaning: Objects with spectral types indicated by M or C, otherwise stars are from Glass et al (1990), triangles are OH/IR stars from Wood et al (1992) and circles are from Wood & Hughes (1992). Outliers not used in the fits are within boxes. Error bars are the standard error on the mean.

The known optical AGB LPVs have continued to be observed by Wood & Hughes (1992), who have been monitoring 60 of the Hughes (1989) LPVs in the LMC, obtaining multi-epoch J, K photometry, with 48 LPVs now having 5 or more epochs. Combining these data with the Glass et al (1990) data used in a similar study by Feast et al (1989), the least squares solution to the $(K, \log P)$ PL relation, for $P < 420$ days, is

$K = 11.10(\pm 0.02) - 3.75(\pm 0.14)[\log P - 2.4]$, $\sigma = 0.13$, for $N = 79$.

Because most of the flux of these optical AGB LPVs is emitted in K, it is not surprising that the slope of this relation is not all that different from that obtained from the bolometric PL relation, which for all AGB LPVs, including the Wood et al (1992) OH/IR variables is found to be

$M_{bol} = -4.42(\pm 0.02) - 3.61(\pm 0.18)[\log P - 2.4]$, $\sigma = 0.27$, for $N = 96$,

where the indicated outliers in Figure 1a are excluded.

The 'faint' outliers in Figure 1a may possibly be semi-obscured stars with high mass loss rates, although a signature of high mass loss is a large K amplitude (Whitelock 1990; Whitelock et al 1991), and the two with the shortest periods have low K amplitudes (~ 0.1 mag). One alternative, that they are background objects is unlikely, for the M star at least, as it has a measured radial velocity (Hughes et al 1991) consistent with LMC membership. Another possibility is that the shorter period objects are RGB stars.

The one 'bright' AGB outlier in Figure 1 is a carbon star, also with an LMC radial velocity. It is one of the shortest period C stars, and may well be in a higher overtone of pulsation, given that its amplitude ($\Delta K = 0.3$ mag) is also low (Wood & Bessell 1985). If it were in the first overtone, then its period of 178 days would imply a fundamental at ~ 356, not too different from a period of 363 predicted by the PL relation from its K mag. Alternatively, as the amplitude is so low, its also possible it has a poorly determined period.

The $(J - K, \log P)$ PC relations (Figure 1b), derived from those LPVs with measured spectral types, are

$J - K = 1.215(\pm 0.014) + 0.37(\pm 0.05)[\log P - 2.4]$, $\sigma = 0.06$, for 21 M stars, and
$J - K = 1.9(\pm 0.9) + 3.3(\pm 7.3)[\log P - 2.4]$, $\sigma = 0.58$, for 10 C stars.

The low dispersion in the PC relation for the M stars is evidence that they obey a PLC relation (see Feast et al 1989), but the divergent PC relation for the C stars indicates this isn't generally applicable to all AGB LPVs.

The three C stars that lie on the M star PC sequence are an anomaly, as their K amplitudes of pulsation are all moderately high (at ~ 0.6 mag), indicating their mass loss loss rates are $\sim 10^{-6} M_\odot$ yr^{-1} (Whitelock 1990). Hence they are likely to be moderately evolved C stars, and should therefore have redder $J - K$ colors.

3. The Galaxy

3.1 The Bulge LPVs

Whitelock et al (1991) have obtained multiple epoch $JHKL$ photometry of a sample of IRAS-identified variable stars in the Galactic bulge. They find that the majority have the IR colors and periods of Mira variables, with very few having $P > 560$. By estimating absolute luminosities from phase-lag and kinematic distances, they derived a bolometric PL relation with a slope of -2.55. Although this is much shallower than for the LMC, a large part of the discrepancy is probably due to the uncertain distances. The $[12]-[25]\mu$m

IRAS colors (indicative of dust, and hence mass loss) were found to be linked to the pulsation amplitude in K. The density distribution was compared to flattened ellipsoidal models (with $a/b = 1$), with a best fit being found for $c/a \sim 0.7$ and $\rho \propto a^{-3.2}$.

In a follow-up analysis, Whitelock & Catchpole (1992) re-modelled the density distribution by a prolate ellipsoid. Their best fit indicates the bulge is likely to be a bar tilted at $\sim 45°$ to the line of sight, in agreement with other recent studies, such as those by Nakada et al (1991) and Weinberg (1992a, 1992b).

3.2 Disk LPVs

Two major studies of the local AGB LPV density distributions have been made by Jura & Kleinmann (1992a, 1992b) and Kershbaum & Hron (1992), using the PL relation to derive distances. Jura & Kleinmann divided their O-rich Miras and semi-regulars (SRVs) into short ($P < 300$) and intermediate ($300 < P < 400$) period groups. They found that the SRVs with $200 < P < 300$ have a similar distribution to the short period Miras, that of a 'thick' disk (scale height ~ 550pc), while the SRVs with $300 < P < 400$ and $100 < P < 150$ have a similar distribution to the intermediate period Miras, that of a 'thin' disk (scale height ~ 240pc). This would imply that most of the SRVs are also AGB stars, different from the Miras only in their pulsation amplitude, and probably being on the lower AGB. It also implies that the $100 < P < 150$ SRVs are of similar mass to the intermediate Miras, but in higher overtones of pulsation (which may explain their sometimes irregular periods).

Kershbaum & Hron found essentially the same results, but derived a thicker 'thick' disk (scale height ~ 780 pc) for the short period Miras. They also found evidence, from $V - [12\mu m]$ colors, for the $P < 150$ SRVs to be 'hot' ($T_{\text{eff}} > 3200$ K).

Jura & Kleinmann (1992a) estimated the number density for the 'thin' disk O-rich Miras to be ~ 210 kpc^{-3}. Mass loss rates of $\sim 10^{-7}$ $M_\odot$ yr^{-1} were derived for both groups of Miras from their IRAS 60μm flux. Mass loss rates for the SRVs were estimated as $10^{-8} - 10^{-7}$ $M_\odot$ yr^{-1}.

Hron (1991) has investigated a potentially important relation between the metallicity of the short period ($P < 200$) Miras and their $(V_{\text{max}} - K)_0$ color, combining a method used by Bessell et al (1986) with the theoretical formulae of Wood (1990a) concerning the relationship between an LPV's luminosity, temperature, mass and metallicity. Although this method is unable to measure an absolute metallicity for any individual Mira, and is limited to LPVs hotter than 3200 K, nevertheless these preliminary results indicate a metallicity difference exists between the AGB LPVs with rotation velocities similar to the thin disk, and those AGB LPVs with a much lower rotation component and a velocity dispersion similar to that of the metal-rich RR Lyraes.

3.3 Nearby LPVs

High resolution imaging with HST, the VLA and mm telescopes, and now increasingly optical imaging from the ground (using both interferometric and aperture synthesis techniques) have resulted in detailed observations of the angular sizes, shapes and even detailed morphology of nearby Miras. Without being exhaustive, the following is a brief summary of papers that have appeared in the last two years concerning R Aquarii and o Ceti.

Paresce et al.(1991), Hege et al.(1991), Hollis et al.(1992), Burgarella & Paresce (1992), Solf (1992) and Lehto & Johnson (1992) have all made detailed studies of R Aquarii, a 387 day Mira in a symbiotic system, producing images of a jet, a possible Herbig-Haro object, and providing evidence for two expanding shells.

Planesas et al (1990a, 1990b) mapped the CO radio emission of o Ceti, finding a circumstellar envelope expansion velocity of only 3 km s^{-1}, implying a mass loss rate of $\sim 1 \times 10^{-7}$ $M_\odot$ yr^{-1}, evidence for bipolar gas outflow, and detecting both SiO and CO emission in the inner regions. Karovska et al (1991) and Haniff et al (1992) obtained optical images of the photosphere and molecular atmosphere of o Ceti. Karovska et al measured departures from circular symmetry (at wavelengths 530 to 850 nm). The direction of the assymetry was found to change with phase, implying that it is related neither to the presence of the companion nor to any rotation of o Ceti itself. Haniff et al found the sizes o Ceti's elongated photosphere (best fitted by a uniformly illuminated disk with $a = 55$ mas and $b/a = 0.82$, at 700.7 nm and phase 0.3) and molecular atmosphere to be consistent with the model atmospheres of Bessell et al (1989), and confirmed the departures from circular symmetry, which they postulated to be due to the presence of a non-radial pulsation mode. Ridgway et al (1992) measured the angular diameter of o Ceti to be 36 mas (best fit of a uniform disk, at phase 0.23 to 0.36). Fitting a limb darkened model gave a diameter of 43 mas, at an effective wavelength of 2.28 μm (equivalent to $\sim$60 mas at 700 nm), implying an effective temperature of 2270 K, consistent with IR colors, but less than the molecular excitation temperature (which would predict a diameter $\sim$half that observed).

Two years ago it seemed that the question as to which mode of pulsation the Miras were in (fundamental or first overtone) had been resolved, with Wood (1990a, 1990b) agreeing with Willson (1982) that the pulsation models for first overtone could not reproduce the observed pulsation velocities, and therefore they had to be fundamental pulsators. In contrast, however, Tuchman (1991) has modelled the pulsation acceleration, and finds the observations are inconsistent with fundamental pulsators, but are in the first overtone. Apart from velocity data, the other method of observationally determining the pulsation mode is to measure the pulsation constant $Q = P(M/R^3)^{1/2}$, where the period P is easily measured, the mass M is reasonably well constrained to be $\sim$1 $M_\odot$, and the radius R is inferred from the luminosity and effective temperature (see eg Wood 1992b). The weakest part of this measure thus lies in the determination of the luminosity and effective temperature. But high resolution studies have the potential to measure R directly (once a reliable model for the effects of limb darkening is established). In the case of o Ceti ($P = 332$), for example, the above diameter of 43 mas, at a parallax distance of 77 pc (Jenkins 1952), gives a radius of 364±17 $R_\odot$. Taking its mass as 1 $M_\odot$, then $Q = 0.048 \pm 0.003$, which the PMR relation of Fox & Wood (1982) would clearly identify as being in the first overtone. Even if the mass of o Ceti was as uncertain as 1.6±1$M_\odot$ (Tuchman 1991), the resultant range in Q (0.06±0.02) still favours the first overtone. However, if the radius predicted by the CO molecular excitation temperature is adopted, then o Ceti would very definitely be in the fundamental mode.

As imaging instrumentation using adaptive optics become available, this field is destined to expand rapidly over the next few years.

At the same time as advances are being made in high resolution imaging, the next few years should also see a vast improvement in the number of Miras with parallax distances. For example, Gatewood (1992) has determined a new parallax for R Leo ($P = 310$), using the Multichannel Astrometric Photometer (MAP) at the Allegheny Observatory, giving a direct trigonometric distance to R Leo of 120±15pc. This is remarkably similar to a PL distance of 117pc, derived by assuming the LMC $K, \log P$ relation with an LMC distance modulus of 18.5, and taking $K = -2.4$ (Robertson & Feast 1981). Such a result disagrees with theoretical calculations by Wood (1990a; 1990b), which predict that Galactic Miras should be about 0.4 mag fainter than an LMC Mira of the same period, due to their presumed metallicity difference.

However, a similar calculation for o Ceti, with a (photographic) parallax distance of 77pc (Jenkins 1952), and $K = -2.5$ (Robertson & Feast 1981), implies a PL distance of ~120pc. Gatewood is currently making MAP observations of o Ceti, and it will be interesting to see if the distance remains the same.

HIPPARCOS is monitoring 245 large amplitude pulsators (Mennessier et al 1992), and it might be hoped there will soon exist a reasonably large sample of absolute Mira distances, which should help to resolve these issues.

4. The Future

Surveys for LPVs in the Magellanic Clouds are being conducted by Moore (1991) in the SMC, and Hughes & Wood (ongoing) have initiated a new UK Schmidt multi-epoch plate survey of the eastern and western edges of the LMC, which should be complete in ~1 year. Another UK Schmidt LPV survey is also being conducted in 5 Galactic regions, 4 of which are in the bulge (Hughes & Whitelock, ongoing).

The results of using HST wide field camera (WFC) exposures to successfully search for Cepheids in M81 (Hughes 1992; Stetson et al 1992, in preparation) and in IC4182 (Sandage et al 1992), proves that HST, even in its present handicapped state, could be used to search for AGB LPVs in external galaxies out to a distance modulus of 27.5 (3.2 Mpc), and with WFPC2 out to 29.5 (7.9 Mpc), providing the potential for studying relationships between galactic environment and AGB and AGB LPV evolution, and for tracing the stellar populations which the AGB LPVs represent in these galaxies.

Wood & Bessell (1985) drew attention to the potential of a PL relation for RSG LPVs as a means of measuring distances to galaxies. Recent CFHT observations by Pierce et al (1992a, 1992b, *these proceedings*) indicate that RSG LPVs can be detected in galaxies as distant as the Virgo cluster. RSG LPVs have the potential to become very useful distance indicators, as they are not only brighter than Cepheids (the brightest Cepheid has an $M_{bol} \sim -6$, compared to $M_{bol} \sim -9$ for RSG LPVs in the LMC), but also emit mostly in the red where uncertainties about internal extinction will be less.

Acknowledgements

My thanks to Peter Wood for all the help over the years, and to Patricia Whitelock and Michael Jura for helpful (e-mail) discussions and providing copies of preprints.

References:

Bessell M.S., Freeman K.C., Wood P.R. 1986 ApJ 310, 710.
Bessell M.S., Brett J.M., Scholz M., Wood P.R. 1989 A&A 213, 209.

Blöcker T., Schönberner D. 1991 A&A 244, L43.
Boothroyd A.I., Sackmann I.-J. 1992, ApJ 393, L21.
Burgarella D., Paresce F., 1992 ApJ 389, L29.
Eggen O.J. 1992 AJ 104, 275.
Feast M.W. 1963 MNRAS 125, 367.
Feast M.W, Glass I.S., Whitelock P.A., Catchpole R. 1989 MNRAS 241, 375.
Fox M.W., Wood P.R. 1982 ApJ 259, 198.
Gatewood G. 1992 PASP 104, 23.
Glass I.S. 1985 Irish AJ 17, 1
Glass I.S., Whitelock P.A., Catchpole R., Feast M.W, Laney C.D. 1990 SAAO Circ. no.14, p 63.
Haniff C.A., Ghez A.M., Gorham P.W., Kulkarni S.R., Matthews K., Neugebauer G. 1992 AJ 103, 1662.
Hege E.K., Allen C.K., Cocke W.J. 1991 ApJ 381, 543.
Hollis J.M., Dorband J.E., Yusef-Zadeh F. 1992 ApJ 386, 293.
Hron J. 1991 A&A 252, 583.
Hughes S.M.G. 1989 AJ 97, 1634.
Hughes S.M.G. 1992 in *IAU Coll. 136: Stellar Photometry*, Dublin August 1992, *in press*.
Hughes S.M.G., Wood P.R. 1990 AJ 99, 784.
Hughes S.M.G., Wood P.R., Reid N. 1991 AJ 101, 1304.
Jenkins L.F. 1952 *General Catalogue of Trig. Stellar Parallaxes* Yale University Press.
Jura M., Kleinmann S.G. 1992a ApJSupp 79, 105.
Jura M., Kleinmann S.G. 1992b ApJSupp *in press*.
Karovska M., Nisenson P., Papaliolios C., Boyle R.P. 1991 ApJ 374, L51.
Kerschbaum F., Hron J. 1992 A&A, *submitted*.
Lehto H.J., Johnson D.R.H. 1992 Nature 355, 705.
Mennessier M.O., Barthés D., Boughaleb H., Figueras F., Mattei J.A. 1992 A&A 258, 99.
Moore G.K.G. 1991 in *IAU Symp. 148. The Magellanic Clouds*, eds. R. Haynes and D. Milne (Kluwer: Dordrecht), p355.
Nakada Y., et al. 1991 Nature 353, 140.
Paresce F., et al 1991 ApJ 369, L67.
Pierce M.J., Welch D.L., McClure R.D., van den Bergh S., Racine R. 1992a *these proceedings*.
Pierce M.J., Crabtree D. 1992b *these proceedings*.
Planesas P., Bachiller R., Martín-Pintado J., Bujarrabal V. 1990a ApJ 351, 263.
Planesas P., Kenney J.D.P., Bachiller R. 1990b ApJ 364, L9.
Ridgway S.T., Benson J.A., Dyck H.M., Townsley L.K., Hermann R.A. 1992 AJ *submitted*, NOAO preprint # 418.
Robertson J.W. 1974 ApJ 191, 67.
Robertson B.S.C., Feast M.W. 1981 MNRAS 196, 111.
Sandage A., Saha A., Tammann, Pangia, Macchetto 1992, *Science with the HST*, ECF conference June 29 - July 7, 1992, Sardinia, in press.
Solf J. 1992 A&A 257, 228.
Tuchman Y. 1991 ApJ 383, 779.
Weinberg M.D. 1992a ApJ 384, 81.
Weinberg M.D. 1992b ApJ 392, L67.
Whitelock P.A. 1990 in *Confrontation between Stellar Pulsation and Evolution*, eds. C. Cacciari and G. Clementini (ASP Conf. Ser. 11), p 365.
Whitelock P.A., Feast M., Catchpole R. 1991 MNRAS 248, 276.

Whitelock P.A., Catchpole R. 1992 in *The Centre, Bulge and Disk of the Milky Way*, ed. L.Blitz (Kluwer: Dordrecht), in press.
Willems F.J., de Jong T. 1988 A&A 196, 173.
Willson L.A. 1982 in *Pulsations in Classical and Cataclysmic Variable Stars*, eds. J.P.Cox and C.J.Hansen (Boulder: JILA), p 269.
Wood P.R. 1990a in *From Miras to PN: Which Path for Stellar Evolution?*, eds. M.O. Mennessier and A. Omont (Gif sur Yvette: Editions Frontières), p 67.
Wood P.R. 1990b in *Confrontation between Stellar Pulsation and Evolution*, eds. C. Cacciari and G. Clementini (ASP Conf. Ser. 11), p 355.
Wood P.R., Bessell M.S. 1985 PASP 97, 681.
Wood P.R., Bessell M.S., Fox M.W. 1983 ApJ 272, 99.
Wood P.R., Hughes S.M.G. 1992, in preparation.
Wood P.R., Moore G.K.G., Hughes S.M.G. 1991 in *IAU Symp. 148. The Magellanic Clouds*, eds. R. Haynes and D. Milne (Kluwer: Dordrecht), p259.
Wood P.R., Whiteoak J.B., Hughes S.M.G., Bessell M.S., Gardner F.F., Hyland A.R. 1992, ApJ *in press (October)*.

Questions:

D. Welch: Robertson (1974) found a number of RSGs in MC clusters. Have there been any additional cluster searches?

Hughes: The I amplitudes of RSG LPVs are generally too low to have been detected in the UK Schmidt I surveys. However, there are at least two groups (Mateo et al and Wood et al, *private communications*) that are currently monitoring several MC clusters with CCD photometry, and these are finding RSG LPVs.

H. Shibahashi: Can the mass and radius of individual stars be determined by comparing the theoretical evolutionary tracks and the observations on the PL diagram?

Hughes: Pulsation masses were derived by Wood, Bessell and Fox (1983) for the MC LPVs, and more recent theoretical pulsation formulae that take into account metallicity effects are given in Wood (1992a, 1992b). Radii can be calculated from the luminosities and temperatures, but would be very uncertain due to the still poorly determined temperatures for these very late-type stars.

Properties of Mira Photospheres

M.Scholz

Institut für Theoretische Astrophysik, Universität Heidelberg, Germany

The first crude pictures of the structure of Mira photospheres were derived from velocity analyses of emission and absorption lines appearing shifted, doubled and doppler-broadened in Mira spectra. Positions and shapes of profiles change with phase and often do not repeat in successive cycles. Non-repeating cycles are also found in colors (e.g., TiO bands in Fig. 9 of Spinrad & Wing 1969) and in light curves and present, of course, a problem in matching specific observed and modeled cycles. The large geometric extension of Mira photospheres was demonstrated by monochromatic radius observations of o Ceti and R Leo by Labeyrie et al. (1977) and Bonneau et al. (1982) which, however, did not systematically follow the stars through different phases and cycles.

The structure of a Mira photosphere is essentially determined by the propagation of shock fronts. A shock front enters the bottom of the photosphere once each cycle and travels outwards. The stellar material is accelerated by the shock wave and falls back after some time, and it is heated by dissipation of shock energy. As a result, the structure of the Mira photosphere differs substantially from that of a non-variable star. (i) There are density discontinuities at the shock positions and very flat density gradients in between, yielding a geometrically very extended configuration. (ii) The photosphere has a pronounced outflow/infall velocity stratification. (iii) There is a thin hot layer just behind the shock front producing typical emission lines plus, depending on how slow or fast relaxation towards equilibrium occurs, a more or less extended non-equilibrium region possibly affecting the absorption line spectrum. See Fig. 1 of Scholz (1992) for a typical stratification where instantaneous relaxation to equilibrium is assumed because of still quite insufficient knowledge of relaxation processes (cf. Beach et al. 1988; Bessell & Scholz 1989).

The predictions presented here of the behavior of typical observable features are based on unpublished self-excited pulsation models of P.R. Wood (cf. Wood 1974, 1990) the photospheric portion of which is a non-grey, spherically extended configuration in local thermodynamic and radiative equilibrium (cf. Scholz 1992) not including the emission-line forming zones. Luminosities L, effective temperature T_{eff} and radii of selected photospheric layers of a typical model sequence covering two successive cycles, with parameters close to o Ceti (1 $M_\odot$, fundamental mode pulsation, M type composition), are shown in **Fig. 1**. Here, the stellar radius R is the Rosseland $\bar{\tau} = 1$ radius which is practically identical with the observable continuum radius. Fig. 1 demonstrates the slow and regular pulsation of the star in terms of R whereas the shock fronts move outward in a fairly erratic manner and do not repeat in successive cycles. The positions of the photospheric surface (*i.e.*, the layers below an outer wind envelope, say around $\bar{\tau} = 10^{-5}$) and of monochromatic $\tau_\lambda = 1$ radii at strong TiO ab-

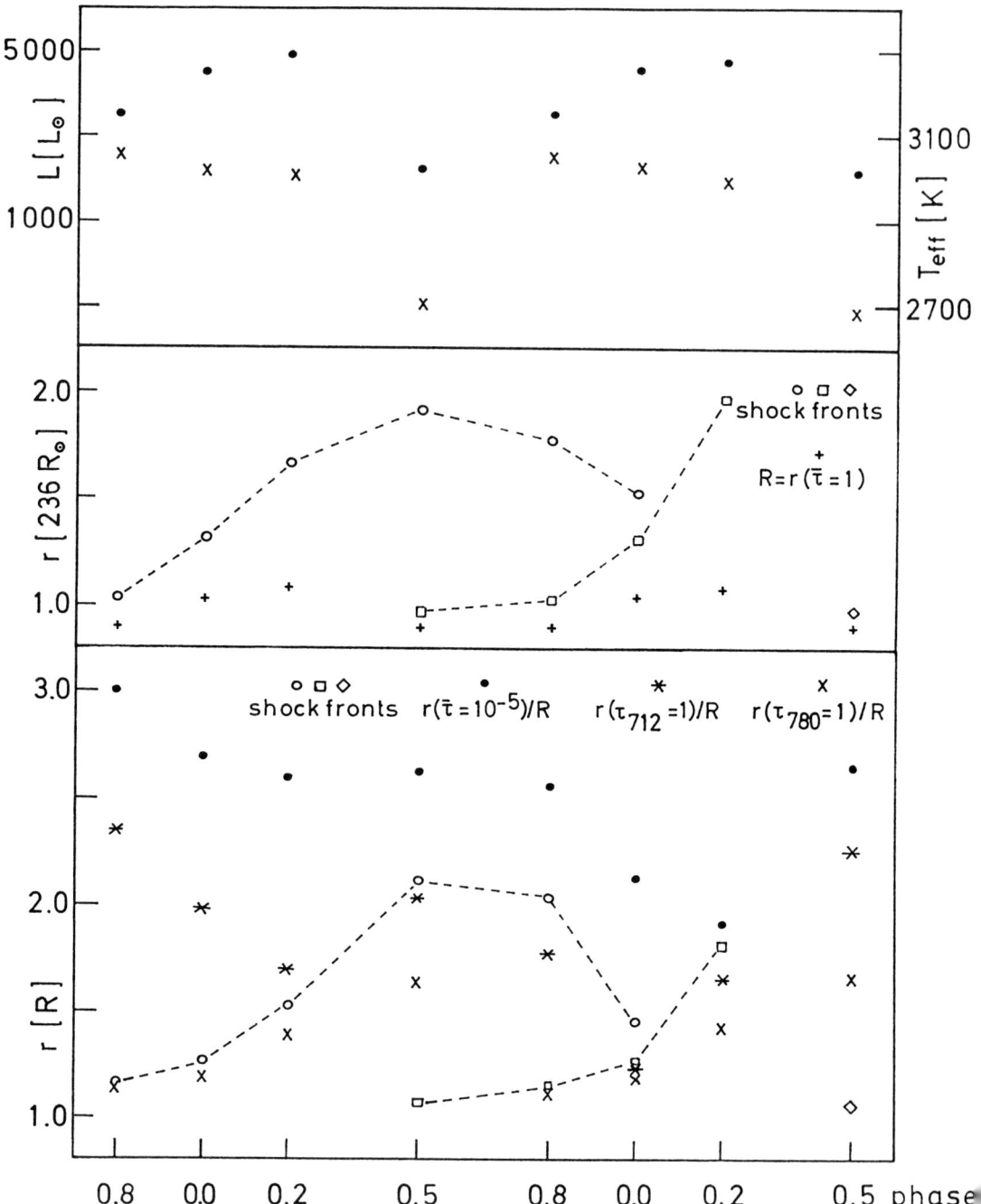

Fig. 1. Luminosities L (dots), effective temperatures T_{eff} (crosses) and radii of selected photospheric layers (symbols given in the panels) of a model sequence covering two successive cycles (see text).

sorption (e.g., 712 nm) are obviously dominated by the details of shock propagation. In contrast, the positions of monochromatic radii at moderate TiO absorption (e.g., 780 nm) vary more slowly and more regularly. As one considers colors measuring the same TiO features, one finds a behavior similar to that given in Fig. 9 of Spinrad & Wing (1969) with a tight correlation with monochromatic radii: the farther above the continuum the feature is formed the stronger the band appears on the details of shock propagation. (Note that the monochromatic radii and colors calculated here should be considered with caution as they are based on frequency-averaged absorption coefficients and as they do not take into account outflow/infall velocity broadening of lines).

The recent model study of Scholz (1992) and high-resolution observations of Mira spectra (see Scholz 1992 for a list of references) have also shown that the profiles of absorption lines of different strengths from selected atoms and molecules react sensitively to the variations of the outflow/infall velocity field from phase to phase and from cycle to cycle. Thus, it is obvious that observations of monochromatic radii and colors probing absorption features of different strengths (including the position and slope of the continuum), covering the phases of successive cycles, provide accurate informaiton about the details of the structure of a Mira photosphere, in particular when these observations are supplemented by simultaneous measurements of absorption-line profiles and of emission-line velocity shifts.

References:

Beach, T.E., Willson, L.A. 1988, ApJ, 329, 241.
Bessell, M.S. & Scholz, M. 1989, in Evolution of Peculiar Red Giant Stars (IAU Colloq. 106), eds. H.R.Johnson & B.Zuckerman, Cambridge University Press, Cambridge, p.67
Bonneau, D. Foy, R., Blazit, A. & Labeyrie, A. 1982, A&A, 106, 235
Labeyrie, A., Koechlin, L., Bonneau, D., Blazit, A. & Foy, R. 1977, ApJ, 218, L75.
Scholz, M. 1992, A&A, 253, 203
Spinrad, H. & Wing, R.F. 1969, ARA&A, 7, 249
Wood, P.R. 1974, ApJ, 190, 609
Wood, P.R. 1990, in From Miras to Planetary Nebulae, eds. M.O. Mennessier & A. Omont, Frontieres, Gif, p.67

Pulsation Properties of Hydrodynamic Models with Dimensional Analysis

Toshiki Aikawa

Tohoku-Gakuin University, Sendai, Japan

Abstract

We demonstrate that the dimension deduced from time series data of hydrodynamic models for chaotic pulsation is a function of luminosity. The dimension is proposed as a good quantity to guess stellar parameters and the physics of stellar envelopes like as the pulsation periods and light curve shapes used for regular variables.

1. Introduction

The dimensional analysis has been applied to time-series data to find chaotic nature and the dimension of the system which generates the time-series data (Atmanspacher et al., 1990). The method may be applied to data generated in theoretical models as well as observational data. In this report, we shall apply the dimensional analysis based on Grassberger and Procaccia (1983) to the data generated from hydrodynamic models for post-AGB stars(Aikawa, 1991, 1992).

The hydrodynamic models are constructed for irregular pulsation with small amplitudes observed in yellow supergiant stars(e.g. 89 Her, HD161796). There is evidence that some of them are post-Asymptotic Giant Branch stars (see, a current review by Fernie, 1992). The stellar parameters for the models are: $M = 0.8 M_\odot, Te = 6300 K$, with the Pop II composition. The luminosity is varied with a range from $3500 L_\odot$ to $7000 L_\odot$ as a control parameter. This sequence has been examined intensively by Aikawa(1992). For lower luminosities, the models show regular pulsation, and, on the other hand, the models with higher luminosities show chaotic pulsation.

We apply the dimensional analysis of Grassberger and Procaccia (1983) to data of time variation of photospheric magnitude of irregular pulsation.

2. Dimensional Analysis

A periodic attractor has the dimension D=1 as the result of dimensional analysis, and a quasi-periodic oscillation with two incommensurable frequencies has D=2. The dimension thus may be an indicator on degree of complexity of oscillations (Moon, 1987). In accordance with Grassberger and Procaccia, we first construct a pseudo-phase space from time-series of photospheric magnitude. Then, we calculate the correlation function in embedding space with dimension m=2 to 8. We use the total number of 8000 points for the time-span of 8000 days for each model and the time delay for the pseudo-phase space is estimated from the value for which the autocorrelation function first passes through zero.

The result is summarized in Table 1 (Embedding dimension 8 is not sufficient to get good convergences for higher luminosities, and so they are marked with :). It demonstrates clearly that the dimension is a function of luminosity. The model $4985L_\odot$ is a regular oscillator just below the transition luminosity from regular to irregular pulsation in the present model sequence. The model $5000L_\odot$ is weakly chaotic, and other higher luminosity models have well-developed chaos.

Table 1

Correlation Dimension	
model($L/L_\odot$)	dimension
4985	1.5
5000	2.4
5500	3.0
6000	5.2:
6500	5.5:

3. Conclusions

We demonstrate that the geometric dimension deduced from dimensional analysis to time-series data for hydrodynamic models may be an indicator on stellar parameters. We may apply the same method to observational data. Fortunately modern data by photoelectric photometry have been accumulated for 89 Her and HD 161796, and within a few years, we will have sufficient data for this purpose for these stars.

References:

Aikawa, T. 1991, Astrophys. J.,**374**,700.
Aikawa, T. 1992, submitted to Mon. Not. Roy. astr. Soc.
Atmanspacher, H., Scheingraber, H. and Voges, W., 1989, in: *Data Analysis of Astronomy III*, eds. Gesú, V. Di. et al., p3.
Grassberger, P. and Procaccia, I., 1983, Phys. Rev. Lett., **50**, 346.
Fernie, J.D., 1992, in :*Luminous High-latitude Stars*, (An International Workshop at CfA), in press.
Moon, F.C., 1987, *Chaotic Vibrations*, John & Sons: New York.

Dust Induced Dynamics of Circumstellar Shells around Long–Period Variables

A. Gauger, A.J. Fleischer, E. Sedlmayr

Institut für Astronomie und Astrophysik, Berlin, Germany

1. Dynamical modelling of circumstellar dust shells around LPVs

The presence of dust essentially affects the optical appearance and the dynamics of circumstellar shells around LPVs, while in turn the formation of solid particles as well as their properties critically depend on the actual local conditions. Thus, a reliable modelling of the structure and dynamics of such shells requires a detailed treatment of the dust complex.

In our hydrocode, which is described in more detail in Fleischer et al. (1992), the time evolution of the dust component is treated consistently by means of a moment method (Gauger et al. 1990). For the solution of the hydrodynamical and thermodynamical problem we essentially follow the approach of Bowen (1988), except for the shocks, which so far are assumed to be isothermal.

2. Dust induced dynamical phenomena

In our models the complex interplay between the dust component and the hydro– and thermodynamical structure of the shell leads to (cf. Fleischer et al. 1991, 1992):

1. An inhomogeneous shell–like distribution of the circumstellar dust, caused by the variations of the temperature, the density and the chemical composition in the dust forming region.

2. Pronounced shocks dominating the shell structure, which are either generated by the interior pulsation and amplified substantially by radiation pressure on newly formed dust grains, or even created by radiation pressure on dust alone.

3. A substantial backwarming effect due to the dust opacity, which increases the radiative equilibrium temperature inside the new dust layer by several hundred Kelvins, and thereby determines the inner boundary of the region where dust can be formed.

The resulting implications for observable quantities are currently under investigation.

References:

Bowen, G.H., 1988, Astrophys.J. **329**, 299
Fleischer, A.J., Gauger, A., Sedlmayr, E., 1991, Astron. Astrophys., **242**, L1
Fleischer, A.J., Gauger, A., Sedlmayr, E., 1992, Astron. Astrophys., in press
Gauger, A., Gail, H.-P., Sedlmayr, E., 1990, Astron. Astrophys., **235**, 345

NLTE Synthetic Spectra of Mira–Type Variable Stars

Donald G. Luttermoser, George H. Bowen, Lee Anne Willson

Iowa State University, Ames, Iowa, USA

Abstract

We present NLTE radiative transfer in hydrodynamic models representative of Mira–type variable stars. Calculations were carried out with the PANDORA code of the Bowen models using a *snapshot* approximation.

Explanation of Observational Characteristics of Miras

Hydrodynamic models produce Balmer lines where $f(H\alpha) < f(H\beta) < f(H\gamma)$. This has been observed in the spectra of Mira–type variables, and in the past, has been attributed to obscuration by overlying absorption. This Balmer line "increment" is the result of two factors in the transfer of radiation in the lines: (1) the optical depth of the line which determines the depth of formation of the line; and (2) the thermalization of the line which determines how closely the source function couples to the Planck function (see Luttermoser & Bowen 1992).

Bowen (1988) has shown that non–dusty dynamic models display a permanent chromosphere (or "calorisphere") that exists throughout the entire pulsation cycle whereas dusty models do not. IUE observations of Miras have shown that the peak of the Mg II flux typically occurs at a photometric phase of 0.3–0.5 (Brugel *et al.* 1988). The non–chromospheric hydro–model has coincident maxima in the Balmer and Mg II lines whereas the chromospheric hydro–model shows a phase shift of ~ 0.4 between the Mg II and Balmer line peak flux which is consistent with the observations. The inner Mg II profile ($\Delta\lambda < 2$ Å) of the non–dusty model forms in the permanent chromosphere. Near pulsation phase 0, the innermost shock (below the chromosphere) enhances the background flux causing the Mg II line to be in absorption. Later as the innermost shock merges with the chromosphere, the background flux decreases and the Mg II line emission increases to form emission lines. In the dusty model, the Mg II emission originates in the innermost shock as does the Balmer lines. This may suggest that optically bright Mira stars (*i.e.*, those observable with IUE) have calorispheres.

References:

Bowen, G.H. 1988, ApJ, 339, 299.
Brugel, E.W., Beach, T.E., Willson, L.A., & Bowen, G.H. 1988, in IAU Coll. #103, The Symbiotic Phenomenon, 67.
Luttermoser, D.G., & Bowen, G.H. 1992, in *Cool Stars, Stellar Systems, and the Sun*, eds. M.S. Giampapa & J.A. Bookbinder (ASP Conf. Series), **26**, 558.

Nonlinear Models of Miras, Including Time-Dependent Convection

Dale A. Ostlie[1] and Arthur N. Cox[2]

[1]*Dept. of Physics, Weber State University, Ogden, UT 84408-2508*
[2]*Los Alamos National Laboratory, Los Alamos, NM 87545*

Abstract

Nonlinear calculations of Mira variable stars of Population I are presented. Each model is 1 $M_\odot$, with a luminosity of 5000 $L_\odot$ and an effective temperature near 3000 K. These models incorporate our theory of time-dependent convection, which is based on a convective phase lag formalism and includes spatial averaging of convective eddies from adjacent zonal interfaces. The theory also includes turbulent pressure, energy, and viscosity terms and allows for negative convective luminosities in subadiabatic regions where overshooting occurs.

Results of the present study suggest that based upon the dynamic behavior of the models, fundamental mode pulsations are the preferred mode of oscillation. In particular, we do not obtain the chaotic behavior that has been noted in previous nonlinear studies of the fundamental mode oscillations of Miras.

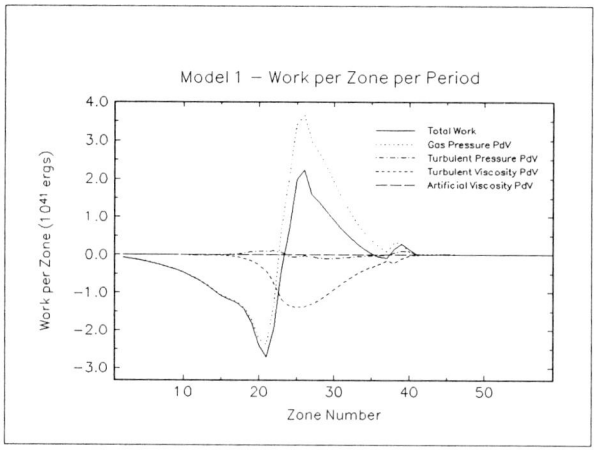

Figure 1 The average amount of work performed per zone per period for a typical fundamental mode model at a limit cycle.

The Evolution of Hα Profiles in S-type Mira Stars

A.W. Woodsworth

Dominion Astrophysical Observatory, National Research Council, Victoria, Canada

Past attempts to model the Hα emission profiles from Mira stars have been hampered by the strong absorption bands produced by TiO (M-type Miras) or carbon-based molecules (C-type Miras). The S-type Miras, however, exhibit relatively clean emission lines. As part of a larger study of Mira variables, I have obtained a series of high-dispersion Hα profiles of S-type Miras at different phases, using the DAO 48" coudé spectrograph. Although most profiles were obtained from different stars, their shapes develop in a regular way with advancing phase, so it appears that the observations can be used together to provide a good data base for the fitting of evolutionary (rather than static) models for the shock waves in S-type Miras.

Initial experiments with fitting simple models to the profiles have been extremely encouraging. The profiles can be represented very well by three Gaussian components. The characteristic asymmetry of the blue side of the profile is reproduced, and the stellar velocity obtained from microwave measurements of circumstellar CO is midway between the red-most component and the new, third component blended with the blue-most component. The accelerations obtained agree with the model of Tuchman (1991), and all the components evolve in a regular way with advancing phase. It is not certain whether these results imply three separate physical components, but they do characterize the line morphology very well and must carry at least some physically meaningful information.

References:

Tuchman, Y., 1991, Astrophys.J., **383**, 779

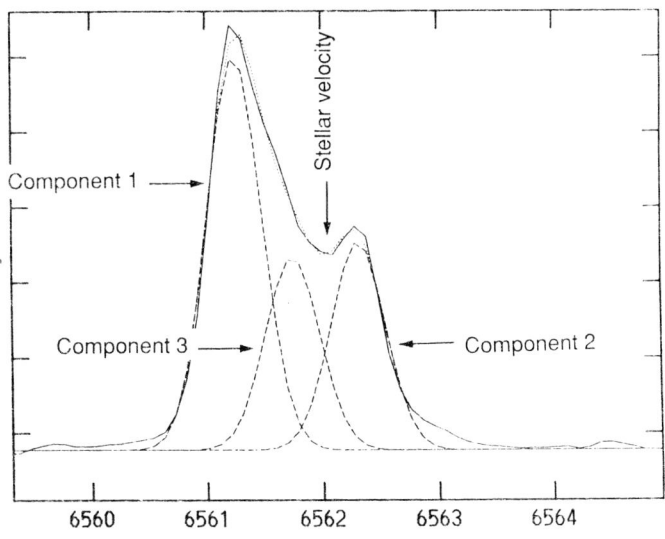

Figure 1 Sample Gaussian fits to S-star Hα emission profiles

Variable UV Line Emission in S Carinae
Miras Do Not Fear Change

E.W. Brugel[1], R. Davis[1], J. Bookbinder[2]

[1] *CASA, University of Colorado, Boulder, CO*, [2] *SAO, Cambridge, MA*

Abstract

The object S Carinae is a bright, relatively hot, short (149^d) period mira. As expected, the Mg II h and k lines dominate the ultraviolet line spectra during a large fraction of the pulsation cycle. There are also numerous emission lines of Fe I and Fe II produced by either fluorescense or possibly collisional excitation. The unifying "theme" for all the ultraviolet line emission is **variability** - *during a pulsation cycle one observes changes in the intensity, the central wavelength and (for the stronger lines) the line profile.*

During the period of 11/87 - 5/91 we obtained 23 high dispersion long wavelength IUE images of S Car. The majority of the data are concentrated during three distinct pulsation cycles. **Cycle 1:** Seven spectra were taken during a 44 day interval covering phases 0.23 to 0.53. **Cycle 2:** Six spectra were taken during a 53 day interval covering phases 0.05 to 0.40. **Cycle 3:** Ten spectra were taken during a 62 day interval covering phases 0.40 to 0.82. Nine of the spectra were taken during a 4 day period, covering phases 0.40 to 0.43.

A general description of these observations follows:

1. **Optical Light Curve**: The optical (i.e. Fes) light curve is fairly constant from cycle to cycle, though the ultraviolet line fluxes change significantly from one cycle to the next.

2. **Mg II Emission Lines**:

    (a) The Mg II flux peaks near phase 0.25.

    (b) Both the Mg II h and k lines are confined to the blueward side of line center in the reference frame of the star.

    (c) The observed ratio of the Mg II k to h line fluxes is always < 0.6. The theoretically expected k/h ratio should lie between two (for an optically thick atmosphere) and unity (for an optically thin atmosphere) (Bookbinder, et al. 1989). Recent models by Luttermoser (private commun.) indicate that "mutilation" of the Mg II k line may be due to radiative transfer effects.

3. **Fe I and Fe II Emission Lines**: In addition to Mg II, there are approximately a hundred unique emission lines detected at the three σ-level. Some lines are detected only once, however the majority are seen on 4 to 25 different spectra. A few of the lines with multiple detections are discussed. Variations in line intensity, velocity (or more precisely, the line centroid) and profiles are seen.

   (a) Fe II UV 1 (2625.669Å) line: This is one of the strongest Fe II lines and is seen at almost all phases (it has also been detected in other miras). The cycle 2 data show a moderate variation in intensity, with the peak flux occurring at approxiamtely the same phase as the peak in Mg II. Cycle 3 data show approximately constant flux over the small phase interval 0.40-0.43, and no detectable flux at phase 0.82. There is however, a significant change in the central wavelength during the cycle. A modest red-shift occurs from phase 0.40 to 0.41 to 0.42, then a dramatic blue-shift of 0.4Å (or $\approx$ 45 km/s) occurs from phase 0.42 to 0.43.

   (b) Fe II UV 32 (2732.446Å, 2759.332Å, and 2775.338Å) lines: As expected, since these line are all from the same multiplet, their behavior is fairly consistent. The cycle 2 data show a continuous increase in flux from phase 0.05 to 0.40 - though the 2775Å line deviates from the pattern. Cycle 3 data show approximately constant flux over the small phase interval 0.40-0.43, and no detectable flux at phase 0.82. Similar to Fe II UV 1, there is a significant change in central wavelength during the cycle. A modest red-shift occurs from phase 0.40 to 0.41 to 0.42, then a dramatic blue-shift of 0.4Å (or $\approx$ 45 km/s) occurs from phase 0.42 to 0.43.

   (c) Fe I UV 44 (2823.276Å) line: This line is produced by fluorescence via pumping from the Mg II k-line (Carpenter et al. 1988).

References:

Bookbinder, J., Brugel, E.W. and Brown, A., 1989 *Astroph. J.*, **342**, 516.

Carpenter, K.G., Pesce, J.E., Stencel, J.E., Brown, A., and Wing, R.F. 1988 *Astroph. J. Supp.*, **68**, 345.

Acknowledgements:

This research was funded by NASA grant NAG5-350 to the University of Colorado and NASA grant NAG5-87 to the Smithsonian Astrophysics Observatory.

Pulsations and declines of RCB stars

P.L. Cottrell[1], W.A. Lawson[2]

[1] Mount John University Observatory, Dept of Physics & Astronomy,
University of Canterbury, Christchurch, New Zealand,
[2] Dept of Physics, University College ADFA, UNSW, Canberra, Australia

Abstract

We have continued to observe many of the R Coronae Borealis (RCB) stars and other related hydrogen–deficient carbon stars. We wish to more fully investigate the photometric and spectroscopic properties of these peculiar stars.

The UBVRI photometric data (acquired at Mount John University Observatory, MJUO) are being used to investigate whether the periods determined by Lawson et al. (1990), based on time intervals of up to 1100 d, are still evident in the larger datasets which now cover time intervals of up to 2200 d.

The long time baseline of photometric observations that we have been able to acquire has enabled us to accumulate an extensive database of decline photometry. These are used to investigate links with the pulsations at maximum light in order to get a better understanding of this remarkable phenomenon.

1. Introduction

Up until a decade ago, the RCB stars had been regarded as an astronomical enigma and objects that were the subject of only occasional and brief investigation during the unusual dimming events that these stars undergo. However, these investigations have often only been undertaken once the object has been in decline for a number of days or weeks. Although this can provide valuable information about the evolution of the obscuring material, any understanding of the triggering mechanism is lost.

2. Pulsations & Declines

(a) Pulsations at maximum light (U Aqr & RY Sgr)

U Aqr was observed to have alternating deep and shallow minima in the data obtained in 1986 and 1987. This effect disappeared during the subsequent 2 years where there was little distinctly periodic photometric variation in the V light curve. The amplitude of the pulsations increased again in 1991. The frequency analysis shows that the 1991 observations are in phase with the previous alternations in light output (see figure 1), similar to the RV Tauri stars (see Pollard et al., this conference).

RY Sgr has continued its semi–regular pulsations, despite the decline during 1990. There are still 2 dominant periods (52–d and 37–d) in the 1991 photometric data (see figure 2). Photometry of RY Sgr obtained at MJUO since 1986 indicates that the amplitude of the 37-d mode has remained essentially constant (0.19±0.03 mag), whereas the amplitude of the 52-d mode varies by a factor of 4 (between 0.06 and 0.24 mag). The varying amplitude of this mode is almost certainly the reason for the semi-regular appearance of the light curve of RY Sgr.

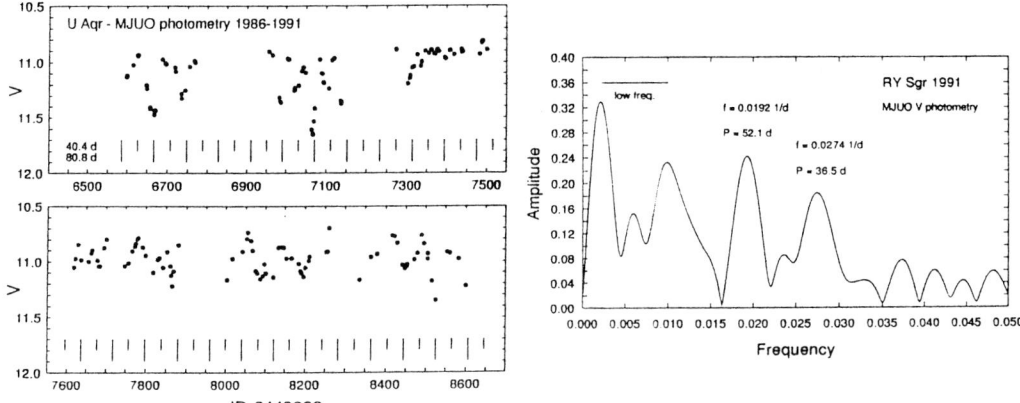

Figure 1 MJUO V photometry of U Aqr obtained since 1986. The 40.4-d and 80.8-d periodicities have been determined from the Fourier analysis of the light curve.

Figure 2 The amplitude spectrum for the 1991 photometry of RY Sgr. The rise from the decline in 1990 shows as low frequency peaks. The other two dominant periodicities, similar to those found in previous years, are indicated.

(b) Pulsation-decline links (RY Sgr & V854 Cen)

Pugach (1977) noted that declines of RY Sgr occurred within a narrow range of phase, near maximum light of the 37-d pulsation mode. A similar relationship was recently found for V854 Cen, which has declines occurring near the maximum of its 43.2-d periodicity (Lawson et al. 1992). These discoveries infer dust formation close to the star (to preserve the phase coherency) and at near-photospheric temperatures. The 'standard' model for mass-loss in RCB stars (see Feast 1990) considers mass-loss via some ejection mechanism with dust formation at $\sim 20\ R_*$. A pulsation-decline link under these conditions would be highly unlikely.

Acknowledgments: We would like to thank Alan Gilmore and Pam Kilmartin for the acquisition of the photometric data. PLC acknowledges the financial support of an Erskine Fellowship from the University of Canterbury and the hospitality of the Department of Physcis, University College ADFA, where this paper was written.

References:

Feast, M.W., 1990, Astr. Soc. of Pacific Conf. Series **11**, 538.
Lawson, W.A., Cottrell, P.L., Kilmartin, P.M., Gilmore, A.C., 1990, MNRAS **247**, 91.
Lawson, W.A., Cottrell, P.L., Gilmore, A.C., Kilmartin, P.M., 1992, MNRAS **256**, 347.
Pugach, A.F., 1977, IBVS No. 1277.

Recent Insights into R Coronae Borealis Stars from Recent UV and Visible Observations

Geoffrey C. Clayton[1], Barbara A. Whitney[2],

[1] CASA, University of Colorado [2] Harvard-Smithsonian Center for Astrophysics

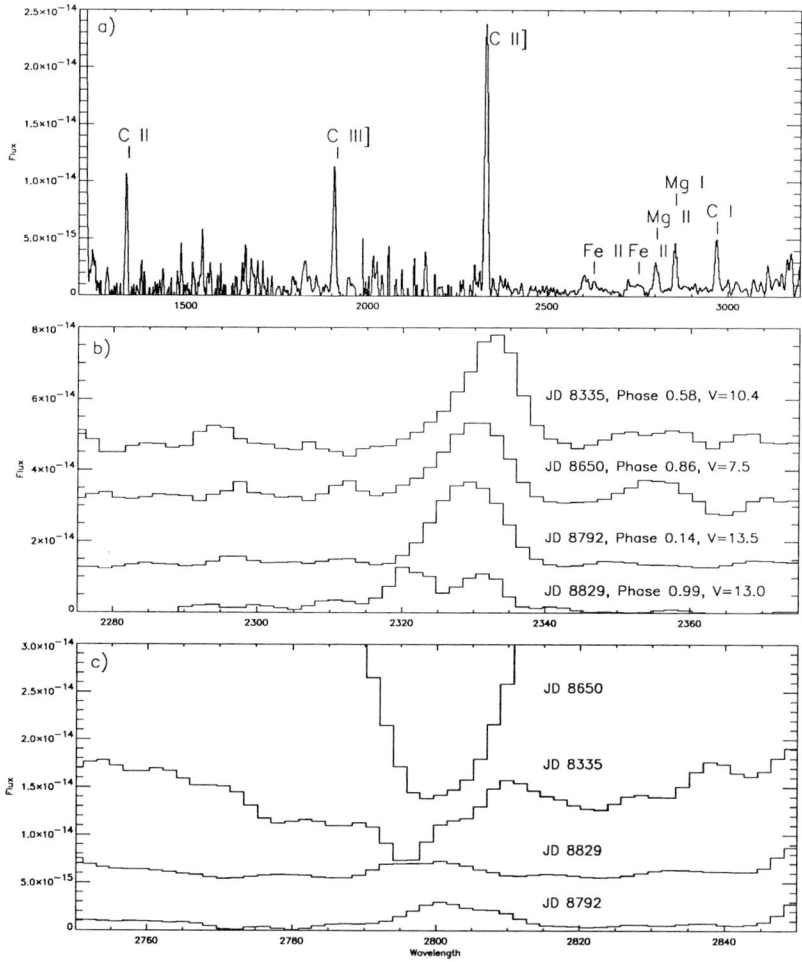

Figure 1 IUE low dispersion spectra of V854 Centauri. Panel a) shows a typical decline spectrum of this unusual R Coronae Borealis (RCB) Star. Note the strong Carbon lines. The C II] and C I lines are not seen in other RCB stars. Panels b) and c) show the spectral regions around C II] 2326 and Mg II 2800 at 4 epochs. Large variations in the line strengths and profiles are seen. These variations may be related to the pulsational phase of V854 Cen. Dust formation in this star takes place at phase 0.0 (Clayton et al. 1992a, ApJ, 384, L19; Whitney et al. 1992, AJ, 103, 1652; Clayton et al. 1992b, ApJ, 397, in press).

A spectroscopic study of RCB stars in the Galaxy and the LMC

Karen Pollard[1], P.L. Cottrell[1], W.A. Lawson[2]

[1] Mount John University Observatory, Department of Physics and Astronomy, University of Canterbury, Christchurch, New Zealand

[2] Department of Physics, University College ADFA, University of New South Wales, Canberra ACT 2600, Australia

Abstract

High resolution échelle spectra have been obtained of two Large Magellanic Cloud (LMC) and one galactic R Coronae Borealis (RCB) star with the Anglo–Australian Telescope. An analysis of these data using He– and C–rich models and the model atmosphere code WIDTH6 of Kurucz indicates that the galactic RCB star SU Tau and the two LMC stars, W Men and HV12842, have similar atmospheric parameters to the warmer galactic RCB and hydrogen–deficient Carbon (HdC) stars e.g., R CrB and XX Cam. Specifically, the new stars have T_{eff} ~7000 K, log g = 0.5–1.0, microturbulent velocities between 6 and 8 km.s^{-1} and C/He ratios from 0.004 to 0.006.

An abundance analysis has been performed on these stars for a wide range of species. Special emphasis has been placed on particular elements H, He, Li, C, N, O, & Fe (see figure 1), which are key indicators of the extent and relative importance of nuclear processes such as the CNO cycle and triple–alpha process, and also of the physical processes such as convective mixing and mass loss. Specific values for the abundances of [H/He], [Li/Fe] and [Fe/total] (expressed relative to the total abundance of all species) in SU Tau and HV12842 are -5.4 & -6.4, +2.5 & +2.4, -0.4 & -0.8.

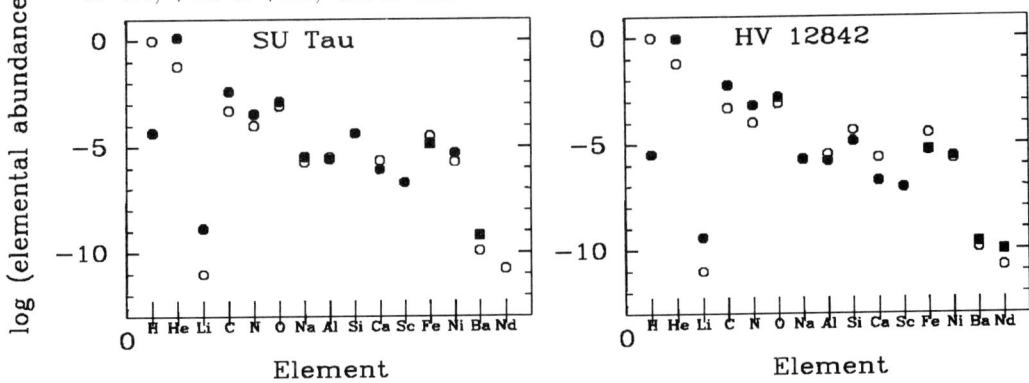

Figure 1 Elemental abundances for (a) SU Tau and (b) HV12842, compared with those for the Sun. The RCB stars are shown with filled symbols and the Sun with open symbols. This figure shows the same general trend of abundances for the two RCB stars, with the large enhancement of Li and N (and of course He and C). Hydrogen is depleted by more than 10^4. Iron is depleted in the LMC object by a factor of ~5.

A Photometric and Spectroscopic Study of Southern RV Tauri Stars

Karen Pollard, P.M. Kilmartin, A.C. Gilmore, P.L. Cottrell

Mount John University Observatory, Department of Physics and Astronomy, University of Canterbury, Christchurch, New Zealand

Abstract

A program to obtain photometric and spectroscopic (high and medium resolution) observations of a number of southern RV Tauri stars has been undertaken over the past two years at the Mount John University Observatory (MJUO). Eleven RV Tauri stars of both RVa (constant mean magnitude) and RVb (varying mean magnitude) photometric type have been chosen as well as normal and weak metal lined RV Tauri stars.

Most program stars display the alternating deep and shallow semi-regular light variations as well as the light curve – colour curve phase lag characteristic of RV Tauri stars. Fourier analyses of the light curves have revealed the dominant periodicities (see figure 1) and allowed phasing of the spectroscopic observations.

High resolution échelle spectra obtained of these stars around the Hα region display the complex emission and absorption structure of the Hα line at various phases. Metallic lines show emission and line doubling or 'splitting' – profiles characteristic of the shock wave that propagates through the line-formation regions of these stars during a pulsational cycle. Spectra at specific phases will be used in an abundance analysis of selected RV Tauri stars.

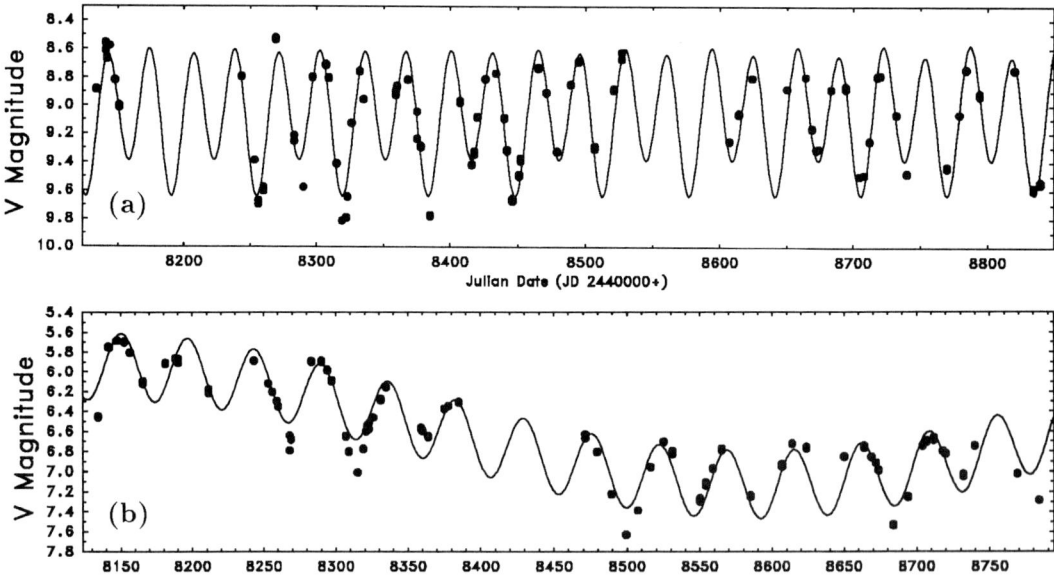

Figure 1 Comparison of the fit of the synthetic light curves (composed of periods determined from the Fourier analyses) to the MJUO photometric data of two southern RV Tauri stars, (a) RU Cen and (b) U Mon.

Hydrodynamical Models of Radially Pulsating Hot Extreme Helium Stars

Yu. A. Fadeyev

Institute for Astronomy of the Russian Academy of Sciences
and
Institute for Mathematics of Vienna University

Abstract

The blue edge of the instability region of radially pulsating helium stars is located on the H-R diagram nearly vertically at luminosities less than $10^4 L_\odot$ and corresponds to $T_{eff} = 7400K$ and $1.1 \cdot 10^4 K$ for stars with mass of $1 M_\odot$ and $0.7 M_\odot$, respectively. The pulsation instability of the stars located along the vertical part of the instability edge is characterised by the increasing order of the pulsation mode with decreasing luminosity. For example, for stars with $M = 0.7 M_\odot$ and $T_{eff} = 10^4 K$ the principal pulsation mode is nearly second overtone for $M_{bol} = -5$ mag and is nearly fourth overtone for $M_{bol} = -4$ mag. For stars brighter than $M_{bol} = -5$ mag the blue edge turns blueward so that the models with T_{eff} as high as $3 \cdot 10^4 K$ are pulsationally unstable. The sequences of the hydrodynamic models characterized by constant luminosity ($M_{bol} < -5$ mag) reveal the decrease of the light amplitude with increasing T_{eff}, whereas the amplitude of the radial velocities of the outer layers is almost independent of T_{eff}. For example, for stars with $T_{eff} > 2 \cdot 10^4 K$ the light amplitude is less than 0.01 mag, whereas the radial velocity amplitude is in the range from 20 to 50 km/s. The pulsation instability of these stars is driven mainly due to the γ-mechanism. Fourier analysis of the hydrodynamic solution shows that the pulsation motions of the hot helium stars can be represented as a superposition of the running waves (in contrast to Classical cepheids where the pulsation motions are described in the terms of superposition of standing waves). The pulsation constant gradually decreases with increasing T_{eff}, down to $Q \approx 0.012$ day at $T_{eff} \approx 3 \cdot 10^4 K$.

Theoretical Breakthroughs

OPAL Opacities

F. J. Rogers and C. A. Iglesias

Lawrence Livermore National Laboratory
Livermore, California 94550

Abstract

We have continued to improve and update the OPAL opacity code. Addition of intermediate coupling has further increased the opacity over earlier LS coupling results. A 'corresponding states' method has been used to extend the tables in both X and Z. This has allowed the calculation and distribution of extensive opacity tables for several different sets of metal abundance.

1. Introduction

The opacity of matter is crucial for studies of stellar evolution and pulsation. Recent improvements in opacity calculations have led to the resolution of a number of long-standing problems. For example, the mechanism for pulsation in β Cephei stars has been identified (Cox *et al.* 1992; Kiriakidis, El Eid & Glatzel 1992; Moskalik & Dziembowski 1992) and the "bump" and "beat mass" discrepancies in Cepheid variables has been removed (Moskalik, Buchler & Marom 1991). Many different mechanisms were introduced over the years to explain these discrepancies, but the resolution has proven ultimately to depend on improved input physics.

The success of the new OPAL opacities in the resolution of these problems has encouraged us to introduce additional improvements in the calculations and to produce updated and expanded tables. The main physics improvement has been the inclusion of spin-orbit effects in the atomic physics calculations. Furthermore, in order to facilitate the calculation of large databases we have developed a corresponding states method. We have also studied the effect of reducing the number of components in the mixture by combining the lesser abundant elements with prominent elements. A brief description of recent developments is given in the following sections.

OPAL is a completely new code having improved equation of state and atomic physics compared to that used to calculate the Los Alamos Opacity Library (Huebner *et al.* 1977, LAOL). It is also different in many respects from the Opacity Project (OP) (Seaton, this volume; 1987; Yu 1992). The equation of state is obtained from a many-body expansion of the grand canonical partition function (Rogers 1991) and the atomic physics is obtained from parametric potentials (Rogers, Wilson & Iglesias 1988). The atomic structure calculations are done on-line and have accuracy similar to single-configuration Hartree-Fock with relativistic corrections. Important improvements in the broadening of spectral lines, inverse bremsstrahlung, and Thomson scattering have also been made. A more detailed description of OPAL is given in Rogers and Iglesias (1992a).

2. Spin orbit interaction

The Los Alamos opacity codes use a detailed configuration accounting method (DCA) to treat bound-state absorption (Cox & Stewart 1965; 1970a; 1970b; Cox & Tabor 1976, Huebner 1986). Introduction of term splitting in the LS coupling scheme was responsible for a major part of the enhanced opacity obtained by OPAL compared to LAOL in the few hundred thousand degree range (Iglesias, Rogers & Wilson 1987; 1990; Iglesias & Rogers 1991a,b; Rogers & Iglesias 1992a). LS coupling is valid for low-Z elements where the electrostatic energy dominates. The spin-orbit interaction is very small for $Z < 10$, but increases with increasing Z leading to lifting of the J degeneracy and to the appearance of intercombination lines. The coupling in this case is intermediate between pure LS and pure jj. For intermediate-Z elements, such as iron, the spin-orbit interaction produces small, but important effects. The different coupling approaches are illustrated in Figure 1 for the case of transitions of the type sp to $p2$. In the DCA approach, there exists a single line. The LS coupling splits this line into three components corresponding to singlets and triplets. In intermediate coupling, the spin-orbit effect further splits the spectrum into 8 lines having $\Delta S = 0$ and 6 intercombination lines; 3 lines having $\Delta S = +1$, and 3 having $\Delta S = -1$. For more complicated configurations, the increase in the number of distinct lines can be much larger.

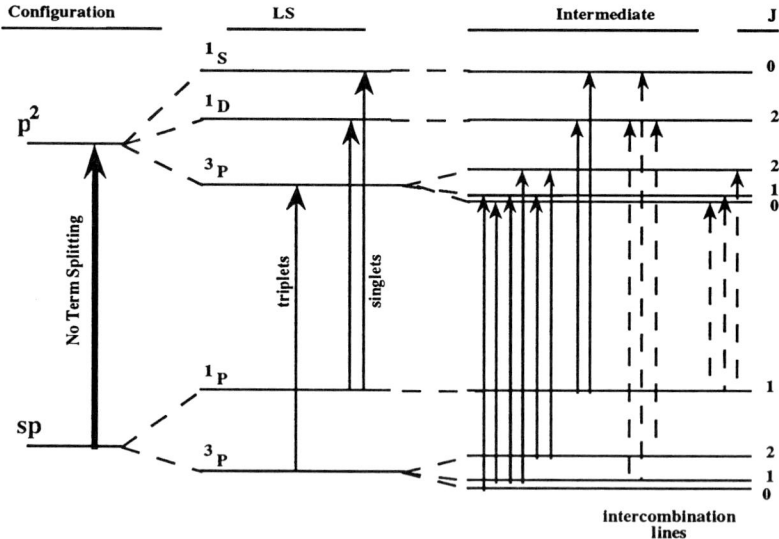

Figure 1. Schematic drawing showing various angular coupling schemes.

The splitting into additional lines distributes the oscillator strength more uniformly over frequencies and is similar to increasing the line width. The introduction of intermediate coupling can, thus, further increase the opacity when the spin-orbit "broadening" exceeds the line broadening. Since the spin-orbit effect does not depend on density, while line broadening does, its effects become more important at low density. This is illustrated in Figures 2(a-c), which compares the OPAL intermediate-coupling result to the LS coupling result for several values of R vs. T ($R = \rho/T_6^3$ and $T_6 = T/10^6$ K). The intermediate-coupling scheme gives an enhancement of 50% at log $R = -5$ and $T_6 = 0.25$, whereas the enhancement is insignificant at log $R = -2$. A more complete description of this work is given in Iglesias, Rogers & Wilson (1992).

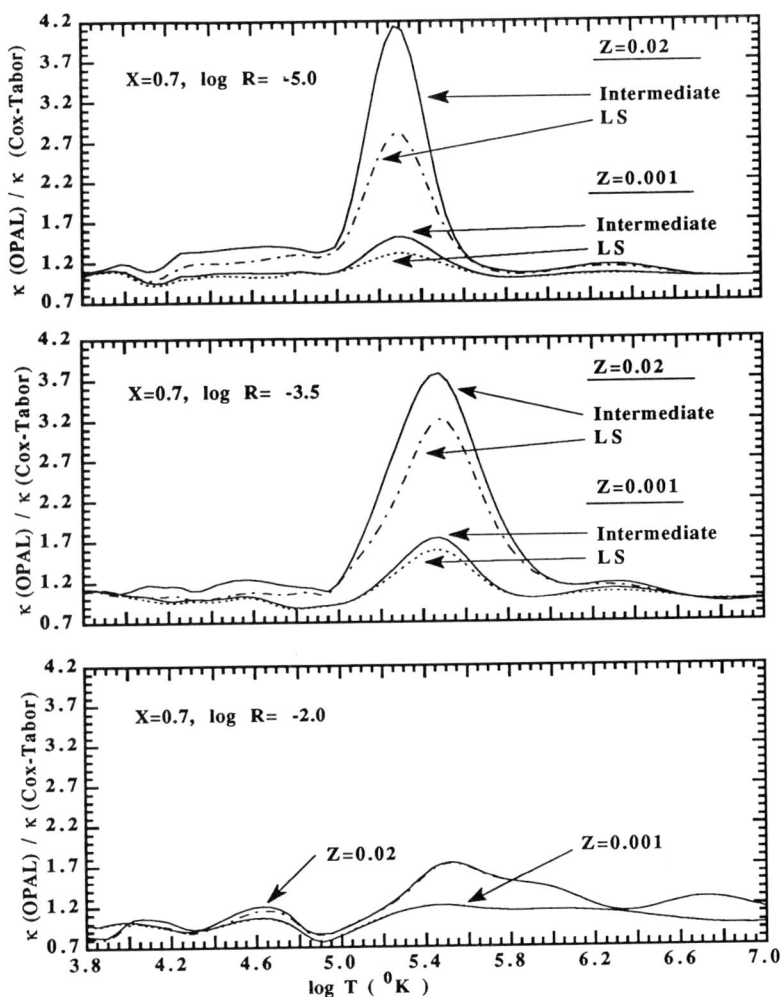

Figure 2. Intermediate coupling effects on the Rosseland mean opacity.

3. Corresponding States

The Los Alamos Opacity Library is a compilation of single-element calculations at specific values of T and electron chemical potential, μ_e. The photoabsorption coefficient for a mixture is obtained by adding these individual element cross-sections at fixed T and μ_e weighted by their abundance. The Rosseland mean opacity is then obtained by integrating the total over photon energy and the density of the resulting mixture is obtained from the ideal gas law of additive volumes.

In our work we calculate the equation of state for the full mixture, including the Coulomb coupling between the various constituents. Consequently, we cannot use the LAOL approach for mixing. However, it is possible to show that the intra-ionic ratios for ions of type i in the various states of ionisation can be the same for mixtures having different elemental mass fractions, χ_i, and densities. The conditions for this to occur is that the temperature and electron number density be the same. It follows that a solution for the equation of state for a mixture characterised by T, n_e, ρ, and χ_i can be used to find the equation of state for a different mixture characterised by T, n_e, ρ', and χ'_i. The electron density for the initial and final mixtures are given by

$$n_e = \rho \sum Q_i^* \chi_i / A_i \quad \text{and} \quad n_e = \rho' \sum Q_i^* \chi'_i / A_i \quad (1)$$

where Q_i^* is the average state of ionization for element i and A_i is the atomic weight. The densities of the two mixtures are thus related according to

$$\rho' = \rho \{\sum Q_i^* \chi_i / A_i\} \{\sum Q_i^* \chi'_i / A_i\}^{-1} \quad (2)$$

Since the relative intra-ionic occupation numbers are the same in the two mixtures, it is possible to rapidly calculate the opacity of the second mixture from the data already calculated and stored for the first mixture. We have used this approach to expand the tables in X and Z (Rogers & Iglesias 1992b). The current tables cover 10 values of X in the range 0 to $1 - Z$ and 13 values of Z in the range 0 to 0.1.

Figure 3 shows log κ_R vs. log T_6 for the extremes of X when $Z = 0.02$. A number of differences related to the H abundance in the mixture are apparent. There is a rapid rise in opacity at low temperatures which is caused by the ionisation of hydrogen when $X = 1 - Z$ and to the ionisation of helium when $X = 0$. In the latter, the higher ionisation potential of He shifts the onset of this rapid increase in κ_R towards higher temperatures. The opacity bump occurring near log $T_6 = -0.7$ is more pronounced for $X = 0$ and is due, in part, to the photoionisation of the K-shell electron in singly-ionized He. In addition, there are relatively more metals by number in the $X = 0$ mixture, further enhancing the bump near log $T_6 = 5.2$. Finally, the lower electron density of the $X = 0$ composition leads to lower photon scattering and free-free absorption contributions; consequently, its opacity is lower at high temperatures.

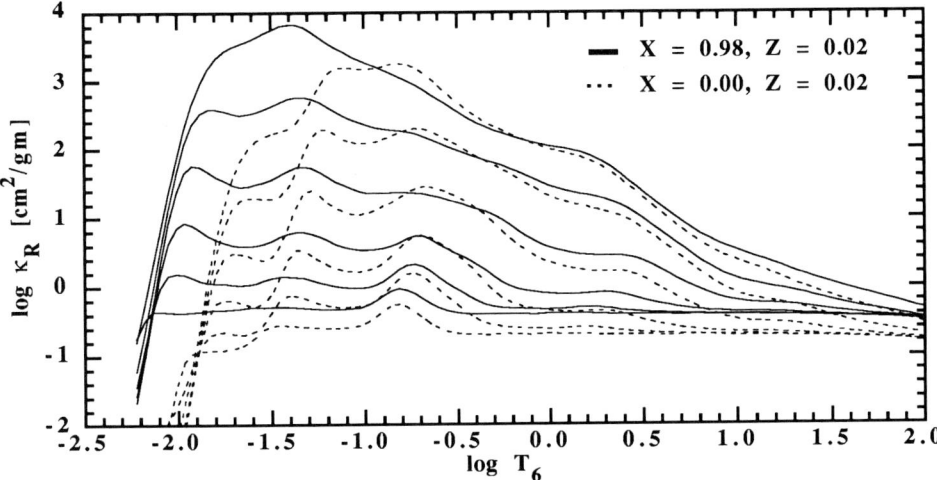

Figure 3. Comparison of κ_R for $X = 0$ and $1 - Z$. The curves are for constant R values: (from top to bottom) $\log R = -1, -2, -3, -4, -5$, and -7.

Due to the versatility of the 'corresponding states' method we can use the stored occupation number and photoabsorption data to rapidly calculate opacity tables covering the same range of X and Z for arbitrary heavy-element abundance. We are currently producing tables for advanced-type stars that have enriched He, C, and O abundance.

4. Abundance Effects

Metal abundance affects opacities in both fundamental and practical ways. The former arise from observational and calculational uncertainties in elemental number fractions. The latter results from the need to conserve computer time. Since some elements have very low abundance compared to their neighbors, it is common practice to lump the abundance of these elements with the more abundant neighbor. The opacity tables given in Rogers & Iglesias (1992a) were for H, He, and a twelve-element composition for Z. In particular, we added Cr and Ni together with Fe. Figure 4 shows the ratio of the OPAL opacities vs. temperature at $\log R = -5$ for a 16-element mixture that treats Cr, Fe, and Ni as distinct components in contrast to the 14-element mixture in which Cr and Ni have been combined with Fe. A substantial enhancement of about 28% occurs around $T_6 = 0.25$. Similar results have also been obtained by the Opacity Project (Seaton, this volume). Even so, this enhancement is small compared to the factor of 4 obtained over LOAL with the introduction of LS coupling.

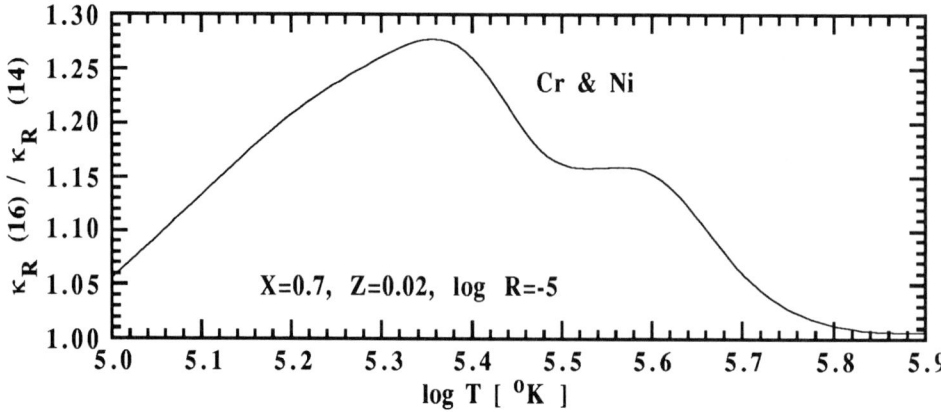

Figure 4. Effect of including Cr and Ni explicitly in mixture.

Figure 5 gives the frequency-dependent absorption cross-section for Cr, Fe, and Ni at $T_6 = 0.2$ and $\log R = -5$. Due to the Z^2 scaling of the energies, valleys in the Fe absorption are filled in by strong features in the Cr and Ni spectrum. In such cases, it is not a good approximation to treat the Cr and Ni as though their spectrum is the same as that of Fe. This was also noted earlier by Rogers & Iglesias (1992a), where only the enhancement due to Ni was considered. Revised tables including Cr and Ni are being calculated.

5. Experimental verification

Measuring the opacity at stellar interior conditions is very difficult and the capability has only recently been developed (Perry *et al.* 1991). The procedure involves use of two laser beams, one to heat a thin metal sample up to several hundred thousand degrees and the other to act as a backlighter source. The measured transmission of the backlighter source, $e^{-\rho L \kappa(\nu)}$, where ρ is the sample density and L the sample thickness, yields the frequency-dependent opacity, $\kappa(\nu)$.

The recent OPAL calculations predict large enhancements to astrophysical opacities due to transitions in the M-shell of iron; particularly the $\Delta n = 0$ lines around 70 eV. So far two experiments have been completed that test the validity of the OPAL code for treating absorption in iron. The first (Da Silva *et al.* 1992) studied pure Fe at $\log T \simeq 5.4$ and $\rho \simeq 0.008$ g/cm^3 and was designed to observe the prominent $\Delta n = 0$ features. Since the calculations predict that these features persist over a wide range of temperature and density, the diagnostics in the experiment were not designed to make accurate measurements of plasma conditions. As a result, the temperature uncertainty is 20% and the density is estimated to within a factor of two.

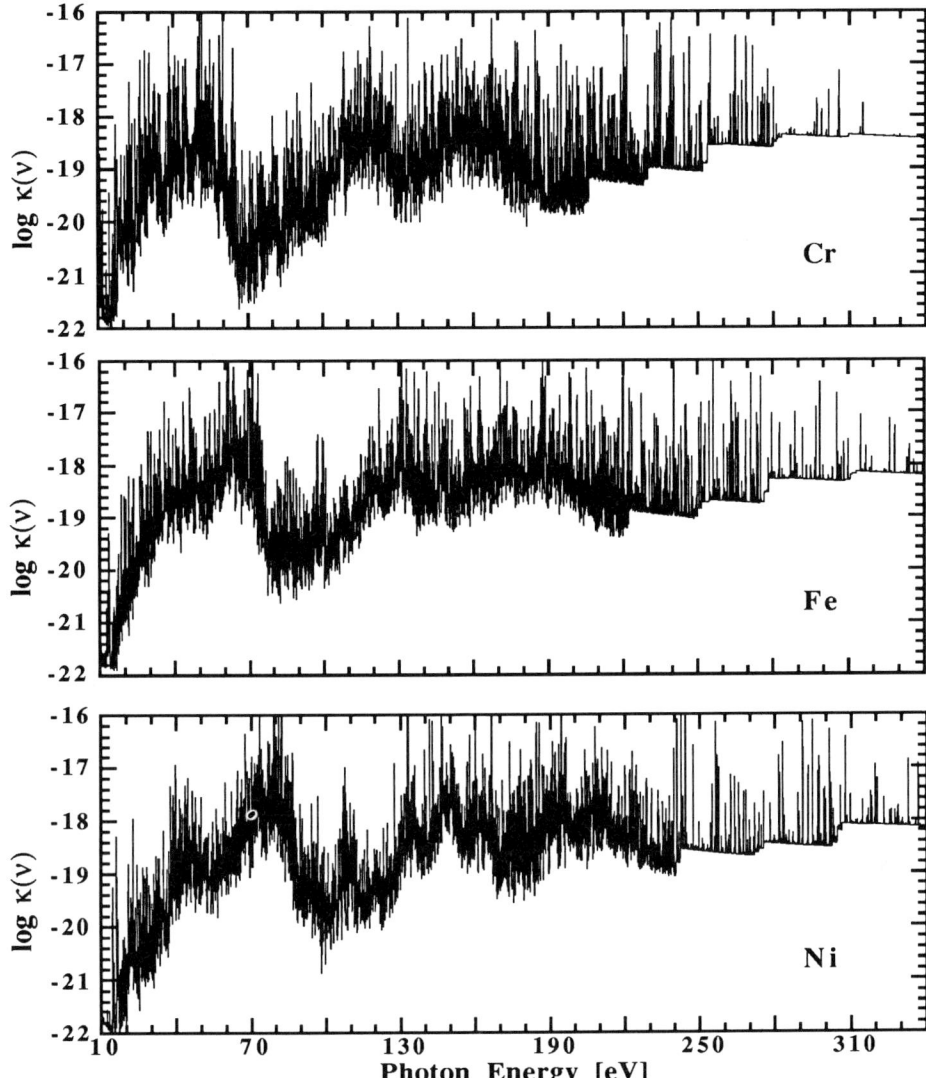

Figure 5. Photoabsorption coefficients (cm²/nuclei) for Cr, Fe, and Ni for a plasma at $\log T = 5.3$ and $\log R = -5$. Note similarity, but also small shift towards higher photon energy with increasing nuclear charge.

The results of the Da Silva *et al.* experiment are compared with two versions of OPAL in Figure 6. It is apparent that the DCA calculation, which approximates the LAOL calculations, has no absorption features in the crucial 70 eV photon energy range; whereas both the OPAL calculation with full intermediate coupling and the experiment show strong absorption features in this range. The experiment thus offers confirmation that the OPAL prediction is qualitatively correct.

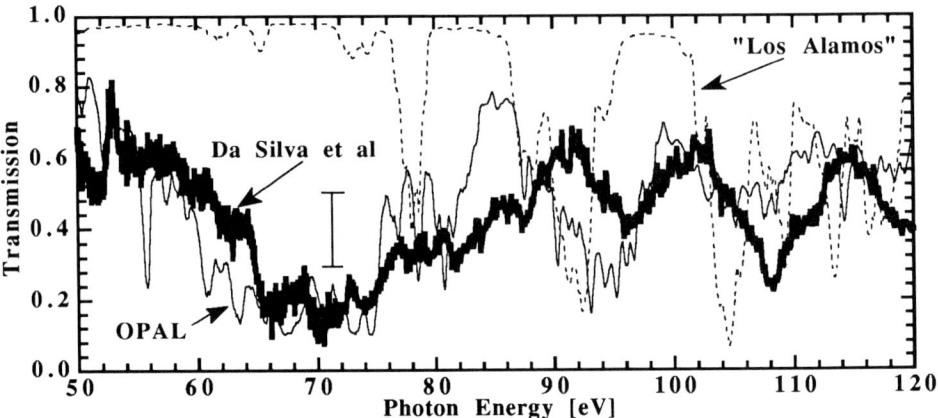

Figure 6. The Fe transmission experiment (heavy solid line) is compared to OPAL with intermediate coupling (thin solid line) and OPAL in DCA with hydrogenic oscillator strengths (dashed line) simulating the LAOL calculations. Experimental error bars for the transmission are indicated in the figure.

The second experiment (Springer et al. 1992) introduces a small impurity of NaF in the Fe sample in order to determine the plasma temperature using an established method (Perry et al. 1991). They also developed a technique to determine the density accurately. The purpose of this experiment was to measure the frequency-dependent absorption over the energy range that contributes most to the Rosseland mean in a well-characterised experiment; the first such experiment in a hot plasma. The conditions were determined to be $kT = 59 \pm 3$ eV and $\rho = 0.0135 \pm 0.0014$ g/cm^3. Integration over the experimentally obtained $\kappa(\nu)$ gave $\kappa_R = 4400 \pm 600$ cm^2/g. The OPAL result for nominal conditions is 4100 cm^2/g. This experiment, thus, also supports the OPAL results.

6. Conclusion

We have continued to improve and update the OPAL opacity calculations. Substantial increases in opacity over our earlier work were obtained with the introduction of intermediate coupling; particularly at low density where the spin orbit splitting can be greater than the spectral line broadening. Significant increases in opacity were also obtained by explicitly including the lowly abundant Cr and Ni in the metal mixture; clearly showing that even elements of very low abundance can make important contributions when their spectrum fills in valleys where no other element has absorption features. This again demonstrates that accurate opacities also require an accurate determination of the stellar element composition.

The use of OPAL opacities has led to the resolution of a number of long-standing discrepancies between theory and observation in pulsating stars. The OPAL opacities have also improved the agreement with observation on a number of other problems (Cox 1991; Kovacs, Buchler & Marom 1991; Stothers 1992; Stothers & Chin 1992; El Eid & Hartman 1992). Recent laser experiments on thin metal foils have also corroborated the large $\Delta n = 0$ contribution from M-shell iron and verified the Rosseland mean at one temperature-density point. This is strong evidence that the OPAL opacities represent a significant improvement over LAOL. The close agreement of OP and OPAL (Seaton, this volume) in the range where OP is valid ($\rho < 0.01$ g/cm^3) is additional strong evidence supporting both efforts.

We are indebted to B.G. Wilson for the angular momentum coupling code in the atomic data generation and R.W. Lee for his linear Stark broadening subroutines. Additional thanks are due to R.W. Lee for providing us with the experimental Fe transmission data. Work performed under the auspices of the U. S. department of Energy by the Lawrence Livermore National Laboratory under contract W-7405-Eng-48.

References:

Cox, A. N. 1991, ApJ, 381, L71

Cox, A. N., Morgan, S. M., Rogers, F. J., & Iglesias, C. A. 1992, ApJ, 392, 272

Cox, A. N., & Stewart, J. N., 1965, ApJS, 11, 22

—. 1970a, ApJS, 19, 243

—. 1970b, ApJS, 19, 261

Cox, A. N., & Tabor, J. E. 1976, ApJS, 31, 271

Da Silva, L. B., MacGowan, B. J., Kania, D. R., Hammel, B. A., Back, C. A., Hsieh, E., Doyas, R., Iglesias, C. A., Rogers, F. J., & Lee, R. W. 1992, Phys. Rev. Lett., 69, 438

El Eid, M. F. & Hartman, D. H., 1992, ApJL (submitted)

Huebner, W. F. 1986, Physics of the Sun, Vol. 1, ed. P. A. Sturrock, E. Holzer, D. M. Mihalas, & R. K. Ulrich (Dordrecht: Reidel), p. 33

Huebner, W. F., Merts, A. L., Magee, N. H., & Argo, M. F. 1977, Los Alamos Scientific Report LA-6760-M

Iglesias, C. A., & Rogers, F. J. 1991a, ApJ, 371, 173

—.1991b, ApJ, 371, 408

Iglesias, C. A., Rogers, F. J., & Wilson, B. G. 1987, ApJ, 322, L45

—. 1990, ApJ, 360, 221

—. 1992, ApJ (October)

Kiriakidis, M., El Eid, M. F., & Glatzel, W. 1992, MNRAS (Letters), 255, 1

Kovacs, G., Buchler, J. R., & Marom, A. 1991, A&A, 25, 685

Moskalik, P., & Dziembowski, W. A. 1992, A&A, 256, L5

Perry, T. S., Davidson, S. J., Serduke, F. J. D., Bach, D. R., Smith, C. C., Foster, J. M., Doyas, R. J., Ward, R. A., Iglesias, C. A., Rogers, F. J., Abdallah, Jr., J., Stewart, R. E., Kilkeny, J. D., Lee, R. W. 1991, Phys. Rev. Lett., 67, 3784

Rogers, F. J. 1991, in High Pressure Equations Of State: Theory and Applications, ed. S. Eliezer & R. A. Ricci (North Holland, New York, 1991)

Rogers, F. J., & Iglesias, C. A. 1992a, ApJS, 79, 507

—. 1992b, ApJ (December)

Seaton, M. J. 1987, J. Phys. B, 20, 6263

Springer, P. T., Fields, D. F., Wilson, B. G., Nash, J. K., Goldstein, W. H., Iglesias, C. A., Rogers, F. J., Swenson, J. K., Chen, M. H., Bar-Shalom, A., & Stewart, R. E. 1992, Phys. Rev. Lett. (submitted)

Stothers, R. B., & Chin, C.-W., 1992, ApJ, 390, 136

Stothers, R. B. 1992, ApJ, 392, 706

Swenson, F. J., Stringfellow, G., & Faulkner, J. 1990, ApJ, 348, L33

Yu, Y. 1992, Rev. Mex. Astron. Astrof., 23, 171

The Opacity Project

M.J. Seaton

Department of Physics and Astronomy, University College London,
Gower St., London WC1E 6BT, UK.

Abstract

The paper gives a brief summary of the work of the Opacity Project.

For the purposes of opacity calculations one may divide stellar interiors into two regions: envelopes with mass-densities, ρ, less than about 10^{-2} g cm^{-3}; and deeper interiors. In stellar envelopes complex atomic systems exist and are not markedly perturbed by the plasma environment. It follows that, for the calculation of envelope opacities, a main need is to have accurate and extensive data for the radiative properties of free atoms and atomic ions. For deeper interiors the nature of the problem changes, one is concerned with simpler radiative processes but plasma perturbations may be of major importance.

The work of the international Opacity Project (to be referred to as OP), has been concerned with the calculation of envelope opacities, which are of particular inportance for studies of stellar pulsations.

The atomic data required are: energy levels; oscillator strengths; photo-ionisation cross-sections; and line-profile parameters. Calculations are made for all cosmically-abundant elements in all stages of ionisation. The main atomic-physics calculations are made using R-matrix methods and a large team of workers has been involved. Table 1 gives a list of papers on the OP atomic-physics work published in the *Journal of Physics B*. That work is discussed further by Seaton *et al.* 1992.

For the calculation of opacities one also requires information concerning the populations of the atomic energy-levels — the problem of the equation of state (EOS). The OP approach is to introduce occupation probabilities $W(i)$ for each level i, such that $W(i) \to 0$ for sufficiently large values of i (highly-excited states), giving convergent partition functions. Table 2 gives a list of papers concerned with the OP EOS work, together with other OP papers published in the *Astrophysical Journal*.

We find that, to a good approximation for envelopes, the level populations depend only on temperature, T, and electron-density, N_e. The OP opacity calculations are done twice, using codes which are largely independent: firstly in Urbana (Illinois) and Columbus (Ohio) using CRAY-YMP machines at the National Center for Supercomputer Applications of the University of Illinois and at the Supercomputer Center of Ohio State University; and secondly in London using the IBM-3090 at the Rutherford and Appleton Laboratory. Close agreement between the USA and UK results gives us added confidence in their being correct. We compute and archive monochromatic opacities for each chemical element on a grid of (T, N_e)-values. These monochromatic opacities can then be added for any required chemical mixture and used to calculate Rosseland-mean opacities, κ_R. We obtain tables giving values of ρ and κ_R as functions of (T, N_e). Interpolation routines give κ_R for any required values of ρ and T. These routines can also be used to produce tables in OPAL format, κ_R as a function of T for fixed values of $\log(R)$ where $R = \rho/T_6^3$, ρ is mass density in g cm^{-2} and $T_6 = 10^{-6} \times T$ with T in K.

A full account of the OP opacity work will be given in a paper by M.J. Seaton, Yu Yan, D. Mihalas & A.K. Pradhan to be submitted to *Monthly Notices of the Royal Astronomical Society*. The main questions of interest for the present meeting are: how do the OP opacities compare with earlier results from the Los Alamos Opacity Library (LAOL, see Weiss, Keady & Magee (1990) which gives references to earlier work); and with more recent results from the OPAL project (see Rogers & Iglesias (1992) and references therein, and the contribution by Rogers in the present volume)? The short answer is that, compared with LAOL, OP gives enhancements at least as large as those postulated by Simon (1982), and that the OP and OPAL results are in good general agreement.

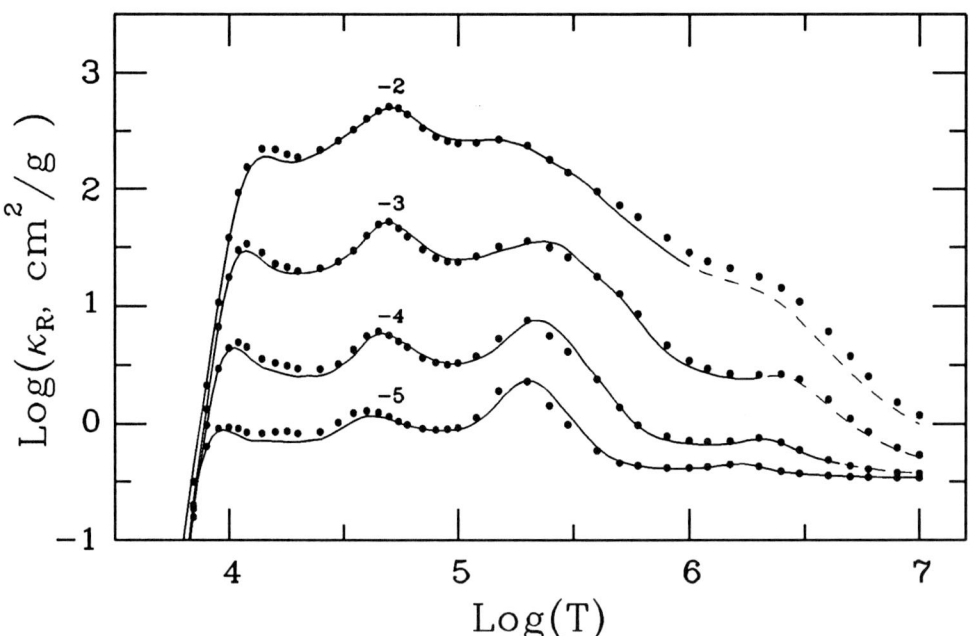

Fig. 1. Rosseland mean opacities for X=0.7, Z=0.02 and relative metal abundances from Anders & Grevesse (1989). Full lines from OP, filled circles from OPA . The four curves are for $\log(R) = -1, -2, -3,$ and -4 where $R = \rho/T_6^3$.

Fig. 1 gives a detailed comparison for just one case, $X = 0.7$, $Z = 0.02$ and relative abundances of "metals" from Anders & Grevesse (1989). For the benefit of users of the opacities I employ the latest results from both opacity projects available at the time of the meeting. Logarithms of Rosseland mean opacities are plotted against $\log(T)$ for four values of $\log(R)$. The following points may be noted.

1. The OP results are plotted with full lines for regions which we define as envelopes, $\log(\rho) \leq -2$, and with dashed lines for higher densities. The OPAL calculations extend to much higher values of $\log(\rho)$.

2. For individual transitions, the atomic data used by OP are more accurate than those used by OPAL. Differences in atomic data may not be so important when one considers the accumulative effects of large numbers of transitions.

3. Both calculations are made with inclusion of fine-structure for transitions in iron. OP uses LSJ coupling while OPAL uses intermediate coupling and hence includes inter-combination lines omitted by OP. At low temperatures ($\log(T)$ of about 4.3), OPAL obtains enhancements, due to inclusion of fine-structure, which are larger than those obtained by OP.

4. The OP calculations have been made with inclusion of the iron-group elements Cr, Mn and Ni. These elements, which have abundances much smaller than that of iron, make contributions to opacities which are small but none-the-less significant. OPAL calculations including these elements have been made only for a small number of temperature-density points.

5. Much of the pulsation work is concerned with values of $\log(R)$ in the range -3 to -5. The main difference between the new opacities, from OPAL and OP, and the old opacities from LAOL is in the magnitude of the maxima which occur at $\log(T) \simeq 5.3$ (the "Z-bump"). The OPAL and OP results are in quite good agreement in that region, although OP gives the maxima shifted to slightly higher temperatures. Some recent calculations show that pulsation results can show surprising sensitivities to the exact values of opacities in the vicinity of the Z-bump.

Other speakers at this meeting discuss the improved agreement between pulsation theory and observations which results from use of the new opacities. A glance at Tables 1 and 2 may convince the reader of the magnitude of the task involved in the opacity work. I think it very fortunate that the new calculations have been made by two entirely independent projects and that the results obtained are in broad agreement. Where there are differences it would be hard to say which results are to be preferred. It may be wise to consider such differences as indicative of the level of uncertainty in both sets of calculations.

I thank all members of the OP team for their cheerful collaboration. I also thank Carlos Iglesias and Forrest Rogers for many helpful discussions and for providing, in advance of publication, their results used in Figure 1.

References:

Anders, E. & Grevesse, N. 1989, *Geochim. Cosmochim. Acta*, **53**, 197.
Rogers, F.J. & Iglesias, C.A. 1992, *Ap. J. Supp.*, **79**, 507.
Seaton, M.J., Zeippen, C.J., Tully, J.A., Pradhan, A.K., Mendoza, C., Hibbert, A. & Berrington, K.A. 1992, *Rev. Mex. Astrn. Astrophys.*, **23**, 19.
Simon, N. R. 1982, *Ap. J.*, **260**, L87.
Weiss, A., Keady, J.J. & Magee, N.H. 1990, *Atomic Data Nucl. Data Tables*, **45**, 209.

DISCUSSION

C.A. IGLESIAS: The uncertainties in the abundances are likely to overwhelm the 20% discrepancies between the OP and OPAL results.

M.J. SEATON: True. But we can always hope that the accuracy of abundances will improve.

N.R. SIMON: How much more are the opacities likely to increase?

M.J. SEATON. Our sins may be of omission rather than of commission. One can go on thinking of processes which should be included and which may lead to further opacity increases. But I do not think that we can give you any further factors of 3 or more.

D. WELCH: Are monochromatic opacity tables available on request?

M.J. SEATON: In principle, yes. But beware, the amounts of data may be very large, despite our "packing" of the archived files (we use the minimum number of frequency points which allow linear interpolations to some prescribed accuracy, usually taken to be 1 or 2%).

R. STELLINGWERF: Do you expect the Rosseland mean opacities to be "smooth" in temperature and density? This is a basic assumption when interpolating in tables.

M.J. SEATON: So far as the tables themselves are concerned, the question of "smoothness" does not arise. It all depends on the interpolation procedure used. Cubic splines will give smoothness in the functions and first derivatives. Something better may be needed. I intend to give some thought to that.

Table 1. OP papers published in the *Journal of Physics B.*

Papers in the series "Atomic data for Opacity Calculations":-

    I. Seaton, General formulation. **20**, 6363, 1987.

    II. K.A.Berrington, P.G.Burke, K.Butler, M.J.Seaton, P.J.Storey, K.T.Taylor & Yu Yan, Computational Methods. **20**, 6379, 1987.

    III. Yu Yan, K.T.Taylor & M.J.Seaton. Oscillator strengths for C II. **20**, 6399, 1987.

    IV. Yu Yan & M.J.Seaton. Photoionisation cross sections for C II. **20**, 6409, 1987.

    V. M.J.Seaton. Electron impact broadening of some C III lines. **20**, 6431, 1987.

    VI. J.F.Thornbury & A.H.Hibbert. Static dipole polarisabilities of the ground states of the helium sequence. **20**, 6447, 1987.

    VII. J.A.Fernley, K.T.Taylor & M.J.Seaton. Energy-levels, f-values and photo-ionisation cross sections for He-like ions. **20**, 6457, 1987.

    VIII. M.J.Seaton. Line-profile parameters for 42 transitions in Li-like and Be-like ions. **21**, 3033, 1988.

    IX. G.Peach, H.E.Saraph & M.J.Seaton. The lithium iso-electronic sequence. **21**, 3669, 1988.

    X. D.Luo, A.K.Pradhan, H.E.Saraph, P.J.Storey & Yu Yan. Oscillator strengths and photoionisation cross sections for O III. **22**, 389, 1989.

    XI. D.Luo & A.K.Pradhan. The carbon iso-electronic sequence. **22**, 3377, 1989.

    XII. M.J. Seaton, Line-profile parameters for neutral atoms of He, C, N and O. **22**, 3603, 1989.

    XIII. M.J. Seaton. Line profiles for transitions in hydrogenic ions. **23**, 3255, 1990.

    XIV. J.A.Tully, M.J.Seaton & K.A.Berrington. The berrylium sequence. **23**, 3811, 1990.

    XV. P.M.J. Sawey & K.A. Berrington, Fe I to IV. **25**, 1451, 1992.

    XVI. H.E. Saraph, P.J. Storey & K.T. Taylor, *Ab initio* calculations for Fe VIII and Fe VII. In press.

Other papers:-

- M.J. Seaton, On the R-matrix method for bound-state calculations: I. General theory. **18**, 2111, 1985.
- K.A. Berrington & M.J. Seaton. On the R-matrix method for bound-state calculations. II Results for energy levels of C;+. **18**, 2587, 1985.
- M.J. Seaton, Outer-region contributions to radiative transition probabilities. **19**, 2601, 1986.

Table 2. OP papers published in the *Astrophysical Journal*

Papers in the series "Equations of State for Stellar Envelopes":-

  I. D.G. Hummer & D. Mihalas. An occupation probability formalism for the truncation of the internal partition function. **331**, 794, 1988.
  II. D. Mihalas, W. Däppen & D.G. Hummer. Algorithms and selected results. **331**, 815, 1988.
  III. W. Däppen, D. Mihalas, D.G. Hummer & B.W. Mihalas. Thermodynamic quantities. **332**, 261, 1988.
  IV. D. Mihalas, D.G. Hummer, B.W. Mihalas & W. Däppen. Thermodynamic quantities and selected ionization fractions for six elemental mixes. **350**, 300, 1990.

Other papers:-

- Däppen, W., Anderson, L.S. & Mihalas, D. "Statistical mechanics of partially ionised stellar plasmas", **319**, 195, 1987.
- D.G. Hummer, "A fast and accurate method for evaluating the nonrelativistic free-free Gaunt factor for hydrogenic ions", **327**, 477, 1988.

The Bump Cepheid Mass Discrepancy Laid to Rest (?)

P. Moskalik[1] and J. R. Buchler[2],

[1]Copernicus Astronomical Center, Warsaw, Poland, [2]University of Florida, Gainesville, USA

1. New Opacities

The beat and bump mass discrepancies have been a long standing, unsolved difficulty in the Cepheid modelling (Cox 1980). The new opacities of Rogers & Iglesias (1992) has provided a partial solution to the problem, bringing the beat masses into good agreement with other mass determinations (Moskalik, Buchler & Marom 1992; herafter MBM). The discrepancy for the bump masses has also been greatly reduced, nevertheless it has not been eliminated entirely.

Recently, the new version of the Livermore tables has been released (Iglesias, Rogers & Wilson 1992). The new calculations take into account previously disregarded spin-orbit coupling in the iron atoms. The inclusion of this effect leeds to a further enhancement of the "metal opacity bump" around 2-5×10^5K. In the following we repeat the calculations of MBM to assess the consequences of this enhancement for the bump mass calibration.

2. New Bump Masses

The Cepheid bump progression has its origin in a 2:1 resonance between the fundamental mode and the second overtone (Simon & Schmidt 1976; Buchler, Moskalik & Kovács 1990). The requirement of placing the center of the progression at $10^d (\pm 0.^d 5)$ is equivalent to the requirement of placing the resonance at this period. Thus, the bump mass problem can be studied with the linear models alone.

Following MBM we start from determining the luminosity of a 10^d Cepheid. Averaging four independent $P-L$ relations (Caldwell & Coulson 1987; Gieren 1988; Walker 1988; Fernie 1992) we obtain $M_V = -4.^m 11$. The largest difference between this value and a prediction of any particular $P-L$ relation (at 10^d) is $0.^m 04$, but we feel that $0.^m 10$ is a more realistic error estimation. Adopting the bolometric correction scale of Gieren (1989) we find $L = 3800 L_\odot \pm 10\%$. Next, we construct the period ratio diagram (Petersen diagram) of P_2/P_0 vs. P_0 for models with $L = 3800 L_\odot$ and with different masses. From that diagram we find that the resonance condition ($P_2/P_0 = 0.5$ at 10^d) is satisfied for $M = 5.90 \pm 0.15 M_\odot$ if $Z=0.02$ is assumed, or for $M = 6.65 \pm 0.15 M_\odot$ if $Z=0.03$ is assumed (quoted errors correspond to the adopted error in luminosity). These are our new bump masses.

3. Baade-Wesselink Masses

Gieren's (1989) average Baade-Wesselink mass of a 10^d Cepheid is $6.52 \pm 0.9 M_\odot$. This value, however, was obtained using the period-radius-mass relation derived with the older Los Alamos opacities. The models constructed with the new opacities have slightly longer periods, and the resulting $P-R-M$ relation (for $Z = 0.02$) is

$$P_0 = 0.026(M/M_\odot)^{-0.68}(R/R_\odot)^{1.70} \tag{1}$$

Repeating Gieren's (1989) procedure with Eq. (1) we find the average trend of the Baade-Wesselink masses with period, which for a 10^d Cepheid gives $M = 6.81 \pm 0.9 M_\odot$. The same excercise repeated for $Z = 0.03$ (slightly different $P-R-M$ relation) leads to $M = 6.92 \pm 0.9 M_\odot$.

4. Conclusions

Our results are summarized in the table below, where we also present the evolutionary masses inferred from the standard $M-L$ relation of Becker, Iben & Tuggle (1977). The new bump masses *do agree* within the error bars with the Baade-Wesselink masses for both $Z = 0.02$ and for $Z = 0.03$, with a better agreement for higher Z. There is still a disagreement, however, between the bump and the evolutionary masses. The increase of metallicity does not help here, because the evolutionary masses also grow quickly with Z. This discrepancy can be reduced, though, by placing the resonance at a longer period, as shown in the last row of the table.

	metallicity	$M_{bump}/M_\odot$	$M_{BW}/M_\odot$	$M_{EV}/M_\odot$
$P_{rez} = 10^d.0$	$Z = 0.02$	5.90 ± 0.15	6.81 ± 0.9	7.04
	$Z = 0.03$	6.65	6.92	8.20
$P_{rez} = 10^d.5$	$Z = 0.02$	6.21	6.94	7.15

References:

Becker, S. A., Iben, I. & Tuggle, R. S. 1977, *Astrophys. J.*, **218**, 633
Buchler, J. R., Moskalik, P. & Kovács, G. 1990, *Astrophys. J.*, **351**, 617
Caldwell, J. A. R. & Coulson, I. M. 1987, *Astron. J.*, **93**, 1090
Cox, A. N. 1980, *Ann. Rev. Astron. Astrophys.*, **18**, 15
Fernie J. D. 1992, *Astron. J.*, **103**, 1647
Gieren, W. P. 1988, *Astrophys. J.*, **329**, 790
Gieren, W. P. 1989, *Astron. Astrophys.*, **225**, 381
Iglesias, C. A., Rogers, F. J. & Wilson, B. G. 1992, *Astrophys. J.* (submitted)
Moskalik, P., Buchler, J. R. & Marom, A. 1992, *Astrophys. J.*, **385**, 685 (MBM)
Rogers, F. J. & Iglesias, C. A. 1992, *Astrophys. J. Suppl.*, **79**, 507
Simon, N. R. & Schmidt, E. G. 1976, *Astrophys. J.*, **205**, 162
Walker, A. R. 1988, in *The Extragalactic Distance Scale*, eds. S. van den Bergh & C. J. Prichet, ASP Conf. Ser 4, p. 89

Discussion

Simon: 1) Evolutionary tracks with OPAL have been done by Stothers & Chin and they find little difference from standard tracks. 2) Using Baade-Wesselink masses is very tricky; they are notoriously uncertain.

Moskalik: I agree that Baade-Wesselink masses usually have large errors. However, they are the closest to the "observational" masses we can get, depending only on the rather well established pulsational $P-R-M$ relation. The evolutionary masses have small formal errors, but they heavily depend on the theoretical $M-L$ relation. Comparing bump masses with evolutionary masses we compare one theory with another. In my mind, the comparison with the Baade-Wesselink masses, which are based primarily on observed quantities, is more fundamental.

Sreenivasan: How can you be certain that there has been no mass loss and that in fact no discrepancy exists between evolutionary masses and pulsation masses (mass loss produces overluminous stars for their mass as compared to conservative mass evolution) ?

Moskalik: I cannot be certain, this is why I prefer the comparison with Baade-Wesselink masses. Another uncertainity in the evolutionary models is the precise amount of convective overshooting. Our bump masses are already higher than the full-overshooting evolutionary masses of Chiosi.

Cox: Do you use convection in your models ? Poster 72 shows that convection in deep layer changes the structure and gives bump masses about $7M_\odot$.

Moskalik: Our models are purely radiative.

Percy: Although you and other theorists construct models with $Z=0.02$ and $Z=0.03$, those who measure abundances of Pop. I objects prefer $Z=0.02$ (actually $Z=0.016-0.020$).

Moskalik: Well, theorists will do anything to make their models pulsate right. Seriously, I think, that it tells us that the true metal opacities are still a little bit higher.

Pel: In relation to the high Z-sensitivity of your bump masses, it would be important to compare the Hertzsprung progression for Galactic and Magellanic Cloud Cepheids. It is known that the "resonance period" of the Hertzsprung progression of Cloud Cepheids has a slightly different value with the respect to the Galactic Cepheids. This way you could make a nice differential check on the Z-dependence of your bump masses.

Comparative Pulsation Calculations with OP and OPAL Opacities

S.M. Kanbur[1], N.R. Simon[1]

[1]*Department of Physics and Astronomy, University of Nebraska-Lincoln, NE 68588-0111*

The OPAL opacities (Iglesias and Rogers 1991; Iglesias, Rogers and Wilson 1992) are recent revisions to the traditional Los Alamos (LA) opacities and have proven very successful in resolving the Cepheid beat mass, and to a lesser extent, the Cepheid bump mass discrepancies (Moskalik, Buchler and Marom 1991, hereafter MBM). MBM showed that for the beat Cepheids the P1/P0 period ratios, calculated with OPAL, yielded solar masses between 4 and 7, in agreement with other mass determinations. MBM also found that the P2/P0 period ratios were reduced, implying higher bump Cepheid masses, though the conflict with standard evolutionary masses was not completely eliminated. We have made linear nonadiabatic pulsation calculations with another set of recently computed opacities, namely those of the Opacity Project (Seaton 1987, 1992, hereafter OP). Since the two teams have employed different techniques in their calculations, it is of interest to compare pulsation results using the two sets of opacities.

Our conclusion is that the OPAL and OP opacities cannot be differentiated on the basis of our calculations. The Becker Iben Tuggle (1977) mass luminosity relation is sufficient for the beat Cepheids, but has too high a period ratio (by 0.01) at a fundamental period of 10 days for bump Cepheids. The mass luminosity relation involving substantial core overshoot (Chiosi 1988) can model both beat and bump Cepheids, but with lower (by about a solar mass) masses.

References:

Becker, S.A., Iben, I. & Tuggle, R.S., 1977, Astrophys. J., **218**, 633.
Chiosi, C., 1989 in 'The Use of Pulsating Stars in Fundamental Problems of Astronomy', Cambridge University Press, p 19.
Iglesias, C.A., & Rogers, F.J., 1991, Astrophys. J., **371**, 408.
Iglesias, C.A., Rogers, F.J., & Wilson, B. 1992, preprint.
Moskalik, P., Buchler, J. R., & Marom, A., 1992, Astrophys. J., **385**, 685 (MBM).
Seaton, M.J., 1987, J. Phys. B., **20**, 6363.
Seaton, M.J., private communication.

Oosterhoff I and II RR Lyrae Variable Masses

Arthur N. Cox

Los Alamos Astrophysics

Abstract

The anomalously low theoretical masses for double-mode RR Lyrae variable stars obtained from accurately observed period ratios recently have been increased by using stellar models constructed with the new Livermore OPAL opacities. These new models appear to the lowest order radial modes to be less concentrated in density, and they thus produce longer periods. Since the fundamental mode is increased in period more than the first overtone, this period ratio from the model is decreased. Cox (1991) showed that now these pulsation-based masses agree with evolution masses, and all RR Lyrae variable stars show little mass loss from the red giant tip in earlier evolution. This investigation has been validated by two additional papers that use more directly the OPAL opacities rather than simple adjustments to the Stellingwerf (1975ab) fit. These new papers (Kovacs, Buchler, and Marom, 1991, and Kovacs, Buchler, Marom, Iglesias, and Rogers, 1992) also display composition effects for Population II mixtures.

My new study uses directly the latest OPAL intermediate coupling opacities for the Oosterhoff I and II (Z=0.0003 and 0.0001, respectively) compositions in frozen-in convection (mixing length/pressure scale height ratio=1.25) models to see if there is any difference between the masses of variable stars in these two globular cluster classes. Derived masses are just over 0.65 and 0.75 $M_\odot$, respectively. The Oosterhoff I mass could be as large as the Oosterhoff II mass only if the iron abundance in the Z mixture is increased by more than 30% over that for the solar mixture.

The Sandage effect of higher periods for variables in Oosterhoff II clusters at lower Z than for the Oosterhoff I clusters might be explained with my new masses by having the Oosterhoff II variables born much to the blue of the instability strip and evolving at increasing luminosity to the red. For Oosterhoff I variables, they can be born at their lower mass and higher Z inside the instability strip at lower luminosity, evolving both to the red and blue with little evolutionary luminosity increase before they become red nonvariables.

References:

Cox, A. N., 1991, Ap. J. Lett., **381**, L71.
Kovacs, G., Buchler, J. R., and Marom, A., 1991, Astron. Astrophys., **252**, L27.
Kovacs, G., Buchler, J. R., Marom, A., Iglesias, C. A. and Rogers, F. J., 1992, Astron. Astrophys., in press.
Stellingwerf, R. F., 1975, Ap. J. **195**, 441.
Stellingwerf, R. F., 1975, Ap. J. **199**, 705.

Radiation Hydrodynamics in Pulsating Stars

Michael U. Feuchtinger & E. A. Dorfi

Institut für Astronomie der Universität Wien
Türkenschanzstraße 17, A-1180 Wien, Austria

Abstract

We present nonlinear radiation hydrodynamical calculations of Cepheid envelopes carried out using the method described in Dorfi & Feuchtinger (1991, A&A 249, 414). The radiative transport is treated within the grey Eddington approximation and all equations are discretised in conservative form on an adaptive mesh which is solved simultaneously with the physical equations. The resulting system of nonlinear algebraic equations is solved implicitly to avoid the restrictive Courant-Friedrichs-Lewy time step condition. In order to treat schock waves we use the artificial tensor viscosity developed by Tscharnuter & Winkler (1979, Comp.Phys.Comm. 18, 171) which is suitable for spherically symmetric problems. For the opacity we employ the latest OPAL tables (Rogers and Iglesias, 1992, ApJ, submitted); the equation of state corresponds to the standard evolution calculations according to Baker & Kippenhahn (1962, Zeitschr. f. Astrophys. 54,114).

We start from an initial hydrostatic model whose parameters correspond to a typical Cepheid ($5\,M_\odot$, $3000\,L_\odot$, $5600\,K$). The subsequent dynamical evolution is initiated by the excitation of oscillations at the ionization zones in the stellar atmosphere where the opacity changes by several orders of magnitude. After an initial relaxation caused by discretisation errors and numerical noise, a periodic pulsation develops. Each pulsation is accompanied by a shock wave which propagates through the stellar atmosphere. A typical pulsation cycle is covered by about 70 time steps and the obtained periods lie in the range of 1 to 40 days.

We discuss in detail the excitation of the oscillations and the structure of the model during a pulsation cycle.

This work is supported by the Österreichischer Fonds zur Förderung der wissenschaftlichen Forschung under project number P8758.

Element Diffusion in Pulsating Variables

Joyce Ann Guzik

Los Alamos National Laboratory, Los Alamos, New Mexico, USA

Abstract

Can pulsation studies yield information on the operation of element diffusion in stars? Can diffusion explain unusual properties of pulsating variables? Element diffusion theory and recent research relevant to these questions will be reviewed, with emphasis on the Sun and δ Scuti stars. High-degree solar p-modes that are sensitive to helium ionization support a reduced convection zone helium abundance consistent with that expected from diffusion. Intermediate-degree p-modes can be used to probe the structure of the convection zone base and constrain possible diffusion-produced composition gradients. δ Scuti variables have shallow convection zones and relatively short diffusion timescales. Helium diffusion may explain unusual period ratios, influence period changes, and affect the amplitudes and light curve shapes of δ Scuti stars.

1. Introduction

There are several ways in which pulsation studies may help us understand diffusion, or diffusion studies may help to explain pulsation observations. For the Sun, white dwarfs, rapidly oscillating Ap stars, and possibly δ Scuti stars, the rich spectrum of nonradial p- and/or g-modes can be used to probe diffusion-produced stratification and composition gradients. Diffusion of helium and heavier elements can affect pulsation driving and instability strip boundaries. Diffusion calculations may provide indirect information on competing mechanisms that could affect pulsation, such as meridional circulation, turbulence, horizontal shear, convection, mass loss, and accretion. Observations of period and amplitude changes, light curve shapes, or episodic pulsation, combined with theoretical modeling may help to constrain/infer diffusion rates.

A number of recent papers discuss diffusion and pulsation for several types of variable stars. Guzik and Cox (1992), Vorontsov et al. (1992), and Dziembowski et al. (1992) use solar p-modes to infer a convection zone helium mass fraction $Y \cong 0.23$-0.24, less than the initial abundance needed to match the solar luminosity by exactly the amount expected from diffusion. For DB white dwarfs, Bradley (1992) finds that g-mode period spacings can be used to find the helium layer mass and probe the He/C transition region. Matthews (1991) suggests that p-modes be used to probe the structure of roAp stars, where diffusion plays a role in observed abundance anomalies. For δ Scuti stars, Cox et al. (1984) suggest that helium diffusion is responsible for the high 1H/F period ratio of VZ Cnc; Poretti and

Antonello (1988) suggest that helium diffusion could produce the unusual light curves of some large-amplitude δ Scuti stars. Saez et al. (1981) propose that a combination of helium diffusion, radiative levitation, and mixing would maintain enough helium in the driving region to allow pulsation and yet maintain mild abundance anomalies observed in δ Del stars. For RR Lyrae stars, Michaud et al. (1983) suggest that helium diffusion may affect the instability strip blue edge, where these stars have smaller envelope convection zones, and that radiative levitation may produce Ca abundance anomalies that should be taken into account in Z determinations from Ca II lines. Cox et al. (1992) propose that radiative levitation of iron provides the κ-effect pulsation driving for β Cephei stars.

2. Diffusion calculations

Michaud and Proffitt (1992) review the status of diffusion calculation methods and applications to solar and stellar evolution. Here we will describe the method of Burgers (1969). The equations of diffusion, heat flow, no net mass flow, and no net current for each ionic species and the electron are given by:

$$\nabla p_i - \frac{\rho_i}{\rho}\nabla p - n_i q_i E = \sum_j K_{ij}(w_j - w_i) + \sum_j K_{ij} z_{ij} \frac{m_j r_i - m_i r_j}{m_i + m_j} \qquad (1)$$

$$\frac{5}{2} n_i k \nabla T = -\frac{5}{2} \sum_{j \neq i} K_{ij} z_{ij} \frac{m_j}{m_i + m_j}(w_j - w_i) - \frac{2}{5} K_{ii} z_{ii}'' r_i$$

$$-\sum_{j \neq i} \frac{K_{ij}}{(m_i + m_j)^2}(3m_i^2 + m_j^2 z_{ij}' + 0.8 m_i m_j z_{ij}'')r_i + \sum_{j \neq i} \frac{K_{ij} m_i m_j}{(m_i + m_j)^2}(3 + z_{ij}' - 0.8 z_{ij}'')r_j \qquad (2)$$

$$\sum A_i n_i w_i = 0; \quad \sum Z_i n_i w_i = 0 \qquad (3;4)$$

These equations are solved for the unknown diffusion velocities w_j, the residual heat flow vectors r_i, and the electric field E. The first two terms on the left hand side of the diffusion equation can be rewritten using the equation of hydrostatic equilibrium and the ideal gas law for a given ion species as:

$$\frac{1}{n_i}(\nabla p_i - \frac{\rho_i}{\rho}\nabla p) = -A_i m_H g - kT \frac{d\ln T}{dr} - kT \frac{d\ln n_i}{dr}$$

The first two right-hand terms including the gravitational acceleration and the temperature gradient are considered the driving terms for "gravitational" diffusion. Radiative levitation can be included by subtracting from the gravitational acceleration the radiative acceleration on the ion species. The last term containing the composition gradient drives "chemical" diffusion. "Thermal" diffusion is accounted for by including the heat flow equations (2). The resistance coefficients K_{ij} and z_{ij}, z_{ij}' and z_{ij}'' are functions of the collision integrals. The best available analytical fits to these integrals are given by Paquette et al. (1986).

The assumptions in this method are: complete ionization; no magnetic fields; all species have Maxwellian velocity distributions at the same temperature; thermal velocities are much greater than diffusion velocities; collisions are dominated by classical interactions between point particles; and the plasma is considered a dilute gas, i.e. the plasma parameter $\Lambda \ll 1$, and the Boltzmann equation is rigorously valid. For stellar interiors, $\Lambda \sim 1$, in the intermediate regime between weak and strong coupling, and the last two assumptions are questionable. Paquette et al. (1986) discuss the potential uncertainties in this treatment due to the breakdown of these assumptions.

The diffusion velocities w_i are used to solve the species equation of motion

$$\frac{\partial n_i}{\partial t} = -\frac{1}{r^2}\frac{\partial}{\partial r}(r^2 n_i w_i)$$

This equation is discretized to 1st order in time and 2nd order in space, and solved by standard matrix techniques. Iben and MacDonald (1985) describe their solution method and implementation in the Iben stellar evolution code.

3. Diffusion in the Sun

Solar evolution calculations including diffusion (Cox, Guzik and Kidman, 1989; Proffitt and Michaud, 1991) show that diffusion can reduce the solar convection zone helium mass fraction by ~10%, and the convection zone Z by 5-10% during the Sun's 4.5 billion year lifetime. Guzik and Cox (1992) find that solar p-modes of degree l=300-600 have large weight functions in the helium ionization region between 50,000 and 300,000 K, and are quite sensitive to the helium abundance. Table I gives the sensitivity of these p-mode frequencies to a reduction of 0.03 in convection zone helium mass fraction from the initial $Y \cong 0.27$ required to match the solar luminosity, and the effects on the observed minus calculated frequencies. The sensitivity for many l=300-600 modes is 5-8 µHz, much larger than the observational uncertainties of 0.2-0.6 µHz. The O-C frequencies are much improved by the reduced helium for exactly those frequencies that show sensitivity, supporting the diffusion results.

Diffusion produces a rather steep Y (and Z) gradient just below the convection zone. Proffitt and Michaud (1991) show that turbulence, possibly induced by convective overshoot or meridional circulation, in the amount required to account for the solar lithium depletion, would smooth this gradient and reduce the amount of helium settling from the convection zone. The helium diffusion composition profiles of Cox, Guzik and Kidman (1989) and of Proffitt and Michaud (1991) for different turbulence treatments are shown in Fig. 1. We discuss next how intermediate-degree p-modes (l=5-30), with large weight functions near the convection zone bottom, can be used to constrain the shape of a diffusion-produced composition gradient.

Figures 2a, b, and c show the observed minus calculated (O-C) p-mode frequencies versus frequency for degrees l=5-60 for several solar models. Lines connect modes with the same degree l. The solar evolution and oscillation procedures are described in Guzik and Cox (1991a). The observations (Libbrecht et al., 1990) have quoted uncertainties of less than 0.1 µHz. The model of Fig. 2a has no diffusion, and (Y, Z) = (0.267, 0.02). The model of Fig. 2b includes Y and Z diffusion, but no turbulence, and so has a steep composition gradient below the convection zone. The initial Y required to match the solar luminosity is 0.264, and the convection zone (Y, Z) after 4.5 Gyr is (0.236, 0.019). The model of Fig. 2c includes diffusion, but has a Y (and Z) composition gradient that is linear

Table I Sensitivity of p-mode frequencies to solar convection zone helium abundance

p-mode		Sensitivity (µHz)	O-C Frequency (µHz)		Observational
l	n	$\nu(0.27)-\nu(0.24)$	$Y_{cz}=0.27$	$Y_{cz}=0.24$	Uncertainty (µHz)
300	1	5.2	-2.3	2.9	0.3
	2	6.1	-4.7	1.4	0.2
	3	2.3	-2.8	-0.4	0.2
	4	0.1	-0.2	-0.1	0.3
395	1	5.7	-4.6	1.1	0.6
	2	7.0	-9.2	-2.2	0.3
	3	7.5	-10.7	-3.2	0.4
400	0	0.0	6.6	6.6	0.6
	1	5.8	-5.4	0.4	0.3
	2	7.2	-8.8	-1.6	0.2
	3	8.0	-9.3	-1.3	0.2
600	1	4.7	-11.7	-7.0	0.8
	2	7.0	-12.4	-5.4	0.3

Observations are the most accurate of either Korzennik (1990) or Libbrecht et al. (1990).

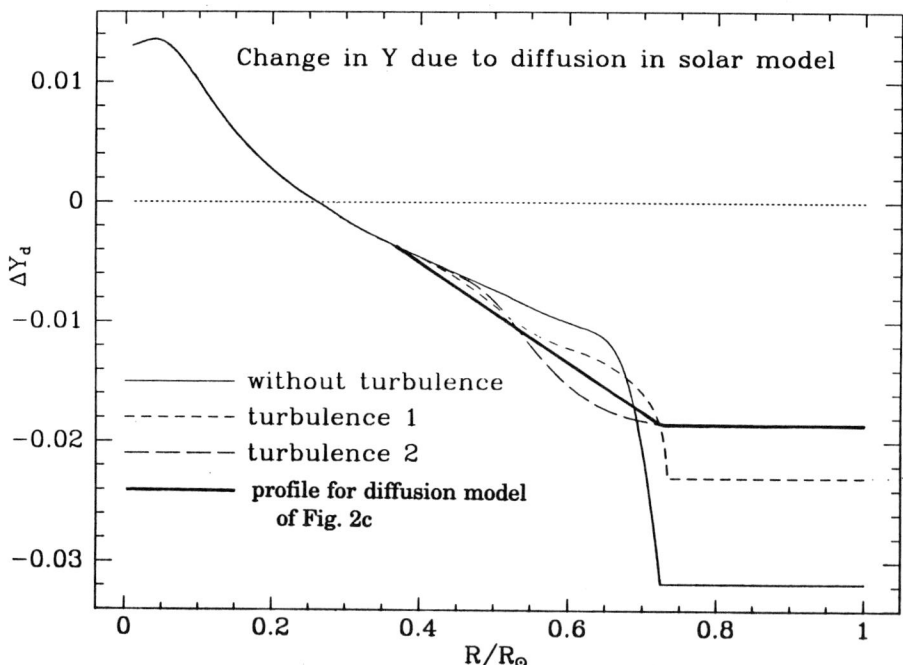

Figure 1 Change in Y due to diffusion in solar models (Proffitt and Michaud, 1991).

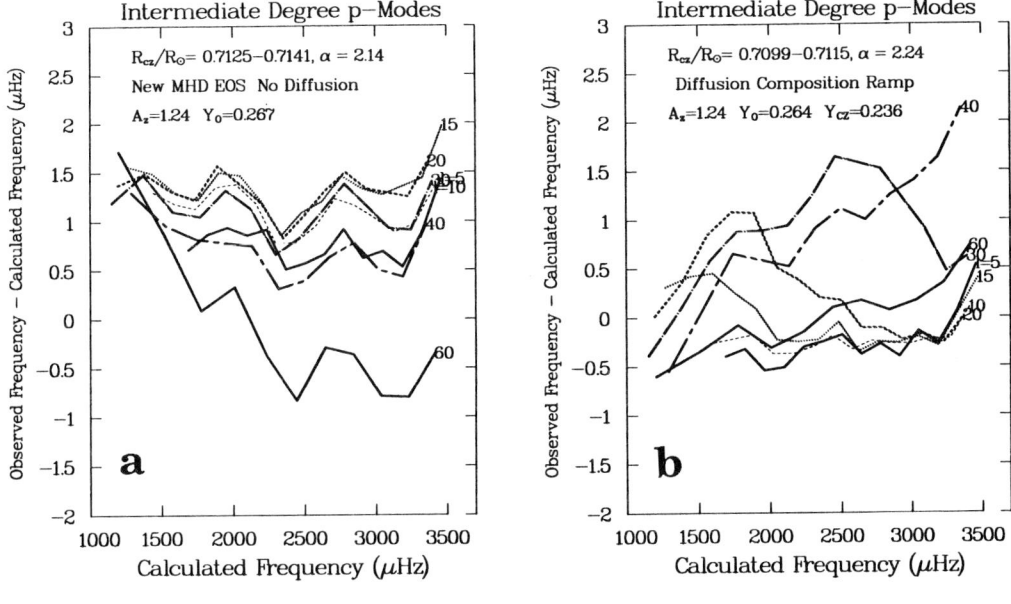
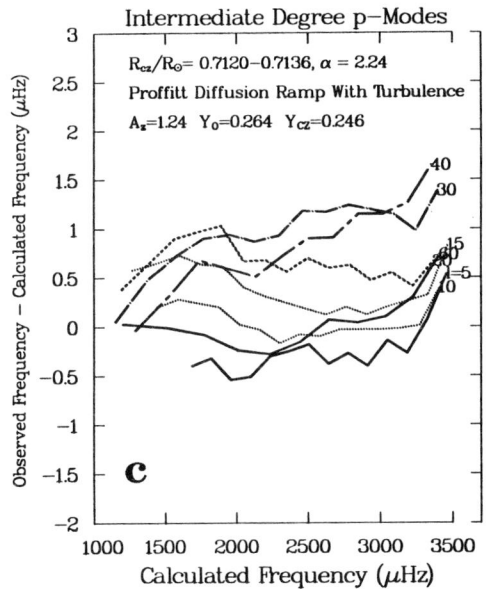

Figure 2 Observed minus calculated vs. calculated p-mode frequencies for solar models without diffusion (a); with diffusion (b); and with the diffusion-produced Y (and Z) composition gradient below the convection zone modified by turbulence (c) as in Fig. 1. Lines connect modes of the same degree l.

with respect to radius, as indicated in Fig. 1, intermediate between the two turbulence models. Turbulence decreases the amount of Y diffusion; Y=0.246 in the convection zone.

The O-C fits in these figures have already been partially optimized (to be generally flat rather than sloping versus frequency), by adjusting the radius of the convection zone bottom via a small change in the mixing length/pressure scale height ratio for the inward integration in our fine-zone model building code. (This procedure preserves solar luminosity and radius). In this way, p-mode observations can be used to find the radius of the convection zone base (0.712 ± 0.002 $R_\odot$; see review by A. N. Cox). The convection zone depth could be adjusted properly in the evolution calculations to agree with the indications from the p-mode observations, by a very small ($\lesssim 1\%$) adjustment in the opacity below the convection zone (within the opacity uncertainties), which would in turn affect the mixing length needed to match the solar radius.

To interpret these plots, a few comments are necessary: The modes of degree l=40 and 60, as well as the low-frequency modes for smaller l, sample the solar structure only within the convection zone, so the O-C differences may be attributed in part to inaccuracies in the solar model structure due to convection treatment. At low frequency (~1000 µHz), all of the plotted modes have some sensitivity to the helium ionization region. The ability of the l=5-30 modes to sense the structure below the convection zone increases with increasing frequency.

A careful study and comparison of these plots reveal: 1) The O-C values at low frequency are ~1.5 µHz positive for the no-diffusion model (Fig. 2a) with highest Y_{cz}, slightly negative for the diffusion model (Fig. 2b) with Y_{cz}=0.236, and slightly positive for turbulence model (Fig. 2c) with Y_{cz}=0.236. Thus a convection zone helium abundance between 0.236 and 0.246 optimizes the low-frequency O-C's. 2) For high-frequency l=5-20 modes with considerable weight below the convection zone, the O-C's of the diffusion model (Fig. 2b) are quite flat with frequency, and coincide for different degrees l. The O-C's of the no-diffusion model (Fig. 2a) vary by about a microhertz over the frequency range considered, and are somewhat dispersed with changing degree l. The O-C fit of the diffusion model (Fig. 2b) is favored. 3) For the O-C's of the model with a composition gradient modified by turbulence (Fig. 2c), the O-C curves are fairly flat in frequency, but are dispersed vertically with respect to l. Thus the O-C fit for the diffusion model with a steep composition profile is favored over that of the model with turbulence.

4. Diffusion in δ Scuti Stars

Cox et al. (1984) suggest that envelope helium depletion could be responsible for the large observed 1st overtone/fundamental period ratio (0.801) of the δ Scuti star VZ Cnc. Breger (1990) lists two Pop I δ Scuti stars, VZ Cnc, and DY Her with reported period decreases $1/P \, dP/dt = -13 \cdot 10^{-8}$/yr and $-6 \cdot 10^{-8}$/yr, respectively. This is surprising, as δ Scuti stars are presumed to evolve across the instability strip from blue to red, resulting in period increases. Could helium diffusion, which would make a star more centrally concentrated, causing the period to decrease, offset the period increase expected from evolution? To answer this question, a 2 $M_\odot$ model was evolved from the zero-age main sequence across the δ Scuti instability strip, including diffusion. The calculations show that helium (and heavier elements, unless radiative levitation is included) diffuse from the upper 10^{-4} $M_\odot$

(T<300,000 K) in ~10^8 yr. This implies that maintaining a nearly Pop. I envelope composition would require some mixing mechanism, or mass loss rates of the order 10^{-12} $M_\odot$/yr.

The evolution timescale across the instability strip is ~$5\cdot10^6$ yr, more than an order of magnitude less than the diffusion timescale. Even if the timescales were comparable, pulsation calculations including a helium-depleted envelope show that the period decrease expected due to diffusion is only ~20% of the period increase resulting from evolution, so diffusion cannot be responsible for the reported period decreases (Guzik and Cox, 1991b).

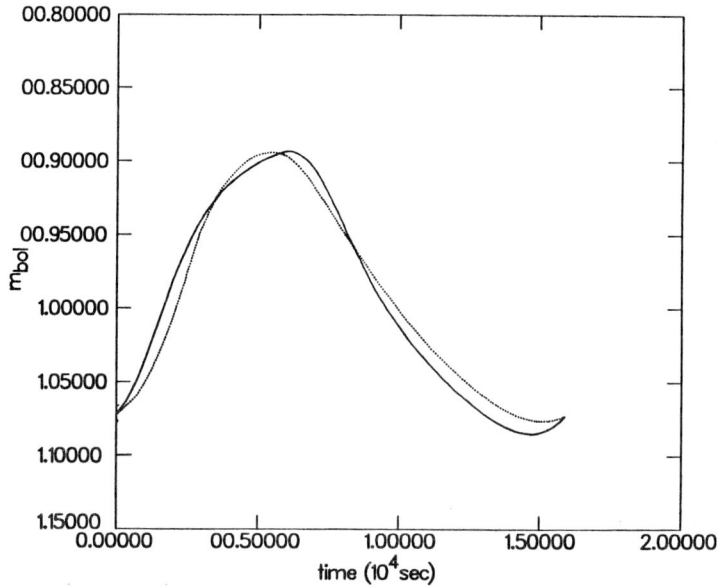

Figure 3 Bolometric magnitude of limiting amplitude solutions for δ Scuti models with envelope helium abundance decreased to Y=0.10 for T≲100,000 K (dotted line) and T≲300,000 K (solid line). These curves show how helium diffusion could alter the light curve shapes of δ Scuti stars.

Poretti and Antonello (1988), and Antonello, Poretti, and Stellingwerf (1988) proposed, based on analysis of one-zone models, that the unusual light curves of the high-amplitude δ Scuti stars V1719 Cyg, V798 Cyg, and V974 Oph can be explained by helium diffusion. The light curves of these stars have descending branches that are steeper than the rising branch. To test this hypothesis, limiting amplitude solutions were calculated for 60-zone radiative 1.8 $M_\odot$ δ Scuti models with and without reduced envelope helium abundances. A decrease in the helium abundance from Y=0.28 to Y=0.10 for temperatures ≲100,000 K decreases the radial velocity amplitude from 40 km/sec to nearer the observed value of ~10 km/sec, but the light curves are still steeper on the rising branch (Fig. 3). When the helium depletion was extended to 200,000 K, the light curve became less steep on the rising branch, and steeper on the descending branch. Extending the helium depletion to 300,000 K further enhances this effect (Fig. 3). The calculated light curve shapes do not quite match those observed for the variables listed above, but the approach appears promising, and additional calculational tests are in progress. We also note that if the helium abundance is reduced too much (to $Y < 0.1$), or for too large a portion of the envelope, the helium-ionization driving is weakened and the models stop pulsating.

5. Conclusions

Pulsation studies are useful to constrain/infer the effects of diffusion, and diffusion may be responsible for some observed properties of pulsating variables. For the Sun, high-degree p-modes support the convection zone helium abundance predicted by diffusion calculations, and intermediate-degree p-modes can be used to determine the convection zone depth and constrain diffusion-produced composition gradients. For δ Scuti stars, helium diffusion can have observable effects on period ratios, period changes, light curve shapes, and amplitudes.

References:

Antonello, E., Poretti, E., and Stellingwerf, R. F., 1988, in: *Multimode Stellar Pulsations*, eds. G. Kovacs, L. Szabados, and B. Szeidl, Kultura, Budapest.

Bradley, P., 1992, "The Potential of Asteroseismology of DB White Dwarfs," these proceedings.

Breger, M., 1990, in: *Confrontation Between Stellar Evolution and Pulsation*, Bologna, Italy, May, 1990.

Burgers, J. M., 1969, *Flow Equations for Composite Gases*, Academic Press, New York.

Cox, A. N., 1992, "Interpretations of Solar Oscillations," these proceedings.

Cox, A. N., Guzik, J. A., and Kidman, R. B., 1989, Astrophys. J. **342**, 1187.

Cox, A. N., McNamara, B. J., and Ryan, W., 1984, Astrophys. J. **284**, 250.

Cox, A. N., Morgan, S. M., Rogers, F. J., and Iglesias, C. A., 1992, Astrophys. J. **393**, 272.

Dziembowski, W. A., Pamyatnykh, A. A., and Sienkiewicz, R., 1992, M.N.R.A.S., in press.

Guzik, J. A. and Cox, A. N., 1991a, Astrophys. J. **381**, 333.

Guzik, J. A. and Cox, A. N., 1991b, Delta Scuti Newsletter, ed. M. Breger, Issue 3.

Guzik, J. A. and Cox, A. N., 1992, Astrophys. J. **386**, 729.

Iben, I. and MacDonald, J., 1985, Astrophys. J. **296**, 540.

Korzennik, S. G., 1990, Ph. D. Dissertation, UCLA.

Libbrecht, K. G., Woodard, M. F., and Kaufman, J. M., 1990, Astrophys. J. Suppl. **74**, 1129.

Matthews, J. M., 1991, Publ. Astron. Soc. Pacific **103**, 5.

Michaud, G., Vauclair, G., and Vauclair, S., 1983, Astrophys. J. **267**, 256.

Michaud, G. and Proffitt, C. R., 1992, in: *Inside the Stars, IAU Colloquium 137*, eds. A. Baglin and W. W. Weiss, Astron. Soc. of Pacific Conf. Series, in press.

Paquette, C., Pelletier, C., Fontaine, G., and Michaud, G., 1986, Astrophys. J. Suppl. **61**, 177.

Poretti, E., and Antonello, E., 1988, Astron. Astrophys. **199**, 191.

Proffitt, C. and Michaud, G., 1991, Astrophys. J. **380**, 238.

Saez, M., Auvergne, M., Valtier, J.-C., Baglin, A., and Morel, P., 1981, Astron. Astrophys. **101**, 259.

Vorontsov, S. V., Baturin, V. A., and Pamyatnykh, A. A., 1992, Nature **349**, 49.

Discussion:

M. Catelan: Could you comment on the constraints imposed on He diffusion for Population II stars by the existence of the Spite Li plateau for such objects? Isn't it true that the presence of stars at the high-temperature end of this plateau provides a strong argument for diffusion being inhibited for such stars?

J. Guzik: Diffusion can explain the Li gap in Pop I F stars, but there is no explanation yet for why diffusion doesn't also produce a Li gap in Pop II stars. This issue is discussed by Michaud and Proffitt (1991).

J. Matthews: Do the most extreme diffusion models you use to try to explain anomalous δ Scuti light curve shapes require mass loss rates comparable to what you quoted when trying to explain the period changes?

J. Guzik: Mass loss rates of 10^{-12} $M_\odot$/yr are required to maintain a nearly normal Pop I surface composition in the presence of diffusion. For surface Y=0.1, the mass loss rates can be an order of magnitude less.

A. N. Cox: I only want to say that John Percy was the first to show that helium settling can stabilize δ Scuti stars.

M. Breger: Your δ Scuti models do not include stellar rotation. Since observations show a strong inverse relation between amplitude and rotational velocity, your large amplitudes computed may not really be anomalous.

E. Antonello: I disagree with Michel (Breger). I do not think the rotational velocity is playing such an important role in δ Scuti stars, even if it known that it affects at some level the pulsational stability. Up to now, no reliable mechanism has been proposed to explain the behavior of the δ Scuti stars as regards their amplitude, and the tests with nonlinear models in the presence of inhomogeneity due to diffusion, even if their results compare only partly well with the observations, are indicating a very promising way for the interpretation of δ Scuti stars.

N. Simon: What is the accuracy of measured p-mode frequencies in the Sun, and to what extent can you simultaneously constrain the various uncertainties: mixing length, opacity, amount of diffusion, etc.?

J. Guzik: The quoted observational uncertainties for the low and intermediate degree p-modes are very small, less than 0.1 µHz, so p-modes can be very effective probes of the solar structure. The mixing length is constrained by the solar radius. The adjustment of the convection zone depth to optimize the agreement between calculated and observed p-modes puts very tight constraints on the opacities at 2-5 million K. The initial solar helium abundance is constrained by the solar luminosity, so the convection zone Y determined from p-modes tells us the amount of diffusion.

Convection in RR Lyrae Stars

R. F. Stellingwerf[1], G. Bono[2]

[1]*Los Alamos National Laboratory Los Alamos, NM,* [2]*Trieste Observatory, Trieste, Italy*

Abstract

Convection undoubtedly plays a strong role in defining the RR Lyrae instability strip, but its effects on amplitude, mode of pulsation, and light curve have always been problematical. One reason is simply that convective models are difficult to compute, and their accuracy is difficult to ascertain. We present results of a new convective survey of RR Lyrae stars that constitute the best models available at this time, with better boundary conditions, physics, zoning, and length of computational run than previous results.

1. Introduction

Attempts at inclusion of convection in pulsating star models date from the early work by Baker and Kippenhahn (1962). They used linear non-adiabatic models in which the convective effects were included in the static structure of the star, but the time dependence was not treated in detail in the pulsation model. Pulsational damping was found for cooler models, but the width of the instability strip as computed was too large, and the phase shifts of the cooler models did not match those of observed stars. This study set the tone of subsequent endeavors in the field of time-dependent convection in the sense that comparison with only a few observational parameters (width of strip, phase of light curve, amplitudes, etc.) comprise the only tests of the theory's validity. Christy (1966) showed that non-convective models seemed to match many of the properties of observed stars, except for the quenching at the red edge of the instability strip. Later work showed that non-convective models tended toward amplitudes larger than observed, and showed difficulties with mode selection - particularly modeling the RRd mixed-mode stars. Convective models, although undoubtedly over-simplifications of reality, may be considered "successful" if some or all of these discrepancies are removed.

Convective theories may be classified according to several characteristics: 1) dimensionality (1D or 2D), 2) effect of the convection on the adjacent regions (either local or nonlocal), 3) amplitude (either linear or nonlinear), and 4) time dependence of the convection (either time-dependent or frozen). Unno (1967) and Baker and Gough (1979) tried linear, local, time-dependent theories, Cox, et al. (1966), and Wood (1979) tried nonlinear, local, time-dependent theories. These attempts all showed promise, but missed significant features of the observations. Deupree (1979) tried a 2D nonlinear theory that matched observations well in many respects. Stellingwerf (1982) derived a 1D nonlinear, nonlocal, time-dependent theory, which is somewhat easier to run and interpret than Deupree's 2-dimensional scheme, and which also seemed to match observation quite well.

This paper presents an overview of a new series of computations using this scheme. For details, see the presentation by Bono in this collection. Here we present a qualitative review of the results and their implications.

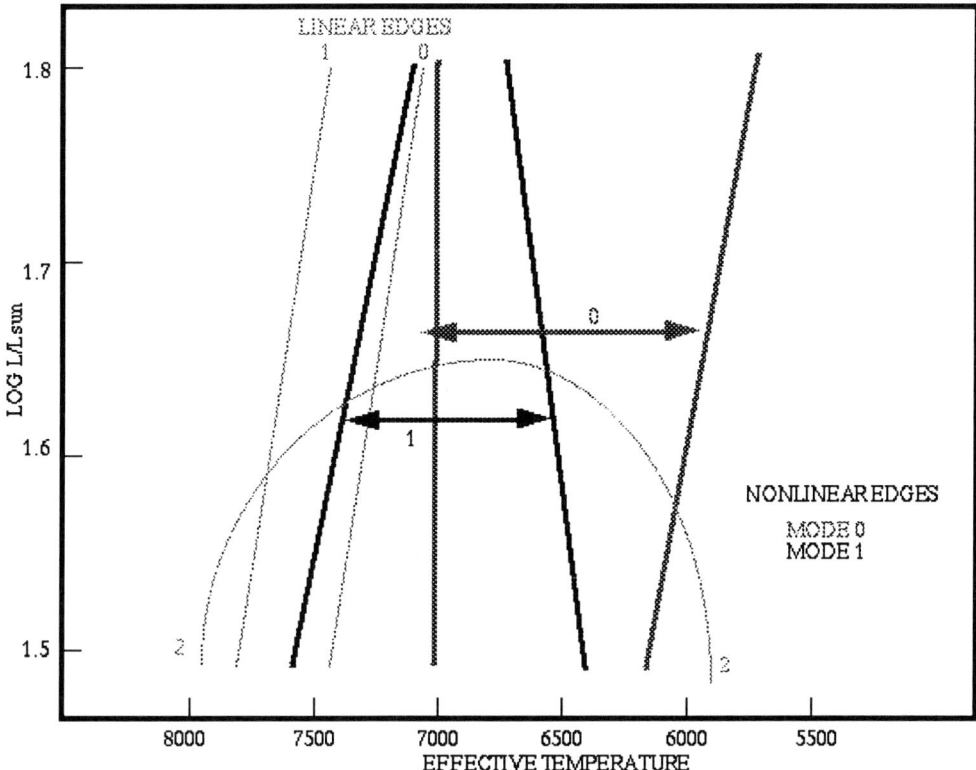

Figure 1: Location of the linear and nonlinear instability strips on the HR diagram.

2. Instability Strip

Figure 1 shows the location of the instability strip of the RR Lyrae stars with mass 0.65 and composition X=0.7, Z=0.001. Linear radiative models were computed for the fundamental (mode 0), first overtone (mode 1), and second overtone (mode 2) modes. The blue edges are shown above as grey lines. Only the second overtone is stabilized toward the red within the limits of this diagram; this is the normal result for non-convective models. These models have about 120 radial zones, which is fine enough to resolve all of the important features of the pulsation.

The models were then repeated using a nonlinear, convective code with the same physics, numerics, and parameters. Only fundamental and first overtone modes have been computed at this time. Models were computed at four different luminosities. The nonlinear models were integrated about 500 periods to determine their final preferred mode. The dark lines shown above show the final position of the nonlinear instability strips for the fundamental and first overtone modes. These lines are smooth approximations to the detailed results, but should be accurate within the uncertainties of the survey. The implications of these results will be discussed below.

Note that the blue edges are considerably shifted toward the red, and the slope of the fundamental blue edge is considerably changed from the linear results. This illustrates the changes in this survey from the "classical" linear, radiative results. These changes represent a combination of the effects of convection and nonlinearities, to be unraveled in the next section. Of course, another change is the appearance of the red edges at the observed locations.

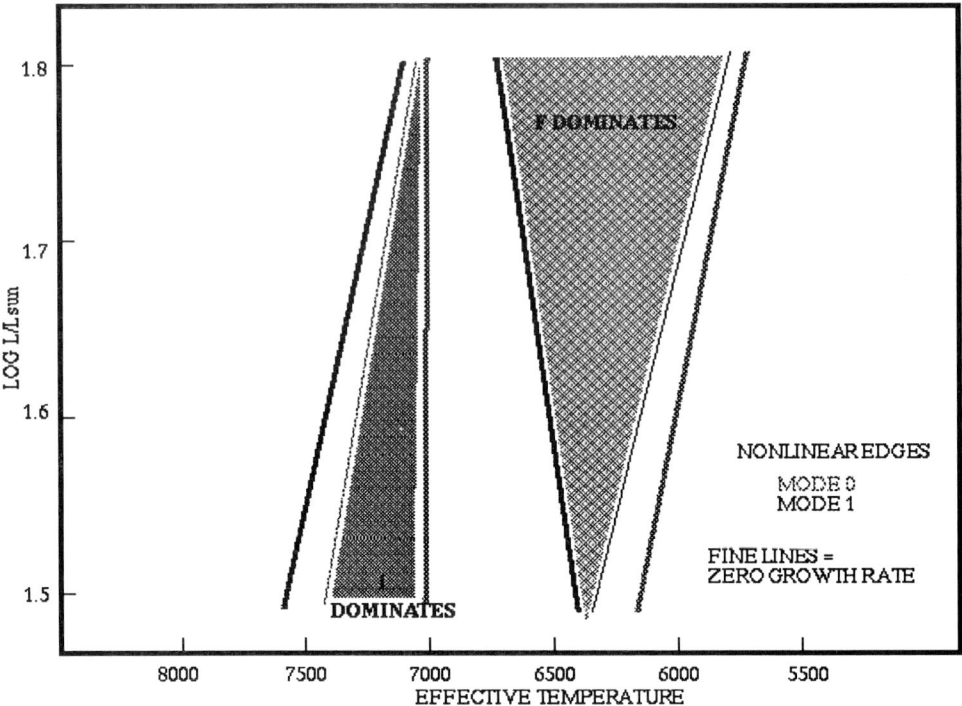

Figure 2: Nonlinear instability strip showing the locations of the "zero growth rate lines" - these are the linear / convective edges of the instability strip as derived from the nonlinear models.

3. Mode Switching

Another piece of information derived from the nonlinear models is the slope of the growth rate at early time during the integration. Since the integration is started at modest amplitude (10 km/s velocity), this gives an approximation to the linear, convective instability strip (this strip was not computed directly because of the difficulty of including the nonlocal effects in the linear analysis). In this way we can separate the effects of the convection from those of the nonlinearity. Figure 2 shows the edges obtained in this manner, as well as the nonlinear results shown in Figure 1. Only the blue fundamental "zero growth" line and the red overtone "zero growth" line are shown, since the other lines coincide with the nonlinear edges. In the cases shown the nonlinear strips begin relatively far from the lines of zero growth, indicating that models in these regions would begin pulsating in one mode, but switch to the other before attaining full amplitude. These "mode switching" regions are shown as shaded regions in Figure 2. The diagonal line at about 6500K is often referred to as the "mode-switch" line. As seen from the figure, however, there are actually two mode switching lines, the usual line at 6500K, which is actually the overtone red edge for the nonlinear models (at this line redward evolving overtone pulsators will switch to the fundamental mode), and one at about 7000K, which is the fundamental nonlinear blue edge (at this line blueward fundamental pulsators will switch to the overtone).

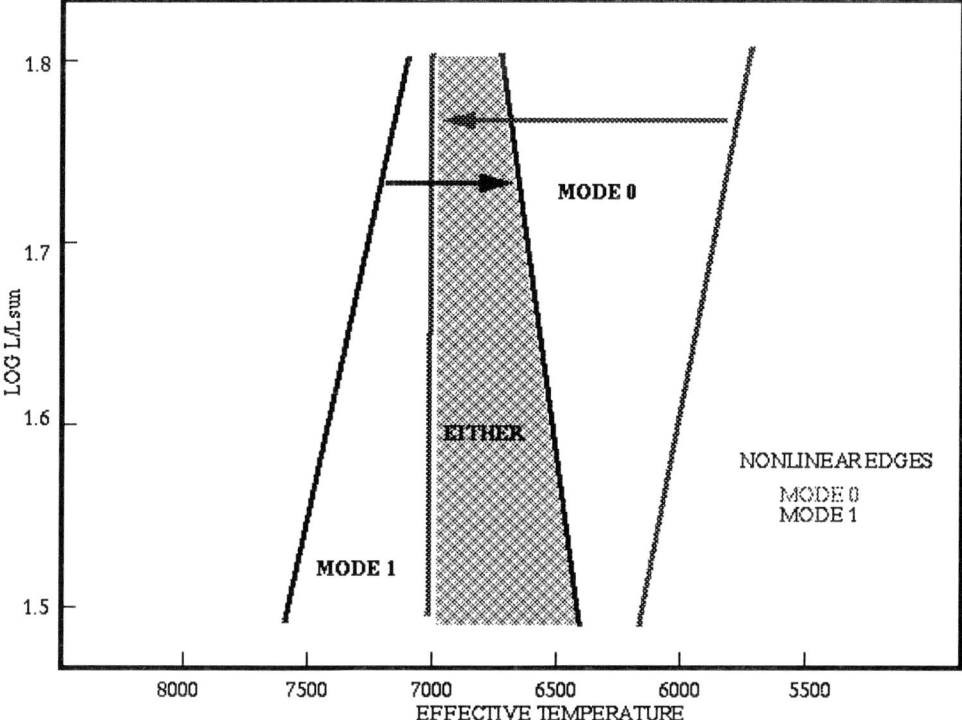

Figure 3: Areas of nonlinear mode preference in the HR diagram.

The region between the two shaded areas in Figure 2 is the "hysteresis region" in which either mode is a possible full amplitude candidate (also called the "either-or" region). This region is shown as a shaded area in Figure 3, above, which also emphasizes the overlapping instability strips for the two modes studied.

Stars evolving to the red will first encounter instability at the overtone blue edge. Overtone pulsations will be stable limit cycles as the star continues to evolve until the far edge of the "either-or" region is reached at the overtone nonlinear red edge. At this point they will switch to the fundamental and continue in this mode until stabilized by convection at the fundamental red edge. Similarly, stars evolving toward the blue will begin pulsating in the fundamental mode at the fundamental red edge and continue until the fundamental blue edge is encountered, this time at the blue edge of the "either-or" region. The star then switches to the overtone until stabilized at the overtone blue edge.

Cluster variables will switch modes at one or the other of the switching lines, depending on the direction of evolution. The fraction of overtone pulsators will depend strongly on the evolution direction and weakly on the temperature of the horizontal branch. Dependences on mass and helium abundance are also expected to be weak, and will be addressed in future work. In Figure 3 at log L = 1.7 there should be roughly equal numbers of overtone and fundamental pulsators for redward evolution, but about 25% overtone pulsators for blueward evolution. This is the sort of difference found between Oosterhoff I and II clusters. Detailed comparisons with specific cluster data are underway.

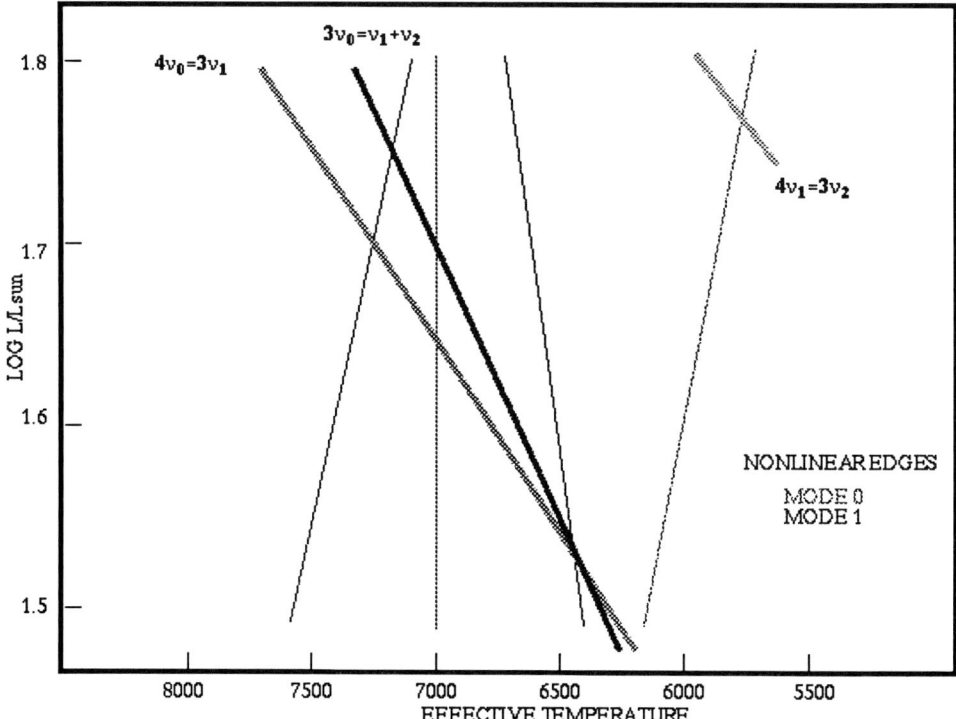

Figure 4: Mode resonance lines in the HR diagram for the current survey.

4. Mixed-Mode RR Lyrae Stars

RR Lyrae stars showing a mixture of modes (RRd stars) were discovered in M15 by Cox, Hodson and Clancy (1983), and have now been found in many other clusters as well as the field. See other papers in this collection for the current observational status of these objects. Although several models have been constructed showing this behavior, the observed location of the stars in a narrow range of temperatures near the center of the instability strip has not been reproduced theoretically. Possible explanations of these stars include 1) mode switching at a transition line (in Figure 4, the lines at 6500K and 7000K), 2) a stable "mixed-mode" region in the HR diagram instead of the "either-or" region found here, and 3) mixed-mode behavior near resonances (see Buchler and Kovacs, 1986). Some likely resonance lines are shown in Figure 4 for the current survey. Linear combinations of these lines are also possible, but should lie in the same general vicinity as the lines shown.

Some preliminary computations have been undertaken with the present models to determine the credibility of these explanations. Only a few models have been computed, so the conclusions are tentative. The computations take the form of time integrations starting with either a relatively pure mode at low amplitude, or a mixture of modes, and continuing until the final nonlinear result can be ascertained. During the calculation, the amplitudes of the components are obtained using the PDM technique (Stellingwerf, 1978). Two models will be discussed here, one at relatively high luminosity (model 2.3), and one at low luminosity (model 4.6).

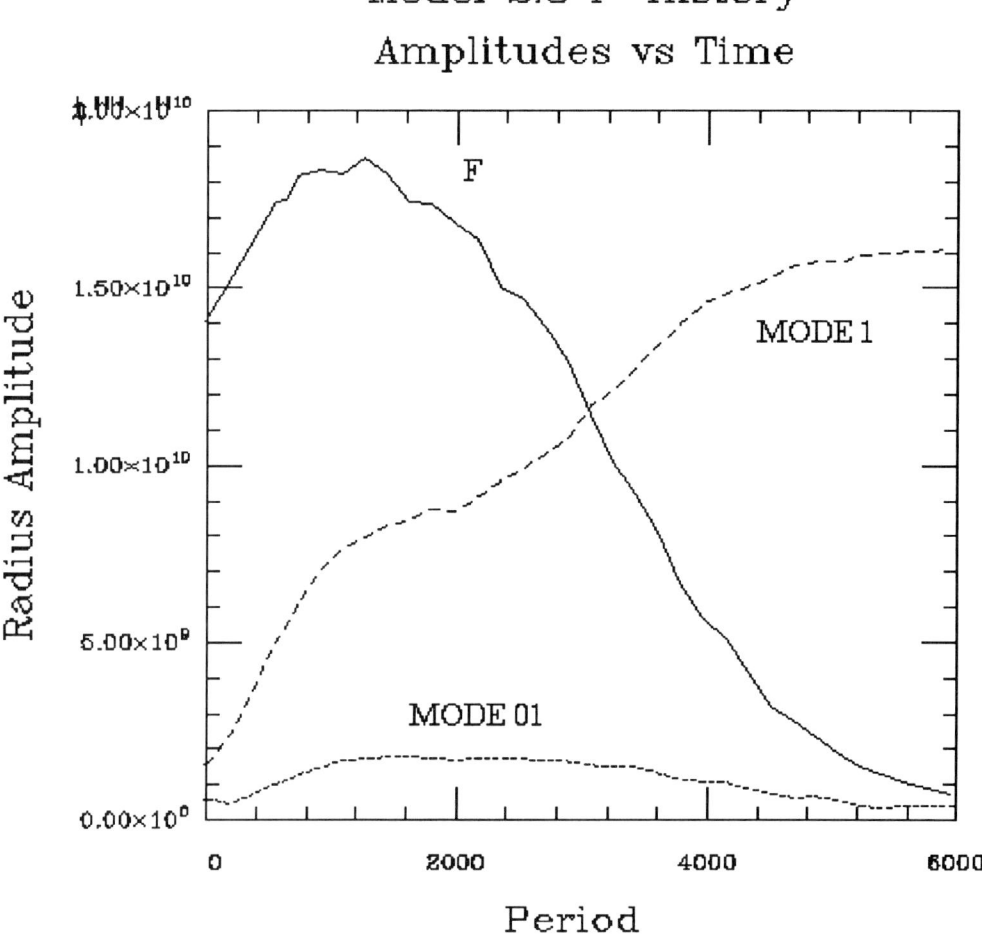

Figure 5: Time history of the radius amplitudes of the fundamental, overtone, and the nonlinear mode coupling term in the spectral analysis of model 2.3.

Model 2.3 has log L = 1.8, T_e = 7100K. Figure 5 shows the evolution of this model when initiated in the fundamental mode. At first both the fundamental and the first overtone grew, until about period 1000, when it appeared that a stable mixture of modes was emerging. Unfortunately, this was not the case since the overtone eventually dominated and quenched the larger amplitude fundamental. Note that the switch took 6000 periods, or about 9 years. Since this model is not exactly at the transition line, an actual mode switch at the line will be much longer (perhaps 100-1000 years, depending on the rate of evolution). Models at lower L will take longer still because of smaller growth rates. These switching times are much longer than those found for purely radiative models. It thus appears that mode switching is a possibility to be considered, and should be checked as the observational time base is extended.

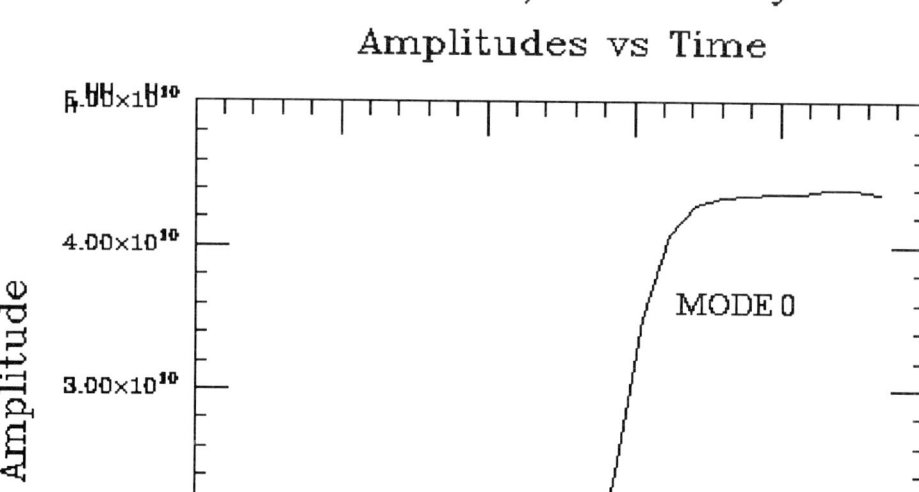

Figure 6: Time history of the first three modes in the development of model 4.6 initiated in the overtone.

Model 4.6 has log L = 1.5 and T_e = 6500K. This model is thus near the two main resonance lines in Figure 4, as well as the cool mode switching line. Also, its linear growth rates are smaller than model 4.6. Figure 6 shows the evolution of the model when initiated in the overtone mode. After 500 periods of integration the overtone establishes itself at a stable limit cycle, and maintains this behavior until period 2500. The light and velocity curves look perfectly regular all during this period, and the model would certainly be classified as a stable overtone with the usual visual analysis. The PDM amplitudes, however, show that the fundamental begins to grow at about period 1000, and is dominant by period 3000. The overtone decays, but does not disappear, and the second overtone also joins in toward the end. The two overtones continue at about a 10% level in radius amplitude, which is about 30% in luminosity. A long term oscillation is also visible, with a period of about 1000 periods (about 1 year).

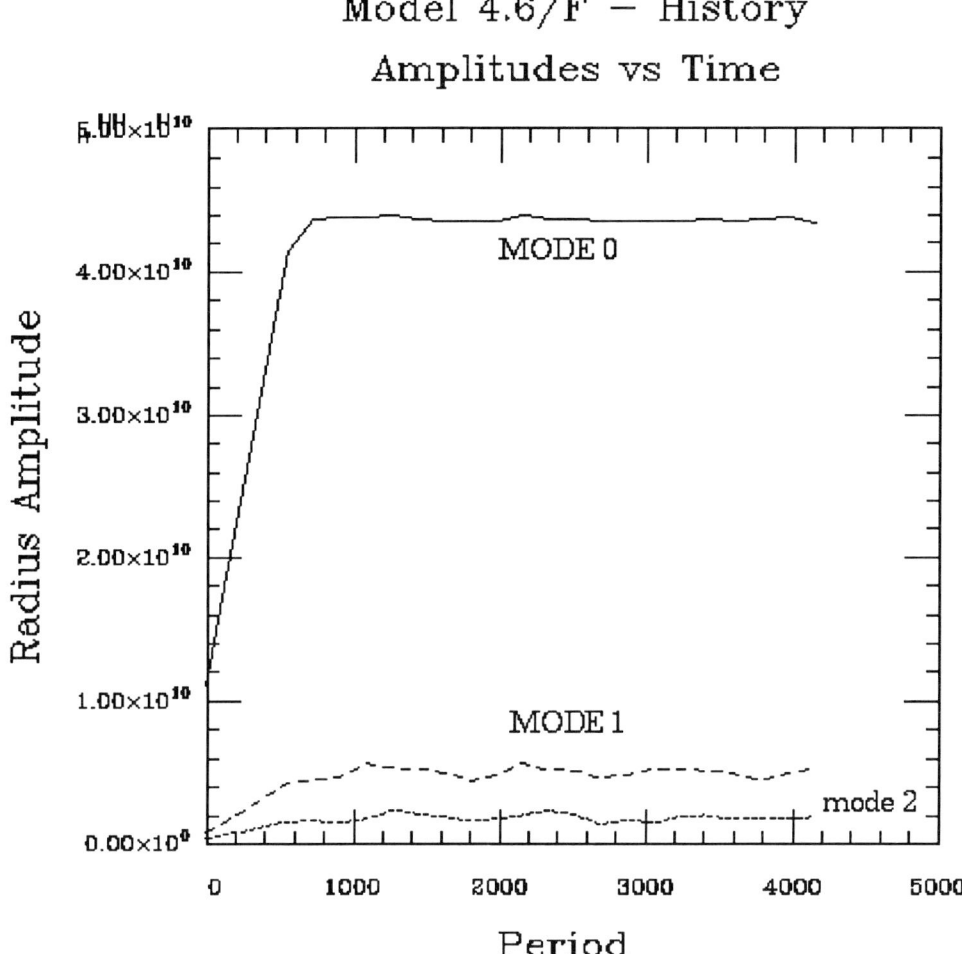

Figure 7: Time history of the first three modes in the development of model 4.6 initiated in the fundamental.

To test whether the long term behavior seen in Figure 6 is really a stable mixed-mode limit cycle, the same model was initialized in the fundamental mode and another long integration undertaken. The result is shown in Figure 7. In this case the model proceeds directly to the same behavior seen after 3500 periods in Figure 6, and is at the final state after about 500 periods. Again the long term modulation is seen in the secondary modes, and both appear to be in phase with about a 1000 period modulation time. This behavior is suggestive of a three mode resonance, but it is not clear whether this phenomenon is identical to that seen in actual RRd stars, or if the resonance actually plays a vital role. In particular, a model at this same luminosity, but at a temperature of 7000K (near the other transition line) also appears to show signs of a mixed mode state, but this time with the overtone dominant. Further tests will be needed to finally resolve these issues.

5. Conclusion

The convective RR Lyrae survey has supplied a new prediction for the instability strip that differs in several significant features from previous surveys. These differences still need to be carefully compared with observational evidence, but indications are that better agreement will be achieved for the RR Lyrae systematics. Amplitudes of the models are generally lower than radiative models, and the growth rates are smaller. This should favor mixed mode behavior, but preliminary results seem to indicate that this is true only at low luminosities. We find that the time required to switch modes is longer than previously derived, however, opening the possibility that the RRd stars are in transit from one mode to another. If this is true, then changes should be seen in the amplitudes of at least some of these objects as a function of time.

References:

Baker, N. H., and Kippenhahn, R., 1962, Zeit. f. Ap. 54, 114.
Baker, N. H., and Gough, D., 1979, Ap. J. 234, 232.
Buchler, J. R., Kovacs, G., 1986, Ap. J. 303, 749.
Christy, R. F., 1966, Ap.J. 145, 337.
Cox, A. N., Brownlee, R. R., and Eilers, D. D.,1966, Ap. J. 144, 1024.
Cox, A. N., Hodson, S. J., and Clancy, S. P., 1983, Ap. J. 266, 94.
Deupree, R. G., 1979, Ap.J. 234, 228.
Stellingwerf, R. F., 1978, Ap.J. 224, 953.
Stellingwerf, R. F., 1982, Ap.J. 262, 330.
Unno, W., 1967, PASJ 19, 140.
Wood, P. R., 1979, Ap.J. 227, 220.

Acknowledgments:

This work has been supported in part by the National Science Foundation, the Department of Energy, and Trieste Observatory.

Questions:

S. R. Sreenivasan: The system of equations you have used is of the "reaction-diffusion" type of equations. They exhibit structural phase transitions, hystersis, and transition to chaos. There is considerable literature on the subject and it might be interesting to make comparisons with your work.

RFS: I agree!

Geza Kovacs: Is not the eddy viscosity just another fudge factor to substitute the artificial viscosity by a "more physical one"?

RFS: We find that the amplitude is only weakly dependent on the eddy viscosity. Remember that the turbulent pressure is one component in the total pressure, and that it is the total pressure that is fixed by the momentum equation. In one test, I turned off the turbulent viscosity completely, but the total work was affected only slightly, although the eddy dissipation was significant. The structure had adjusted to maintain nearly the same total pressure. The eddy term will have a secondary effect on the structure, which is weaker than it looks. In most of our models, limiting amplitude is determined by strong convective quenching, primarily at the phase near minimum radius, rather than by viscous effects.

Convection and the Bump Cepheid Resonance

Arthur N. Cox

Los Alamos Astrophysics

Abstract

Bump Cepheids display a resonance between the fundamental and second overtone modes in the form of a bump on descending light for periods less than 10 days and on the ascending light curve for longer periods. A long-standing problem has been how to explain this resonance for stellar masses consistent with evolution theory, rather than significantly lower ones. New Livermore OPAL intermediate coupling opacities now produce stellar models that are less density concentrated in the outer 1/4 of the radius, and all radial mode periods are increased. The second overtone to fundamental mode period ratio is reduced. This reduction then requires larger masses for the bump resonance, much closer to those from the old Becker, Iben, and Tuggle and the new Stothers and Chin stellar evolution results with no or very little core convection overshooting and the standard (X=0.70 and Z=0.02) composition. This achievement of the OPAL opacities is not adequate, though, because the period ratios have not quite decreased enough. There is another missing ingredient suggested to me by Norman Simon. A second deep "iron line" convection zone now appears between 145,000 and 205,000 K for higher Cepheid luminosities and masses like 7 and 8 $M_\odot$. This convection is necessary to transport the higher luminosities that cannot be carried by radiation alone. An increase in the ratio of the mixing length to the pressure scale height to about 1.5 for 7 $M_\odot$ and 2.0 or more for 8 $M_\odot$ can give an appropriate convection efficiency and the required deep convection zone structure. Thus OPAL opacities, when used with the full physics of convection models all across the instability strip and nonadiabatic pulsation analyses, actually do explain the bump Cepheid puzzle. Further, the unknown convection efficiency for intermediate mass yellow giants can apparently be calibrated as a function of stellar mass.

Bump Cepheid Models Period Ratios

$\ell/Hp=$	0.0		1.0		1.5		2.0	
T_e (K)	Π_0	Π_2/Π_0	Π_0	Π_2/Π_0	Π_0	Π_2/Π_0	Π_0	Π_2/Π_0
7 $M_\odot$								
5350	9.81	0.511	9.93	0.510	10.07	0.500	10.61	0.464
5700	7.92	0.524	7.98	0.524	8.03	0.519	8.13	0.511
6000	6.66	0.533	6.68	0.534	6.70	0.533	6.74	0.530
8 $M_\odot$								
5350	13.67	0.497	13.81	0.496	13.97	0.486	14.43	0.465
5700	11.01	0.509	11.06	0.510	11.13	0.506	11.24	0.499
6000	9.24	0.519	9.24	0.519	9.28	0.519	9.31	0.517

A survey of RR Lyrae models

G. Bono[1], R.F. Stellingwerf[2]

[1] Trieste Astronomical Observatory, Italy [2] Los Alamos National Laboratory, NM

Abstract

An extensive grid of non-linear pulsating models of RR Lyrae stars have been computed. To simulate the outer regions of these variables a non-local and time-dependent treatment of convective transport has been adopted. In this poster we briefly describe some new features of the instability strip (IS).

1. Discussion

In order to derive the IS for the fundamental (F) and first overtone (IO) three different sequences of models have been computed at $logL/L_o = 1.8, 1.7, 1.5$. The steps in temperature lie between 100 and 300 K. The chemical composition (X=.7, Z=.001) and the mass ($M = .65 M_o$) are fixed for all models. For all cases the pulsation has been followed until the limiting amplitude behaviour of the model can be surely identified (the number of periods directly integrated ranges between 500 and 6000). The Figure 1 shows the non-linear instability strip for the F and IO. This figure presents some interesting results: 1) the large difference between the non-linear radiative IO blu edge and the convective one confirms that the convection shifts the blu boundaries toward lower temperatures; 2) the extension of IO region (only the IO results unstable) and of the OR region (the F and IO are unstable) is more in agreement with the observational counterpart; 3) the red edge of the IO has been computed at limiting amplitude for the first time, confirming that the convection is the quenching mechanism of the pulsation both for the F and IO; 4) the new location of the F blue edge and of the IO red edge could be the key parameter for several open astrophysical problems like the Oosterhoff dicotomy and the Sandage period-shift effect.

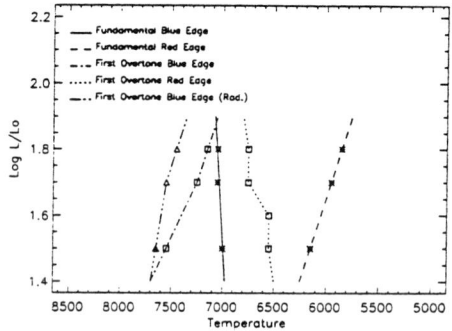

Figure 1 Non-linear instability strip, for more detalis see text.

Higher Vibrational Modes In RR Lyrae Stars

S. A. Glasner[1], J. R. Buchler[2],

[1] *The Racah Institute of Physics, Jerusalem, Israel,*
[2] *Physics Department, University of Florida, Gainesville, Florida, USA*

Abstract

A simple method for computing the full spectrum of the linear (nonadiabatic) radial modes is implemented and tested on RR Lyrae models. The growthrates of the vibrational modes display unexpected, but physically correct undulations as a function of period, with the 8th through 10th overtones almost unstable. Caution must be exercised when only a small number of meshpoints is used, or when the meshpoints are not well distributed, as some of these overtones may become unstable, clearly an artifact of the differencing. The role that such unstable high overtones can play in hydrodynamic calculations is demonstrated.

The lowest order vibrational modes are frequently computed with the Castor (1971) method. Unfortunately this approach fails for the higher vibrational modes as well as for the thermal modes. To overcome this difficulty here, we transform the linearized system of hydrodynamics into a standard eigenvalue problem of the form $\mathbf{A} \cdot \mathbf{z} = \sigma \, \mathbf{z}$, where $\mathbf{A}$ is a $3N \times 3N$ constant matrix and $\mathbf{z} \in \mathbf{R}^{3N}$ is the vector of components $\{\delta R_k, \delta u_k, \delta T_k\}$, $k = 1, N$. To achieve this form we have made use of the continuity equation to replace $p\, dV/dt$ in the energy equation in terms of the velocity u. Standard procedures exist for finding the eigenvalues of such a general (non Hermitean) real matrix.

As an application we have computed the spectrum of vibrational modes for a few RR Lyrae models, *viz.* $0.6 M_\odot$, $60 L_\odot$ with T_{eff} in the range 6800K to 7600K. The composition in all models is $X=0.7$ and $Z=0.001$. Because of the sensitivity of the growthrates to zoning we have used up to 1400 zones. The Figure displays the resultant relation between the frequencies f_k and the relative growthrates $\eta_k = 2\kappa_k/f_k$ for the lowest vibrational modes in the $T_{eff}=7300$ K model which is typical of all the RR Lyrae models. The surprising result is that the growthrates do not level off with increasing mode number, but that instead they show an undulatory behavior. The first excursion is in fact quite substantial with the result that the 8th through 10th overtones are only very weakly damped. The undulations are a result of the phase relationship between δp and $\delta \rho$ which varies with the spatial structure for the successive overtones. The first large decrease in driving (culminating with the 4th overtone) is due to an adverse phase relationship which occurs in the He partial ionization region. For the higher modes the partial He ionization region no longer contributes any driving, which comes entirely from nonadiabatic effects in the partial H ionization region. This driving is particularly efficient for the 9th overtone.

The existence of marginally stable overtones suggests that it may be possible to detect these frequencies in RR Lyrae if they are stochastically excited (in a fashion similar to that of the solar p-modes). This raises the exciting prospect of being able to do astro-seismology on the classical variable stars.

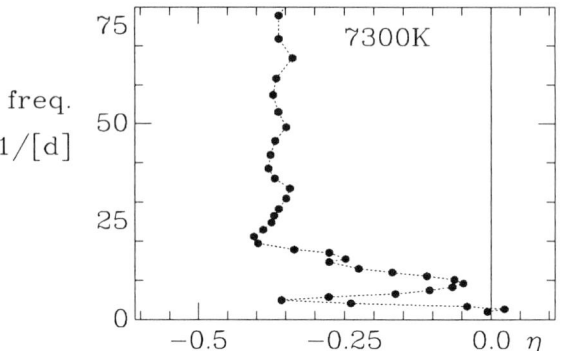

Figure: η vs frequency for an RR Lyr model with T_{eff}= 7300 K

Our study reveals a strong sensitivity of the linear growthrates to the numerical mesh, especially for the higher vibrational modes. For unfortunate choices of the numerical mesh the decrease of stability around overtones 8 – 10 can give rise to unstable modes ($\eta > 0$) and to unphysical effects in the numerical hydrodynamic computations.

In order to illustrate the phenomenon we have constructed models with T_{eff} = 7300K this time with a coarser grid of 120 zones, suitable for hydrodynamical computations. One model, e.g. has 30 equally spaced mass zones up to 11,000 K, followed by geometrically increasing zones up to 2 MK. Another model has again 30 mass zones up to 11,000 K, but with a geometric progression of the mass zone, followed by a geometrically increasing mesh as for the first model. In the two models the first 5 overtones have almost the same properties as those of the finely resolved model. The spectrum of eigenvalues for the higher overtones also has a *qualitatively* similar behavior. However, it is interesting that the 9th overtone for the second model is actually vibrationally *unstable*, while it remains stable in the first model.

Since only the first overtone is linearly unstable in the first model the latter exhibits only stable first overtone limit cycle pulsations. On the other hand the nonlinear behavior of the second model which is linearly unstable in two modes is different. When the model is destabilized with a more or less arbitrary kick the pulsations contain a strong admixture of the linearly unstable 9th overtone which persist for a long time and show up as wiggles with a frequency $f_9 \approx 3.5 f_1$. In addition, there are now at least two stable limit cycles (1st overtone and 9th overtone) with a unstable mixed mode state. These results therefore suggest that some caution needs to be exercised when an initial model is constructed to ensure that the differencing does not introduce unphysical unstable overtones and unphysical attractors.

This work has been supported by NSF.

A Full-Amplitude Nonlinear Model for RR Lyr: Pulsations, Shock Waves and Hα Peculiarities

Andrew Fokin

Institute of Astronomy, Russian Academy of Sciences
Ul. Pyatnitskaya 48, Moscow 109017, Russia

Abstract

The shock phenomena and Hα formation in the atmosphere of RR Lyrae are investigated by means of numerical simulations. The full-amplitude hydrodynamical model is generated, adopting $M = 0.578 M_\odot$, $L = 62 L_\odot$, $T_{eff} = 7175 K$ and Population II composition. The Hα profiles are obtained by solution of the non-LTE line transfer problem for the multilevel hydrogen atom. In the course of pulsations an extended (about 16 static scales) low-density atmosphere is produced with nearly exponential density decrease. Two shocks, propagating outwards, are successively generated during one period. The shock amplitude reaches 140 km/s. No mass loss has been found. These results confirm the earlier results of S. Hill (1972). The main shock develops very high, at mass depths near 0.001 g per area, whereas computations for W Vir yield 100. An analysis of the Fe I and Ti II lines shows that this difference is sufficient to explain the lack of metallic line-doubling in RR Lyrae stars. The bump on the light curve is due to a weak shock, propagating below the photosphere, which is generated with the early shock at the stage of expansion at the H-recombination front. The central intensities, Dopplerian shifts, amplitudes and phases of splitting of the computed Hα profiles agree well with the observed values (Gillet & Crowe 1988). The lack of strong emission also fits the observations, resulting from strong scattering processes near the shock. The predicted Lyman lines show strong emission, reaching maximum at the maximum light.

Nonlinear Radiative RR Lyrae Models: A Search for Double-Mode Behavior

Joyce A. Guzik and Arthur N. Cox

Los Alamos National Laboratory, Los Alamos, New Mexico, USA

Cox (1991), Kovacs, Buchler, and Marom (1991), and Kovacs et al. (1992) showed that opacity changes indicated by the Livermore OPAL opacities increase the pulsation masses of double-mode RR Lyrae variables. We calculated limiting amplitude solutions for radiative RR Lyrae models of 0.75 $M_\odot$, 51 $L_\odot$, and Z=0.0001 (Oosterhoff II) to investigate the effects of the mass increase and opacity changes suggested by the OPAL opacities. In particular, we modified the Stellingwerf analytical fit (1975) to the Cox-Tabor (1976) tables to decrease the opacity by 20% between 20,000 and 30,000 K. The Stellingwerf periodic relaxation method was used to converge the models to a limit cycle, and the Floquet matrix analysed to search for a tendency of the fundamental mode to grow from the full-amplitude overtone solution, and the overtone to grow from the full-amplitude fundamental mode solution, thereby predicting double-mode behavior.

Models with T_{eff} < 7000 K have positive fundamental-mode growth rates in the overtone solution, in contrast to the models with older Los Alamos opacities of Hodson and Cox (1982). Models with T_{eff} > 7000 K have positive 1st overtone growth rates in the fundamental-mode solution. However, no double-mode candidate models were found. An opacity decrease over a wider temperature range (15,000-70,000 K), as suggested by the most recent OPAL opacities with intermediate coupling, increases the switching rates, but still does not produce double-mode behavior near 7000 K.

References:

Cox, A. N. and Tabor, J. E., 1976, Astrophys. J. Suppl. **31**, 271.
Cox, A. N., 1991, Astrophys. J. **381**, L71.
Hodson, S. W. and Cox, A. N., 1982, in: *Pulsations in Classical and Cataclysmic Variable Stars*, eds. J. P. Cox & C. J. Hansen, U. of Colorado, Boulder, p. 201.
Kovacs, G. D., Buchler, J. R., and Marom, A., 1991, Astron. Astrophys. **252**, L27.
Kovacs, G. D., Buchler, J. R., Marom, A., Iglesias, C. A., and Rogers, F. J., 1992, Astron. Astrophys. (in press).
Stellingwerf, R. F., 1975a, Astrophys. J. **195**, 441.
Stellingwerf, R. F., 1975b, Astrophys. J. **199**, 705.

Modelling Cepheids and RR Lyrae Stars

G. Kovács

University of Florida, Department of Physics, Gainesville
and
Konkoly Observatory, Budapest, Hungary

Abstract

We review the recent nonlinear hydrodynamical results and the numerical problems to be solved by the next generation of codes. The behavior of the Fourier parameters and double-mode RR Lyrae pulsation will be discussed in detail. We emphasize the importance of the implementation of the various adaptive-mesh schemes for stellar pulsations. A further goal is the substitution of the artificial viscosity with modern shock capturing methods and a better treatment of radiative transfer. Convection however, must ultimately be included in order to supply the necessary physical mechanism for dissipation.

1. Introduction

The application of the full set of nonlinear equations of radiative hydrodynamics to the modelling large-amplitude stellar pulsations has a history of almost 30 years. Since the pioneering work of Christy (1964) there have been a number of attempts to improve the original idea. The particular problems to tackle in nonlinear stellar pulsation are as follows: (1) there is a very steep temperature and density gradient at the hydrogen ionization zone which moves according to the local ionization conditions, therefore, does not follow the motion of the mass elements; (2) under certain conditions shocks may develop which also move according to the local thermo- and hydrodynamical situation; (3) the exact solution of the radiative transfer problem coupled with the hydrodynamics is very difficult throughout the entire stellar envelope and atmosphere, because of the basic change of the radiative properties in these regions; (4) convection should play a role in the pulsation, which poses the Herculean task of simulating the three-dimensional turbulent flow by a one-dimensional model; (5) since we are interested in the long-term behavior, the code should be very stable numerically and exact total energy conservation is a must. Past code developments handled some of these questions with some success. For reference we mention Davis (1972), Stellingwerf (1974, 1975, 1984), Castor, Davis and Davidson (1977), Aikawa and Simon (1983), Bowen (1988). Since, and parallel with these works, there have been a lot of progress made in solving one- and multi-dimensional hydrodynamical problems in various fields of physics and engineering. A realization of this fact stimulated the organization of a recent workshop on this topic (Buchler 1990).

The purpose of this paper is threefold: (i) to review the major recent results and problems in modelling Cepheid and RR Lyrae pulsations; (ii) to emphasize the importance and present a possible mechanism for multimode stellar pulsation among classical variables; (iii) to give a flavour of the latest efforts in code developments.

2. Stellar masses with the new opacities

During the last two years the widely used Los Alamos opacities have undergone substantial changes following the bold suggestion of Simon (1982). The OPAL opacities of Rogers and Iglesias (1992) cured the basic discrepancy of the beat Cepheids (Moskalik, Buchler and Marom 1991). A similar, but maybe less accommodating conclusion can be drawn by the application of the OP opacities (Seaton; Kanbur, these proceedings). For bump Cepheids, on the other hand, the opacity enhancement did not prove to be completely sufficient (Moskalik et al. 1991). On the other hand, a subsequent improvement of the OPAL opacities with the consideration of the spin-orbit interaction (Iglesias, Rogers and Wilson 1992) further decreased (in our opinion eliminated) the difference between bump and evolutionary masses (Moskalik, these proceedings).

While solving the Cepheid mass discrepancies, the OPAL opacities led to an unexpected sensitivity to metal abundance and chemical composition, and confused the situation of RRd stars (Kovács, Buchler and Marom 1992; Kovács et al. 1992). Considering only luminosity and iron abundance effects, for solar composition the ranges of RRd masses are $0.75-0.8$ and $0.7-0.9 M_\odot$ for M15 (Oo II) and IC 4499 (Oo I) respectively. By taking the 'standard' values of the luminosities, iron abundances and individual periods, there seems to be no systematic difference in mass between the two Oosterhoff type RRd stars. The ambiguities in the observed chemical composition and metal abundances, however, do not permit us to draw any firm conclusion yet.

3. Fourier decomposition

3a. Bump Cepheids

One important utilization of nonlinear hydrodynamics is to compare the theoretical pulsation profiles with the observed ones. A natural frame of reference is the set of Fourier parameters obtained from a fit to the pulsation cycle. Before considering the ultimate goal of fitting the light and velocity curves of the individual stars we attempt to verify the overall trend in the progression of the Fourier parameters.

Because of the new opacities, the center of the $2\omega_0 = \omega_2$ resonance for bump Cepheids is located at 10 days, therefore, the hydrodynamical results can be compared directly with the observations. Such a comparison shows that the overall agreement with the observations is very good, at least for the Fourier parameters up to the third order (Moskalik et al. 1991). There are, however, some systematic differences, especially in ϕ_{21} for the light curves of the low-period stars. Further characteristic features which have not yet been modelled, are the break in the ϕ_{31} and ϕ_{41} progression at ≈ 7 days for the light curves and the strict linear relation between ϕ_{31} and ϕ_{41} after 10 days (Simon and Moffett 1985). There is a possibility that the feature at 7 days is associated with the $3\omega_0 = \omega_4$ resonance (Antonello, these proceedings).

Clearly, understanding these characteristics is an important task for future hydrodynamical simulations. Further detailed discussions of the bump Cepheid progression are in Buchler, Moskalik and Kovács (1990) and Moskalik *et al.* (1991) and references therein.

3b. s-Cepheids

Using the Fourier decomposition technique, Antonello and Poretti (1986) found that a large fraction of the small amplitude Cepheids below $\approx$ 5 days follow a progression in their Fourier parameters different from that of the Classical Cepheids. Because of their sinusoidal light curves and low amplitudes, a common understanding is that s-Cepheids are first overtone variables. An apparent break in their ϕ_{21} could then be accounted for by a $2\omega_1 = \omega_4$ resonance, very similarly to that of the bump Cepheids. Unfortunately, there is very little direct observational evidence so far that s-Cepheids are indeed first overtone pulsators. Though nonlinear results suggest something similar to that what we observe, they are not convincing either. We refer to Antonello's paper of these proceedings for the present status of these variables. It is clear that a more detailed study of the SMC first overtone Cepheids would help to clarify the situation of the s-Cepheids (Smith *et al.*, these proceedings).

3c. RR Lyrae stars

Unlike Classical Cepheids, RRab stars do not show any systematics in the variation of their Fourier parameters as a function of the period (Simon and Teays 1982). Similarly to the bump Cepheids, there are systematic differences in the low-order Fourier phases (Simon 1985). Our unpublished tests confirmed this discrepancy basically independently of the opacity and artificial viscosity (see also Kovács 1990). The obvious agents to resolve these discrepancies are convection and more accurate treatment of radiative transfer. Both of these were absent from the codes used in the above computations.

RRc stars, on the other hand, seem to follow some trend in their Fourier parameters as a function of the period. Also, nonlinear models are in a fair agreement with the observations. The roughly linear relation between the period and ϕ_{31} prompted Simon (1989) to use this phase as a diagnostic tool to obtain stellar parameters. Further discussions of the RRc stars in this context are given by Simon and by Cacciari and Bruzzi, these proceedings.

4. Mode selection, double-mode pulsation

One of the main goal in modelling nonlinear stellar pulsation is the determination of the long-term behavior of the model. In this respect the application of the relaxation technique (Stellingwerf 1974) to find strictly periodic solutions and their stability is extremely important. In the case of coexisting stable single- and multimode, or unstable single-mode solutions however, the direct (*i.e.* 'brute force') time integration cannot be avoided. In this situation the final state can be assessed only by 'long enough' integration combined with a precise analysis of the time-series.

A further major complication in the mode selection is the important role of various viscosity parameters (artificial and eddy). There is probably an optimal tradeoff in the amount of viscosity in order to get a result which is not very sensitive to these parameters, but at the same time to maintain the numerical stability and observationally acceptable amplitude. Until further improvement in the shock treatment and convection theory, statements about mode selection remain somewhat uncertain, especially close to the mode transition (*i.e.* at the bifurcation).

Modelling double-mode pulsation has been a serious challenge for nonlinear hydrodynamics for decades. Here we would like to give only the highlights of the very encouraging developments of this year. A more comprehensive review and details of our double-mode models are given in Kovács (1992); Kovács and Buchler (1992).

The most important constraints for any model compatible with the observations are : (a) periods, (b) amplitudes, (c) amplitude stability. Although previous modelling has already produced double-mode RR Lyrae pulsation in the vicinity of the $2\omega_0 = \omega_3$ resonance (Kovács and Buchler 1988), the requirement of matching the observed periods was not met. Since no low-order resonance could fit the period and period ratio at the same time, we tried to approach the question from another angle. As is well known, viscosity plays an important role in some circumstances in controlling mode selection. To avoid this, we started a survey of low dissipative RR Lyrae models in 1991. The basic question we would like to answer is: how would a purely radiative RR Lyrae model pulsate if we could get rid of all the unnecessary dissipation? These models have higher amplitudes (especially velocities) than those given by the observations, but we defer this question to a later time (see also Section 6). Quite surprisingly, but the low-viscosity models showed a window of unstable fundamental pulsation, which for somewhat cooler models overlapped with the 'fundamental only' region and resulted in a double-mode pulsation.

Many models have been thoroughly studied by limit cycle analysis and by very long direct time integrations. New features of pulsation were detected, like periodically varying mode amplitudes and coexisting double- and single-mode states. Though the periods were matched and the double-mode state was steady, the amplitude ratios were reversed compared to what is observed. It turned out however, that this problem can be cured by allowing the models to be a little more dissipative. The higher dissipation led to the generation of another double-mode state, now with the right, directly observable quantities.

It is, of course, very important to confirm and further refine our finding by independent computations which use different numerical schemes and perhaps more complete physics (most importantly convection). Marvelously enough, it was at this conference that such a supporting new results were presented for the first time. Though at different stellar parameters, new convective models of Bono and Stellingwerf show the same type of double-mode behavior as our radiative minimal viscosity models. It is remarkable too that their double-mode model also appears at the neighbourhood of the same type of resonance (namely $3\omega_0 - \omega_1 = \omega_2$) what we have claimed to be responsible for the double-mode pulsations in our models.

An even more recent development is a test of one of our double-mode model with a new radiative adaptive code (Marom, private communication). The model settled down with the new code to a very similar double-mode state found by our Lagrangean code.

Though it is clear that further work is necessary on the convective models to get better agreement with the observations, we think that this first result is very encouraging. By using updated opacities and extending the survey to a larger parameter space, convective models might cure the problem of low luminosity and temperature of our radiative models.

5. Stochastic effects in classical variables

There is no star in the classical instability strip which avoids convection. As a consequence, turbulence (*i.e.* stochastic forcing) is a natural ingredient of pulsation. Convective pulsation models usually take into consideration that part of the stochastic effect which have nonzero ensemble average (*e.g.* pressure, flux). The remaining fluctuating part has so far been neglected in the studies of the classical variables, because of the assumed small effect of the random perturbations which cancel out on the average. As is well known, this part of the turbulent convection plays an important role in the theory of the stochastic mode excitation of the 5 min. solar oscillations (Goldreich and Keeley 1977). In that theory the overall linear damping and the turbulent intensity which excite the modes balance each other, because both depend mostly on the same physical mechanism, *i.e.* turbulent convection. We would like to indicate here that in the case of limit cycle pulsation the situation is quite different, which may lead to much higher amplitudes of the stochastically excited modes.

Let us consider a single periodic pulsation close to the transition region. Because the pulsation is stable, all modes are damped in the limit cycle. Usually, as in the case of non-resonant pulsation, the linearly damped modes are even more damped in the limit cycle. The linearly excited modes, however, are *arbitrarily* mildly damped, and at the transitions they become actually excited with a rate which is less than their linear excitation rates. There are two possibilities to consider: (a) the model is in the 'one limit cycle only' region, or (b) the model is in the 'either-or' region. In case (a) the modes can reach in principle arbitrarily large amplitudes depending on the intensity of the noise. In case (b), however, there is an upper limit on the amplitudes of the noise generated (*precursor*) modes, because increasing the noise causes not only an amplitude increase, but also the chance that the model switches to the other, more stable limit cycle. Both situations are generic, but in the stellar situation we consider case (b) more likely to lead to observable precursor oscillations, because in principle, close to the transition arbitrary small noise may lead to observable amplitudes. Work is in progress to estimate the mode amplitudes and lifetimes.

An important question is the intensity of the noise. An *ab initio* estimation of this quantity is very complicated and risky (Goldreich and Keeley 1977; Goldreich and Kumar 1990). From the observation of the long-term phase variation, however, one may get some direct estimation on the size of the noise. Very often cited features of the O-C diagrams are the irregular and 'faster than the evolutionary time scale' variations. We think that at least a part of these problems can be explained by *stochastic phase diffusion*, which has basically a *non-stationary, random walk* character.

We think that a search for low-amplitude pulsation in classical variables would be extremely useful. If such a search were positive, we could map the period ratio diagrams or help to estimate masses and chemical compositions much more accurately than the RRd stars allow (see Kovács and Buchler, these proceedings).

For further discussion of the above topic, the general formalism and some interesting dynamical effects of noise, we refer to Kovács and Buchler (in preparation); Buchler, Goupil and Kovács (1992) ; and Buchler and Kovács (1992).

6. The new generation of pulsation codes

The basic numerical technique of nonlinear stellar pulsation is more than 20 years old (Fraley 1968). Though together with Stellingwerf's (1974) relaxation method the scheme is a very powerful tool in studying nonlinear stellar pulsation, obvious improvements are necessary. We emphasize here the importance of the further sophistication of the radiative hydrodynamics. For the problem of convection we refer to Stellingwerf (these proceedings).

As it was mentioned in Section 1, the main problems to be solved in the radiative nonlinear codes are: (i) proper tracing of the hydrogen ionization zone; (ii) substituting the artificial viscosity method by some modern, less dissipative shock capturing scheme; (iii) more accurate solution of the radiative transfer problem throughout the entire envelope.

In the last two years some progress has been made in tackling most importantly, the tracing of the hydrogen ionization. Problems (ii) and (iii) have not yet received proper attention perhaps of the possibly greater importance of problem (i) (see however Buchler and Whalen 1990; Dorfi and Feuchtinger 1991; Gehmeyr 1991).

The first adaptive pulsation code was constructed by Castor *et al.* (1977). Aiming specifically to resolve the hydrogen ionization zone, Aikawa and Simon (1983) opted for a scheme in which the updated temperature plays the role of the independent spatial variable. The new scheme of Gehmeyr (1991) is in principle able to resolve any feature specified in the structure function. Another new technique was suggested by Dorfi and Gautschy (1990) and later implemented by Dorfi and Feuchtinger (1991). In all these works the total energy conservation, a very important quality of nonlinear pulsation, was not guaranteed by the schemes. In addition, in the limit of no spatially resolved structures, the above schemes go to an Eulerian coordinate system instead of a Lagrangean one, which is a more natural system for a radially pulsating star. The new adaptive code developed by Buchler and Marom (1992) has this additional useful feature and furthermore, exactly conserves the total energy.

Beside the very smooth variation of the physical quantities, adaptive codes, in general, also seem to give smaller amplitudes than those of the Lagrangean codes (see however Gehmeyr 1991). Radial velocities usually decrease more than luminosity amplitudes (Kovács 1990; Marom, private communication). Together with the more accurate solution of the hydrodynamical equations, the amplitude decrease and the smooth variation shows that adaptive codes provide a more sound modelling of the pulsation and may even suggest that (at least at the hotter side of the instability strip) convection might be not all that important dynamically.

7. Conclusion

Substantial progress has been made in modelling classical stellar pulsations during the last two years. The revised opacities have led to the elimination of the frustrating Cepheid mass discrepancies. New thorough survey of radiative RR Lyrae models with minimal viscosity has produced the first physically sound double-mode models, thereby breaking the dead-lock of search for RRd models and stimulating further studies in this field. A new generation of nonlinear pulsation codes are under development, which aim it is to solve the complex problem of radiative hydrodynamics very accurately.

We hope that these developments will ultimately lead to a more fruitful application of nonlinear pulsation results, like fitting the theoretical pulsation profiles to the observations of the individual stars. Understanding mode selection, multimode behavior and long-term amplitude modulation (*i.e.* Blazhko effect) of course remain top priority issues.

Acknowledgements

The author is grateful for the financial support of the IAU and of the Institute for Fundamental Theory in the Physics Department of the University of Florida. Fruitful discussions with Robert Buchler and Ariel Marom are greatly appreciated. This work was supported by NSF (AST 89-14425) and by an RCI grant through IBM and the NER Data Center at the University of Florida.

References:

Aikawa, T. and Simon, N. R. 1983, *Ap. J.*, **273**, 346.
Antonello, E. and Poretti, E. 1986, *Astr. Ap.*, **169**, 149.
Bowen, G. H. 1988, *Ap. J.*, **329**, 299.
Buchler, J. R. 1990, in *The Numerical Modelling of Nonlinear Stellar Pulsations; Problems and Prospects*, Kluwer, Dortrecht; Ed.: J. R. Buchler, p. 1.
Buchler, J. R., Goupil, M.-J. and Kovács, G. 1992, *Astr. Ap.*, submitted.
Buchler, J. R. and Kovács, G. 1992, *Physica D*, submitted.
Buchler, J. R. and Marom, A. 1992, in preparation.
Buchler, J. R., Moskalik, P., and Kovács, G. 1990, *Ap. J.*, **351**, 617.
Buchler, J. R. and Whalen, P. 1990, in *The Numerical Modelling of Nonlinear Stellar Pulsations; Problems and Prospects*, Kluwer, Dortrecht; Ed.: J. R. Buchler, p. 315.
Castor, J. I., Davis, C. G. and Davison, D. K. 1977, *Los Alamos report LA-6664*.
Christy, R. 1964, *Rev. of Modern Physics*, **36**, 555.
Davis, C. G. 1972, *Ap. J.*, **172**, 419.

Dorfi, E. A. and Gautschy, A. 1990, in *The Numerical Modelling of Nonlinear Stellar Pulsations; Problems and Prospects*, Kluwer, Dortrecht; Ed.: J. R. Buchler, p. 289.
Dorfi, E. A. and Feuchtinger, M. U. 1991, *Astr. Ap.*, **249**, 417.
Fraley, G. S. 1968, *Ap. Space Sci.*, **2**, 96.
Gehmeyr, M. 1991, *On Non-Lagrangian Computations of Convective RR Lyrae Stars*, Ph. D. Thesis, University of New Mexico.
Goldreich, P. and Keeley, D. A. 1977, *Ap. J.*, **212**, 243.
Goldreich, P. and Kumar, P. 1990, *Ap. J.*, **363**, 694.
Iglesias, C. A., Rogers, F. J and Wilson, B. G. 1992, *Ap. J.*, submitted.
Kovács, G. 1990, in *The Numerical Modelling of Nonlinear Stellar Pulsations; Problems and Prospects*, Kluwer, Dortrecht; Ed.: J. R. Buchler, p. 73.
Kovács, G. 1992, in *Nonlinear Phenomena in Stellar Variability*, IAU Coll. 134, Eds.: J. R. Buchler, M. Takeuti, in press.
Kovács, G. and Buchler, J. R. 1988, *Ap. J.*, **324**, 1026.
Kovács, G. and Buchler, J. R. 1992, *Ap. J.*, in press.
Kovács, G., Buchler, J. R., and Marom, A. 1991, *Astr. Ap. (Letters)*, **252** L27.
Kovács, G., Buchler, J. R., Marom, A., Iglesias, C. A. and Rogers, F. J. 1992, *Astr. Ap. (Letters)*, in press.
Moskalik, P., Buchler, J. R. and Marom, A. 1991, *Ap. J.*, **385**, 685.
Rogers, F. J. and Iglesias, C. A. 1992, *Ap. J. Suppl.*, **79**, 507.
Simon, N. R. 1982, *Ap. J. (Letters)*, **260**, L87.
Simon, N. R. 1985, *Ap. J.*, **299**, 723.
Simon, N. R. 1989, *Ap. J. (Letters)*, **343**, L17.
Simon, N. R. and Moffett, T. J. 1985, *PASP*, **97**, 1078.
Simon, N. R. and Teays, T. J. 1982, *Ap. J.*, **261**, 586.
Stellingwerf, R. F. 1974, *Ap. J.*, **192**, 139.
Stellingwerf, R. F. 1975, *Ap. J.*, **195**, 441.
Stellingwerf, R. F. 1984, *Ap. J.*, **284**, 712.

DISCUSSION

M. FEUCHTINGER: What kind of rezoning is used in your code?
G. KOVÁCS: The method is very similar to the one what you use, but in the limit of no spatial resolution it goes to a Lagrangean instead of an Eulerian system like in your case.
N. SIMON: To what extent can your new code resolve shocks?
G. KOVÁCS: I let Robert Buchler answer this question.
R. BUCHLER: In principle you can resolve any physical feature, provided you use enough meshpoints. For a given number of meshpoints, however, you have to make a balance between the features you deem most important to be resolved.

Double-mode RR Lyrae models

G. Bono[1], R.F. Stellingwerf[2]

[1] Trieste Astronomical Observatory, Italy [2] Los Alamos National Laboratory, NM

Abstract

Double-mode RR Lyrae stars (RRd) represent a fundamental testing ground for the theory of stellar pulsation because they play an important role in understanding the interaction between different pulsation modes. In spite of the fact that from an observational point of view the identification of the fundamental (F) and first overtone (IO) periods needs only few periods of light curve coverage (Sandage et al. 1981, Nemec 1985), the theoretical models of double-mode and mixed-mode variables dating back to the original papers of Cox et al. (1983) and Stellingwerf (1975) were not integrated for long enough times to ensure proper mode identification. Recently Kovacs and Buchler (1992) using a standard radiative hydrodynamical code investigated the theoretical properties of RRd stars in more detail. This paper represents a brief description of the results of few cases computed adopting a non-linear, non-local treatment of convective transport, and long integration times to enable positive mode identification.

1. Discussion

The cases discussed, are model 2.3 ($log L/L_o = 1.8, T_e = 7100K, P_0 = .5386d, P_1 = .4010d, P_1/P_0 = .7446$) and model 4.6 ($log L/L_o = 1.5, T_e = 6500K, P_0 = .4097d, P_1 = .3083d, P_1/P_0 = .7525$). Both have been computed with $M = .65M_o$ and King IA chemical composition (for mor details see Bono and Stellingwerf 1991, Stellingwerf and Bono 1992). The first one is located between the IO and the F blue edges, the latter one very close to the IO red edge. The period dispersion minimization technique (PDM, Stellingwerf 1978) has been applied to the radial variation of the surface zone for 20 consecutive periods to derive the amplitudes and periods of different modes. Along the run of models this analysis has been performed every 300 periods. It is worth emphasizing that this approach gives reliable results only if the physical and numerical approximations adopted in the code retain the exact energy conservation so that gains or losses from numerical perturbations a re avoided. In our case generally the energy conservation ranges between $10^{-8} \div 10^{-9}$ the initial value of the energy. Panel a of fig.1 shows the switching from the F to the IO of model 2.3. The formula derived by Stellingwerf (1975b) and Cox et al. (1976) has been used to evaluate the switching time. For case 2.3 the switching rate ranges between .001 and .002 1/(yr K), while the evolutionary time that a star with the same mass and almost the same chemical composition spends inside the instability strip is approximately 15000 yr/K

(Castellani et al. 1991). Therefore a rough evaluation of the switching time is of the order of thousands of years (5000 ± 1000 yr). This value is more in agreement with the relatively large number of RRd stars found in galactic globular clusters and in the field. This value is an order of magnitude larger than the old estimation derived by Stellingwerf (1975). This discrepancy is essentially due to the decrease of the switching rate and partially to a small increase in the evolutionary time. Case 4.6 represents the first firm evidence of multimode stability in Pop. II variables. As a matter of fact the panels b and c of fig. 1 clearly show the stability of the first three modes obtained initiating the model in the F mode with a surface amplitude of 10 km/s. To check the final amplitude and stability of these modes, the calculations have been repeated perturbing the IO linear radiative eigenfunctions with the same surface amplitude. Although, the amplitude of the IO remains constant over a long time, it is very comforting to note that the final amplitudes of the three modes are exactly the same. Unfortunately no firm conclusion can be reached concerning the location of RRd stars inside the stability strip and in particular about the physical mechanisms that govern the stability of these objects. Model 4.6 is located at the intersection between the resonance lines ($3\nu_0=\nu_1+\nu_2$) and ($4\nu_0=3\nu_1$) therefore it will be necessary to extend the calculation along the resonance lines and the F blue edge to account for the double-mode pulsation of RR Lyrae.

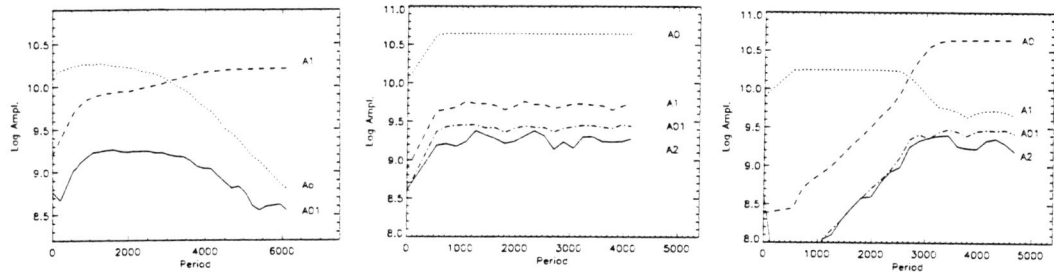

Figure 1 Radius amplitudes of cases 2.3 and 4.6. A0, A1, A2, A01 are the amplitudes of first three modes and the non-linear term.

References:

Bono, G., Stellingwerf, R.F., 1991, Mem. Soc. Astron. It., in press.
Castellani, V., Chieffi, A., Pulone, L., 1991, Ap. J. Suppl. **76**, 911.
Cox, A.N., Hodson, S.W., Davey, W.L., 1976, In "Proceedings Solar and Stellar Pulsation Conference", ed. A.N. Cox, R.G. Dupree.
Cox, A.N., Hodson, S.W., Clancy, S.P., 1983, Ap. J. **266**, 94.
Kovacs, G., and Buchler, R.J., 1992, Preprint.
Nemec, J.M., 1985, Astron. J. **90**, 204.
Sandage, A., Katem, B., Sandage, M, 1981, Ap. J. Suppl. **46**, 41.
Stellingwerf, R.F., 1975a, Ap. J. **195**, 441.
Stellingwerf, R.F., 1975b, Ap. J. **199**, 705.
Stellingwerf, R.F., 1978, Ap. J. **224**, 953.
Stellingwerf, R.F., and Bono, G., 1992, This proceedings.

A Survey of BL Herculis-Type Models

J. Robert Buchler and Pawel Moskalik

*Department of Physics, University of Florida,
Gainesville, Florida 32611, U.S.A.*

Abstract

We have studied the nonlinear behavior of several sequences of BL Herculis-type models. The question arose whether the 2:1 resonance between the fundamental mode and the second overtone would cause the same systematic variation of Fourier parameters of the pulsation cycle with period ratio P_2/P_0 as was seen in the classical Cepheids. We find that for the BL Her stars, the behaviour of the light-curve Fourier phases is markedly different from the Cepheid case. In particular, ϕ_{21} exhibits essentially a featureless, monotone increase throughout the range of P_2/P_0, which is in qualitative agreement with the observed trend (Petersen & Diethelm 1986). In the velocity curves, on the other hand, the 2:1 resonance is a dominant feature and the progression of the Fourier phases and the amplitude ratios is similar to those witnessed in the Cepheids. However, here the sensitivity to the stellar masses and luminosities is significantly stronger. Our results show that radial velocity observations of the BL Her stars would pinpoint the resonance and put important new constraints on the models.

Theoretical Implications of Triple-Mode RR Lyrae Pulsations

G. Kovács[1,2], J. R. Buchler[1]

[1]*Department of Physics, University of Florida, Gainesville*
[2]*Konkoly Observatory, Budapest, Hungary*

Abstract

We argue that triple-mode RR Lyrae pulsation with low amplitudes might be quite common. It is shown that until very accurate abundance data become available, triple-mode RR Lyrae stars are the only hope to estimate reliable stellar parameters from the periods alone.

1. Introduction

The motivation of this study comes from the following observations: (1) most of the radiative RR Lyrae models are *linearly excited* in some range of temperature in the three lowest order radial modes; (2) many RR Lyrae stars show *non-repetitive* light variation. Accurate photoelectric observations might reveal some low-amplitude mode contamination (*e.g.* Fernley *et al.* 1990); (3) marginally stable modes in a limit cycle may oscillate at low amplitudes because of *stochastic mode excitation* (Kovács and Buchler, in preparation). Some high-amplitude δ Scuti stars seem to support the existence of such a multimode pulsation (*e.g.* Walraven *et al.* 1992).

We address the following questions in this note: (a) How much the knowledge of the periods of the first three radial modes constrains the derived stellar parameters? (b) Is there any limit cycle in which the other two modes are simultaneously marginally stable?

2. Determination of M and Z

A large number of RR Lyrae models have been computed in the range $0.6 - 0.9$, $40 - 70$, $6000 - 8000$ for M, L and T_{eff} respectively. Two sets of models have been computed for $Z = 0.0001$ and $Z = 0.001$, both with $X = 0.7$ and with solar (*i.e.* Anders-Grevesse) mixtures. We used the opacities published by Rogers and Iglesias (1992).

The novel sensitivity of the period ratios on Z due to the revised opacities seriously jeopardizes the applicability of the single period ratio method for the mass determination. For three periods, however, the effect of Z can be largely eliminated. The $P_1/P_0 - P_2/P_0$ *versus* P_0 diagram shows very little dependence on Z (Fig. 1).

In conclusion, if three periods are known, then the $P_0 \to P_1/P_0 - P_2/P_0$ diagram can be used for a mass determination. With this mass, the $P_0 \to P_2/P_0$ diagram constrains the metal abundance.

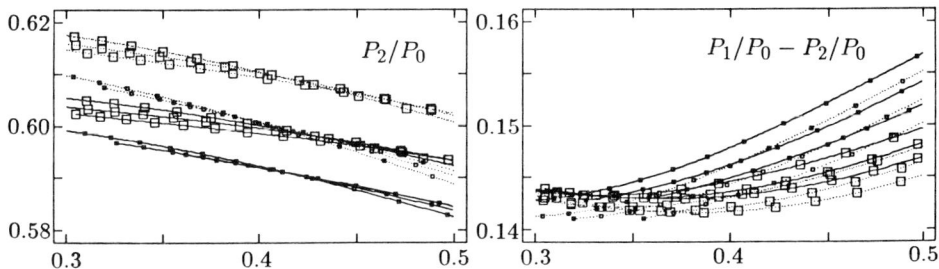

Figure 2. Period ratio diagrams. *Dotted lines:* $Z = 0.0001$, *solid lines:* $Z = 0.001$; *small squares:* $M = 0.6$, *large squares:* $M = 0.8$. At each mass three luminosity values are plotted with $L = 40, 50, 60$. Horizontal axis: P_0 in [day].

3. Limit cycle stability

One possible scenario for triple-mode RR Lyrae pulsation is that the marginally stable modes in a limit cycle get stochastically excited due to turbulent convection. A limit cycle analysis of a sequence of RR Lyrae models showed that *simultaneous marginal stability* is possible only in the *second overtone* limit cycle. Predictions of evolution calculations do certainly allow (re)entering into the instability strip from the very hot side, where only the second overtone could be excited.

4. Conclusions

Theoretical considerations suggest that triple-mode RR Lyrae pulsation is possible. If it is observed, the periods put further significant constraints on the stellar parameters (especially on M and Z) over those of the traditional single period ratio method. So far only AC And (Fitch and Szeidl 1976) could be possibly identified as a triple-mode RR Lyrae star pulsating in the first three radial modes. For recent analyses of two RRc stars we refer to Peniche *et al.* (1989) and Hobart *et al.* (1991).

Based on the commonly observed non-repetitive variation of the light curves and on the theoretical possibility of triple-mode pulsation, we think it would be very important and profitable to search for low-amplitude pulsations in RR Lyrae stars.

References:

Fernley, J.A., Skillen, I., Jameson, R.F. and Longmore, A.J. 1990, *MNRAS*, **242**, 685.
Fitch, W.S. and Szeidl, B. 1976, *Ap. J.*, **203**, 616.
Hobart, M.A., Peña, J.H. and Peniche, R. 1991, *Rev. Mexicana Astr. Ap.*, **22**, 275.
Peniche, R., Gomez, T., Parrao, L. and Peña, J.H. 1989, *Astr. Ap.*, **209**, 59.
Rogers, F.J. and Iglesias, C.A. 1992, *Ap. J. Suppl.*, **79**, 507.
Walraven, Th, Walraven, J. and Balona, L.A. 1992, *MNRAS*, **254**, 59.

On the explanation of the Sandage effect

M. Catelan

Instituto Astronômico e Geofísico, Universidade de São Paulo, São Paulo, Brazil

There are two problems with the Lee et al. (1990, LDZ; Lee 1990, 1991) scenario for the Sandage effect (Sandage 1990 and references therein) which may still warrant further detailed analyses: 1) on the basis of standard tracks, RR Lyrae stars in OoII, M15-like globular clusters can only be well-evolved objects from the blue ZAHB *for unrealistically low helium abundance (Y) values*; 2) in comparison with the models by Sweigart (1987, SW87), the evolutionary tracks of Lee & Demarque (1990, employed by LDZ) tend to overestimate the effect of evolution away from the ZAHB for the RR Lyrae stars, despite the similar assumptions involved in the calculations of the two sets of models. (See Catelan 1992a, C92A.) It is indeed possible to conclude that the claims by LDZ that the Sandage effect has been explained arise not only from the inclusion of the final 5-10% of HB evolution in their case, but also from the fact that their tracks do not agree very well with those of SW87, for both high and low Z [as an example, $\Delta \log P(T_{\rm eff})$(M15 - NGC 2808/6864) increases by $\simeq$ 10-30% due to the former effect, and by $\simeq$ 25-45% due to the latter – see Catelan 1992b for details]. At any rate, if the RR Lyrae variables in OoII clusters *are* well-evolved objects from the blue ZAHB, an alternative way to obtain HB tracks with the necessary morphological characteristics (evolution predominantly redward for low metallicities) is urged, for $Y_{\rm MS} \simeq 0.20$ (or equivalently $Y_{\rm HB} \simeq 0.21$ - 0.22) – as presently required – is a value that seems both *cosmologically too low* (Olive et al. 1990), and unjustified in terms of independent observational and/or theoretical evidence (C92A). I suggest that a "mimicking effect" may be operating, in which HB stars do not have abnormally low Y, but instead enhanced core-mass values (with respect to the standard ones): $\Delta M_{\rm c} \approx +0.02 \, M_\odot$. Several advantages are inherent to this scenario, such as a natural explanation of the results by Walker (1992) concerning the distance modulus to the LMC, and an increase in evolutionary RR Lyrae masses ("α-enhanced" evolutionary values are lower than the latest double-mode ones – see C92A).

References:

Catelan, M., 1992a, Astron. Astrophys., in press. (C92A)
Catelan, M., 1992b, in preparation.
Lee, Y.-W., 1990, Astrophys. J., **363**, 159.
Lee, Y.-W., 1991, Astrophys. J., **367**, 524.
Lee, Y.-W., Demarque, P., 1990, Astrophys. J. Suppl., **73**, 709.
Lee, Y.-W., Demarque, P., Zinn, R., 1990, Astrophys. J., **350**, 155. (LDZ)
Olive, K. A., Schramm, D. N., Steigman, G., Walker, T., 1990, Phys. Lett. B, **236**, 454.
Sandage, A., 1990, Astrophys. J., **350**, 631.
Sweigart, A. V., 1987, Astrophys. J. Suppl., **65**, 95. (SW87)
Walker, A., 1992, Astrophys. J., **390**, L81.

Hydrogen Emission Lines from Extended Pulsating Atmospheres

P. de Laverny[1,2], C. Magnan [1,2,3]

[1] GRAAL, cc72, Université Montpellier II, F-34095 Montpellier cedex05, France.
[2] Institut d'Astrophysique de Paris, France [3] Collège de France, Paris, France

Abstract

In order to study the basic radiative mechanisms in the extended envelopes of evolved stars (e.g. Long Period Variables), we determine the spectrum emerging from a very optically thick Non-LTE hydrogen layer surrounding a core of high temperature (see figure 1). Such a model was first proposed by Sobolev (1960) and Menzel (1946). It is consistent with the fact that LPVs are likely evolving toward planetary nebulae. The main parameters of the model are the temperature of the illuminating star (T_r), the dilution factor (ω) and the Lyman continuum optical depth of the envelope (τ_{1c}). We examine their influence upon the emerging intensities in the lines and in the continua.
The model could explain some observations related to evolved stars and especially the presence of emission features in an otherwise cold environment. Furthermore atmospheric pulsation phenomena produce large variations in the emerging spectra: reddening, emission/absorption transition of the hydrogen lines, profile deformations...

We assume that the only energy deposited in the envelope is the ionizing radiation of the stellar core. We reject the Local Thermodynamical Equilibrium hypothesis in the medium. The lines thus result from purely radiative processes via a fluorescence mechanism. We solve the radiative transfer equations in the lines (including the Lyman transitions) and in the continua together with the equations of statistical equilibrium of a hydrogen atom with 4 levels and a continuum. This system is linearized and solved by the method of addition of layers in an iterative way. (Gros and Magnan, 1981; Magnan, 1992). A typical model of a mira is obtained with a core of temperature $T_r = 30000K$, a dilution factor of 10^{-2} and an envelope optical depth in the Lyman continuum $\tau_{1c} = 100$. The geometric thickness of the envelope is thus $\simeq 1 R_\odot$ for a star which radius is close to $200 R_\odot$. The calculated emerging spectrum shows that the illuminating radiation is extremely reddened by the envelope. Moreover the Lyman and Balmer lines are strong in emission with central reversal profiles due to a strong decrease of the source function in the most external layers of the envelope.
We mimic an atmospheric pulsation by varying the envelope radius. If δR is assumed to be constant, a variation of R leads to a change of the parameters $\omega \simeq R^{-2}$ and $\tau \simeq R^{-2}$ since $n_e^2 R^2 \delta R$ is constant. In figure 2 is shown the effects of a +/- 25%

variation of the envelope radius on the emerging $H\alpha$ profile. We may notice that the higher is the envelope radius, the stronger are the lines in emission, the wider are the lines, and the bluer is the emerging spectrum.

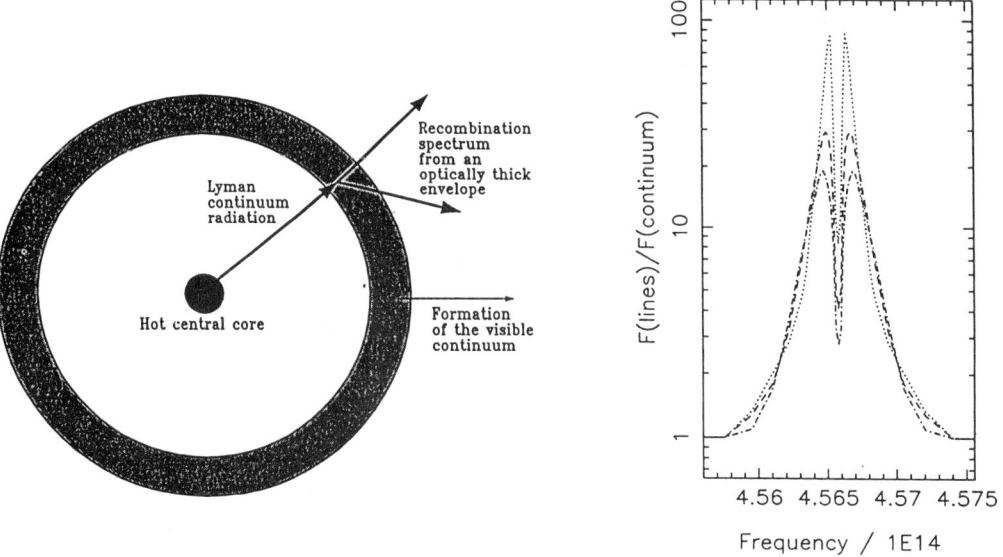

Figure 1 (left) Schematic modeling of the radiative transfer in an evolved star.
Figure 2 (right) H_α emerging profiles from 3 envelopes of different radius.

This study suggests that emission lines in diluted envelopes could result from a recombination mechanism and not from thermal processes through a temperature rise. Moreover atmospheric pulsations produce changes in the emerging spectrum. They lead to variations of the intensity of the emerging lines with corresponding emission/absorption passages, to variations of the emerging profiles, and to a reddening of the emerging continuum.

This model can be applied to evolved pulsating stars like the Mira variables which show similar variations during their cycle (for more details see de Laverny and Magnan, 1992).

References:

Gros M., Magnan C., 1981, A&A, 93, 150.
de Laverny P., Magnan C., 1992, in preparation for A&A.
Magnan C., 1992, submitted to A&A.
Menzel D.H., 1946, Physica, XII, n.9-10, 768.
Sobolev V.V., 1960, in "Moving Envelopes of Stars", Harvard University Press.

Propagation of Radial Pulsation Modes in the Outer Atmosphere of Arcturus: First Results

M. Cuntz

High Altitude Observatory, National Center for Atmospheric Research[1], Boulder, CO 80307-3000, USA

Abstract

I present first results of a study about propagating radial pulsation modes in the outer atmosphere of Arcturus (K1.5 III). Mechanical energy input is explicitly taken into account by treating shock wave dissipation. I investigate the influence of different wave frequencies on the mass loss behavior of the star. I show that significant time–averaged mass loss can only be produced when periods larger than 5 x 10^5 s ($\sim$ 1 week) are employed. The initial atmosphere I use extends from 1.2 R_* up to 11.8 R_*. All wave models are adiabatic. I found that the mass loss rates and the final flow speeds obtained are extremely sensitive to the wave periods. In a certain regime the effect of increasing the period by a factor of 4 is to increase the corresponding mass loss rate by four orders of magnitude. I note that the mass loss rate and final flow speed of the wind for a 5.6 x 10^5 s period wave are somewhat close to the observed values. A more complete discussion regarding mass loss generation in Arcturus has been presented by Cuntz (1990). Recent observations of low–amplitude radial velocity variations in the photosphere of Arcturus provide evidence that the theoretically predicted mass loss frequencies might exist. Belmonte et al. (1990) presented evidence for a $\sim$8.3 d period with an amplitude of $\sim$50 m s^{-1}, which they attributed to the fundamental radial pulsation mode. Further studies are in progress (Larson et al. 1992). I note that if the 8.3 d period is real than it would be sufficiently large to support continuous mass loss. Judge & Stencel (1991) argued that the mechanical energy in the observed disturbances could be $\sim$15 times greater than the energy required to drive the wind. For other stars than Arcturus the required minimum mass loss periods can be estimated as $P_{ML} \sim R_*$.

References:

Belmonte J.A., Jones A.R., Pallé P.L., & Cortés T.R.: 1990, ApJ, 358, 595
Cuntz M.: 1990, ApJ, 353, 255
Judge P.G., & Stencel R.E.: 1991, ApJ, 371, 357
Larson A., Irwin A.W., Yang S., Goodenough C., Walker G.A.H., Bohlender D.A., & Walker A.R.: 1992, this meeting

[1] The National Center for Atmospheric Research is sponsored by the National Science Foundation

Pulsations of Proto-Giant-Planets

G. Wuchterl[1]

Institut für Theoretische Astrophysik der Univ. Heidelberg, Germany and Institute for Theoretical Physics, UCSB, Santa Barbara, California

Abstract

Nonlinear oscillations of proto-giant-planets have been found in recent numerical calculations relevant to planetary formation. Pulsations are excited in two phases of the protoplanetary evolution. (a) In an 'instability strip' at core masses of typically $0.2 M_\oplus$ ($M_\oplus$ is the earth mass). Perturbations grow into the nonlinear domain and saturate into perodic variations with relative luminosity-amplitudes of $0.2^{\rm m}$ (b) At the so called *critical mass* (typically at $M_{\rm core} \approx 15 M_\oplus$). There the pulsations drive a strong mass loss. A large portion of the envelope is ejected. Then the mass loss fades and the envelope settles into a new quasi-equilibrium. This remnant — a *post nucleated instability protoplanet* — has a compact envelope and is in core and envelope mass similar to Uranus and Neptune.

Nonlinear Hydrodynamics of Protoplanetary Pulsations

Using the methods described be Wuchterl 1991a,b, I followed the evolution of protogiant planets for the 'Jupiter' to 'Neptune' conditions in a standard solar nebula model. The evolutionary sequences where obtained by solving the non-linear, time-dependent equations of radiations hydrodynamics in the grey Eddington approximation. Calculations starting at planetesimal-sized cores show static solutions up to core-masses of $0.2\,M_\oplus$. For the 'Uranus' and 'Neptune' cases pulsations are excited at this core mass as described in the abstract. In the 'Jupiter' and 'Saturn' cases the protoplanetary envelopes remain static until quasi-hydrostatic contraction sets in at core-masses of $\approx 8\,M_\oplus$.

To study the nature and outcome of the *nucleated instability* the hydrodynamic evolution starting at the so called critical mass (at $M_{\rm core} \approx 15\,M_\oplus$) was studied for the same set of solar nebula conditions as above. The onset of mass loss driven by pulsations of the protoplanetary envelope is found in all cases. After a large portion of the envelope mass has been ejected, new more tightly bound envelopes are formed. The excitation mechanism is a κ-mechanism operating at the Silicate-dust- and H$^-$-opacity feature (cf. Wuchterl 1990).

References:

Wuchterl, G. 1990, *Astron. Astrophys.*, **238**, 83–94.
Wuchterl, G. 1991a, *Icarus* **91**, 39–52.
Wuchterl, G. 1991b, *Icarus* **91**, 53–64.

---

[1] *This work was supported by the Deutsche Forschungsgemeinschaft, DFG, project No. Ts 17/2-2 and in part by the National Science Foundation under Grant No. PHY89-04035.*

Classical Models of Cepheid Stars

C.Chiosi,

Department of Astronomy, University of Padova, Padova, Italy

Abstract

In this paper we summarize recent work on Cepheid stars by Chiosi *et al* (1992a), who calculated large grids of models at varying mass, effective temperature, mass-luminosity (ML) relation or equivalently underlying evolutionary scheme for intermediate mass stars, i.e. classical models, models with mild core overshoot, and models with large core overshoot, and finally initial chemical composition. The chemical parameters bracket the abundances of the Cepheid stars in the Milky Way and Magellanic Clouds. First, we present the theoretical results limited to the analytical fits of the fundamental period-mass-radius (PMR) relation and the blue and red edges of the instability strip for the fundamental mode and first overtone. Second, we discuss the data converted to magnitudes and colours of the BVRcIc passbands with particular attention to the period-luminosity-colour (PLC) and period-luminosity (PL) relations. Finally, we briefly summarize three studies based on these models, which aimed to explain the shape and width of the observed instability strip (Chiosi *et al* 1992b), and the main reason for the mass discrepancy of the Cepheid stars (Chiosi *et al* 1992c) by means of the mass equivalency (ME) method applied to clusters of the Large Magellanic Cloud (LMC). Because the ME method turned out to be sensitive to the distance modulus, the dependence of this on the metallicity has been the subject of the last study by Bertelli *et al* (1992).

1. Introduction

In recent years, there have been a considerable number of photometric studies of Cepheids in the field of the LMC and SMC (eg. Caldwell & Coulson, 1986, and references therein) and there is currently much observational effort being put into the search for, and study of, Cepheids in the rich star clusters of the Magellanic Clouds (eg. Mateo *et al* 1990a,b; Welch *et al* 1991). Because the lie at the same distance, the Cepheids in the Magellanic Cloud clusters are basic to two important topics of astronomy, the understanding of pulsation theory itself and stellar evolution theories in general, and the establishment of the cosmic distance scale through the calibration of the PLC relation (Schmidt 1984; Feast & Walker, 1987; van den Bergh 1989; Madore & Freedman 1991). Furthhermore, modern observations are done increasingly towards the red using BVRcIc photometry rather than the more traditional BV photometry (see Madore & Freedman 1991). Most of the theoretical modelling of the Cepheid pulsation rests on the pioneer work of Iben & Tuggle (1972a,b; 1975) and

Becker et al (1977). See the reviews by Becker (1985) and Chiosi (1989, 1990). However, those calculations covered a limited range of masses, were based on old models for intermediate mass stars, and were made mostly at solar metallicity (Z=0.02). The recent work by Chiosi et al (1992a), who produced large grids of Cepheid models covering the appropriate metallicity range and considering the different possible scenarios for the evolution of intermediate mass stars (see below), has much extended this subject. In addition to this, those models, suitably converted into observable photometric quantities, were used to address a few questions, such as the shape and width of the instability strip (Chiosi et al (1992b), and the cause of the discrepancy between the evolutionary and pulsational mass of the Cepheid stars (Chiosi et al 1992c, Bertelli et al 1992). In this paper we shall briefly touch upon these topics.

2. The Cepheid Models

All details concerning the physical assumptions of model calculations (e.g. the definition of the growth rate, the boundary conditions, the coupling between pulsation and convection, the mixing length parameter) can be found in Fox & Wood (1982) and Chiosi et al (1992a), to whom the reader should refer. It suffices to recall that:
i) the abundances adopted in those computations were Y=0.25 and Y=0.30, and Z=0.016, Z=0.008, and Z=0.004, corresponding approximately with the solar, LMC, and SMC abundance scales, respectively (eg. Russell & Bessell 1989);
ii) the opacities for these mixtures were computed using the Los Alamos Opacity Library (Huebner et al 1977). The radiative opacities are supplemented by the molecular contribution (Alexander 1975; Alexander et al 1983) according to the prescription by Bessell et al (1989) and the revision by Wood (1990, unpublished);
iii) over the mass range spanned by the models (3 to 12 $M_\odot$), three values of the luminosity were associated to each value of the mass in order tobracket the range of luminosities expected from stellar evolution models with and without convective overshoot;
iv) the luminosities and T_{eff}'s of the models were transformed into absolute BVRcIc magnitudes and colors by means tables of colors and bolometric corrections as a function of the effective temperature, gravity, and chemical composition. Two sources of colours and bolometric corrections were used: the Green et al (1987) data, otherwise known as the 87 release of the Yale isochrones, and based on theoretical model atmospheres calculated by Buser (1989), and otherwise referred to as the Padova scale. These latter was compared with the observational calibrations of the (V-I)- and (B-V)-$logT_{eff}$ relations for giants (McWilliam 1990) and dwarfs (Bessell 1979). It turned out that Buser's model colors for gravity $logg = 4.5$ closely agree with the dwarf temperature scale. Similarly, the bolometric corrections for the cepheid models agree with the the bolometric correction scale of Flower (1977). within less than 0.05 magnitudes.

The detailed results of the model cepheid calculations are not given here for the sake of brevity. They can be found in Chiosi et al (1992a). Nevertheless, it it worth recalling that while the blue edge of the instability was easy to identify, the red edge was more uncertain. Chiosi et al (1992a) adopted as red edge the T_{eff}'s at which the growth rate reached its maximum value. As a result of it the instability strip could be wider than assumed (see Chiosi et al 1992a for more details). The discussion below will be limited to the Padova scale.

From the results of the linear calculations linear relationships have been constructed, which represent the numerical results with a sufficient degree of accuracy. For use in the analytical expressions below, we define the following quantities $L' = LogL/L_\odot$, $T' = LogT_{eff}$, $M' = LogM/M_\odot$, and $P'_i = LogP_i$, where the periods are in days and i=0,1 denote the harmonic under consideration.

Period-Mass-Radius Relation. The theory of stellar pulsation reveals that the mass, radius, luminosity, effective temperature, and period of a Cepheid are ideally related by the relations

$$P \alpha L^{0.75} T_{eff}^{-3} M^{-0.5} \qquad (1)$$

In practice, the exponents in equation (1) differ from the ideal values. If we write equation (1) in the slightly more general logarithmic form

$$P' = A + BL' + CL'L' + DM' + ET' \qquad (2)$$

we get the following equations:
Fundamental

$$P' = 12.063 + 0.668L' + 0.027L'L' - 0.691M' - 3.551T' \qquad (3)$$

First Overtone

$$P' = 10.219 + 0.785L' + 0.000L'L' - 0.527M' - 3.153T' \ . \qquad (4)$$

As there is little dependence of the coefficients on the chemical composition, a mean fit to all abundances is given. The term in $L'L'$ was necessary in order to get a good fit. The rms errors in the fits to $logP$ are all less than 0.005.

The Instability Strip in the H-R Diagram: The Blue Edge. Along the blue edge of the instability strip, suitable abundance-dependent expressions relating the effective temperature, luminosity, and mass are:
Fundamental

$$T'_{BE} = [3.99 + 0.07Y - 0.54Z] + 0.01M' - 0.09L' + [0.007 + 0.004Y - 0.009Z]L'L' \quad (5)$$

$$L'_{BE} = [2.111 + 0.302Y - 2.364Z] + 0.738M' + [0.971 + 0.211Y - 0.595Z]P'_o \quad (6)$$

First Overtone

$$T'_{BE} = [3.86 + 0.06Y + 0.69Z] + 0.11M' - 0.01L' + [-0.015 + 0.014Y - 0.152Z]L'L' \quad (7)$$

$$L'_{BE} = [2.174 + 0.489Y - 1.748Z] + 0.813M' + [0.877 + 0.326Y - 2.916Z]P'_1 \quad (8)$$

The Instability Strip in the H-R Diagram: The Red Edge. Similarly for the red edge we get:

Fundamental

$$T'_{RE} = [3.95 + 0.12Y - 0.41Z] + 0.04M' - 0.10L' + [0.009 - 0.005Y - 0.074Z]L'L' \quad (9)$$

$$L'_{RE} = [1.913 + 0.237Y - 2.433Z] + 0.862M' + [0.962 - 0.037Y - 2.314Z]P'_o \quad (10)$$

First Overtone

$$T'_{RE} = [3.79 + 0.15Y + 0.09Z] + 0.06M' - 0.01L' + [-0.006 - 0.006Y - 0.081Z]L'L' \quad (11)$$

$$L'_{RE} = [2.088 + 0.349Y - 1.447Z] + 0.746M' + [1.075 - 0.174Y - 2.067]P'_1 \quad (12)$$

The $P - M - M_V - (B - V)$ Relation. This relation is the analog of the MPLC relation of equation (1). Translated into observational quantities it becomes:

Fundamental

$$\begin{aligned}P' + 0.691M' =\ & -(0.549 + 0.007Z_1) - (0.335 - 0.009Z_1)M_V \\ & + (1.107 + 0.798Z_1^2)(B - V) - (0.063 + 0.677Z_1^2)(B - V)^2\end{aligned} \quad (13)$$

First Overtone

$$\begin{aligned}P' + 0.527M' =\ & -(0.749 + 0.004Z_1) - (0.310 - 0.006Z_1)M_V \\ & + (1.294 + 0.904Z_1^2)(B - V) - (0.307 + 0.938Z_1^2)(B - V)^2\end{aligned} \quad (14)$$

The maximum rms error in any of the fits is 0.0071 and the maximum deviation of any point from any of the fits is 0.022.

The PLC Relation. The PLC (and PL) relation can be obtained from the MPLC relation above with the aid of a ML relation provided by stellar evolution theory. Since the ML relation is a function of the amount of convective overshoot assumed on the main sequence and chemical parameters, the PLC relation will also exhibit such dependences. The ML relation of classical models (without convective overshoot) can be derived from a number of recent evolution calculations which use input physics similar to those used here for the pulsation calculations, in particular, from models using the Huebner *et al* (1977) opacity. Recent classical evolutionary are by Castellani *et al* (1990), Lattanzio (1991), and Alongi *et al* (1992). To obtain the ML relation for models with convective overshoot on the main sequence, an increment f_{ov} in $log L$ can be simply added to ML relation of classical models. Clearly, $f_{ov} = 0$ for classical models, while $f_{ov} = 0.25$ is appropriate for models with mild overshoot models and $f_{ov} = 0.5$ corresponds to models with full overshoot. We note that f_{ov} is approximately equal to the parameter d_{over}/H_P of Maeder & Meynet (1989) or half the parameter λ of Bertelli *et al* (1985). The full ML relation is thus

$$L' = -0.015 + 3.14Y - 10.0Z + 3.502M' + f_{ov}. \quad (15)$$

Writing the PLC relation in the usual form

$$M_V = \alpha log P + \beta(B - V) + \phi \quad (16)$$

we find for the fundamental mode

$$\alpha = \frac{-5.31}{[1 + 0.115L']},\qquad(17)$$

and

$$\beta = \frac{[(110.01 + 22.81Z_1) - (28.58 + 5.92 * Z_1)T' + 18.86/(1 + 0.115L')]}{[(81.1 + 0.54Z_1) - 20.4T']},\qquad(18)$$

where, $Z_1 = log(Z/0.016)$, $T' = logT_{eff}$ and $L' = log(L/L\odot)$. The coefficients α and β depend on abundance either explicitly or implicitly through the dependence of L' and T' on abundance. For Cepheids in the SMC and LMC, Caldwell & Laney (1991) find α = -3.75 and -3.72 and β = 2.63 and 2.45, respectively, for ridge line solutions. Taking values of T'=3.7 and L'=3.9, which correspond to values near the middle of the period range for the LMC and SMC data (ex. Caldwell & Coulson 1986), and a metal abundance of Z=0.008 for the LMC and Z=0.004 for the SMC, equations (17) and (18) give α = -3.67 for both the LMC and SMC and β = 3.16 in the SMC and 3.12 in the LMC. The observed and theoretical values of α agree well but the observationally determined value of β is significantly smaller than the theoretical value. We note, however, that theoretical determinations of β generally lie in the range 3.1±0.1 (Stothers 1988; Stift 1990) and that an analysis of LMC observational data by Stift (1990) also finds a value $\beta \approx 3$. In order to obtain zero points for the PLC relations, and to derive PL relations, a different fit of the theoretical data has been tried in which the explicit dependence on the composition and overshoot parameter are included in the coefficients α, β, and ϕ. The coefficient α is assumed to vary as

$$\alpha = a_1 + a_2Y + a_3Z_1 + a_4f_{ov}\qquad(19)$$

and similarly for β and ϕ (see Table 13 of Chiosi et al 1992a). For LMC and SMC abundances (Y=0.27 and Z_1 = -0.3 and -0.6, respectively) and no overshoot, α = -3.69 and β = 3.46 for the LMC and α = -3.68 and β = 3.59 for the SMC. The α values agree well with the observed ones and the analytic values derived above. However, the β values are larger than the observed or analytic values. It appears that the determination of β is quite sensitive to the sample of Cepheids used in the analysis and the method used. We note that Fernie (1990) was unable to find any color term in a PLC relation for Galactic Cepheids. The zero point of the PLC relation provides a test of the amount of overshoot occurring during the main sequence evolutionary phase of Cepheid stars. The procedure works as follows. The zero points of the PLC relation is obtained for abundances appropriate for the Galaxy, the LMC and the SMC, and for no overshoot (f_{ov} = 0.0) and full overshoot (f_{ov} = 0.5). The zero points are then determined from observational PLC relations of the Galaxy, the LMC and the SMC, assuming distance moduli of 18.5 and 18.9 for the LMC and SMC, respectively. These distance moduli are within a few hundredths of the values derived by Caldwell & Laney (1991) using three separate methods. Finally, the theoretical and observational zero points are compared. The analysis indicates that for the Galactic Cepheids zero to mild amounts of convective overshoot ($f_{ov} < 0.25$ or $d_{over}/H_P < 0.25$ or $\lambda < 0.5$) seem to be appropriate. For the LMC and SMC, the theoretical and observed values agree best with no overshoot. The final point to be considered regarding the PLC relation is the abundance dependence of the zero points. This can be well represented by the relation

$$\delta M_V = 2.15\, \delta Y - 0.72\, \delta log Z \;. \tag{20}$$

For abundances in the range solar to 1/4 solar (Galactic to SMC) these abundance dependences are similar to those given by Stothers (1988).

The PL Relation. The theoretical results indicate that the PL relation for models along the blue edge is abundance independent, while that for the red edge models is only slightly affected. In particular, abundance effects are small compared to changes that result from plausible changes in the amount of overshoot.

In order to compare the observed and theoretical PL relations, some estimate of the way in which Cepheids populate the instability strip is required. Here we assume a uniform population, so that the mean theoretical PL relation lies midway between the blue and red edge sequences. The theoretical PL relation therefore depends on having a good estimate of the red edge, as well as of the blue edge. The theoretical results show that the width of our instability strip is not underestimated by more than $\approx 20\%$ and that the blue edge is in good agreement with observations, at least for the Galactic Cepheids. The comparison of the theoretical PL relations with the observational ones for the Galaxy (Wilson *et al* 1991) and for the SMC and LMC (Caldwell & Laney 1991, and assumed distance moduli of 18.9 and 18.5, respectively) leads to the following conclusions: the theoretical PL relation defined by models with no overshoot is ≈ 0.35 magnitudes too bright. The theoretical and observational PL relations agree only with mild to large amounts of overshoot.

Limiting the discussion to the case of the fundamental mode and writing the PL relation in the form

$$M_V = A log P + B \tag{21}$$

where the coefficients A and B are assumed to have the functional dependence

$$A = A_1 + A_2 Y + A_3 Z + A_4 f_{ov} \tag{22}$$

and are calculated for the mean PL relation defined to be mid-way between the red and blue edge relations. The coefficients are: A_1=-3.092, A_2=-0.019, A_3=7.799, A_4=0.003, B_1=-1.595, B_2=1.018, B_3=-4.057, and B_4=0.702.

5. A Few Selected Topics and Conclusions

Shape of the Instability Strip. The blue edges of the instability strips of Chiosi *et al* (1992a) agree with the corresponding ones of Iben & Tuggle (1972a,b; 1975), whereas the red edges have a different inclination whose slope varies with the metallicity. Red edges not running parallel to the blue ones have been suggested by Fernie (1990) for the galactic Cepheids and are perhaps confirmed by the observational study of Mateo et al (1990a and references) of Cepheids in LMC clusters. Chiosi et al (1992b) compared the empirical instability strip by Fernie (1990) with the theoretical predictions from the Chiosi et al (1992a) Cepheid models, and by means of the synthetic CMD technique showed that both the edges and the distribution of stars within the strip could be reproduced. In particular, they found that the number frequency-period distribution of Cepheid stars can be better accounted for by adopting models with overshoot.

Mass Discrepancy of the Cepheid Stars. It has long been debated whether the masses determined from stellar evolution theory agree with those derived from pulsation theory (see Iben 1974; Iben & Tuggle 1972a,b; 1975; Cox 1980, 1985). In general, pulsational masses (M_{pul}) are estimated to be 30 % to 40 % lower than evolutionary masses (M_{evol}) of the same luminosity. Various causes have been proposed to solve the mass discrepancy problem, each of which would affect the masses in question in a different way (see the reviews by Becker 1985; Cox 1980, 1985; Pel 1985). In the following we would like to call the attention on the role played by convective overshoot as a possible candidate. As already amply discussed, convective overshoot alters the ML relation of core He-burning models. Thus, at any given initial mass, the tracks cross the instability strip at higher luminosity than classical models, or conversely, at any given luminosity the correspondent Cepheid mass is significantly lower (Matraka et al 1982; Bertelli et al 1985). This topic has been examined in a great detail by Chiosi et al (1992c) using the Cepheid stars and CMD of the LMC cluster NGC 2157 (Mateo et al 1990a,b). In fact, the star clusters of LMC with Cepheids are the ideal workbench because all stars lie at the same distance and membership is less of a problem compared to the case of Galactic clusters. On the one hand, the fit of the CMD with theoretical simulations based either on classical models or models incorporating core overshoot leads to accurate determination of the M_{evol} of the Cepheid stars, together with the age and chemical compositions. In particular, M_{evol} will turn out to be a function of the underlying distance modulus. On the other hand, the use of the MPLC for Cepheid stars with the chemical composition suited to the cluster in question, allows a good determination of M_{pul}. Once again M_{pul} is a sensitive function of the underlying distance modulus. By imposing that $M_{evol}=M_{pul}$, one may solve for the distance modulus. The study of the Cepheid stars of NGC 2157 indicated that only using models with convective overshoot the distance modulus could agree with other independent estimates. Specifically, the distance modulus to LMC turned out to be $(m-M)_o = 18.5 \pm 0.1$ in agreement with the recent determination by Panagia et al (1991) based on the circumstellar ring observed by HST around the supernova 1987A in the LMC. When classical models were used, the resulting distance modulus was unacceptably too high. In general, the analysis clarified that the problem of the mass discrepancy likely originates from the adoption of classical models, i.e. without overshoot, to derive M_{evol}, and from the lack of sufficient accuracy in the determination of the distance which bears on both M_{pul} and M_{evol}.

The Distance Modulus to the LMC. The above method (thereinafter the mass equivalency method, ME) has been applied by Bertelli *et al* (1992) to study the Cepheid stars in two other rich clusters of the LMC, NGC 1866 and NGC 2031, to constrain the cluster distances and the sensitivity to the metallicity. Using evolutionary models incorporating a mild amount of core and envelope overshooting along with the present pulsational models, they derived distance moduli of 18.51 ± 0.21 and 18.32 ± 0.20 for NGC 1866 and NGC 2031, respectively. The quoted errors are dominated by the uncertainties in the heavy element abundances of the clusters (assumed to be 0.3 dex for both clusters), with a smaller contribution due to the apparently intrinsic spread in the masses of the Cepheids in each cluster. For the ME method, they found that $\Delta(m-M)_0/Z_1 = 0.69$, where $Z_1 \equiv \log(Z/0.016)$. This result implies that the cluster distances can be determined to better than ± 5% if the cluster abundances can be measured to better than ∼ ± 0.15 dex. The distance moduli derived for NGC 1866 and NGC 2031 are consistent with other recent results for the LMC, meaning that the models used in their analysis avoid the classical evolutionary/pulsational Cepheid mass discrepancy. The results of Bertelli's *et al* (1992) study were based on models using the Los Alamos opacities; had the authors used models incorporating the new Livermore opacities instead, the cluster distance moduli would be *larger* by at most 0.1 mag.

Radiative Opacities. Studies of the effects of the new radiative opacities by Iglesias & Rogers (1991a,b) and Rogers & Iglesias (1992) on the pulsational properties of the Cepheid stars are under way. Preliminary test calculations done for intermediate mass stars with solar composition indicate period variations less than 4% (Capitanio *et al* 1992).

References:

Alongi, M., Bertelli, G., Bressan, A., Chiosi, C., Fagotto, F., Greggio, L., Nasi, E., 1992, Astron. Astrophys. Suppl. in press.

Alexander, D. R., 1975, Astrophys. J. Suppl. **29**, 363.

Alexander, D. R., Johnson, H. R., Rympa, R. C., 1983, Astrophys. J. **273**, 773.

Becker, S. A., 1985, in *Cepheids: Theory and Observations*, ed. B. F. Madore, Cambridge University Press, p. 104.

Becker, S. A., Iben, I. Jr., Tuggle, R. S., 1977, Astrophys. J. **218**, 633.

Bertelli, G., Bressan, A., Chiosi, C., 1985, Astron. Astrophys. **150**, 33.

Bertelli, G., Bressan, A., Chiosi, C., Mateo, M., Wood, P. R., 1992, Astrophys. J. submitted.

Bessell, M.S., 1979, Pub. Astron. Soc. Pacific **91**, 589.

Bessel, M. S., Brett, J. M., Scholz, M., Wood, P. R., 1989, Astrophys. J. Suppl. **77**, 1.

Buser, R., 1989, private communication.

Caldwell, J. A. R., Coulson, I.M., 1986, Mont. Noc.Roy. Astr. Soc. **218**, 223.

Caldwell, J.A.R., Laney, C.D., 1991, in *The Magellanic Clouds*, ed. R. Haynes and D. Milne, Kluwer, Dordrecht, p. 249.

Capitanio, N., Chiosi, C., Wood, P.R., 1992, in preparation.

Castellani, V., Chieffi, A., Straniero, O., 1990, Astrophys. J. Suppl. **74**, 463.

Chiosi, C., 1989, in *The Use of Pulsating Stars in Fundamental Problems of Astronomy*, ed. E. G. Schmidt, Cambridge University Press, p.19.
Chiosi, C., 1990, in *Confrontation between Stellar Pulsation and Evolution*, ed. C. Cacciari and G. Clementini, Astr. Soc. of Pacific Conf. Series **11**, p. 158.
Chiosi, C., Wood, P. R., Capitanio, N., 1992a, Astrophys. J. Suppl. in press.
Chiosi, C., Wood, P.R., Bertelli, G., Bressan, A., 1992b, Astrophys. J. **387**, 320.
Chiosi, C., Wood, P.R., Bertelli, G., Bressan, A., Mateo, M. 1992c, Astrophys. J. **385**, 205.
Cox, A.N., 1980, Ann. Rev. Astron. Astrophys. **18**, 15.
Cox, A. N., 1985, in *Cepheids: Theory and Observations*, ed. B. F. Madore, Cambridge University Press, p. 126.
Feast, M.W., Walker, A.R., 1987, Ann. Rev. Astron. Astrophys. **25**, 345.
Fernie, J.D., 1990, Astrophys. J. **354**, 295.
Flower, P.J., 1977, Astron. Astrophys. **54**, 31.
Fox, M.W., Wood, P.R., 1982, Astrophys. J. **259**, 198.
Green, E.M., Bessel, M.S., Demarque, P., King, C.R., Peters, W.L. 1987, in *The Revised Yale Isochrones and Luminosity Functions*. Yale University, Observatory, New Haven.
Huebner, W.F., Mertz, A.L., Magee, N.H. Jr, Argo, M.F., 1977, Astrophysical Opacity Library, UC-34b.
Iben, I. Jr., 1974, Ann. Rev. Astron. Astrophys. **12**, 215.
Iben, I. Jr., Tuggle, R.S., 1972a, Astrophys. J. **173**, 135.
Iben, I. Jr., Tuggle, R.S., 1972b, Astrophys. J. **178**, 441.
Iben, I. Jr., Tuggle, R.S., 1975, Astrophys. J. **197**, 39.
Iglesias, C., Rogers, F.J., 1991a, Astrophys. J. **371**, 408.
Iglesias, C., Rogers, F.J., 1991a, Astrophys. J. **371**, L73.
Lattanzio, J.C., 1991, Astrophys. J. Suppl. **76**, 215.
Madore, B. F., Freedman, W. L., 1991, Publ. Astron. Soc. Pacific **103**, 933.
Maeder, A., Meynet, G., 1989, Astron. Astrophys. Suppl. **210**, 155.
Mateo, M., Olszewski, E., Madore, B.F., 1990a, Astrophys. J. **107**, 203.
Mateo, M., Olszewski, E.W., MadoreB.F. 1990b, in *"Confrontation between Stellar Pulsation and Evolution"*, ed. C. Cacciari and G. Clementini, Astr. Soc. of Pacific Conf. Series **11**, p. 241.
Matraka,B., Wassermann, C., Weigert, A., 1982, Astron. Astrophys. bf 107, 283.
McWilliam , A., 1990, Astrophys. J. Suppl. **74**, 1075.
Panagia, , N., Gilmozzi, R., Macchetto, F., Adorf, H. M., Kirshner, R. P., 1991, Astrophys. J. **380**, L23.
Pel, J. W., 1985, in *Cepheids: Theory and Observations*, ed. B. F. Madore, Cambridge University Press, p. 1.
Rogers, F.J., Iglesias, C., 1992, Astrophys. J. Suppl. in press.
Russell, S.C., Bessell, M.S., 1989, Astrophys. J. Suppl. **70**, 865.
Schmidt, E. G., 1984, Astrophys. J. **285**, 501.
Stift, M.J., 1990, Astron. Astrophys. **229**, 143.
Stothers, R.B., 1988, Astrophys. J. **329**, 712.
van den Bergh, S, 1989, Astron. Astrophys. Rev **1**, 111.
Welch, D. L., Mateo, M., Cote', P., Fisher, P., Madore, B., 1991, Astron. J. **101**, 490.
Wilson, T.D., Barnes, T.G., Hawley, S.L., Jefferys, W.H., 1991, Astrophys. J. **378**, 708.

Evolutionary Models of RR Lyrae Stars

Young-Wook Lee
Department of Astronomy, Yale University, New Haven, CT, USA

1. Introduction

As tracers of old stellar populations and as primary Population II standard candles, RR Lyrae stars have played an important role in the development of modern astronomy. Our knowledge of stellar evolution has identified these variable stars in a core helium burning phase of low-mass star evolution, the horizontal-branch (HB) phase. Consequently, not only to understand fully the nature of RR Lyrae stars, but also to apply them correctly as population probes and distance indicators, we must understand the underlying evolutionary effect of HB stars.

In this paper, I briefly review the most important properties of RR Lyrae stars predicted from the HB evolutionary models, and present many pieces of supporting evidence for these models. For the implications of these models on the chronology of the Galactic formation and on the cosmological distance scale, the reader is referred to several recent publications by Lee (1992a,b,c,d).

2. Properties of RR Lyrae Stars Predicted from HB Population Models

The construction of HB population models is straightforward, once we have all the evolutionary tracks for HB stars of different masses for a given composition. It is generally assumed that the distribution of mass on the HB is Gaussian, resulting from variable amounts of mass-loss on the giant-branch. For each HB star, the time elapsed since the zero-age HB (ZAHB) is obtained by using a uniform random number generator. When the HB stars fall into the instability strip, we treat them as RR Lyrae variables. The HB population models yield the luminosity, mass, and effective temperature of each star within the instability strip; hence it is straightforward to calculate the periods of model RR Lyrae stars from the period-mean density relation (e.g., van Albada and Baker 1971). I refer the reader to Lee et al. (1990) and Lee (1990) for details of the model constructions.

As we have seen in this meeting, there is still some debate concerning the correlation between the luminosity of RR Lyraes and metallicity and other related problems, such as the Sandage period-shift effect. I believe that the reason for this annoying situation is nothing but due to the ignorance of the effect of post ZAHB evolution in some investigations. The importance of the effect of post ZAHB evolution on the properties of RR Lyrae stars is demonstrated in Figure 1, where we have two HB models of the same metallicity but different HB morphology. The model (a) is for the globular cluster like M3, which

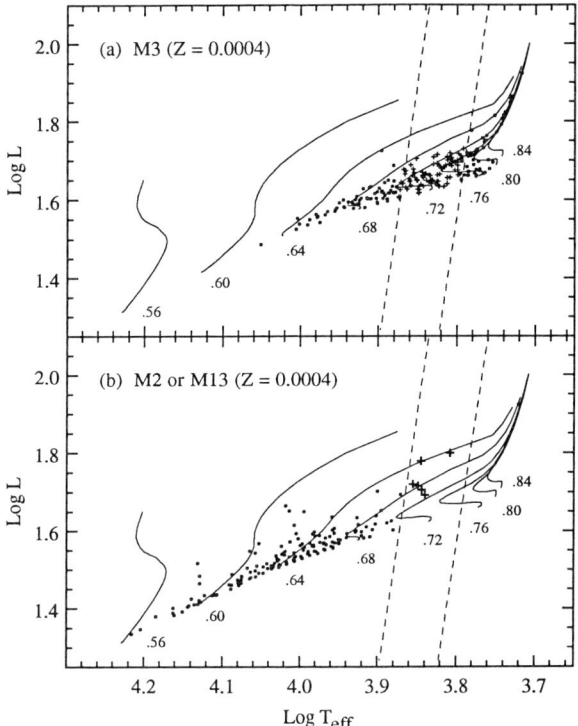

Fig. 1. The HB population models for M3 and M13 (or M2) with HB evolutionary tracks ($Y_{HB} \approx 0.22$). The instability strip is represented schematically by dashed lines, and each track is labeled by its total mass in solar units (from Fig. 4 of Lee 1990).

possesses nearly equal numbers of red and blue HB stars, while model (b) is for the globular cluster like M13 or M2, whose HB is almost entirely to the blue of the instability strip. We can see in model (a) that most RR Lyraes are near the ZAHB; hence the effect of evolution is relatively small. In model (b), however, almost all RR Lyraes are highly evolved stars from the blue HB, and thus the mean RR Lyrae luminosity ($<M_v^{RR}>$) for model (b) is significantly brighter than that for model (a), even though they have the same metal abundance. Note also that the variables in model (b) are less massive than the ones in model (a).

Extensive model calculations by Lee (1990, 1992b) demonstrate this effect more clearly. Figure 2 presents the results of such calculations for $<M_v^{RR}>$, mean mass of RR Lyraes (<Mass>), and period-shift (at fixed T_{eff}) as functions of my HB morphology index under various assumptions regarding the metal abundance. We can see here clearly that $<M_v^{RR}>$ and <Mass> depends not only on metal abundance (as widely assumed) but also on HB morphology due to the effect of post ZAHB evolution. According to the period-mean density relation, the periods of variable stars are proportional to their luminosity, but are inversely proportional to their mass. Consequently, the period-shift of RR Lyraes

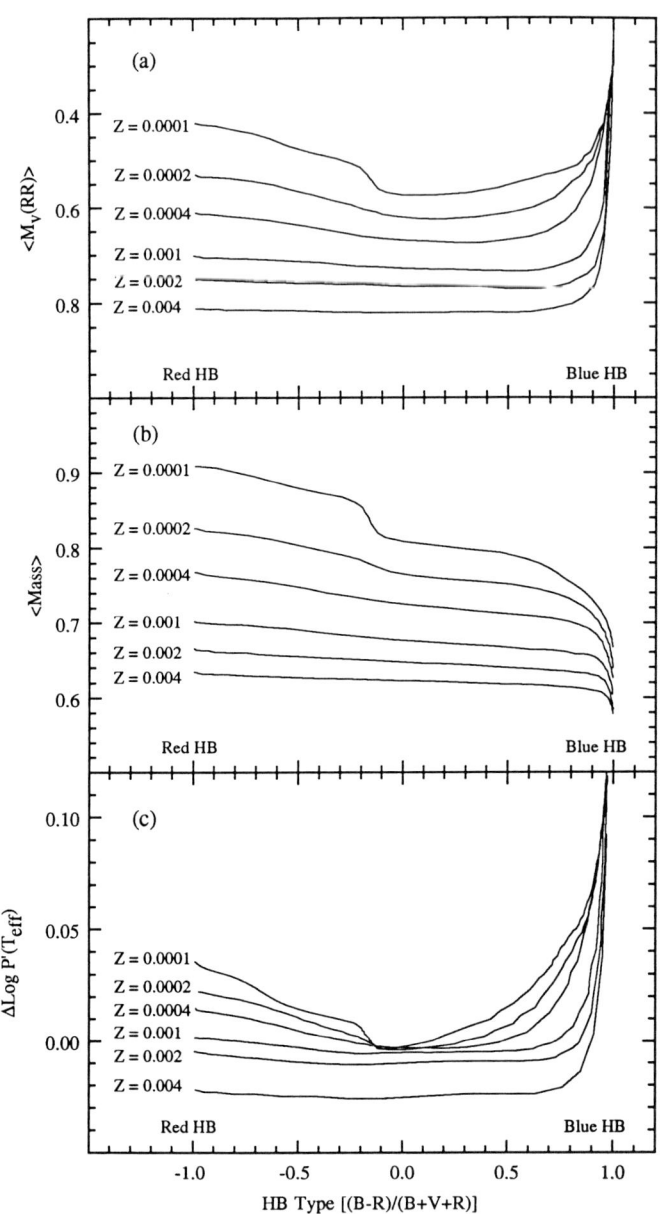

Fig. 2. The HB population model calculations for $\langle M_v^{RR} \rangle$, mean mass of RR Lyraes ($\langle Mass \rangle$), and period-shift (at fixed T_{eff}) as functions of HB type and metallicity ($Y_{HB} \approx 0.22$, $\sigma_M = 0.01 M_\odot$). The quantity (B-R)/(B+V+R) is a new index devised by Lee (1989) to characterize the morphology of the HB (B, V, and R are the numbers of blue HB, RR Lyrae variable, and red HB stars, respectively).

measured at a given effective temperature [ΔLog P'(T_{eff}) = Log P' - log P'$_{M3}$ at Log T_{eff} = 3.83] is much more sensitive to the HB morphology (see Fig. 2c). Similarly, we can also calculate the mean period of type ab RR Lyraes as functions of HB type and metallicity, once some reasonable assumption is made regarding the transition temperature between the first overtone and fundamental mode of pulsation (see Lee and Zinn 1990 and below).

3. Supporting Evidence

Because of their impacts on the formation chronology of the Galaxy (see Lee 1992a,c) and on the cosmological distance scale (see Lee 1992b,d), it is of considerable importance to determine whether these new evolutionary models are correct. Below, I list many pieces of supporting evidence for these models.

3.1. Period Changes of RR Lyrae Variables

Since the periods of RR Lyrae variables should be either increasing, if the stars evolve from blue to red in the HR diagram, or decreasing, if they evolve from red to blue (Eddington 1918), the secular period changes of RR Lyrae stars provide a decisive test of the evolutionary models. If, as predicted by the Lee (1990) HB evolutionary models (see Fig. 1b), most RR Lyraes stars in blue HB clusters pass through the instability strip from blue to red toward the end of core helium burning phase, we expect large positive period changes for these RR Lyraes. In fact, Lee (1991a) has shown that the mean rate of period change in RR Lyrae stars depends sensitively on HB type due to the effect of post ZAHB evolution. In particular, he has shown that the observed period changes for the five best-studied clusters are consistent with those expected from the evolutionary models.

3.2. The RR Lyrae Stars in the Globular Cluster ω Centauri

As discussed above (see Fig. 2a), the HB evolutionary models have shown that $<M_v^{RR}>$ depends sensitively on HB morphology as well as metallicity because of the effect of post ZAHB evolution. This has an important effect on the mean correlation between $<M_v^{RR}>$ and metallicity among the Galactic globular cluster system, because the HB morphology itself varies with [Fe/H]. As illustrated in Figure 3, the observed HB type gets bluer with decreasing [Fe/H], and in the mean, the clusters of intermediate metallicity (-1.9 < [Fe/H] < -1.4) have the bluest HBs (e.g., M2, M13, M22). But, as [Fe/H] decreases still further, the HBs move back through the instability strip, producing the clusters with slightly redder HB types (e.g., NGC 5466, M53, M15). This implies that the HB morphology is not a monotonic function of [Fe/H] (Lee 1990, 1991b; see also Renzini 1983). When this effect is combined with the model calculations for $<M_v^{RR}>$ in Fig. 2a, the models predict a *mean* correlation between $<M_v^{RR}>$ and metallicity for the Galactic halo population that looks like a step function (see solid line in Fig. 4; see also Lee 1991b for details).

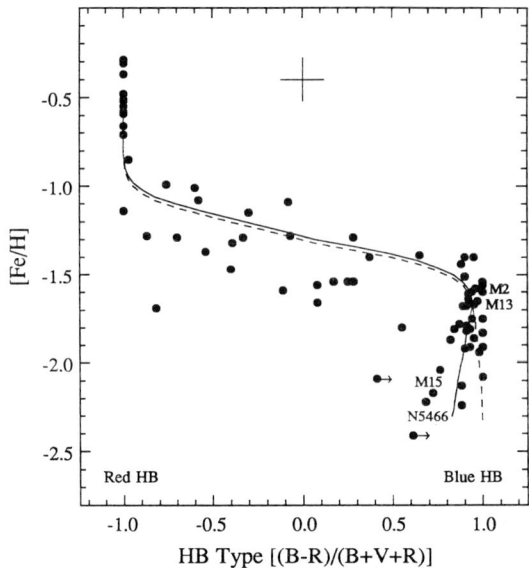

Fig. 3. HB type index is plotted against [Fe/H] for Galactic globular clusters ($R_G < 40$ kpc; from Lee 1991b). The solid line is a model locus with nonlinear mass-loss - [Fe/H] relation, and can be viewed as a mean correlation between HB type and [Fe/H] for halo population. The dashed line is a model locus with fixed mass-loss, which produces too blue HB types for clusters having [Fe/H] < -1.6 (see Lee 1991b for details). This difference has profound effects on the values of $<M_{bol}^{RR}>$, $<P_{ab}>$, and period-shift (see solid and dashed lines in Figs. 4 and 6 for corresponding correlations in different diagrams).

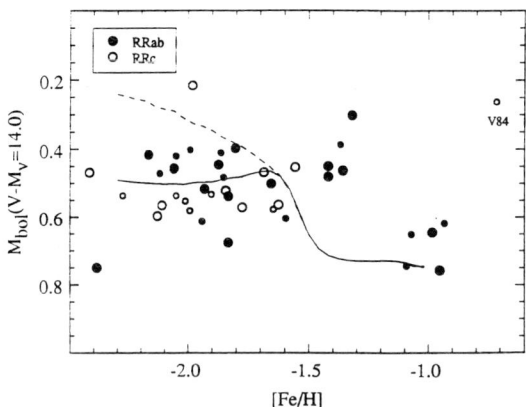

Fig. 4. Observed correlation between $<M_{bol}^{RR}>$ and [Fe/H] for RR Lyrae stars in ω Centauri. The solid and dashed lines are model loci under two different assumptions regarding the HB type - [Fe/H] correlations (see Fig. 3). Note that the correlation is not linear because $<M_{bol}^{RR}>$ depends sensitively on HB morphology as well as metallicity (from Fig. 1 of Lee 1991b).

Lee: Evolutionary Models of RR Lyrae Stars

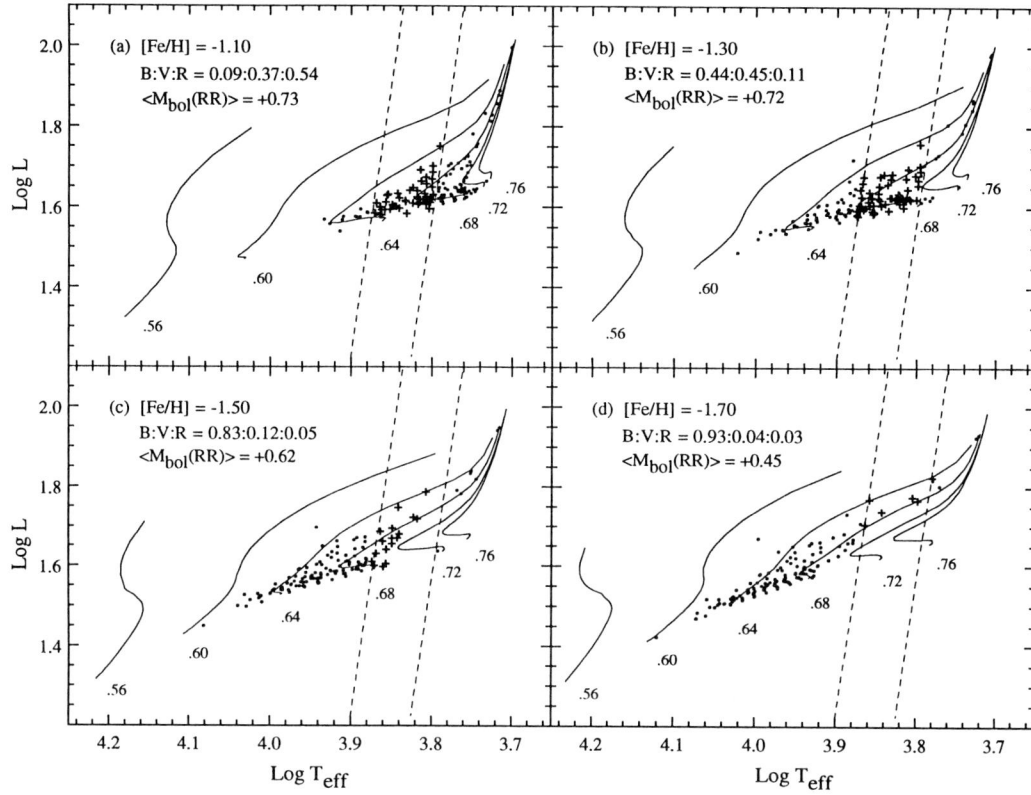

Fig. 5. Same as Fig. 1, but these models are constructed specifically to explain the origin of sudden upturn in $\langle M_{bol}^{RR} \rangle$ and $\langle P_{ab} \rangle$ at $[Fe/H] \approx -1.5$ (see Figs. 4 and 6).

To test this model prediction, we need a large sample of RR Lyrae stars, spanning a large range of [Fe/H], for which precise measurements of relative luminosity and [Fe/H] exist. The RR Lyrae stars in ω Cen, which span a large range in [Fe/H], are the most ideal sample for this test. In particular, the relative values of M_v^{RR} for RR Lyraes in ω cen can be inferred straightforwardly from the apparent visual magnitudes because they are all located at the same distance and are all reddened by the same magnitudes. The observational data for the RR Lyraes in ω Cen are compared with the model prediction in Fig. 4, where one can see that the correlation becomes nonlinear at approximately [Fe/H] = -1.5, in reasonable agreement with the models.

The origin of this sudden upturn in $\langle M_v^{RR} \rangle$ at $[Fe/H] \approx -1.5$ is explained by a series of HB population models in Fig. 5, where one can see how sensitively the population of the instability strip changes with decreasing [Fe/H]. As [Fe/H] decreases, HB morphology gets bluer, but between models (a) and (b), the increase in mean RR Lyrae luminosity is small because most RR Lyraes are still near the ZAHB (note that the evolution is slow near the ZAHB). However, as [Fe/H] decreases further, there is a certain point where the zero age portion of the HB just crosses the blue edge of the instability strip [see models (c) and (d)]. Then, only highly evolved stars from the blue HB can penetrate back into the instability strip, and the mean RR Lyrae luminosity increases abruptly. The increase in luminosity would continue if the HB morphology moves to blue monotonically with decreasing [Fe/H] (see dashed line in Fig. 4). However, because of the nonmonotonic behavior of the HB morphology with decreasing [Fe/H], the HB morphology gets slightly redder after model (d) and this effect offsets further increase in luminosity with decreasing [Fe/H], producing the overall correlation that looks like a step function.

The fact that the correlation between $<M_v^{RR}>$ and metallicity (and hence between period-shift and metallicity; see below) is not linear would clarify some of the disagreements with other investigators because fits of straight lines to different data sets produce significantly different slopes.

3.3. The Oosterhoff Period Dichotomy among Globular Clusters

The very same effect provides a natural explanation of the one of the long-standing problem in modern astronomy, the Oosterhoff (1939) period dichotomy among Galactic globular clusters. In Figure 6a, we plot the observed mean period of type ab RR Lyraes ($<P_{ab}>$) in globular clusters as a function of [Fe/H]. The solid line is a correlation predicted from the model calculations. Because the periods of RR Lyrae variables are proportional to their luminosity, it is not surprising to see again that the correlation looks like a step function. Intermediate values of $<P_{ab}>$ between the two groups can be produced in principle, but only by clusters within a very narrow range of HB type and [Fe/H] (see Lee and Zinn 1990). This explains why ~ 50 clusters in the Galaxy containing sufficient numbers of RR Lyrae variables fall into one or the other Oosterhoff group. Following the analyses of Lee et al. (1990), these calculations assume a small amount of hysteresis in mode switching, which is equivalent to a shift of ~100 K in the T_{eff} of the transition edge, depending on the direction of evolution. While this assumption is necessary if the models are to match the observed difference in $<P_{ab}>$ between the Oosterhoff groups (see Lee et al. 1990), it is important to note that it cannot by itself produce the dichotomy.

Renzini (1983), Castellani (1983), and more recently Sandage (1990) have suggested that the Oosterhoff period dichotomy is due to the absence of RR Lyraes in clusters of intermediate metallicity caused by the nonmonotonic tracking of HB morphology with decreasing [Fe/H]. As noticed by Sandage (1990) himself, however, this cannot be the whole explanation because there are clusters on each side of the empty period range that have almost the same values of [Fe/H] ([Fe/H] ≈ -1.5), yet have very different values of $<P_{ab}>$. This sudden jump in $<P_{ab}>$ at [Fe/H] ≈ -1.5 is now understood as discussed above (see Fig. 5).

As first noticed by Lee (1990), the same effect is also seen in the correlation between the period-shift (at fixed T_{eff}) and [Fe/H] for field RR Lyrae stars (see Fig. 6b; see also Caputo and De Santis 1992). Again, the fact that the correlation between period-shift and metallicity is not linear would clarify some of the disagreements with other investigators because fits of straight lines to different data sets produce significantly different slopes.

Lee: Evolutionary Models of RR Lyrae Stars 301

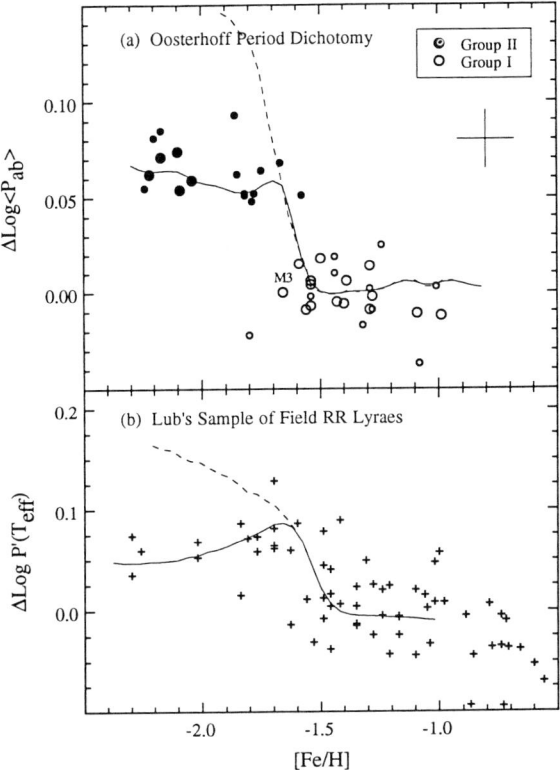

Fig. 6. (a) The Oosterhoff period dichotomy among Galactic globular clusters. (b) The correlation between period-shift (at fixed T_{eff}) and metallicity for Lub's sample of field RR Lyrae stars. The solid and dashed lines are model loci (see Fig. 3 for corresponding model loci in HB type - [Fe/H] plane). Note again that the correlations look like a step function (see text).

3.4. Masses of the RR Lyrae Variables

There has been some discrepancy between the masses of the double-mode RR Lyrae stars as determined from pulsation calculations and from stellar evolution models. The sense was that the masses derived from the Petersen diagram (pulsation calculations) are systematically lower than those indicated by the standard stellar evolution calculations by 0.10 - 0.15 $M_\odot$ (see Lee et al. 1990). Sandage (1990), among others, considered this as evidence against the Lee et al. (1990) HB evolutionary models.

However, the most recent stellar pulsation calculations based on new Livermore OPAL opacities suggest that the double-mode masses are about 0.1 - 0.2 $M_\odot$ larger than those using earlier opacities (Cox 1991; Kovacs et al. 1991; see also Kovacs, this volume). To be consistent with the stellar pulsation calculations, we have used the same Livermore OPAL opacity tables in the calculation of HB evolutionary tracks (Yi, Lee, and Demarque 1992). The result is presented in Figure 7, where one can see that, unlike the

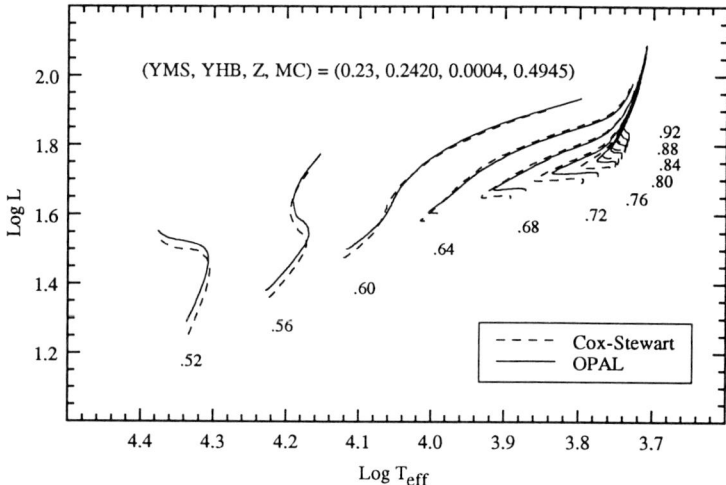

Fig. 7. The HB evolutionary tracks constructed with two opacity tables (from Yi, Lee, and Demarque 1992). Each track is labeled by its total mass in solar units.

pulsation calculations, the effect of new opacity is small in evolutionary calculations. In fact, the HB population models based on these tracks suggest that the decrease in mean RR Lyrae mass due to the new opacity is less than 0.01 $M_\odot$. Therefore, with the new Livermore OPAL opacities, both the stellar pulsation and evolution calculations predict approximately the same masses for the RR Lyrae stars. This is a strong independent check on the evolutionary models.

3.5. Baade-Wesselink Measurements of Field RR Lyrae Stars

One of the strongest observational evidence in favor of the HB evolutionary models is provided by three recent investigations of field RR Lyrae variables using the Baade-Wesselink method (Cacciari et al. 1989; Liu and Janes 1990; Carney et al. 1992). These observations have yielded a relationship between $<M_v^{RR}>$ and metallicity that has approximately one-half the slope of Sandage's (1982) relation (see also Sandage and Cacciari 1990), but is in excellent agreement with the Lee et al. (1990) HB evolutionary models. It is important to note that the analysis of Lee et al. (1990) did not include the extremely blue HB clusters of intermediate metallicity because in general they contain fewer variables. Consequently, their $<M_v^{RR}>$ - metallicity correlation appears to be linear (see Fig. 10 of Lee et al. 1990). The sample of RR Lyrae stars used in the recent Baade-Wesselink analysis also lacks such highly evolved stars of intermediate metallicity, and therefore it is not surprising to see the excellent agreement between these two results. Clearly, it is of great importance to continue the Baade-Wesselink measurements of RR Lyrae stars in this metallicity range to see whether the correlation becomes nonlinear due to the effect of evolution (as is the case in the ω Cen). Note that Clementini and Cacciari (1990) already report two such highly evolved stars in this metallicity range (see also Cacciari, Clementini, and Fernley 1992).

Support for this work was provided by NASA through a grant HF-1014.01-90A awarded by the Space Telescope Science Institute, which is operated by the Association of Universities for Research in Astronomy, Inc., for NASA under contract NAS5-26555.

References

Cacciari, C., Clementini, G., and Fernley, J. A. 1992, ApJ, in press
Cacciari, C., Clementini, G., Prevot, L., and Buser, R. 1989, A&A, 209, 141
Caputo, F., and De Santis, R. 1992, AJ, 104, 253
Carney, B. W., Storm, J., and Jones, R. V. 1992, ApJ, 386, 663
Castellani, V. 1983, Mem. Soc. Astron. Italiana, 54, 141
Clementini, G., and Cacciari, C. 1990, in Confrontation between Stellar Pulsation and Evolution, ed. C. Cacciari (San Francisco: ASP), p. 109
Cox, A. N. 1991, ApJL, 381, L71
Eddington, A. S. 1918, MNRAS, 79, 2
Kovacs, G., Buchler, J. R., and Marom, A. 1991, A&A, 252, L27
Lee, Y.-W. 1989, Ph.D. thesis, Yale University
Lee, Y.-W. 1990, ApJ, 363, 159
Lee, Y.-W. 1991a, ApJ, 367, 524
Lee, Y.-W. 1991b, ApJL, 373, L43
Lee, Y.-W. 1992a, PASP, in press (September)
Lee, Y.-W. 1992b, Mem. Soc. Astron. Italiana, in press
Lee, Y.-W. 1992c, AJ, in press (November)
Lee, Y.-W. 1992d, in Variable Stars and Galaxies, ed. B. Warner (San Francisco: ASP), in press
Lee, Y.-W., Demarque, P., and Zinn, R. 1990, ApJ, 350, 155
Lee, Y.-W., and Zinn. R., 1990, in Confrontation between Stellar Pulsation and Evolution, ed. C. Cacciari (San Francisco: ASP), p. 26
Liu, T., and Janes, K. 1990, ApJ, 354, 273
Oosterhoff, P. Th. 1939, Observatory, 62, 104
Renzini, A. 1983, Mem. Soc. Astron. Italiana, 54, 335
Sandage, A. 1982, ApJ, 252, 553
Sandage, A. 1990, ApJ, 350, 631
Sandage, A., and Cacciari, C. 1990, ApJ, 350, 645
van Albada, T. S., and Baker, N. 1971, ApJ, 169, 311
Yi, S., Lee, Y.-W., and Demarque, P. 1992, in preparation

The Red Edge of the Cepheid Instability Strip

Yan Li

Yunnan Observatory, Chinese Academy of Sciences
P.O. Box 110, Kunming 650011, P.R. China

Abstract

Recent results of investigations of the nonequilibrium effects between gas and radiation on the red edge of the Cepheid instability strip are presented.

1. Introduction

Many Cepheid investigations focus on the red edge of the instability strip. Deupree (1977) and Stellingwerf (1982) confirm that if time-dependent convection is considered, stars that are sufficiently cool return to a stable state. However, many uncertainties lie in the treatment of convection and we are not completely sure that this treatment correctly describes the interaction of convection with pulsation. Therefore other explanations of the red edge are of great significance. Presented here are our recent results of investigations of the non-equilibrium effects between gas and radiation on the red edge of the Cepheid instability strip.

2. Results

Two stellar models with masses of 9 $M_\odot$ and 5 $M_\odot$ and Population I chemical composition $(Y,Z) = (0.24, 0.021)$ were evolved to the instability strip. The linear stability analysis was carried out for the first three radial modes using Li's (1992a) method, a treatment based on the frozen convection approximation ($\nabla \cdot F'_C = 0$) which precisely describes the process where the gas absorbs or emits radiation.

Numerical results show that all of the modes that have been considered are pulsationally unstable in a certain range of temperature with definite blue and red edges. Fig. 1 gives the instability strip of the fundamental mode. The blue edge obtained is in good agreement with that obtained by Iben & Tuggle (1975). The most striking characteristic is the emergence of the red edge. This contradicts the common opinion that unless a proper treatment of convection is included the red edge is not found. The first and second overtones also have complete instability strips, and the blue edge occurs at progressively higher temperatures for successively higher order modes while the red edge is at almost the same place for all of the three modes.

The work function of the first overtone is displayed in Fig. 2 for the model of $9M_\odot$ at the center of the instability strip during the second crossing. The pulsation is excited by the frozen temperature mechanism (Li 1992b) operating in the hydrogen and helium ionization zones. In the above regions, the ionization restrains the variation of gas temperature. Therefore the gas absorbs or emits radiation upon compression or expansion as the gas temperature is lower or higher than the radiation temperature respectively. As the absorbed energy is more than the emitted one, the difference is transformed into pulsation energy.

We suggest that the red edge of the instability strip results from the emergence of convection in the stellar envelope. With the cooling of the star, the successively increasing convection brings more and more energy and gradually replaces the radiative transfer as the primary means of energy transportation. The sharp decline of the radiative luminosity reduces the supply of energy to be absorbed by the gas and then transformed into pulsation energy. As a result, the variation of radiative temperature is restricted, and the efficiency of the frozen temperature mechanism goes down. This effect can be seen in Fig. 2, which shows as well the work function of the same mode for the same model but at the red edge of the instability strip. The contribution of the second helium ionization driving is depressed at the center of the ionization zone, where convection is most developed. In the hydrogen and first helium ionization zones, convection is so strong that the driving effects only appear at the edges of the ionization zones. When the star is even cooler, there are no driving effects left in the ionization zones and the star returns to the static state.

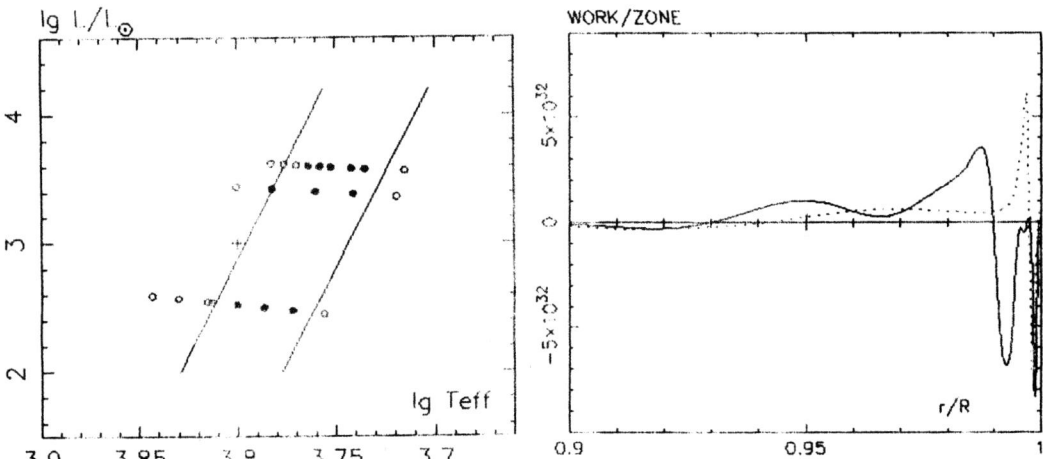

Figure 1. (Left) Instability strip of the fundamental mode: open circles – stable; filled circles – unstable; plus sign – Iben & Tuggle (1975). **Figure 2.** (Right) Work function for the first overtone: dotted curve – at the center of the instability strip; solid curve – at the red edge of the instability strip.

3. Conclusions

Taking into account the non-equilibrium effects between gas and radiation we obtain the red edge of the instability strip without considering the interaction between convection and pulsation. Therefore, we do not need to use very crude treatment of convection-pulsation interaction to account for the red edge of the Cepheid instability strip.

I would like to thank Dr. J. Nemec and Dr. D.L. Welch for their cordial invitation to this colloquium, and for the IAU travel grant. This research has been supported by the National Natural Sciences Foundation grant IEC(92)352.

References:

Deupree, R.G. 1977, ApJ **211**, 509.
Iben, I., Jr. & Tuggle, R.S. 1975, ApJ **197**, 39.
Li, Y., 1992a, A&A **257**, 133.
Li, Y., 1992b, A&A **257**, 145.
Stellingwerf, R.F. 1982, ApJ **262**, 339.

The Cepheid Instability "Wedge"

Siobahn M. Morgan

University of Northern Iowa, U. S. A.

For the most part it had been assumed that the red and the blue edges of the Cepheid Instability region were parallel. However, previous work by Pel and Lub (1978) and recent work by Fernie (1990) seems to reveal a rather interesting structure to the shape of the Cepheid Instability region. Figure 1 shows the shape defined using the data from Fernie (1990) and the observational data of Gieren (1989). It is apparent that the edges defined by the distribution of these points are not parallel.

I have calculated a series of pulsation models that included varying values of the mixing length to try and produce a distribution of Cepheids as seen in Figure 1. Calculations were done using the methods outlined by Castor (1971) with a Linear Non Adiabatic pulsation code to determine the characteristics for a given model. The masses range in value from 5 to 12 $M_\odot$, while the luminosities extended over a range of 3.0 to 4.8 in $\log L/L_\odot$. Each series of models was calculated over a temperature range of 4500 to 7000 K or until the blue edge of the Cepheid Instability Region was obtained. It is found that for different masses of Cepheids, different values of the mixing length are required to produce the empirical red edge. Values for the mixing length are close to $2.2H_p$ for the low mass Cepheids, and they are near $1.4H_p$ for the higher mass Cepheids.

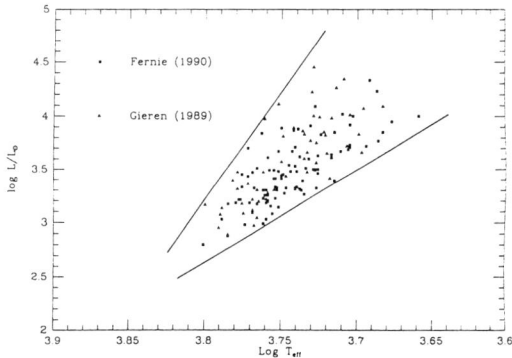

Figure 1 Cepheids from Fernie (1990) and Gieren (1989) are shown along with the approximate location of the red and blue edges of the Instability region.

References:

Castor, J. I. 1971, Astrophys. J., **166**, 109.
Fernie, J. D. 1990, Astrophys. J., **354**, 295.
Gieren, W. P. 1989, Astron. Astrophys., **225**, 381.
Pel, J. W. and Lub, J. 1978, in *The HR Diagram, IAU Symp. number 80*, eds. A. G. Davis Philip and D. S. Hayes, Reidel, Dordrecht Holland, p. 229.

Time Dependent Convection and the Pulsations of Polaris

Siobahn M. Morgan[1], Arthur N. Cox[2]

[1]*University of Northern Iowa, U. S. A.,* [2]*Los Alamos National Laboratory, U. S. A.*

The low amplitude radially pulsating classical Cepheid Polaris (α UMi) has well observed decreasing radial velocity and light amplitudes and an increasing period. Recent work indicates that Polaris may stop pulsating by 1995 (Dinshaw et al, 1989), which would make it an excellent indicator for the red edge of the Cepheid instability strip in the Hertzsprung-Russell diagram.

We have calculated Population I envelope models with 200 mass shells of a 5.5 $M_\odot$ star near $T_{eff} = 5700$ K and $\log(L/L_\odot) = 3.11$ which match Polaris' characteristics (Arellano Ferro, 1984). Models were calculated using the Linear Non Adiabatic pulsation analysis code of Castor (1971). The effects of convection on the pulsational stability were calculated with the time dependent convection theory of Cox, Brownlee and Eilers (1966) as discussed by Cox and Guili (1968). The characteristics of the models were altered until the appropriate period, temperature and luminosity were obtained.

Without including the effects of time dependent convection, we can not find models which are stable against pulsation and which have Polaris' physical characteristics. Using the linear time dependent convection theory we obtain models which are stable against pulsation at temperatures cooler than the red edge. The effect of the mixing length in the standard convection theory on the red edge of the Cepheid instability strip is examined and values for the mixing length of $l = 2.2 H_p$ are required to obtain models with the period, luminosity and temperature of Polaris. Such values for the mixing length are not unusual considering the values needed for solar models and for Mira variables. Also recent work of the shape on the Cepheid instability region appear to require values of the mixing length close to $2.0 H_p$ for stars with masses similar to Polaris' (Morgan, 1992).

References:

Arellano Ferro, A. 1984, M. N. R. A. S., **209**, 481.
Castor, J. I. 1971, Astrophys. J., **166**, 109.
Cox, A. N., Brownlee, R. R. and Eilers, D. D. 1966, Astrophys. J., **144**, 1024.
Cox, J. P. and Giuli, R. T. 1968, Principles of Stellar Structure, Gordon and Breach, New York, p. 1045.
Dinshaw, N., Matthews, J. M., Walker, G. A. H., and Hill, G. M. 1989, Astronom. J., **98**, 2249.
Morgan, S. M. 1992, *these proceedings.*

Stellar Structure and RR Lyrae Masses

Ben Dorman

Astronomy Department, U. Virginia.

Abstract

I describe qualitatively the constraints on RR Lyrae masses that arise from consideration of the interior structure of Horizontal Branch (HB) stars. Briefly, the stellar models lying within the temperature range of the instability strip are slightly less massive than those with deep exterior convection zones. I discuss therefore what brings about the existence of these zones in the models.

1. Horizontal Branch Stellar Structure and the Instability Strip

The object of this paper is to present a simple argument showing how the masses of stellar models that appear in the instability strip depend on the physics of stellar interiors. The discussion here follows after a detailed treatment of HB stellar structure given in Dorman (1992a, hereafter D92). Recall first that the HB is understood to be a set of helium core-burning stars which have (to a good approximation) the same helium rich core masses and are inferred to have a range of total mass. This mass range is responsible for the colour spread to which the HB owes its name.

For fixed composition, models of decreasing mass are located successively further to the blue, and they are distinguished by successively less luminous hydrogen burning shells. In particular, for most compositions, the reddest models of each Zero Age HB sequence have exterior convection zones. These models also have the shortest blueward loops in the HR diagram because if the convection zone persists throughout evolution the surface temperature changes very little. The mechanisms thought to be responsible for RR Lyrae pulsations cannot operate in the presence of a stellar convection zone deep enough to engulf the H and He ionization zones. Hence the red edge of the instability strip lies at some point on the ZAHB blueward (*i.e.* on the less massive side) of the point beyond which these extensive exterior convection zones are not present. The figure shows HB evolutionary tracks from Dorman (1992b). The lines show, very schematically, the limit of models with exterior convection on the ZAHB (solid line), with the fundamental red edge slightly blueward of this limit. For sufficiently low envelope mass, the ZAHB model and most of its evolutionary track will be located blueward of the instability strip and its hydrogen burning shell will be much dimmer, the outer layers being hot and dense but strongly radiative. One therefore needs to understand what produces models with exterior convection zones, and how the masses of such models change with assumed composition.

An important exception to the argument presented here concerns stars which only enter the instability strip at a late stage of evolution, on the way to the AGB; no relevant constraints exist for their masses. For these stars there is also no mass-metallicity relationship.

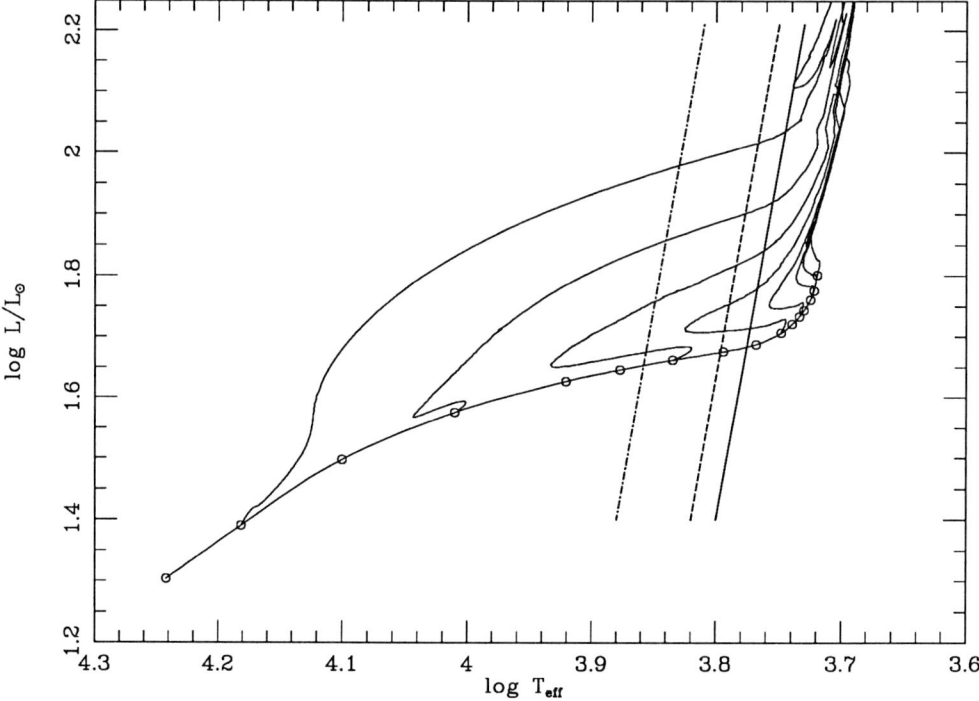

Figure. Evolutionary tracks are shown for [Fe/H] = −1.66, [O/Fe] = 0.63. The solid line represents the limit of models with convective envelopes; the dashed and dashed-dotted lines represent, roughly, the edges of the instability strip.

2. How to Obtain Exterior Convection Zones in Stellar Models

It is useful to separate the effects of the CNO elements, which act as nuclear catalysts, and other heavy elements which only contribute to the opacity. One can think of the CNO abundance as giving the hydrogen shell a potential to burn, whose magnitude is regulated by the opacity sources in the envelope above it. If the CNO abundance is increased, everything else being fixed, the most important things that happen to the deep interior are (i) the hydrogen burning shell (and thus the model luminosity) becomes brighter, and (ii) the temperature gradients in the hydrogen rich envelope become steeper, *lowering the temperature of the outermost layers*. This is demonstrated by explicit calculation (D92, Figure 4). The fact that at fixed mass the outer layers are at lower temperature of course implies a deeper convection zone, and from there we see easily that the models without convection zones will have lower masses as the CNO abundance is increased.

The D92 study also showed that if Z was high, the masses of models in the instability strip would also be small, irrespective of the CNO abundance. This occurs because the opacity clamps down the temperature of the hydrogen burning shell, regulating the luminosity but also ensuring that smaller masses have exterior convection zones (D92, Fig. 6). Clearly, however, this 'clamping' is an effect caused by the opacity at high temperature. The recent superior opacity treatments by the OPAL group and the Opacity Project have not so far produced large changes at these temperatures. Numerical experiments show that the properties of models are quite sensitive to small changes ($\sim$ 10%) in the opacity for $T > 10^6 K$. However for the most metal poor compositions the metal contribution is a small proportion of the total opacity, so that the remaining uncertainties are expected to be small.

3. Masses of Metal Poor Variables

The mass of RR Lyrae variables in the cluster M15 has been the subject of much debate in the last year (Kovács et. al. 1991, Cox 1991, Simon 1992). With the new OPAL opacities, the most recent pulsational estimates for the mass are around 0.80 $M_\odot$. Note that the model red-giant tip mass is about 0.80 $M_\odot$ for a cluster age of 15 Gyr. The pulsational masses are thus only compatible with the evolutionary models (a) if the abundances are in solar proportion and (b) if there are many stars which undergo little mass loss at the red-giant tip. The former is uncomfortable because of a large body of evidence suggesting that [O/Fe] $\sim 0.2 - 0.4$ in globular clusters. For M15 in particular, there are a set of apparently unevolved HB stars (Buonanno et. al. 1985; Stetson 1991) which are too red for the scaled-solar ZAHB. The latter (little mass loss) presents difficulties if it also implies that bright AGB stars which are indistiguishable from red giants in their exterior characteristics also may not lose mass. However, for [O/Fe] as high as 0.75, the inferred evolutionary masses are around 0.72 $M_\odot$. A smaller assumed oxygen enhancement gives an intermediate value, probably still discrepant with the new pulsational estimates by a few hundredths of a solar mass.

References:

Buonanno, R., Corsi, C.E., and Fusi Pecci, F. 1985 A&A 145, 97

Cox, A.N. 1991 ApJ, 381, L71

Dorman, B. 1992a ApJS, 80, 701

Dorman, B. 1992b ApJS, 81, 221

Kovács, G., Buchler, J.R., and Marom, A. 1991 A&A 252, L27

Simon, N.R., 1992, this volume

Stetson, P.B. 1991 in *Precision Photometry: Astrophysics of the Galaxy*, eds. A.G.D. Philip, A.R. Upgren & K.A. Janes (New York: Davis), p.69

Mass-Loss During the RR Lyrae Phase of the HB:
Mass Dispersion on the HB and RR Lyrae Period Changes

Rebecca A. Koopmann, Young-Wook Lee, Pierre Demarque, and Jamie M. Howard
Department of Astronomy, Yale University, New Haven, CT, USA

Horizontal branch (HB) models were evolved using the Yale stellar evolution code, YREC, to test the possibility that mass loss during the RR Lyrae phase is able to produce the observed color (mass) dispersion on the HB (Willson and Bowen 1984) and the anomalous period changes in RR Lyrae stars (Laskarides 1974). Models of total mass 0.64, 0.66, 0.68, 0.70, and 0.72 $M_\odot$ (Y_{MS} = 0.23, Z = 0.001) were evolved with constant mass loss rates of 0, 10^{-10}, and 10^{-9} $M_\odot$ yr^{-1}. Mass loss was assumed to occur only in the RR Lyrae phase, and the instability strip was defined by 3.800 < log T_{eff} < 3.875.

HB stars which lose mass evolve further to the blue. Low mass loss rates do not affect the shape of the tracks significantly. Stars, which without mass loss could not become blue HB stars, were able to emerge from the instability strip on the blue side. The trapping phenomenon predicted by Willson and Bowen (1984) was present in the case of the 0.68 $M_\odot$ star with a 10^{-9} $M_\odot$ yr^{-1} mass loss rate. Trapping occurs when mass loss, which causes a star to move blueward, is balanced by the redward evolutionary trend. Synthetic HB's were calculated from the tracks with and without mass loss. The addition of mass loss does produce a slightly larger spread in mass, but these HB's do not resemble observed HB's (Koopmann et al. 1993, in preparation). We conclude that mass loss in the RR Lyrae phase cannot explain the dispersion in color on the HB.

Periods of RR Lyrae were computed using the period-density relation derived by van Albada & Baker (1971). We find that the magnitude of period changes in stars without mass loss is similar to those with mass loss. Positive period changes have somewhat smaller magnitudes when mass loss is included. This was expected since the effect of mass loss is to impede the evolution to the red. Though mass loss in principle should increase the rate of evolution to the blue (and therefore increase the magnitude of negative period changes), the blueward portion of the evolution is so slow that mass loss rates tested here did not greatly speed the blueward evolution. We conclude that much larger rates of mass loss would be necessary to cause large negative period changes observed in some RR Lyrae stars.

In summary, mass loss on the HB of the magnitude tested here is not capable of explaining either the color dispersion in the HB or the large negative period changes in RR Lyrae stars. Larger mass loss rates would have an observable effect on HB morphology and are therefore unlikely.

Laskarides, P. G. 1974, Astrophys.&Sp.Science, 27, 485
van Albada, T. S. & Baker, N. 1971, Ap.J., 169, 311
Willson, L. A. & Bowen, G. H. 1984, Nature, 312, 429

Windows on the Instability Strip

The Masses and Luminosities of Globular Cluster RRc Stars

Norman R. Simon[1] and Christine M. Clement[2]

[1] *Department of Physics and Astronomy, University of Nebraska – Lincoln*
[2] *Department of Astronomy, University of Toronto*

Abstract

A large number of hydrodynamic pulsation models are converged in the first overtone, and their mean properties compared with observations of RRc stars in six globular clusters. The two observed quantities, period and Fourier parameter ϕ_{31}, lead, via the models, to inferred values for mean mass, luminosity and temperature of the RRc sample in each cluster. We find a narrow range in intracluster RRc luminosity and temperature, but a wider range in mass. At the same time, the intercluster spread is wide in all three parameters. A full discussion of our techniques and results will be given elsewhere (Simon and Clement, in preparation).

1. Introduction

We begin with the technique of Fourier decomposition (e.g., Simon 1988), in which a Fourier series,

$$\mathrm{mag} = A_0 + \Sigma_{j=1}^{n} A_j \cos(j\omega t + \phi_j) \qquad (1)$$

is fitted to the observed magnitudes of a pulsating star, and the shape of the light curve quantified in terms of the low-order coefficients, viz., $R_{j1} = A_j/A_1$, $\phi_{j1} = \phi_j - j\phi_1$, $(j = 1, 2, 3, 4)$. A recent study by Clement, Jankulak and Simon (1992; hereafter CJS) applied this technique to globular cluster RRc stars, focusing on the phase parameter ϕ_{31}. Figure 1, borrowed from CJS, summarises the results of this study in the form of a plot of ϕ_{31} vs. log(period). Here each symbol represents an RRc star and the differing symbols denote different clusters as indicated in the figure caption, with the cluster's [Fe/H] value given in parentheses.

Despite considerable scatter, the data in Figure 1 show two clearly discernable trends: 1) there is an increase of ϕ_{31} with period within each cluster; and 2) the clusters are segregated according to metallicity, with the metal-rich clusters lying higher and the metal-poor clusters lower. In what follows we shall use linear and hydrodynamic pulsation models to interpret these data, deriving mean masses and luminosities for the various clusters and reproducing the intracluster relation between ϕ_{31} and period. We shall see that a considerable range of masses is indicated, both in a given cluster and among the clusters.

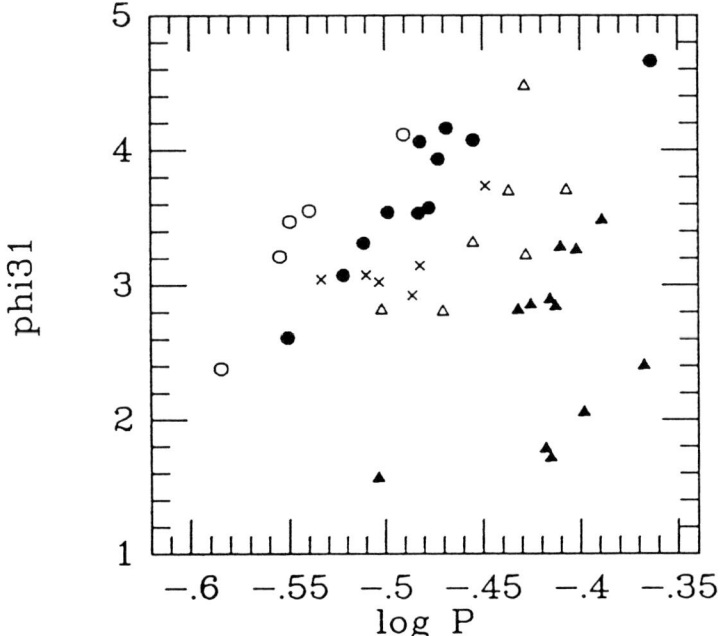

Figure 1. ϕ_{31} vs. log(period) for five globular clusters: open circles – NGC 6171 (−0.99); filled circles – M5 (−1.40); crosses – M3 (−1.66); open triangles – M53 (−2.04); closed triangles – M15 (−2.17). Numbers in parentheses indicate published values of [Fe/H].

2. The models

Hydrodynamic pulsation models were perturbed in the first overtone and integrated from 100 to 300 periods to allow the light curves to settle down to their limiting form. An eighth-order Fourier fit was then performed according to Eq. (1), and the value of ϕ_{31} extracted for each model. The calculations are very similar to those described by Simon (1990a; hereafter S90a) except that OPAL opacities (Rogers and Iglesias 1992) were employed. The chemical composition, dictated by then available opacity tables, was $X = 0.7$, and $Z = 0.001$ or 0.0001 (Anders–Grevesse mixture).

Table 1 presents the results of these calculations, giving for each mass (solar units), the values of log L (solar units), T_{eff}, metal abundance (see code beneath table), pulsation period (days), and Fourier parameters R_{21}, ϕ_{21}, R_{31} and ϕ_{31}. The structure of the present models in general, and the ϕ_{31} values in particular, is very similar to that found by S90a using Los Alamos opacities. The opacity law does not play a crucial role in determining ϕ_{31}. Neither does the metallicity. Figure 2 shows a plot of ϕ_{31} ($Z = 0.001$) vs. ϕ_{31} ($Z = 0.0001$) for pairs of models with the same values of M, log L and T_{eff}. A linear fit to these points yields a slope 1.000 and zero point 0.019. It seems likely from Figure 1 that the differences in ϕ_{31} are random, probably caused by slightly different convergence properties in the models, i.e., ϕ_{31}

Table 1 Hydrodynamic Models for the RRc Stars (X = 0.70)

log L	Te	Z(*)	P1	R21	φ21	R31	φ31
M = 0.55							
1.58	7450	1	0.253	0.222	4.43	0.043	2.72
1.58	7450	3	0.250	0.280	4.33	0.056	2.84
1.58	7300	1	0.270	0.232	4.51	0.060	2.93
1.58	7300	3	0.268	0.283	4.44	0.077	3.09
1.62	7400	1	0.278	0.187	4.52	0.045	3.04
1.62	7400	3	0.276	0.240	4.51	0.058	3.15
1.62	7250	1	0.298	0.179	4.54	0.081	3.07
1.62	7250	3	0.295	0.247	4.67	0.081	3.13
1.66	7400	1	0.300	0.129	4.54	0.030	3.46
1.66	7400	3	0.297	0.170	4.67	0.036	3.64
1.66	7250	1	0.321	0.128	4.39	0.104	3.21
1.66	7250	3	0.318	0.170	4.75	0.064	3.49
1.70	7350	1	0.330	0.117	4.45	0.045	4.19
1.70	7350	3	0.327	0.131	4.69	0.045	4.04
1.70	7200	1	0.354	0.092	4.38	0.074	4.04
1.70	7200	3	0.351	0.102	4.79	0.069	3.66
1.74	7300	1	0.361	0.096	4.77	0.033	4.65
1.74	7300	3	0.373	0.072	4.23	0.023	4.47
1.74	7250	1	0.369	0.087	4.84	0.024	4.66
1.74	7250	3	0.410	0.063	3.45	0.068	4.63
1.74	7050	1	0.407	0.023	4.47	0.077	4.49
1.74	7050	2	0.406	0.027	5.12	0.080	4.35
1.78	7150	1	0.422	0.089	3.40	0.055	4.60
1.78	7150	3	0.417	0.032	3.80	0.056	4.22
M = 0.65							
1.66	7450	1	0.265	0.207	4.26	0.036	2.11
1.66	7450	3	0.263	0.303	4.03	0.052	2.71
1.66	7200	1	0.296	0.341	4.35	0.102	2.64
1.70	7200	3	0.294	0.391	4.16	0.142	2.59
1.70	7400	1	0.291	0.226	4.39	0.038	2.41
1.70	7400	3	0.289	0.286	4.27	0.046	2.68
1.70	7200	1	0.319	0.255	4.46	0.077	2.71
1.70	7200	3	0.316	0.349	4.39	0.115	2.81
1.74	7350	1	0.321	0.189	4.47	0.038	2.70
1.74	7350	3	0.318	0.239	4.46	0.043	2.95
1.74	7100	1	0.360	0.168	4.53	0.121	2.79
1.74	7100	3	0.357	0.252	4.73	0.082	2.85
1.78	7350	1	0.346	0.119	4.51	0.014	3.16
1.78	7350	3	0.343	0.174	4.59	0.024	3.50
1.78	7200	1	0.370	0.152	4.37	0.110	3.12
1.78	7200	3	0.367	0.200	4.68	0.078	3.06
1.82	7200	1	0.399	0.130	4.29	0.098	3.77
1.82	7200	3	0.396	0.133	4.67	0.075	3.31
1.82	7000	1	0.439	0.082	4.25	0.090	4.22
1.82	7000	3	0.435	0.084	4.82	0.095	3.69

Table 1 (continued)

log L	Te	Z(*)	P1	R21	φ21	R31	φ31
M = 0.75							
1.70	7400	1	0.268	0.313	4.03	0.042	2.00
1.70	7400	3	0.266	0.303	3.87	0.076	1.94
1.70	7200	1	0.293	0.441	4.03	0.169	2.35
1.70	7200	3	0.290	0.432	3.84	0.236	1.99
1.74	7350	1	0.295	0.321	4.18	0.064	2.19
1.74	7350	3	0.292	0.366	3.95	0.091	2.28
1.74	7200	1	0.315	0.394	4.19	0.117	2.51
1.74	7200	3	0.313	0.433	3.98	0.196	2.30
1.78	7350	1	0.317	0.238	4.31	0.044	2.18
1.78	7350	3	0.315	0.318	4.13	0.049	2.52
1.78	7100	1	0.355	0.330	4.43	0.113	2.48
1.78	7100	3	0.352	0.395	4.27	0.139	2.61
1.82	7300	1	0.349	0.227	4.42	0.047	2.46
1.82	7300	3	0.346	0.297	4.32	0.057	2.70
1.82	7050	1	0.391	0.234	4.52	0.130	2.51
1.82	7050	3	0.388	0.315	4.57	0.096	2.72
1.86	7250	1	0.385	0.190	4.48	0.052	2.81
1.86	7250	3	0.381	0.213	4.52	0.031	2.89
1.86	7050	1	0.422	0.173	4.37	0.133	2.85
1.86	7050	3	0.418	0.237	4.71	0.088	2.73
1.90	7250	1	0.415	0.130	4.43	0.038	3.42
1.90	7250	3	0.411	0.170	4.62	0.029	3.33
1.90	7000	1	0.466	0.138	4.28	0.113	3.69
1.90	7000	3	0.462	0.135	4.75	0.100	3.04
M = 0.85							
1.70	7200	1	0.272	0.410	3.91	0.145	1.96
1.70	7200	3	0.270	0.413	3.73	0.226	1.49
1.74	7150	1	0.299	0.446	3.96	0.171	2.13
1.74	7150	3	0.297	0.432	3.78	0.247	1.76
1.78	7150	1	0.322	0.453	3.98	0.185	2.23
1.78	7150	3	0.320	0.385	3.98	0.118	2.22
1.82	7150	1	0.346	0.416	4.15	0.124	2.41
1.82	7150	3	0.344	0.433	3.97	0.172	2.29
1.86	7150	1	0.373	0.368	4.27	0.105	2.46
1.86	7150	3	0.370	0.410	4.11	0.142	2.48
1.90	7150	1	0.402	0.303	4.38	0.092	2.52
1.90	7150	3	0.399	0.358	4.29	0.105	2.68
1.94	7100	3	0.439	0.296	4.48	0.093	2.73

(*) 1: Z = 0.001; 2: Z = 0.0003; 3: Z = 0.0001

approaching its limiting value by a slightly different path. If this is the case, then the mean difference between the ϕ_{31} pairs (which turns out to be $|\delta\phi_{31}| = 0.2$) may be taken as a measure of the uncertainty in determining ϕ_{31} for a given model: namely, $\Delta\phi_{31} = 10.1$. This number is in accord with error estimates made in a different way by S90a.

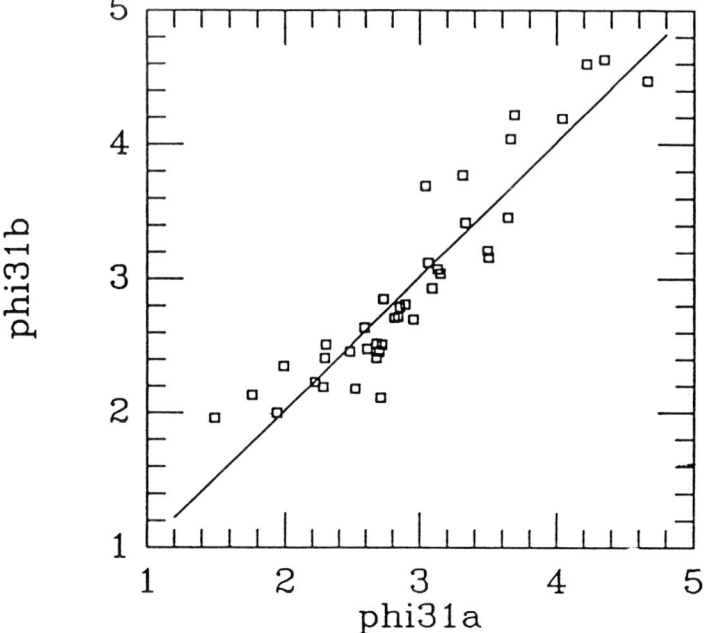

Figure 2. ϕ_{31} values for pairs of hydrodynamic models with the same M, L and T_{eff} but two different values of the metallicity: "phi31a" corresponds to $Z = 0.0001$; "phi31b" to $Z = 0.001$. The solid line is a least-squares fit to the points.

The model grid in Table 1 was chosen in a similar manner to that in S90a. For each mass and luminosity, models were calculated near the first-overtone blue edge and near the fundamental-mode blue edge, thus spanning the region where first-overtone pulsation is expected. An examination of Table 1 discloses trends in the models which are reminiscent of those in the observations: 1) for given mass there is an increase of ϕ_{31} with period; and 2) the higher the mass, the lower the values of ϕ_{31}.

Finally, we point out the crucial feature that emerges from both the present calculations and those of S90 — over 120 hydrodynamic models, in all: The phase parameter ϕ_{31} is determined essentially by mass and luminosity and is hardly sensitive to metallicity, helium abundance, effective temperature and the choice of opacity law (see also Simon 1989; Simon 1990b). It is this property that shall allow us to use ϕ_{31} to determine masses and luminosities for the RRc stars.

3. Theory vs. observations

In order to make the comparison with observations, we shall describe the average properties of the calculations by means of parameter fits to the model sample. Although the models of Table 1 were calculated for a single value of the helium parameter, $Y \simeq 0.30$, the properties of ϕ_{31} described just above make it possible to use linear calculations to generalise the models to other values of Y. This has been done, resulting in a grid of 140 models, each characterized by six parameters: M, $\log L$, Y, T_{eff}, ϕ_{31} and the first-overtone period, P_1. Parameter fits to these models then yield relations of the following form:

$$\log T_{\text{eff}} = 3.265 - 0.3026 \log P_1 - 0.1777 \log M + 0.2402 \log L \qquad (2)$$

$$\log Y = -20.26 + 4.935 \log T_{\text{eff}} - 0.2638 \log M + 0.3318 \log L \qquad (3)$$

$$\log M = a_0 + a_1 \log P_1 + a_2 \phi_{31} + a_3 \log Y \qquad (4)$$

$$\log L = b_0 + b_1 \log P_1 + b_2 \phi_{31} + b_4 \log Y \qquad (5)$$

Equation (2) is a version of the familiar period/mean-density relation, the so-called "pulsation law," while Eq. (3) describes how the mean location of the first-overtone instability strip changes with the helium parameter, Y. We have rendered Eqs. (4) and (5) schematically, since there are a number of different ways in which these fits can be made. The fitting procedure, along with the generalisation of the hydrodynamic models, mentioned above, are discussed in detail elsewhere (Simon and Clement, in preparation; hereafter SCprep).

We note that the parameter list in Eqs. (2) through (5) includes Y, but omits the metal abundance, Z. This is because, for given mass and luminosity, Z is unimportant in determining either ϕ_{31} or P_1, whereas Y influences the location of the instability strip and thus the mean periods of the models (see also, Simon 1990b). However, it is well known that the calculated blue edges are also influenced by certain (problematical) characteristics of the models, e.g., the boundary conditions and the treatment of convection. We have chosen to subsume all of these uncertainties into the quantity Y. Thus, when, below, we infer a large (small) value of Y, this could mean either a high (low) abundance of helium in the ionisation zones, or that proper physics and numerics in the code would result in hotter (cooler) blue edges, or both. This question is discussed further by SCprep.

Equations (2) through (5) constitute a set which may be solved for the derived variables $\log M$, $\log L$, $\log T_{\text{eff}}$ and $\log Y$ in terms of the observables $\log P_1$ and ϕ_{31}. The following relations result (see SCprep):

$$\log M = 0.39 + 0.52 \log P_1 - 0.11 \phi_{31} \qquad (6)$$

$$\log L = 2.41 + 1.04 \log P_1 - 0.058 \phi_{31}, \qquad (7)$$

whereupon $\log T_{\text{eff}}$ and $\log Y$ may be calculated from Eqs. (2) and (3).

Because these relations are all linear they may be applied equally well to individual RRc stars or to cluster averages. In Table 2 we show average RRc masses, luminosities, and temperatures for six clusters — five from CJS, plus M68 (Clement,

Ferance and Simon, in preparation). Derived values of the helium parameter, Y, are also given, along with published cluster metallicities. One sees a general increase of mass and luminosity, and a fall in temperature and in Y as [Fe/H] gets smaller, with M15 and M68 containing by far the most massive and brightest stars and exhibiting the coolest temperatures. In fact, if the values given in Table 2 are correct, it means that the Oosterhoff effect is due mainly to differences in temperature. That is, the periods in M68, for example, are longer than those in, say, M5 largely because the RRc stars in the former are considerably cooler. In that case, any formulation which attempts to express a period shift at constant temperature (e.g., Sandage 1982) is clearly unjustified.

Table 2

Cluster	[Fe/H]	Y	$<M>$	$<\log L>$	$<T_{\text{eff}}>$
NGC 6171	-0.99	0.30	0.57	1.65	7420
M5	-1.40	0.28	0.56	1.70	7300
M3	-1.66	0.28	0.63	1.71	7280
M53	-2.04	0.27	0.62	1.75	7190
M68	-2.09	0.24	0.79	1.82	7070
M15	-2.17	0.24	0.80	1.83	7050

Table 3 shows the results of applying our equations to the individual stars in each cluster. All of the observations are photographic, but those for NGC 6171, M5 and M68 were obtained at a better site and are probably more accurate (CJS). One notices for these clusters (excluding the anomalously long-period star #76 in M5), a very tight range of intracluster luminosity and temperature, but a large range of mass. Some of the mass scatter may be explained by errors in measuring ϕ_{31} [an error of 0.3 in ϕ_{31} leads to an error of about 0.03 in the inferred value of $\log M$ from Eq. (6)], but much of it must be real.

To show this, we have made linear fits to the observed ϕ_{31} vs. log(period) relations for NGC 6171, M5 (excluding star #76) and M68, and obtain slopes of 17.4, 16.2, and 17.7, respectively, with an average value, $d\phi_{31}/d\log P_1 = 17.1$. We note that Eq. (7), obtained strictly from theoretical models, reproduces this value very well, provided that $\log L \sim$ constant. However, it is also clear that Eq. (6) cannot yield the observed slope if $\log M \sim$ constant. In fact, a decrease of $\log M$ with increasing period is required. Using an expression of the form, $\log M = a\log P_1 + b$, in Eq. (6) and requiring the observed value $d\phi_{31}/d\log P_1 = 17$, one easily finds $a = 1.3$. Given a typical intracluster period spread (e.g., M5 without star #76) of $|\Delta\log P_1| \sim 0.1$, we find $|\Delta\log M| \sim 0.13$ or $|\Delta M| \sim 0.15 M_\odot$.

It should be pointed out that, while this range is close to what we have actually obtained for M5 (see Table 3), the spread may still be exaggerated. This is due both to errors in measuring ϕ_{31} and to the fact that Eq. (6) describes a mean line through the theoretical models (with standard deviation 0.03 in log M) and must thus be uncertain in any individual case. Nonetheless the argument given above makes it clear that some spread is required — perhaps (guessing on the basis of our error estimates) of the order $0.1 M_\odot$. Thus the following picture emerges of the RRc domain on the horizontal branch within a given cluster: narrow in temperature, narrow in luminosity, but with a considerable range in mass, which accounts for the observed range in period. When one turns to the family of clusters, Table 2 shows a substantial spread in intercluster temperature and luminosity as well as in mass.

Finally, we shall briefly mention three independent tests that argue for the validity of Eqs. (6) and (7): 1) Eq. (7) reproduces the observed hierarchy of relative luminosity among a large sample of RRc stars in ω Centauri; 2) for M68 and M15, Eq. (6) yields mean RRc masses ($M \sim 0.80 M_\odot$) which agree quite well with the mean masses obtained for the RRd stars in those clusters; and 3) Eq. (7) gives an LMC distance modulus in agreement with that emerging from other methods. These tests shall be described in detail in SCprep.

4. Questions and comments

A. N. COX: What is the uncertainty in your RRc masses inferred from ϕ_{31}?

N. R. SIMON: I would not like to state a formal uncertainty. An error of, say, 0.3 in measuring ϕ_{31} leads to an error of 0.03 in log M. In addition, the theoretical calculations have been described by mean functions fit to the model sample. The standard deviation of these fits is also about 0.03 in log M. Thus I would say that the uncertainty in individual masses is large — say 10.05, or even worse in some cases. However, since the uncertainties I mentioned seem to be strictly random, the mean RRc mass derived for a given cluster should be much more accurate.

A. SANDAGE: Am I correct in believing that your RRc masses agree with the RRd masses recently derived by Art Cox (using OPAL opacities), at least in the crude sense that the Oo I stars turn out to be less massive than the Oo II stars, i.e., that the trend of the original Petersen diagram is confirmed?

N. R. SIMON: Yes, I think that is correct.

G. KOVACS: Can you comment on the systematic difference between your luminosities and the evolutionary luminosities for M15?

N. R. SIMON: I have not looked at this carefully yet. I believe that the masses are also involved here, and that the answer depends upon what you take for the O/Fe ratio, whether you think the stars are evolved, etc.

G. KOVACS: Combining the theoretical relation for the position of the blue edge with the probably more solid relations involving ϕ_{31} might be dangerous because of the large uncertainties involved in the blue-edge position.

N. R. SIMON: That is why we have subsumed all uncertainties in the absolute blue edge location into the helium parameter, Y. The results we have obtained won't change much unless the blue (or red) edges turn out to have much stronger (and weirder) dependences on L and M than present calculations show.

D. WELCH: I don't really know how to say this diplomatically, so I'll just say it. It is unconscionable, in view of the importance of the RR Lyrae luminosities and their [Fe/H] dependence, to introduce photographic data into the discussion. RRc stars in the rich fields of globular clusters require modern detectors and reduction techniques.
N. R. SIMON: I would be delighted to see such data, provided that the phase coverage is good. A filled-in photographic light curve is, for our purposes, superior to a CCD light curve with holes. This is an important point that I hope the observers will note.
J. O. PETERSEN: Concerning photographic vs. modern CCD observations: the old photographic data for ω Centauri published by Martin give very accurate Fourier parameters because Martin had 350–400 observations of each variable.

J. O. PETERSEN: In your beautiful analysis you have used only one Fourier decomposition parameter, ϕ_{31}. One should expect that other Fourier parameters also provide valuable information. Have you tried to use two or more parameters simultaneously?
N. R. SIMON: Unfortunately, the present nonlinear pulsation codes are not up to this task. While they seem to model ϕ_{31} rather well for the overtone mode (RRc stars), the codes are less successful for ϕ_{21}. In the case of the fundamental mode (RRab stars), the calculations seem deficient for both ϕ_{21} and ϕ_{31}. With regard to higher-order terms, I don't think the observations are yet precise enough to compare with models.

J. NEMEC: Do you have an explanation for why the longest-period c-type RR Lyrae stars in ω Cen have periods as long as 0.485 (V47) or 0.534 (V68)? And, why none of the c-type stars with periods longer than 0.35 day are double-mode RR Lyrae stars?
N. R. SIMON: No.
E. BELSERENE: In response to Nemec's question about the long periods of RRc stars in ω Cen: doesn't it help that the RR Lyrae stars in ω Cen are not ZAHB stars at all, but evolved stars?
J. NEMEC: While it is probably true that the ω Cen RR Lyrae stars are evolved, isn't it also true that the M15 RR Lyrae stars are also evolved, and many RRd stars are present in M15?
N. R. SIMON: I want to point out that there is a body of opinion (to which I subscribe) holding that the M15 RR Lyraes cannot constitute a largely evolved population, simply because there are so many of them.

References:

Clement, C. M., Ferance, S. and Simon, N. R., in preparation.
Clement, C. M., Jankulak, M. and Simon, N. R. 1992, Ap. J. 395, 192.
Rogers, F. J. and Iglesias, C. A. 1992, Ap. J. Suppl. 79, 507.
Sandage, A. 1982, Ap. J. 252, 553.
Simon, N. R. 1988, in Pulsation and Mass Loss in Stars, ed. R. Stalio and L. A. Willson (Dordrecht: Reidel), p. 27.
———. 1989, Ap. J. (Letters) 343, L17.

———. 1990a, M.N.R.A.S. 246, 70.

———. 1990b, Ap. J. 360, 119.

Simon, N. R. and Clement, C. M., in preparation (SCprep).

Table 3 Measured and Derived Parameters for Individual Stars

Star No.	P1	φ31	M	log L	Te
NGC 6171					
4	0.282	3.47	0.54	1.64	7448
6	0.260	2.38	0.68	1.66	7442
15	0.289	3.55	0.54	1.64	7430
19	0.279	3.21	0.57	1.65	7439
23	0.323	4.11	0.50	1.66	7356
M5					
15	0.337	3.93	0.53	1.69	7297
31	0.301	3.07	0.62	1.69	7345
35	0.308	3.31	0.59	1.69	7339
40	0.317	3.54	0.57	1.69	7327
44	0.330	4.06	0.51	1.67	7331
55	0.329	3.53	0.58	1.70	7288
62	0.281	2.61	0.67	1.69	7377
66	0.351	4.07	0.52	1.70	7266
73	0.340	4.16	0.50	1.68	7306
76	0.432	4.66	0.50	1.76	7095
79	0.333	3.57	0.58	1.71	7278
M3					
37	0.327	2.92	0.67	1.73	7245
56	0.330	3.14	0.64	1.73	7253
75	0.314	3.02	0.64	1.71	7294
85	0.356	3.73	0.57	1.73	7222
86	0.293	3.04	0.61	1.68	7372
107	0.309	3.07	0.63	1.70	7316
M53					
19	0.391	3.70	0.61	1.77	7120
21	0.339	2.80	0.70	1.76	7197
23	0.366	3.69	0.59	1.74	7190
35	0.373	4.47	0.49	1.70	7235
36	0.373	3.22	0.67	1.78	7130
40	0.315	2.81	0.67	1.72	7274
47	0.351	3.31	0.63	1.74	7203
M68					
1	0.350	2.15	0.84	1.81	7110
8	0.390	2.58	0.80	1.83	7031
13 (*)	0.362	1.76	0.94	1.85	7043
18	0.367	2.67	0.76	1.80	7101
24	0.376	2.67	0.77	1.81	7076
33	0.391	2.98	0.72	1.81	7063
3	0.391	2.95	0.73	1.81	7060
M15					
3	0.389	3.28	0.67	1.79	7093
4	0.314	1.56	0.92	1.80	7174
5	0.384	1.71	0.98	1.88	6977
10	0.386	2.84	0.74	1.81	7063
14	0.382	1.78	0.96	1.87	6989
17	0.429	2.40	0.88	1.89	6921
24	0.370	2.81	0.73	1.80	7106
31	0.408	3.48	0.65	1.80	7059
35	0.384	2.89	0.73	1.81	7073
38	0.375	2.85	0.73	1.80	7094
43	0.396	3.26	0.68	1.80	7072
54	0.400	2.05	0.92	1.88	6965

(*) Included according to criterion of CJS, but crowding effects may be significant here (see Clement, Ferance and Simon, in preparation).

Masses of c-type RR Lyrae Variables in Globular Clusters

C. Cacciari[1], A. Bruzzi[2]

[1]*Bologna Observatory, Italy,* [2]*University of Bologna, Italy,*

The mass of RR Lyrae variables has been a controversial problem for about a decade: while the stellar evolution theory predicts masses ranging between 0.65 and 0.75 $M_\odot$ for Oosterhoff type I and II clusters respectively, the stellar pulsation theory predicts smaller masses (0.55 and 0.65 respectively) using the double-mode pulsators. Simon (1990, M.N.R.A.S. 246, 70), comparing hydrodynamical models with observed stars by means of Fourier parameters, has found relations between the stellar mass and its luminosity, pulsation period, Helium content and Fourier parameter ϕ_{31}. Combining his equations we obtain:

$$log M = 0.590 log L - 0.052 \phi_{31} - 0.056 log P - 1.068 \qquad (1)$$

from which one can estimate the stellar mass by using the observable quantities P and ϕ_{31} and a luminosity scale, e.g. the one derived by Cacciari, Clementini and Fernley (1992, Astrophys. J. in press). We have selected the clusters with a good number of c-type variables and available photometry in the literature, except for M3 for which we have used our own CCD photometry (Carretta 1991, Thesis, University of Bologna): the list includes M3, M5, M62, NGC3201, NGC4147 and NGC6171 in the Oosterhoff I group, and M15, M53, M68 and NGC4833 in the Oosterhoff II group, for a total of 76 and 74 stars respectively. For each star the parameter ϕ_{31} was calculated and the stellar mass was derived. The average values of mass were 0.53 ± 0.07 $M_\odot$ (OoI) and 0.58 ± 0.09 $M_\odot$ (OoII), with M $\sim -0.05[Fe/H] + 0.47$ (typical error of 15% on the average mass per cluster). The main conclusions can be summarized as follows:

- These values of mass are a direct consequence of the adopted luminosity scale. They appear rather low, and in order to reconcile them with the larger values predicted by the evolution theory the zero-point of the luminosity scale should be brighter: an increase in mass by $\sim 10\%$ would increase the luminosity by ~ 0.2 mag. This is in agreement with the suggestion made by several other authors in this conference. The difference in mass between OoI and II groups is smaller than predicted by the classical pulsation and evolution theories, and more in agreement with the results of calculations which make use of enhanced (metal) opacities.

- Simon's ϕ_{31} method, although potentially very interesting when applied on a statistically significant sample of stars, is not sufficiently accurate for deriving reliable masses of individual stars.

CCD Photometry of RR Lyrae Stars in M3

C. Cacciari[1], E. Carretta[2], F. Ferraro[1], F. Fusi Pecci[1], G. Tessicini[1], J.M. Nemec[3], A.R. Walker[4]

[1]*Bologna Observatory, Italy*, [2]*University of Padova, Italy*, [3]*Washington State University, Pullman, USA*, [4]*University of British Columbia, Victoria, Canada*

Abstract

New CCD BVI observations of RR Lyrae variables in M3 are presented. Mean magnitudes and colours are derived, as well as their relations with periods and amplitudes, and comparisons are made with previous data (Sandage 1981, 1990). Preliminary results are presented on the temperature distribution of the variables and the period-shift effect with respect to M15 and M68.

1. Introduction

The globular cluster M3 has traditionally been assumed as the typical Oosterhoff type-I cluster, and therefore it has been used as the reference cluster for all problems related to globular clusters in general and RR Lyraes in particular, e.g. Oosterhoff dichotomy, period-shift effect, etc.

The data base, however, is still the B and V photographic photometry by Roberts and Sandage (1955), Baker and Baker (1956) and Sandage (1959), later re-analysed by Sandage (1981, 1990). The need was felt for more recent and accurate observations, in order to verify the assumptions and conclusions derived during the past decade. A project was thus undertaken by Buonanno et al. (1986, 1992) with the aim of obtaining photographic and CCD BVI accurate photometry and an independent absolute photometric calibration, as well as infrared JHK CCD photometry, and hence the best possible information on all parts of the CMD.

Different topics (i.e. characteristics and morphologies of the various branches in the CMD) will be treated elsewhere. Here we focus on the RR Lyrae stars only.

2. The Data Base

Our data have been collected at two different sites:

LOIANO: 152cm telescope + RCA 320x512 pixels, scale=0.5 arcsec/px, FOV= 2'40"x4'16", two non-overlapping fields (NW and SE) were observed. 65 frames were taken in each colour (BVI) covering the NW field, and 69 frames in each colour were taken on the SE field. Of these, respectively 44 and 53 have been reduced. A total of 95 variables have been measured, for several of them new or improved periods have been determined. Here we present the preliminary results

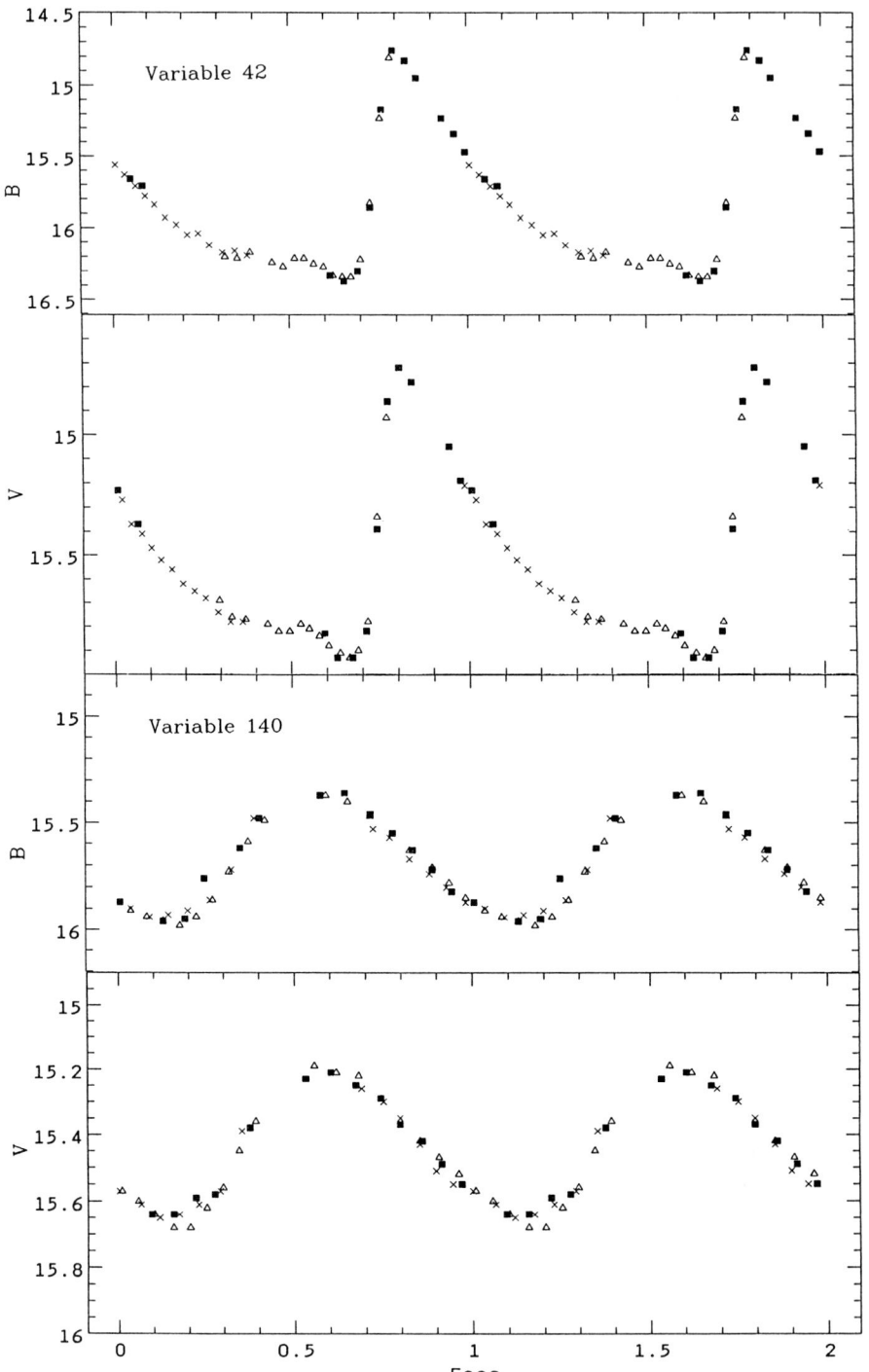

Figure 1 B and V light curves for the variables V42 and V140 in M3.

only for the best 38 stars that were measured in the NW field, based on 44 frames (i.e. the data base is still incomplete at the present time). The absolute calibration, based on 15 Landolt standard stars plus 6 stars in M92, has been performed for the NW field only, the calibration work is still in progress for the SE field. The photometric measurements were obtained using the reduction packages ROMAFOT and DAOPHOT, optimized for working in crowded areas, and the mean estimated rms error on the single measure are $\sim$ 0.01-0.02 mag.

PALOMAR: 152cm telescope + RCA 800x800 pixels (rebinned down to 400x400 px), scale=0.9 arcsec/px, FOV= 6' 16" square, four overlapping fields (NE, NW, SE, SW) were observed. Most frames were aimed at detecting Blue Straggler stars (and actually several variable Blue Stragglers have been found), only about 15-20 frames per colour (B and V) per field are suitable for RR Lyrae studies. 137 known or candidate variables have been measured. The calibration of these data is based on Sandage and Katem (1982) photometry, and the data reduction has been performed using the package DoPHOT, which is faster than DAOPHOT but slightly less accurate in crowded areas, the mean estimated rms error on the single measure being $\sim$ 0.05-0.06 mag.

In Figure 1 we show the B and V light curves of the variables 42 and 140 as an example. The preliminary results on our new photometric calibration confirm Sandage (1981, 1990) data within 0.01-0.02 mag: the possibility of a blue shift by $\sim$ 0.02 mag in B-V cannot be ruled out at present, but needs confirmation from the final calibration work.

A detailed study of the HB morphology will be made when the complete sample of variable and non-variable HB stars will be available. Some information on the period-shift effect, however, can be derived also using the partial sample of stars presented here.

3. The Colour-Temperature Transformation

The "Period-Shift Effect" (PSE) means that the period-temperature relations in clusters of different metallicities are shifted as a function of metallicity, in the sense that more metal-poor clusters have tendentially longer periods for the same temperatures. Therefore colours must be transformed to temperatures in order to investigate this issue. However, several problems, both observational and theoretical, affect this transformation:

1. Definition of the "best" mean colour. There are several possibilities: $<$B–V$>$, $<$B$> - <$V$>$, (B-V)$_{eq}$ = 2/3$<$B–V$>$ + 1/3 ($<$B$> - <$V$>$), (B-V)$_S$ = $<$B$> - <$V$>$+C(A_B) (Sandage 1990). These same types of indices can be defined for V-I, or for any other colour. They are all affected by some degree of inaccuracy, the effect being strongest on bluer colours, since they average LTE with non-LTE (shock waves at maximum light) values. The colours $<$V–K$>$ or $<$V$> - <$K$>$ are in principle better temperature indicators since they

have a longer baseline, are less affected by non-LTE problems and are less sensitive to metallicity.

Finally, another question which affects the definition of the "best" mean colour is: intensity or magnitude means?

2. From the observational point of view, the reddening represents a serious problem: as we shall see later, even a difference of 0.02-0.03 mag, which is of the same order as the accuracy of most reddening determinations, may make a significant difference in the analysis and interpretation of the PSE (Caputo 1987).

3. The temperature scale.
   a) By definition, the equilibrium temperature of a star is given by the Stefan-Boltzmann law, but radius and luminosity must be known. For RR Lyrae stars this is presently possible only for the ~ 25 stars analysed with the Baade-Wesselink method (Jones et al. 1992; Cacciari, Clementini and Fernley 1992), but there may be some problem with the zero-point of the B-W results, which appears to be ~ 0.2 mag too faint, as suggested by several authors also in this conference (e.g. Fernley, Saha, Cacciari, Dorman, Freedman, Longmore, Walker).
   b) A relation was found between equilibrium temperature, pulsation period, metallicity and amplitude of the B light curve (Carney, Storm and Jones 1992), which however does not appear to be very accurate.
   c) A different set of problems is related to the use of purely theoretical models (Kurucz, Bell-Gustafsson, VandenBerg-Bell, etc.). Kurucz models, for example, have been found to be somewhat incorrect, especially in the bluest part of the spectrum, and possibly in correlation with metallicity.
   d) Empirical scales can be used as an alternative: however, they are derived only for Population I main sequence stars where there is a reasonable number of stars with measured angular diameters. For more evolved or more metal-poor stars the transformations make use again of model atmospheres differentially.

The problem of a correct temperature determination is a very important one, but will not be addressed here in any further detail. For the sake of comparison with Sandage's previous work on M3 and with other clusters, we shall use consistently the colours $<B> - <V> + C(A_B)$ and Kurucz models, and $<V> - <K>$ and Fernley (1989) calibration when infrared colours are available.

4. The Period-Shift Effect

First we compare the present data (using only the 38 variables we have accurately measured in the NW field of M3) with the previous data obtained by Sandage (1981, 1990) in the logP-logT$_e$ plane. The B-V colours are actually <B> − <V>+C(A$_B$), where C(A$_B$) is determined according to Sandage (1990), and the Kurucz models have been interpolated for the cluster metallicity and logg = 2.75. The reddening for M3 has been assumed to be zero. Figure 2 shows our data and, superimposed, the solid line representing the best fit to Sandage (1990) data. The two distributions are consistent, and would coincide if our data were red-shifted by about 0.02 mag. The same indication comes from the comparison with the M5 data using the mean V magnitudes by Storm et al. (1991) and Goranskj (1982), assuming E(B−V) = 0.03 for M5. Since our calibration is still in progress, we provisionally apply a red-shift of 0.02 mag to our colours.

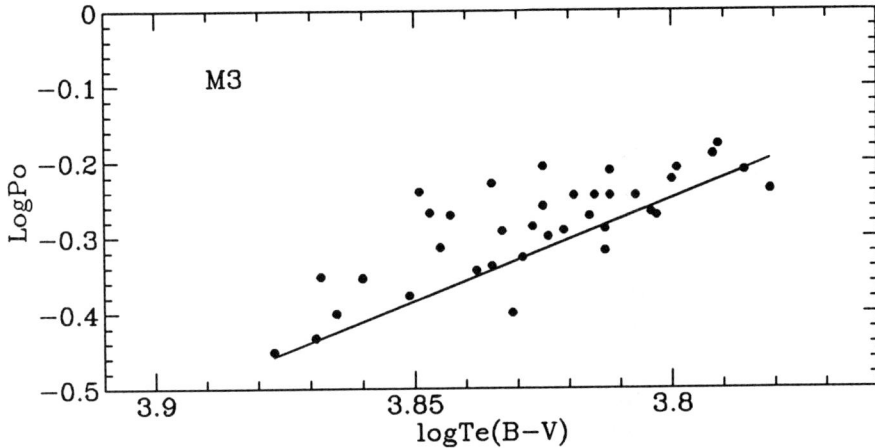

Figure 2 Data points from our new photometry. The logP$_0$ represents the fundamental period (0.125 was added to the c-type variables). The solid line is the best fit to Sandage (1990) data.

If we now compare the M3 data with the data obtained for M15 using Bingham et al. (1984) photometry and E(B−V)=0.10, and with M68 using unpublished photometry by Andrews and E(B−V)=0.03, we find a period shift $\Delta \log P/\Delta[Fe/H] \sim -0.12$ and −0.07 respectively. These values would coincide (i.e. ∼ −0.07) if the reddening for M15 were 0.07, as suggested by a number of independent studies, especially on UV data.

Since for all these clusters, except M68, K photometry is available (Longmore et al. 1990), it is worth repeating this comparison using V–K colours and Fernley's (1989) semi-empirical calibration:

$$log T_e = -0.125(V - K) - 0.003[Fe/H] + 3.934 \quad (1)$$

which is valid for RR Lyraes at temperatures around 6500K and all metallicities. The smaller slope of the colour-temperature relation decreases the effect of possible errors in the colours and improves the accuracy of the comparison: again the period-shift between M3 and M15 leads to $\Delta \log P/\Delta[Fe/H] \sim -0.07$ or -0.12 if a reddening of 0.07 or 0.10 respectively are assumed for M15.

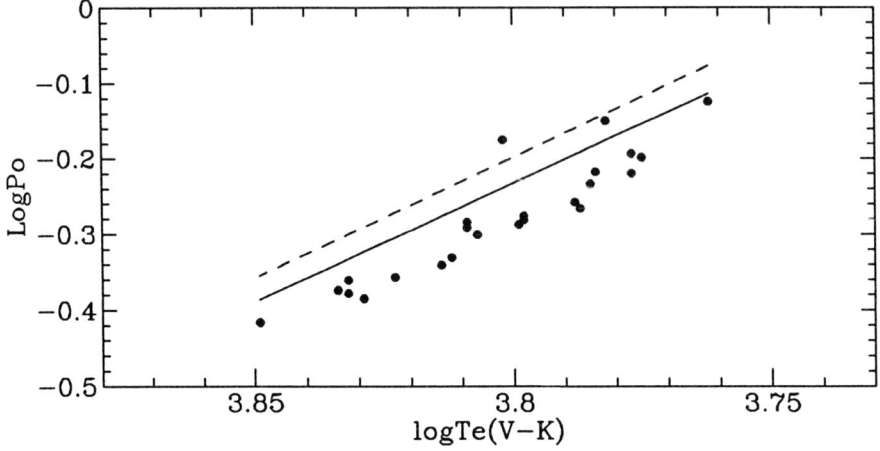

Figure 3 Our data for the variables in M3. The lines represent the best fit to the M15 variables using $E(B-V) = 0.10$ (dashed line) and 0.07 (solid line).

5. Summary and Conclusions

(a) For the "standard" Oosterhoff I cluster M3 a new photometric data base (CCD BVI) has been obtained. A preliminary independent calibration confirms Sandage (1981, 1990) results within about 0.02 mag. Work is still in progress for a final accurate calibration.

(b) Referring to RR Lyraes only: the problem of the PSE depends mainly on: a) the colour - temperature transformation; b) Off-ZAHB evolution; c) the reddening.

(c) The reddening plays a key role: if $E(B-V) = 0.07$ for M15, which gives consistent results with other clusters and is suggested by UV studies, then the period-shift between M3 & M5 and M15 & M68 is about 0.03-0.04. Therefore the period-shift dependence on metallicity is about 0.06-0.07, as suggested independently by other studies (e.g. Fernley elsewhere in this conference).
If, on the contrary, the reddening of M15 is 0.10, as suggested by optical studies, then evolution off the ZAHB for most of the RR Lyraes in M15

could explain the high slope of the period shift-metallicity relation (and hence of the luminosity-metallicity relation), as already suggested by Lee, Demarque and Zinn (1990).

References:

Baker, R.H., and Baker, H.V., 1956, Astron. J., **61**, 283.
Bingham, E.A., Cacciari, C., Dickens, R.J., and Fusi Pecci, F., 1984, M.N.R.A.S. **209**, 765.
Buonanno, R., Buzzoni, A., Corsi, C.E., Fusi Pecci, F. and Sandage A., 1986, Mem. Soc. Astr. It. **57**, 535.
Buonanno, R., et al., 1992, in preparation
Cacciari, C., Clementini, G., and Fernley, J.A., 1992, Astrophys. J., Sept. 1 issue, in press.
Caputo, F., 1987, in *Stellar Evolution and Dynamics in the Outer Halo of the Galaxy*, Ed.s M.Azzopardi and F.Matteucci, ESO Conf. Proc. No.27, p.321.
Carney, B.W., Storm, J., and Jones, R.V., 1992, Astrophys. J., **386**, 663.
Fernley, J.A., 1989, M.N.R.A.S., **239**, 905.
Goranskij, V.P., 1982, Astr. Tsirk. **1207**, 4.
Jones, R.V., Carney, B.W., Storm, J., and Latham, D.W., 1992, Astrophys. J., **386**, 646.
Lee, Y.W., Demarque, P. and Zinn, R., 1990, Astrophys. J., **350**, 155.
Longmore, A.J., Dixon, R., Skillen, I., Jameson, R.F., and Fernley, J.A., 1990, M.N.R.A.S., **247**, 684.
Roberts, M.S., and Sandage, A., 1955, Astron. J., **60**, 185.
Sandage, A., 1959, Astrophys. J., **129**, 596.
Sandage, A., 1981, Astrophys. J., **248**, 161.
Sandage, A., 1990, Astrophys. J., **350**, 603.
Sandage, A., and Katem, B., 1982, Astron. J., **87**, 537.
Storm, J., Carney, B.W., and Beck, J.A., 1991, P.A.S.P. **103**, 1264.

Discussion

COX: Could the -0.07 coefficient of the period shift - metallicity relation be within the errors of the Sandage value near -0.1?

CACCIARI: As I said before, the effect of reddening is quite important on this issue, and reddening is not determined with a precision better than 0.02-0.03 mag, especially when it's large. This error is sufficient to change the coefficient from -0.12 to -0.07 even in the case of M15, whose reddening is not very high.

FERNLEY: I have a comment: just to add another piece to the issue I find a slope $d\log P/d[Fe/H] = -0.07$ from field RR Lyrae stars (see talk in these proceedings).

Period Shifts in RR Lyrae Stars

J. Fernley

IUE–Vilspa, P.O. Box 50727, 28080 Madrid, Spain

Using recently published infrared photometry of RR Lyrae stars from both the field (Fernley et al. 1992) and globular clusters (Longmore et al. 1990), period shifts have been calculated using the mean $(V-K)$ colour rather than the mean $(B-V)$ colour employed in most previous analyses (e.g., Sandage 1990 and references therein). The advantage of using $(V-K)$ rather than $(B-V)$ is the reduced sensitivity to, firstly, metallicity and, secondly, non-LTE radiation. This latter occurs in many RR Lyrae stars near maximum light and clearly should not be included when calculating the mean colour and hence, mean temperature. This, and all other aspects of the present paper, are discussed more fully in Fernley (1992a).

The mean $(V-K)$ colour of both the field and cluster stars were de-reddened using the maps of Burstein and Heiles (1982) and converted to effective temperature using the calibration of Fernley (1989). The resulting period $-< T_{\text{eff}} >$ plots are shown in Figure 1 for the field stars, where – as can be seen in Figure 1 – the stars have been divided into four metallicity bins. The metallicities of both the field and globular cluster stars have been taken from the literature. They are on the Preston (1959) ΔS scale as calibrated into [Fe/H] by Butler (1975).

It can be seen in Figure 1 that there are several discrepent stars and these have been excluded from the least-squares fits. Of particular interest is BB Vir which, on the basis of its colour being too blue and it having a large ultraviolet excess, has been proposed as a binary with a blue-horizontal-branch companion (Kinman and Caretta 1992; Fernley 1992b).

Similar plots were made for the RR Lyrae stars in M3, M5 and M15. For both the field and cluster stars, the period at $\log T_{\text{eff}} = 3.81$ was read off and those periods are shown plotted against [Fe/H] in Figure 2. It can be seen that the periods of the field stars are systematically displaced above the globular cluster stars; the effect being particularly large at the metal-poor end.

Similar plots were made for the RR Lyrae stars in M3, M5 and M15. It has been suggested recently by Carney et al. (1992) that there is a bias in the field stars in the sense that an unrepresentative number of evolved stars are included. This bias arises because field stars were selected by the original observers (e.g., Fitch et al. 1966; Lub 1977) to cover a wide range of periods at a given metallicity. To test this further, we have calculated the mean period at a given metallicity for RR Lyrae stars from three samples: (1) the globular cluster stars used in this analysis, (2) the field stars from this same analysis, and (3) the field stars discovered in the Lick Survey (e.g., Suntzeff et al. 1991 and references therein). These mean periods are shown in Table 1, where it can be seen that the suggestion of Carney et al. (1992) is confirmed.

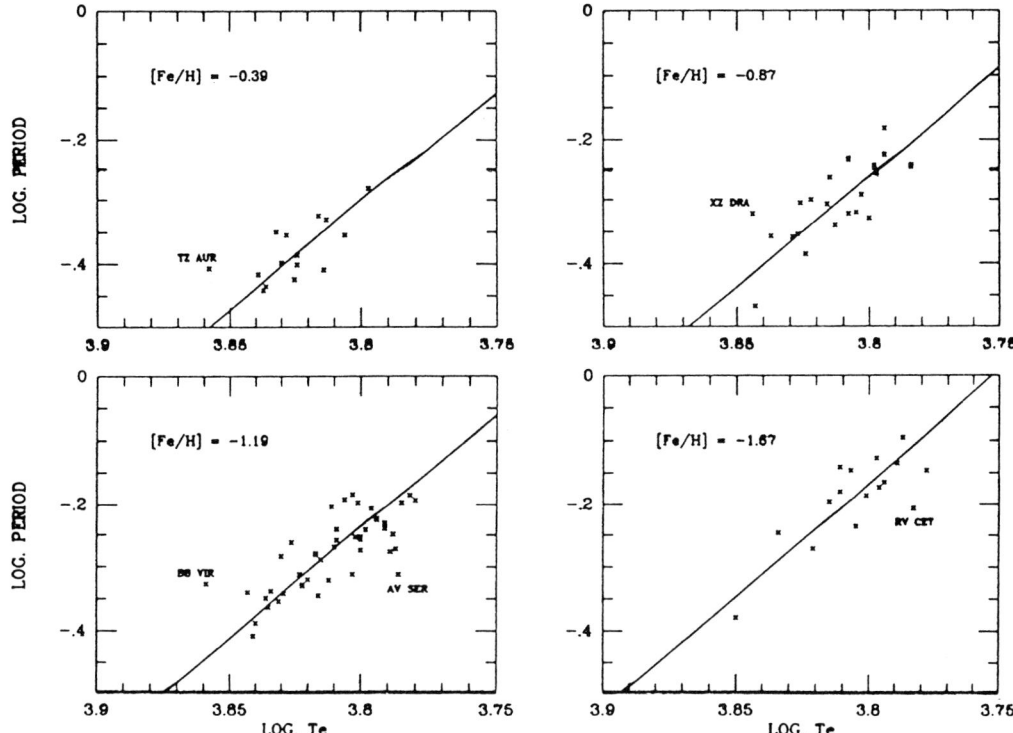

Figure 1. Period–colour diagrams at four different metallicities.

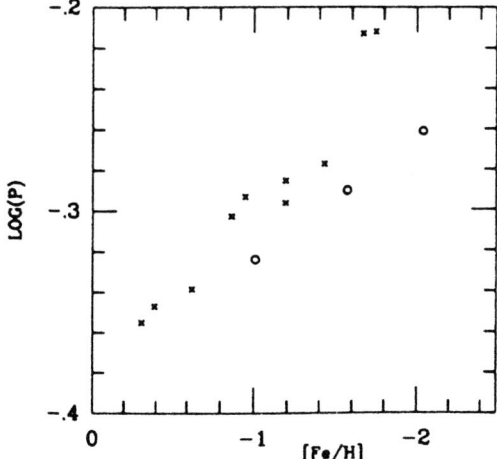

Figure 2. Period shifts (crosses = field stars; open circles = cluster stars)

Table 1. Mean periods at different metallicities

[Fe/H]	Cluster Stars	Field Stars (This Paper)	Field Stars (Lick Survey)
−1.03	0.518 (19)	0.515 (30)	0.532 (36)
−1.51	0.546 (22)	0.569 (39)	0.560 (44)
−1.83	0.566 (16)	0.654 (16)	0.576 (22)

Note: The number of stars used to derive the mean period is shown in brackets.

Omitting therefore the point from the most metal-poor stars, the best-fit lines to the data in Figure 2 give

$$\frac{d(\log P)}{d[\text{Fe/H}]} \propto -0.073 : \text{field stars} \quad (1)$$

$$\frac{d(\log P)}{d[\text{Fe/H}]} \propto -0.062 : \text{cluster stars} \quad (2)$$

References:

Burstein, D., Heiles, C., 1982, AJ 87, 1165

Butler, D., 1975, ApJ 200, 68

Carney, B.W., Storm, J., Jones, R.V., 1992, ApJ 386, 663

Fernley, J., 1989, MNRAS 239, 905

Fernley, J., 1992a, A&A, in press

Fernley, J., 1992b, Observatory, submitted

Fernley, J., Skillen, I., Burki, G., 1992, A&AS, in press

Fitch, W.S., Wisniewski, W.Z., Johnson, H.L., 1966, Comm. Lunar & Plan. Lab. 71, 5

Kinman, T.D., Caretta, E., 1992, PASP 104, 111

Longmore, A.J., Dixon, R., Skillen, I., Jameson, R.F., Fernley, J., 1990, MNRAS 247, 684

Lub, J., 1977, Ph.D. Thesis, University of Leiden

Preston, G.W., 1959, ApJ 130, 507

Sandage, A., 1990, ApJ 350, 631

Suntzeff, N.B., Kinman, T.D., Kraft, R.P., 1991, ApJ 367, 528

New results on field and cluster RR Lyraes

G. Clementini[1], R. Merighi[1], M. Tosi[1], R. Gratton[2], E. Carretta[3]

[1]*Osservatorio Astronomico, Bologna, Italy*, [2]*Osservatorio Astronomico, Padova, Italy*, [3]*Dipartimento di Astronomia, Padova, Italy*

Abstract

RR Lyrae stars both in the field and in clusters can be used to derive the metal abundance of the regions and systems where they are found.
(1) New data have been collected on a sample of field ab-type RR Lyraes with the aim of studying the composition of the halo and the disk of the galaxy, (Clementini et al. 1992a, in preparation), using the relation found by Clementini et al. (1991), (hereafter CTM91), between [Fe/H] and the equivalent width of the Ca II K-line W'(K). (2) A quantitative chemical abundance analysis of the ab type RR Lyrae (V29) in the globular cluster M4 has been performed using high resolution, high S/N spectroscopy. We obtain [Fe/H]$=-1.3\pm0.2$ and the $\alpha-$ elements (Mg and Ti) are overabundant by 0.6 dex. These results are in good agreement with determinations from high resolution spectra of giants and blue horizontal branch stars (Clementini et al. 1992b, in preparation).

1. Abundances of field RR Lyrae stars

To study the composition of the halo and the disk of our Galaxy, and to get hints on the existence of a thick and/or thin galactic disk we have selected 7 field ab-type RR Lyraes which, from data in the literature, appear to be located at distances from the galactic plane varying in the range $0< |Z| <9$ kpc and that according to their pulsational properties (periods and Bailey type) are expected to have high metal abundance (Castellani et al., 1981). Photometric and spectroscopic data have been collected to check the pulsational properties of these stars and to derive their metal abundance using the relation [Fe/H] $= 0.53(\pm0.09)$W'(K) $- 3.08(\pm0.22)$ (CTM91). The new data show that 5 out of 7 of the stars in our sample had improperly been classified as Bailey type-ab RR Lyraes and that among the remaining two ab-type stars only one, (RS Boo), has short period and high metallicity ([Fe/H]$=-0.10$), but this star is also the closest to the galactic plane. We conclude that the selected sample was not the proper one to study the [Fe/H] *versus* $|Z|$ variation and that caution should be taken when selecting samples from data that may be not updated.

2. Chemical abundances in a globular cluster RR Lyrae

Echelle spectra of the variable V29 in M4 were obtained with the 3.6 m ESO telescope, at La Silla, Chile, equipped with the CASPEC spectrograph. Eight spectra (for a total exposure time of about 4 hr) taken between phase 0.65 and 0.8 (close to the minimum of the light curve, where the temperature is constant within ±100 K and the effective gravity varies less than 3 %) were coadded after appropriate shifts to account for the different radial velocities. The resulting spectrum has a $S/N > 50$ and a resolution of $R = 15,000$. Equivalent width (EW)s for about 40 lines due to various elements were measured using a special gaussian fitting routine. The abundances we derive are shown below.

[A/H]	No. lines	$T_{\text{eff}} = 6,000$ K	$T_{\text{eff}} = 6,200$ K
[Mg/H]	1	−0.7	−0.6
[Sc/H]	1	−1.5	−1.4
[Ti/H]	10	−0.85	−0.75
[Cr/H]	5	−1.2	−1.15
[Mn/H]	1	−1.25	−1.05
[Fe/H] I	13	−1.5	−1.35
[Fe/H] II	9	−1.3	−1.3
[Ni/H]	1	−1.1	−1.0
[Y/H]	1	−1.1	−1.0
[Ba/H]	1	−1.5	−1.3

Our abundances compare very well with other determinations from high resolution spectra of M4 stars, both giants (Gratton et al. 1986) and blue HB stars (Lambert et al. 1992). A comparison with other abundance determinations (ΔS index), is also quite satisfactory, given the uncertainties and the large scatter in abundances obtained from low dispersion spectra.

References:

Castellani, V., Maceroni, C., Tosi, M. 1981, Astron. Astrophys. **102**, 411.
Clementini, G., Merighi, R., Tosi, M. 1992a, in preparation.
Clementini, G., Merighi, R., Gratton, R., Carretta, E. 1992b, in preparation.
Clementini, G., Tosi, M., Merighi, R. 1991, Astron. J **101**, 2168 (CTM91).
Gratton, R.G., Quarta, M.L., Ortolani, S. 1986, Astron. Astrophys. **169**, 208.
Lambert, D.L., McWilliam, A. and Smith, V.V. 1992, Astrophys. J. **386**, 685

Discussion:

KOVACS : How large is the dependence of your [Fe/H] values on the pulsation phase?

CLEMENTINI : We have used only spectra taken at phase corresponding to minimum light so I don't expect our determinations to be phase-dependent.

SMITH : The ΔS results for M4 are made uncertain by the strong interstellar K-line in the direction of M4. This may introduce some scatter in the comparison of ΔS and echelle abundances for this cluster.

FERNLEY : Why do you not use the 'c' type field RR Lyraes in your analysis?

CLEMENTINI : For the c-type RR Lyraes it is more difficult to properly take into account the effect of temperature on abundance determinations from intermediate resolution spectroscopy.

Search for Variables in the Galactic Globular Clusters NGC 6544 and NGC 6642

Martha L. Hazen[1],

[1]*Harvard-Smithsonian Center for Astrophysics, Cambridge, MA, USA*

A search for variable stars in the globular cluster NGC 6544 has revealed only one possible short period variable within the tidal radius of the cluster. A search in NGC 6642 yielded 16 new RR Lyrae stars within the tidal radius and 5 new field RRs. The previously discovered (Hoffleit 1972) V1 is a slow variable, and V2 is an RR Lyrae star. Photometry of the variables within the tidal radius gives a mean B for the horizontal branch of $<B> = 17.0$ mag. With $E(B-V) = 0.37$ mag and $(B-V) = 0.35$ mag for RR Lyraes, a value for $V(\text{HB}) = 16.3$ mag is derived. This is about one mag fainter than previous estimates (Webbink 1985), and places NGC 6642 at a distance of approximately 7.9 kpc.

References:

Hoffleit, D., 1972, IBVS No. 660.
Webbink, R.F., 1985, in: *Dynamics of Star Clusters*, eds. J. Goodman and P. Hut, Reidel, Dordrecht, p 541.

CCD Photometry of RR Lyrae Stars in NGC 6388 and M15

N.A. Silbermann[1], H.A. Smith[1], M. Bolte[2],

[1]*Dept. of Physics and Astronomy, Michigan State University, East Lansing, Michigan, USA* [2]*Lick Observatory, UCSC, USA*

Abstract

We present preliminary results of a program of CCD photometry of RR Lyrae variable stars in the globular clusters NGC 6388 and M15.

We have begun an investigation of the RR Lyrae stars in the globular clusters NGC 6388 and M15 based on CCD photometry. Hazen and Hesser (1986) investigated the variable stars of NGC 6388 and noted that, although several studies place NGC 6388 as metal-rich as 47 Tuc, it may contain a number of RR Lyrae stars. We obtained new B, V, and R CCD observations of this cluster in 1987, 1988, and 1989 with the 0.9m and 4m telescopes at CTIO. We identified eight variable stars, two of which, v17 and v20, were found previously by Hazen and Hesser. Their v20 is badly blended but is probably an RR Lyrae star with a period of approximately 0.45 days. The six new variables are relatively close to NGC 6388 and most appear to be RR Lyrae stars. We have been able to determine a reliable period for only one of the new variables. The data on the eight variables are given in the table below.

Data for M15 were obtained with the 2.3m telescope at WIRO and, for the most part, the 0.6m telescope at MSU. Reductions of the CCD frames for M15 are still incomplete: so far, only 85 of the V frames obtained with the MSU 0.6m have been reduced with Stetson's DAOPHOT program. Over 40 RR Lyrae stars fall within the CCD frames for this cluster. We expect to eventually obtain approximately 200 V and 200 R magnitudes for each M15 variable, with a somewhat smaller number of B magnitudes.

References:

Hazen, M.L., Hesser, B.H., 1986, Astronomical J., **92**, 1094.

NGC 6388 RR Lyrae Variable stars

Star	Type	Period	ΔV	ΔB	$<B>$
v17	RRab	0.603	15.6-17.3	16.6-18.2	17.40
v27	RRc	0.361	16.7-17.2	17.2-17.9	17.55
v28	RR	0.3 or 0.5	—	—	—
v29	RR?	short	—	—	—
v30	RRc?	0.343?	16.6-17.3	17.1-17.6	17.35
v31	RR?	short	—	—	—
v32	RR	0.4 or 0.8	16.6-17.4	17.3-18.0	17.65

Amplitude of RR Lyrae Star Light Curves: Comparison between Observations and One–Zone Model Predictions

E. Antonello, S. Cernuti

Osservatorio Astronomico di Brera, Via E. Bianchi 46, 22055 Merate, Italy

Stellingwerf's one–zone model is a simple and useful tool for reproducing the main observed pulsational characteristics of RR Lyrae and high amplitude δScuti stars, in particular their light and color curves (Stellingwerf et al., 1987, *Ap.J.* **313**, L75; Antonello, 1990, *Astr. Ap.* **230**, 127). In the present poster we show in better detail a comparison of the observed *amplitudes* of the light curve at various wavelengths with those predicted by the one–zone model; a preliminary result on this subject was reported in the short note by Grieco and Antonello (1990, in *Confrontation between Stellar Pulsation and Evolution* p. 101). As in the previous applications, here we use the published grids of atmospheric models (Kurucz, 1979, *Ap.J.Suppl.* **40**, 1) and we do not consider possible shock effects.

The amplitudes predicted by the model as a function of wavelength are in good qualitative agreement with the spectrophotometric and photometric data of RR Lyrae stars available from the literature. The predicted amplitudes depend on the adopted equilibrium values of T_e, $\log g$ and on $[Fe/H]$. A high T_e gives a large amplitude mainly in the UV region, while a different $\log g$ gives a slightly different amplitude only in the blue–visual region. The amplitude in the UV region is particularly sensitive to $[Fe/H]$. Moreover, we find an interesting feature: at the wavelength corresponding to the CaII K line the amplitude appears to be independent on T_e, weakly dependent on the gravity and strongly dependent on $[Fe/H]$.

The model prediction of a sensitivity of the amplitude of the light curve at the various wavelengths to the parameters T_e, $\log g$ and $[Fe/H]$ is checked by means of a comparison with the observed data of some RR Lyrae stars (Liu and Janes, 1990, *Ap.J.* **354**, 273). The differences between the observed amplitudes in the bands U, B and V, that is $(\Delta U - \Delta V)$ and $(\Delta B - \Delta V)$, are computed and compared with the corresponding differences predicted by the model. The differences are plotted as a function of T_e, or $\log g$ or $[Fe/H]$. There is a qualitative agreement between the observed trends and the predicted ones. The are some small (0.05 mag) systematic differences between observations and models for $(\Delta B - \Delta V)$. The largest discrepancy is for the amplitude difference $(\Delta U - \Delta V)$, and this is tentatively interpreted as due to shock effects, which are expected to be larger in the U band than in the V band.

Pulsation Variables in the AF stars of the Case Low-Dispersion Survey

T.D.Kinman

Kitt Peak National Observatory, P.O.Box 26732, Tucson, AZ 85732, U.S.A.

Abstract

The AF stars are those of spectral types A and F that have been discovered on objective-prism plates taken with the Burrell Schmidt in the Case Low-Dispersion Northern Sky Survey (Pesch and Sanduleak, 1983). In SA 57, this survey is complete to V = 16.5. In this field and also in the Lick Astrograph RR Lyrae survey field RR 7 (in the anticenter) the AF stars comprise (a) blue horizontal branch stars and RR Lyrae stars of the halo and (b) stars which have the higher surface gravities of main sequence stars. The two groups can be separated primarily by their differing Balmer jumps and Balmer line-widths. The latter group (which may well include blue stragglers of both Pop I and Pop II) extends some 10 kpc above the galactic plane and shows a wide range of metallicity. Photoelectric photometry of this AF star sample has allowed the detection field RR Lyrae stars of lower-amplitudes than could have been found by conventional blinking techniques; this has led to a significant increase in the number of RR Lyrae stars that are known in SA 57 and RR 7. The cooler main sequence and/or blue straggler AF stars lie in the zone of pulsational instability and one higher-amplitude δ-Scuti star was detected in field RR 7. It is suggested that these AF stars provide a good sample for studying the incidence of pulsation in the population of older stars that extends beyond the thin disk.

1. Introduction

The Case Low-Dispersion Northern Sky Survey (Pesch and Sanduleak, 1983 and Pesch, 1991) is being made with the 0.6-m/0.9-m F/3.5 Burrell Schmidt telescope using a 1.8° UV transmitting prism ($\sim$1000 Å mm^{-1} at H and K). The AF stars (stars with spectral types A and F) are those which are described as being in Category IV, IV-V or V in the Case survey. It was originally thought that Category IV and V corresponded respectively to main sequence and to horizontal branch stars. Later work (Sanduleak, 1989; Kinman, 1992 and MacConnell, Stephenson and Pesch, 1992) has shown, however, that this distinction cannot be made unambiguously from the relatively low S/N objective-prism spectra of the fainter stars. Ultraviolet-excess objects of stellar appearance (such as white dwarfs and quasars) can, however, be distinguished in these spectra and so excluded from Categories IV and V. Lists of both Categories of these AF stars are available for regions near the North Galactic

Pole (NGP) (Sanduleak, 1988 and MacConnell et al., 1992). Category V stars alone have been published for a large region between the Anticenter and the NGP (Pesch and Sanduleak, 1989). I was first interested in the AF stars because they included nearly all the RR Lyrae stars (brighter than B~17) that had been discovered by blinking plates taken with the Lick Astrograph at the NGP (Kinman, Wirtanen and Janes, 1966). Sanduleak kindly made available to me the identifications of the stars of both categories in SA 57 at the NGP (which is also Lick Astrograph field RR 4) and also in the Lick Astrograph field RR 7 (l =183° ;b=+37°) in the Anticenter. The field of SA 57 is important because it has been included in a number of other surveys — in particular the multicolor survey of Stobie and Ishida (1987). A comparison with this survey showed that the Case survey in SA 57 is complete to V = 16.5 within the range $0.0 \leq (B-V) \leq 0.32$ (Kinman, 1992). More recent work suggests that the red limit of the Case survey may vary from region to region and this is also probably true of the bright limit.

The importance of the Case AF stars is that they provide a moderately deep survey for probing both the halo (with the RR Lyrae stars, blue horizontal branch stars and possibly the halo blue stragglers) and also the stars of the extended or "thick" disk. *In the case of the halo, the use of the AF stars allows one to determine the horizontal-branch morphology from the relative numbers of bhb to RR Lyrae stars and so puts an interesting constraint on the nature of this population.*

2. Observational techniques for identifying blue horizontal branch (bhb) stars

The characteristic variability of the RR Lyrae stars — the bhb stars that lie in the zone of pulsational instability ($0.20 \leq (B-V)_0 \leq 0.40$) — makes them relatively easy to identify. It is those with $(B-V)_0 \leq 0.20$ and particularly those with $(B-V)_0 \sim 0.0$ that are more difficult to classify with certainty. The two fields (SA 57 and RR 7) contain some 150 AF stars and the majority of these are fainter than V = 14. The well-known technique of separating bhb stars from higher-gravity main-sequence stars using the c_1 index of Stromgren photometry was not practical for such faint stars with the equipment available. A hybrid (u−B) index (where u is a Stromgren filter) can, however, be used to measure the Balmer jump and the separation can be made in a (u−B) vs. (B−V) plot (Kinman, 1992). Bidelman (1992) has noted that, in addition to the bhb stars, somewhat evolved normal early-type stars can also have a high c_1 index. Corbally and Gray (1992) have used high S/N CCD spectra to classify 67 of the field horizontal branch stars listed by Philip (1984) and conclude that *only two are actually field horizontal branch stars.* This conclusion may need modification. Their *spectral classification* of the classic prototype field horizontal branch star HD 161817 (Oke, Greenstein and Gunn, 1966) which has a radial velocity of −363 km/s and [Fe/H] = −1.7 (Adelman and Hill, 1987), as an A8 dwarf with weak metal lines does not agree with our assessment. HD 161817 has a $(B-V)_0$ of ~0.13 which is close to that of an A3 V star, but its Balmer lines are much narrower and its ultraviolet continuum quite different from such a star (Fig. 1). The subject needs further review; confusion is possible if only one parameter (e.g. the c_1 index, (u−B) or *a fortiori* (U−B)) is used as a discriminant. The confusion will be greatest in the solar neighbourhood where disk stars greatly outnumber those of the halo. It should be less in samples that are far from the plane where the proportion of halo to disk stars is much higher.

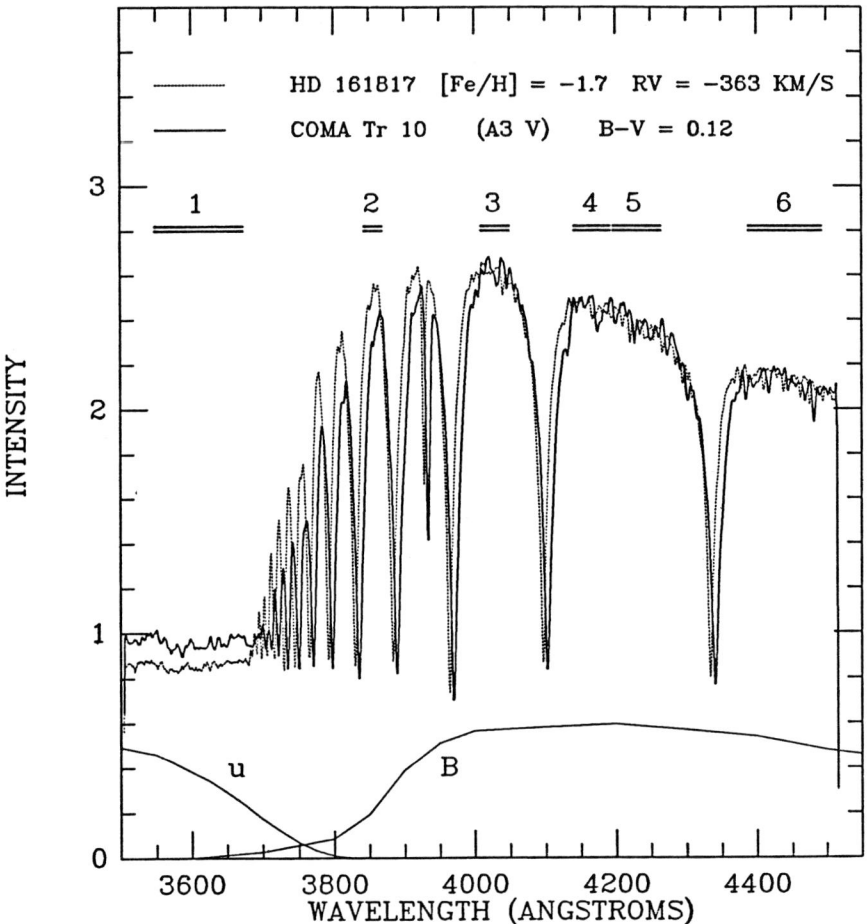

Figure 1 A comparison of spectrophotometric scans made with the Goldcam CCD spectrograph by Kinman, Kraft and Suntzeff of HD 161817 and the A3 V star Trumpler 10 in Coma. Both stars have similar B−V colors and spectral gradients measured by wavebands 3, 4, 5 and 6. HD 161817 has much narrower Balmer lines and weaker metal lines. It also has a larger Balmer jump as measured by the ratio of the fluxes in wavebands 1 and 2 or the (u−B) index.

Additional discriminants are needed. A spectrophotometric survey of these stars was therefore undertaken in collaboration with R.P. Kraft and N.B. Suntzeff using the Goldcam CCD spectrograph on the KPNO 2.1-m telescope (resolution $\sim$4 Å). These scans provide:

(a) A more direct measurement of the Balmer jump by a comparison of the fluxes in wavebands centered on λ3610 and λ3860,

(b) The widths ($D_{0.2}$) at the 20% depth (cf Pier, 1983) of the Balmer lines Hδ and Hγ and

(c) Equivalent widths of the Balmer lines Hδ and Hγ.

These parameters all allow some separation of the bhb from higher-gravity stars in plots against either (B−V) or PCG (a pseudo-continuum gradient in the wavelength range $\lambda\lambda$ 4010 to 4500 which correlates very well with B−V).

Parameters which can be used for the *statistical* separation of halo from disk stars include rotation (Peterson, Tarbell and Carney, 1983), radial velocity and metallicity; the last of these will be considered in the next section.

3. The metallicity of the AF stars

The prototype halo objects, the halo globular clusters, have [Fe/H] $\leq$−1.0. According to Morrison, Flynn and Freeman (1990) stars with disk kinematics can have [Fe/H] as low as −1.6. Unfortunately, A stars are well known for the peculiarities of their spectra. Further, most of the metal lines in these stars are rather weak for accurate measurement in low resolution scans. The following lines are strong enough for consideration:

Ca II λ3933: Pier (1983) has discussed the use of the K-line for determining [Fe/H] in A stars. The main disadvantages are (a) the unknown interstellar component can cause large uncertainties for the hotter stars (B−V $\leq$ 0.10), (b) the line is susceptible to anomalous behaviour (Am stars) and (c) it is quite sensitive to temperature.

Fe I λ4271: A broad blend that is therefore not too sensitive to rotation, but sensitive to temperature (Gray and Garrison, 1987).

Mg II λ4481: Morgan (1933) noted that this line shows increasing luminosity sensitivity as one goes to later spectral types. Gray and Garrison (1989), however, note only a mild sensitivity to luminosity in F stars and also to dilution effects in shells stars (Struve and Wurm, 1938). Being a narrow line it is sensitive to rotation effects. In his rotation studies, Slettebak (1954) noted that in the spectral range B7 V to A5 V, it was strong in supergiants and Am stars and weakened in Ap stars and weak-line stars. It is greatly weakened in λ Boo stars (Population I).

The blends at $\lambda\lambda$ 4173−4179 and $\lambda\lambda$4383−4385 are strongly luminosity sensitive and are not considered further.

It is clear that no line or blend of lines is an ideal abundance indicator for low resolution spectra. It may be noted that the pseudo-continuum in such spectra is rather strongly influenced by weaker absorption lines, so that the equivalent widths are not necessarily comparable to those obtained at other resolutions. Under these circumstances, it seems reasonable to derive a separate metallicity from each line and also to be aware that abundance ratios of Ca/Fe and Mg/Fe may not be a constant for all stars.

A provisional analysis showed 15 stars in which there seemed to be a significant difference between the metallicity derived from the K-line and that derived from the Mg II λ4481 line. In 13 of these cases, the metallicity derived from the Fe I λ4271 line was in better agreement with that found from the Mg II line than that derived from the K-line.

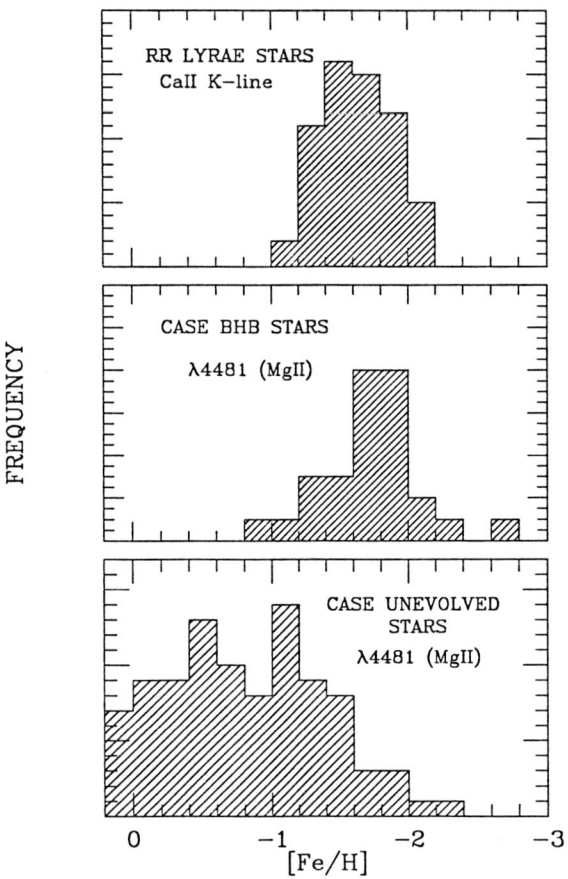

Figure 2 The upper panel shows the distribution of [Fe/H] derived from the Ca II K-line (ΔS method) for the RR Lyrae stars outside the solar circle (Suntzeff, Kinman and Kraft, 1990). The middle panel shows the distribution of [Fe/H] for the Case AF stars that are judged to be bhb stars on the basis of their (u−B) index; and the lower panel gives the distribution for those that are not. In both lower and middle panels, [Fe/H] is derived from provisional measurements of the Mg II λ4481 line.

The distribution of the values of [Fe/H] for the AF stars in the two fields (SA 57 and RR 7) derived from these provisional measurements of the Mg II λ4481 line are shown in Fig 2. The middle panel shows the distribution for the bhb stars; the lower panel shows the distribution for the remaining (unevolved) stars. Two stars for which the Mg II line gives [Fe/H] of +0.5 and +0.7 have been omitted from the histogram in the lowest panel; one of these stars is a known Am star. The top panel in Fig. 2 shows the distribution of [Fe/H] for the RR Lyrae variables outside the solar circle (Suntzeff, Kinman and Kraft, 1990). The RR Lyrae abundances were derived from the Ca II K-line using Preston's ΔS method. Bearing in mind the measuring errors that must be present in the equivalent widths of the Mg II line, the agreement between the [Fe/H] distributions of the two halo samples is as satisfactory as could be expected. Very provisional radial velocities were determined for the stars in SA 57. In this field the radial velocity dispersion is nearly the same as the z-motion dispersion and is ± 136 km/s for the bhb stars. For the remaining or "unevolved" AF stars, those with [Fe/H]≥ -1.0 have a dispersion of ± 48 km/s and those with [Fe/H]≤ -1.0 a dispersion of ± 70 km/s. These velocity dispersions are probably inflated by significant errors but valid enough to show that the stars with weak Mg II $\lambda 4481$ which have been assigned [Fe/H]≤ -1.0 are not low-velocity Population I λ Boo stars.

4. Pulsation variables in the AF stars

Some of the AF stars populate the instability strip. Photoelectric monitoring of a sample of AF stars affords a practical way of picking out the pulsation variables even

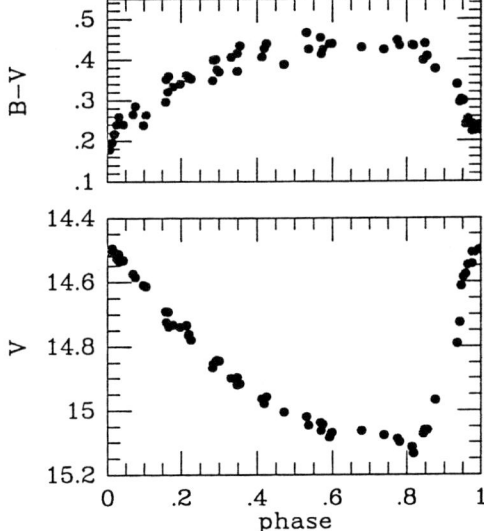

Figure 3 The light curves of the Case star AF 167 (Sanduleak, 1988). This type ab RR Lyrae star has a period of 0.640648 days.

if they have quite small amplitudes. After discovery, the periods and light curves of the variables can be obtained by further photoelectric observations or by measuring the stars on existing survey plates such as those taken with the Lick Astrograph. This is a much more effective way of finding RR Lyrae variables — particularly those of lower amplitude — than by blinking. An example of such a variable is the Case star AF 167 (Sanduleak, 1988). This star lies in the field RR 3 and close to the edge of the overlapping field RR 4 and was missed from their survey by Kinman, Wirtanen and Janes (1966). It is interesting that it is listed (NSV 6031) in the catalogue of stars of suspected variability; it was identified as a possible RR Lyrae star by Kurochkin (1960). It has a blue amplitude (see Fig. 3) of 0.75 magnitudes. This is the amplitude limit at which the surveys with the Lick Astrograph are thought to be complete. Variables with lower amplitude were found — but only with partial completeness. Surveys such as those made with the Lick Astrograph will therefore be incomplete for the lower-amplitude c-type variables. The ratio of the c-type to ab-type variables is one of the parameters that characterizes the Oosterhoff type, and like the ratio of the numbers of RR Lyrae stars to bhb stars is essentially a measure of the way the stars are distributed along the horizontal branch. Twelve new RR Lyrae variables have been discovered so far in SA 57 and RR 7 and work is in progress to determine their periods and light curves.

Amongst the "unevolved stars", a δ Scuti variable with a B-amplitude of 0.25 magnitudes (period 0.093 days) was found in field RR 7 (GSC 2985 01044 in the Space Telescope Guide Star Catalog). The star is relatively bright (B$\sim$ 12.2) and its images on the survey plates taken with the Lick Astrograph are very strong; it is therefore not surprising that its variability went undetected. It was possible, however, to make iris astrophotometer measurements of GSC 2985 01044 on these plates with enough accuracy to produce a light curve. The Astrograph images are of sufficient quality to derive light curves even though the amplitudes are too small for the variables to have been discovered by blinking the plates.

5. The nature of the non-variable AF stars

The bhb stars can be convincingly separated out from the other AF stars by observational parameters (Balmer jump, Balmer line widths, etc.) that are related to their lower surface gravity compared with main sequence stars. Undoubtedly there are stars in the disk with high c_1 index that may mimic these bhb stars. At the height above the plane of most of our bhb star candidates, the ratio of main sequence to halo stars is much smaller than in the solar neighbourhood and so this kind of contamination in our sample is likely to be minimal. We find a metallicity distribution for our bhb sample that is much like that found for RR Lyrae stars in the outer halo and we also (provisionally) find an appropriate velocity dispersion.

The nature of the remaining non-variable stars is less clear and it could well be that they comprise several populations. The merging of other small galaxies with our own is one possiblity (Freeman, 1990). We should also consider mechanisms such as those involving collisions within high-velocity HI clouds (Dyson and Hartquist, 1983) and the ejection of stars from open clusters (Leonard and Duncan, 1990) which are invoked to explain runaway B stars and may also explain some of the extended disk of

A stars. It has also been suggested that field blue stragglers are merged stars that have been ejected from clusters (Leonard, 1989). Over twenty years ago, Eggen (1969) and later Bond and MacConnell (1971) suggested that the blue stragglers of the old open clusters and the globulars clusters have counterparts in the field; Shields and Twarog (1988) estimated that these field blue stragglers should exist in substantial numbers. That many of these non-bhb AF stars are blue stragglers associated the older stellar populations agrees qualitatively with their having an extended distribution in z and a wide distribution in metallicity. According to Buonanno, Buzzoni, Corsi, Fusi Pecci and Sandage (1988) the number of blue stragglers and horizontal branch stars in the globular cluster M 3 are rather similar. If this ratio also applies to our sample in SA 57 we should expect to find 3 or 4 halo blue stragglers with $15.0 \leq V \leq 16.5$. There are 7 non-bhb stars with [Fe/H]≤-1.4 in our sample in this magnitude range. Nemec (1989) has reported that 25% of globular cluster blue stragglers are variable and that most of these are SX Phe stars. It would therefore be interesting to search for this type of variability among the metal-poor non-bhb AF stars.

References:
Adelman, S.J. and Hill, G., 1987, M.N.R.A.S., **226**, 581.
Bidelman, W.P., 1992, in: *Luminous High-Latitude Stars*, Astr. Soc. of Pacific Conf. Series, to be printed, ed. D. D. Sasselov, Chelsea MI, Astr. Soc. of the Pacific.
Bond, H.E. and MacConnell, D.J., 1971, Astrophys. J., **165**, 51.
Buonanno, R., Buzzoni, A., Corsi, C.E., Fusi Pecci F. and Sandage, A.R. 1988, in: *I.A.U. Symposium 126*, ed. J.E. Grindley and A.G.D. Philip, Dordrecht, Kluwer, p 621.
Corbally, C.J. and Gray, R.O., 1992, in: *Peculiar versus Normal Phenomena in A-type and Related stars*, ed. F. Castelli and M.M. Dworetsky.
Dyson, J.E. and Hartquist, T.W., 1983, M.N.R.A.S., **203**, 1223.
Eggen, O.J. 1969, Publ. Astr. Soc. Pacific, **81**, 741.
Freeman, K.C., 1990, in: *Dynamics and Interactions of Galaxies*, ed. R. Wielen, Springer, Berlin, p. 36.
Gray, R.O., and Garrison, R.F., 1987, Astrophys. J., **65**, 581.
Gray, R.O., and Garrison, R.F., 1989, Astrophys. J., **69**, 301.
Kinman, T.D., 1992, in: *Variable Stars and Galaxies*, Astr. Soc. of Pacific Conf. Series, **30**, ed. Brian Warner, Chelsea MI, Astr. Soc. of Pacific, p. 19.
Kinman, T.D., Wirtanen, C.A., and Janes, K. 1966, Astrophys. J. Suppl., **13**, 379.
Kinman, T.D., Mahaffey, C. and Wirtanen, C.A. 1982, Astron. J., **87**, 314.
Kurochkin, H.E., 1960, Variable Stars, **12**, 409.
Leonard, P.J.T., 1989, Astron. J., **98**, 217.
Leonard, P.J.T. and Duncan, M.J., 1990, Astron. J., **99**, 608.
MacConnell, D.J., Stephenson, C.B. and Pesch, P., 1992, Astron. J., (submitted).
Morgan, W.W., 1933, Astrophys. J., **77**, 291.
Morrison, H., Flynn, C. and Freeman, K., 1990, Astron. J., **100**, 1191.
Nemec, J.M., 1989, in: *The use of pulsating variables in fundamental problems in astronomy*, ed. E.G. Schmidt, Cambridge Univ. Press, Cambridge, p. 215.
Oke, J.B., Greenstein, J.L. and Gunn, J., 1966, in: *Stellar Evolution*, ed. R.F. Stein and A.G.W. Cameron, Plenum Press, New York, p. 399.

Pesch, P., 1991, in: *Objective-Prism and Other Surveys: A Meeting in Memory of Nicholas Sanduleak*, ed. A.G.D. Philip and A.R. Upgren, L. Davis Philip Press, Schenectady, p. 3.
Pesch, P. and Sanduleak, N., 1983, Astrophys. J. Suppl., **51**, 171.
Pesch, P. and Sanduleak, N., 1989, Astrophys. J. Suppl., **71**, 549.
Peterson, R.C., Tarbell, T.D. and Carney, B.W., 1983, Astrophys. J., **265**, 972.
Philip, A.G.D., 1984, Van Vleck Obs. Contrib., No 2, p. 1.
Pier, J.R., 1983, Astrophys. J. Suppl., **53**, 791.
Sanduleak, N., 1988, Astrophys. J. Suppl., **66**, 309.
Sanduleak, N., 1989, Astrophys. J. Suppl., **71**, 713.
Shields, J.C. and Twarog, B.A., 1988, Astrophy. J., **324**, 859.
Slettebak, A., 1954, Astrophy. J., **119**, 146.
Stobie, R. and Ishida, K., 1987, Astron. J., **93**, 624.
Struve, O. and Wurm, K., 1938, Astrophys. J., **88**, 84.
Suntzeff, N.B., Kinman, T.D. and Kraft, R.P., 1991, Astrophys. J., **367**, 528.

Discussion:

M. Breger: You mentioned that the amplitudes of the RR Lyrae stars detected in your survey fields are very low. Are the amplitudes unusually low relative to other field or cluster variables?

T.D.Kinman: Field RR Lyrae stars are usually discovered by blinking survey plates taken with wide-angle Astrographs or Schmidt telescopes which have plate scales of 50 to 100 arcsec mm^{-1}. With this equipment, variables of lower amplitude (say B-amplitude less than 0.75 magnitudes) are not found with any completeness. Surveys for RR Lyrae stars in globular clusters are usually made on reflector plates of better plate scale and variables of lower amplitude can be detected; in this case most of the c-type variables will probably be discovered. The technique of using the AF stars allows us to acheive a completeness for the discovery of the low amplitude variables in the field which should be at least as good as that in the well-searched globular clusters.

Fundamental Mode and First Overtone Mode Cepheids in the Small Magellanic Cloud

H.A. Smith[1], N.A. Silbermann[1], S.R. Baird[2], J.A. Graham[3]

[1]*Dept. of Physics and Astronomy, Michigan State University, USA* [2]*Dept. of Physics and Astronomy, Benedictine College and the University of Kansas, USA*
[3]*Dept. of Terrestrial Magnetism, Carnegie Institution of Washington, USA*

Abstract

We report results of a new photographic survey of variable stars in a 1 x 1.3 degree region near the Northeast Arm of the Small Magellanic Cloud. We have discovered 133 new variable stars in this field and have determined periods and B lightcurves for 78 new and 72 previously known variables. At periods shorter than about 3 days, the Cepheid period-luminosity relation splits into two sequences. The brighter sequence is believed to be populated by stars pulsating in the first overtone radial mode, whereas the fainter sequence is populated by fundamental mode pulsators. The peak in the Cepheid period-frequency distribution occurs near a period of 1.8 days. The surface density of RR Lyrae stars in this field is comparable to that in an outlying SMC field near NGC 121.

1. Introduction

In the eight decades since Henrietta Leavitt (Pickering 1912) discovered that Cepheid variable stars in the Small Magellanic Cloud obey a period-luminosity relation, there have been numerous studies of Magellanic Cloud Cepheids. Most of these studies, however, have concentrated upon the brighter Cepheids, leaving the fainter Cepheids and RR Lyrae stars rather neglected. We have carried out a new photographic survey of variable stars in the SMC designed in part to remedy this neglect.

Our survey is based upon 29 B-plates obtained between 2 July and 29 September 1970 with the 1.5-m telescope at CTIO. These plates cover a 1 x 1.3 degree region and have a limiting magnitude near B = 21. Centered at 1975 coordinates $\alpha = 1^h02^m$ $\delta=-71°30'$, this photographic field is located near the Northeast Arm of the SMC and includes the cluster NGC 361 (Fig. 1). For brevity, we shall refer to this region as the NGC 361 field.

This investigation is a companion study to that reported by Graham (1975), who used similar plate material to study variable stars in an outlying field near the old SMC cluster NGC 121. Graham found many RR Lyrae stars but few Cepheids in the field around NGC 121.

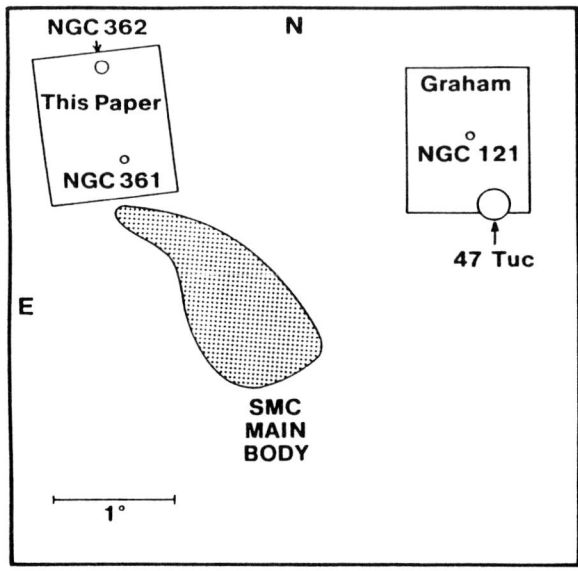

Figure 1. Diagram of the Small Magellanic Cloud, showing the locations of the outlying field studied by Graham (1975) and the NGC 361 field.

2. Variable Stars in the NGC 361 Field

By visually blinking 8 plate pairs, and by intercomparing scans of an additional plate pair obtained with the Automatic Plate Scanner at the University of Minnesota, we identified 276 candidate variables in the NGC 361 field. 209 of these were later confirmed as definitely or probably variable. 133 of these were new variable stars.

Candidate variable stars were scanned with the PDS machine at the Dominion Astrophysical Observatory. Standard stars selected from those observed by Harris (1982) and Da Costa & Mould (1986) were also scanned and used to calibrate the photographic photometry. Periods and lightcurves were determined for 78 new and 72 previously known variables.

In Figure 2 we plot $<B>$ against the logarithm of the period in days for all of the variables for which we determined lightcurves. Four groups of variables are visible in this diagram. Near 20th magnitude and having periods shorter than one day are the field RR Lyrae stars which belong to the SMC. A second group of variables with RR Lyrae-like periods extends nearly vertically from $<B>=15.5$ to 18. The brighter of these are likely foreground RR Lyrae stars in the galactic halo, though the status of the fainter members of this group is more ambiguous. Finally, there are the Classical Cepheids of the SMC which, at periods shorter than about 3 days, seem to divide into two distinct sequences of period versus luminosity.

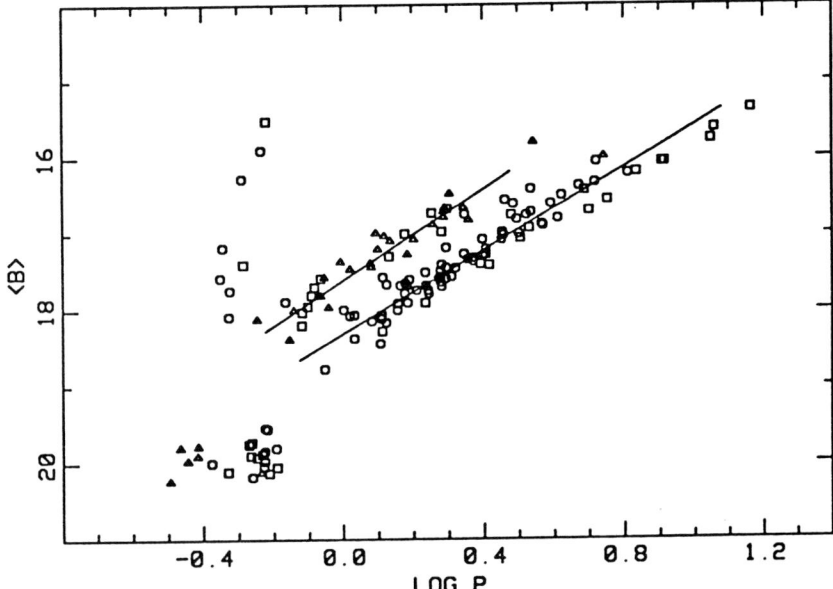

Figure 2. Period-apparent magnitude relation for variable stars in the NGC 361 field. Symbols are explained in the text.

Cepheids in the two P-L sequences are distinguished not only by differing luminosities at equal periods, but also by lightcurve shape. This difference is well illustrated in Figure 3, which depicts the lightcurves of two Cepheids of nearly equal period - one from the upper and one from the lower sequence. HV1871, the lower sequence variable, has a greater amplitude and more asymmetric lightcurve than HV11449, the upper sequence variable. In Figure 2 we have plotted as circles those Cepheids which have relatively large-amplitude, asymmetric lightcurves similar to those of ab-type RR Lyrae stars. Cepheids with smaller amplitude, more symmetric lightcurves similar to RR Lyrae stars of type c have been plotted as triangles. Stars with significant gaps in phase coverage or otherwise difficult to classify on this scheme have been plotted as squares.

By analogy with the RR Lyrae stars, we might suppose that Cepheids in the brighter sequence are pulsating in the first overtone radial mode, while Cepheids in the fainter sequence pulsate in the fundamental mode. This would be consistent

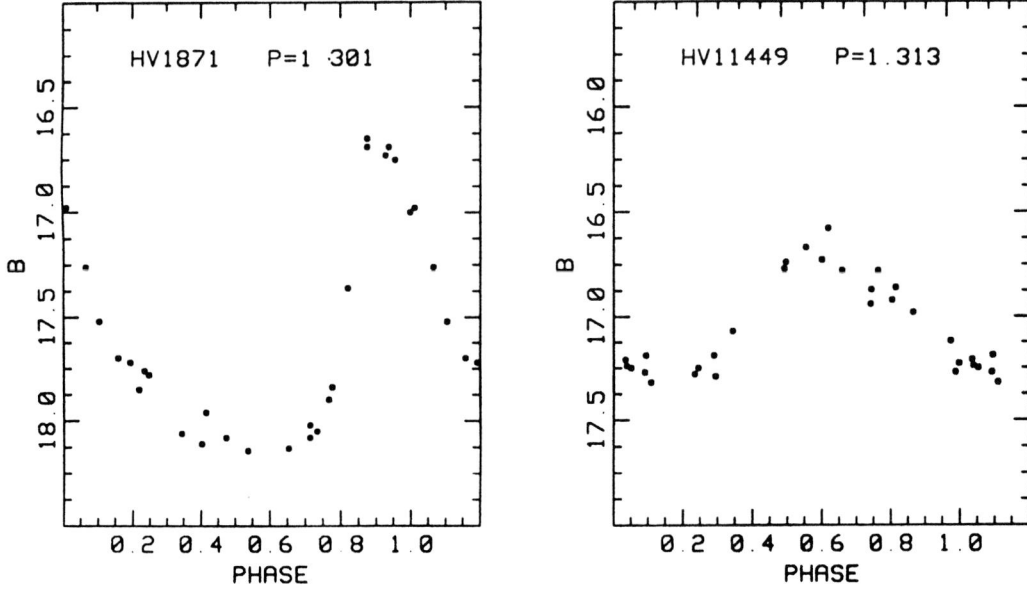

Figure 3. Lightcurves of the upper sequence Cepheid HV11449 and the lower sequence Cepheid HV1871. Both have periods near 1.3 days.

with the shift in period between the two sequences, approximately $P_1/P_0 = 0.72$. That some of the shortest period SMC Cepheids were first overtone pulsators has been suspected at least since the work of Arp (1960) and Payne-Gaposchkin & Gaposchkin (1966). That case now seems very strong.

3. Period-Luminosity Relations

Linear least squares fits to the two Cepheid sequences give the relations

$$<B> = 18.29(\pm 0.04) - 2.71(\pm 0.09) \log P_0 \qquad (1)$$

$$<B> = 17.61(\pm 0.04) - 2.98(\pm 0.17) \log P_1 \qquad (2)$$

which are shown in Figure 2. The standard deviations around relations (1) and (2) are about 0.18 mag. This relatively small scatter is indicative of small differential reddening and modest back-to-front depth among Cepheids in the NGC 361 field.

Previous work indicates that reddening is relatively small in the NGC 361 field (Da Costa & Mould 1986, Harris 1982). We assume $E(B-V) = 0.06 \pm 0.03$. RR Lyrae stars in the SMC appear to be about 0.4 ± 0.1 mag fainter than those in the LMC (this paper, Graham 1975, 1977, Nemec & Hazen 1992, Walker 1991). If this difference can be entirely attributed to distance, and if we adopt an LMC distance modulus of 18.5, we can transform apparent to absolute magnitudes for the SMC variables. The observed P-L relations then become

$$< M_B > = -0.87(\pm 0.21) - 2.71(\pm 0.09) \log P_0 \quad (3)$$

$$< M_B > = -1.55(\pm 0.21) - 2.98(\pm 0.17) \log P_1 \quad (4)$$

Relations (3) and (4) remain, however, somewhat tentative.

Nemec et al. (1988) found that the so-called anomalous Cepheids in dwarf spheroidal galaxies fell along two distinct period-luminosity relations. They interpreted these as indicating the presence of both first overtone and fundamental mode pulsators among the anomalous Cepheids. The period-luminosity relations for these anomalous Cepheids (Nemec 1989) are not, however, identical to those found here for the SMC classical Cepheids. Because most stars in dwarf spheroidal systems are very old, it has been suggested that the anomalous Cepheids, with masses near 1.5 solar masses, are not evolved single stars, but are instead coalesced binaries (Renzini, Mengel, & Sweigart 1977). A range of ages, from old to young, is present in the SMC, so that no such hypothesis is necessary to explain the large majority of the short period Cepheids in the NGC 361 field.

4. Period-Frequency Distribution

The period-frequency distribution for variables which are believed to be

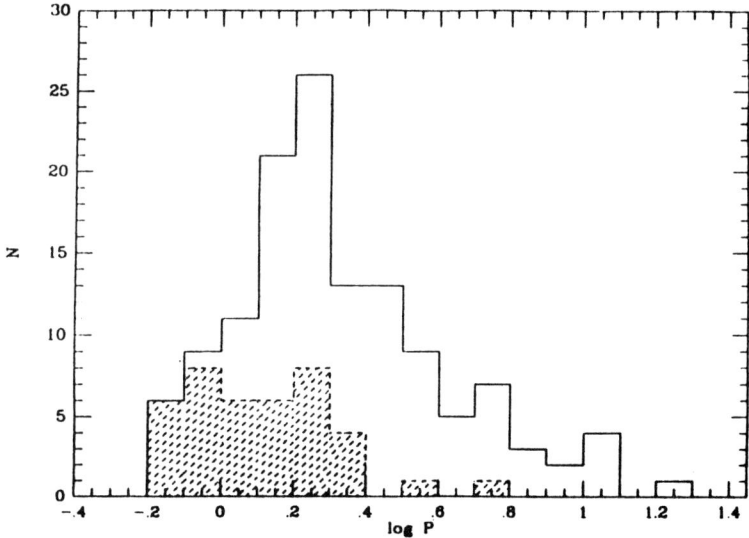

Figure 4. Period-frequency distribution for Cepheids in the NGC 361 field. The hatched area denotes Cepheids which are belived to pulsating in the first overtone mode

SMC Cepheids is shown in Figure 4. Incorporated into this figure are data for 12 additional Cepheids which fall within the NGC 361 field but which, for various reasons, were not measured by us. For these stars, the decision as to whether a Cepheid is a first overtone or fundamental mode pulsator is based upon the lightcurves in Payne-Gaposchkin & Gaposchkin (1966).

The peak in Figure 4 occurs near a period of 1.8 days. This is consistent with the studies of Dessy (1959) and Wesselink & Shuttlewoth (1965), which found similar peaks near a period of 1.5 days. It is, however, somewhat at odds with the extensive survey of Payne-Gaposchkin & Gaposchkin (1966). The Payne-Gaposchin & Gaposchkin survey did find a secondary peak in the SMC period-frequency distribution near 1.6 days, but also found a higher peak near 3 days. It seems quite possible that the 3 day peak is not real, but is instead a consequence of the limiting magnitude of the Harvard plate material used by Payne-Gaposchkin & Gaposchkin. With a limiting magnitude near B = 18, much of the Harvard plate material may not go deep enough to reveal the faintest SMC Cepheids, thus underrepresenting the frequency of short period Cepheids.

5. RR Lyrae Stars

We have identified 42 probable RR Lyrae stars of the SMC within the NGC 361 field, although we have been able to determine periods for only 22 of these. Of those with periods, 17 appear to be RRab variables, while only 5 are RRc stars. The period-amplitude diagram for these RR Lyrae stars is similar to that found by Graham for RR Lyraes in the more outlying NGC 121 field.

Our survey of RR Lyrae stars in the NGC 361 field is probably seriously incomplete. An approximate correction for this incompleteness indicates that the NGC 361 field actually contains perhaps 80-85 RR Lyrae stars. This number is comparable to that obtained by Graham (1975) for RR Lyraes in the more outlying NGC 121 field. Excluding RR Lyraes which are members of NGC 121 itself, and those stars with RR Lyrae-like periods which are probably short period Cepheids, Graham's results indicate that there are about 90-95 RR Lyraes in the NGC 121 field. We thus confirm Graham's (1975) conclusion that RR Lyrae stars are not strongly concentrated to the main body of the SMC.

Future Observations

This study leaves many questions unanswered: What are the colors of the short period Cepheids? Do the fundamental mode and first overtone mode Cepheids overlap in color at a given luminosity? What are the periods of the variables identified, but still without reliable lightcurves? In an attempt to answer some of these questions, Alistair Walker and H. Smith plan to obtain B and V CCD photometry of most of the variables in the NGC 361 field. This photometry will be obtained with a CCD on the Schmidt telescope at CTIO.

Acknowledgements

This work has been supported in part by a Gaposchkin Research Award of the AAS and by the National Science Foundation under grant AST89-15422.

References:

Arp, H., 1960, Astron. J, **65**, 404.
Da Costa, G.S. & Mould, J.R., 1986, ApJ **305**, 214.
Dessy, J.L., 1959, **71**, 435.
Graham, J.A., 1975, PASP, **87**, 641.
Graham, J.A., 1977, PASP **89**, 425.
Harris, W.E., 1982, Ap.J. Suppl. **50**, 573.
Hazen, M.L. & Nemec, J.M., 1992, Astron. J. **104**, 111.
Nemec, J.M., 1989, IAU Colloq. No. 111, *The Use of Pulsating Stars in Fundamental Problems of Astronomy*, ed. E.G. Schmidt, Cambridge University Press, Cambridge, p. 215.
Nemec, J.L., Wehlau, A., & Mendes de Oliveira, C. 1988, Astron. J. **96**, 528.
Payne-Gaposchkin, C. & Gaposchkin, S. 1966, Smith. Contr. Ap. **9**, 1.
Pickering, E.C., 1912, Harvard Coll. Obs. Circ. No. 173, 3.
Walker, A.R. 1991, IAU Symp. 148, *The Magellanic Clouds*, ed. R. Haynes and D. Milne, Kluwer, Dordrecht, p. 307.
Wesselink, A.J., & Shuttleworth, M., 1965, MNRAS **130**, 443.

Discussion

G. KOVACS: Did you observe any bump progression among the fundamental mode Cepheids? Because of the resonance, a comparison with the Galactic Cepheid bump progression would constrain the chemical composition and M-L relation substantially.

SMITH: There is indeed some evidence for a progression of lightcurve shape with period among the fundamental mode Cepheids. However, some of our lightcurves still have gaps, and we would like to obtain additional data before drawing conclusions from this. It is also worth remembering that the Payne-Gaposchkin & Gaposchkin (1966) survey contains excellent lightcurves for SMC Cepheids.

D. WELCH: The intrinsic width of your <B> P-L relation for fundamental mode stars is so small that the line-of-sight depth in this field must be very small, which would be in contradiction to the large bulge giant scatter seen in the northeast portion of the SMC by Hatzidimitriou and Hawkins (MNRAS **241**, 645).

SMITH: This contradiction might have several explanations. The fields examined for bulge giants, although in the northeast part of the SMC, are somewhat farther from the main body of the SMC than is the NGC 361 field. Moreover, the bulge giants may belong to an older population than most of the Cepheids we have observed. The issue will need further observations to resolve.

S. HUGHES: Will you be obtaining I-band CCD frames in order to estimate individual extinctions to the Cepheids?

SMITH: It would indeed be very interesting to have I-band photometry for the Cepheids. However, our first CCD observations will be limited to the B- and V-bands.

D. TURNER: Presumably there are also a few double-mode Cepheids in the survey field. Could these stars lie in your sample of objects for which no definitive period was found?

SMITH: That is very likely the case, given our relatively small number of observations.

Resonance Effects in Fundamental Mode, First-Overtone, and Double-Mode Cepheids

Elio Antonello

Osservatorio Astronomico di Brera,
via E. Bianchi 46, I-22055 Merate (CO), Italy

Abstract

Linear adiabatic periods, period ratios and frequencies of Cepheid models with P_0 less than about 10 days have been computed, taking into account standard and nonstandard mass–luminosity relations and the new (or augmented) opacities. A comparison of the results with the observed properties of Cepheids has yielded the following conclusions:

(a) A non-standard mass–luminosity relation is needed in order to have good agreement between observed stars and models on the temperature–period (or luminosity) diagram; 'standard' models have much lower temperatures and luminosities than observed stars.

(b) The linear models predict the following resonances: $f_2/f_0 = 2$ (well known) at $P_0 \sim 10$ days, $f_4/f_0 = 3$ at $P_0 \sim 6.8$ days, $f_4/f_1 = 2$ at $P_1 \sim 3.2$ days, and $f_1 + f_0 = f_3$ at $P_0 \sim 6.5$ days.

(c) All these resonances yield effects which have been observed in light curves of Cepheids *at the predicted periods*: $f_2/f_0 = 2$ and $f_4/f_0 = 3$ in classical Cepheids; $f_4/f_1 = 2$ in first-overtone Cepheids; and $f_4/f_1 = 2$ and $f_1 + f_0 = f_3$ in double-mode Cepheids.

The discussion will include comparison with the results obtained with nonlinear models by Moskalik *et al.* (1992) for fundamental-mode Cepheids, and by Aikawa (1992) for first-overtone Cepheids.

References:

Aikawa T. 1992, in *Nonlinear Phenomena in Stellar Variability*, eds. M. Takeuti & J.R. Buchler, I.A.U. Coll. 134 (Mito, Japan), in press.

Moskalik P., Buchler J.R. & Marom A. 1992, ApJ, 385, 685.

Variable Stars in the Sculptor Dwarf Galaxy

C. G. Goldsmith

York University, North York, Ontario, Canada

1. Introduction

This project was initiated in 1985 by James Nemec (University of Washington) and Nicholas Suntzeff (C.T.I.O.). The goal was to study the system of ~600 variable stars in the Sculptor dwarf galaxy. In 1987 the author became the recipient of the plate collection, which formed the basis for his Ph.D. dissertation. In this paper preliminary results are presented. Briefly, 612 stars were studied, of which 432 are van Agt (1978) stars and 180 are newly discovered variable stars. A total of 381 stars are confirmed variables. Most of these are RR Lyraes, but many anomalous Cepheids and some candidate eclipsing variables were also found. Several candidate double-mode RR Lyrae stars were also identified. The mean period of the ab-type RR Lyrae stars is 0.60±0.08 day, and the mean period of the c-type stars is 0.35±0.03 day, not unlike the mean periods of other nearby dwarf galaxies.

2. Period-Amplitude Diagram

Figure 1 shows the period-amplitude diagram for the RR Lyrae stars in Sculptor. The ab-types show the characteristic decrease in amplitude with increasing period, the longest period being ~0.80 day, characteristic of a very low metal abundance. Also, the P-A_B relationship is significantly wider at a given amplitude than that for globular clusters (Sandage 1981), probably largely due to a range in metallicity for the stars. The slope of P-A_B relation appears to be shallower than that for globular clusters. Several stars may also be misclassified in this diagram.

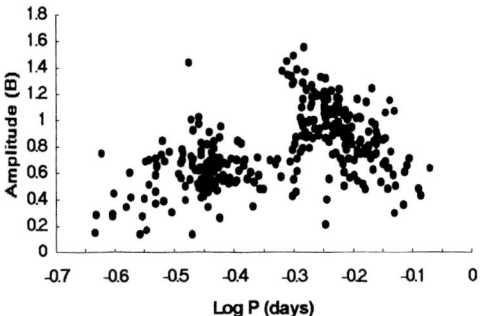

Figure 1. Period-amplitude diagram for the Sculptor variables.

References:

Sandage, A., 1981, Astrophys. J. **248**, 168.

van Agt, S., 1978, Publ.David Dunlap Observatory, Vol.3 (number 7).

Cepheids in Magellanic Cloud Clusters

D.L. Welch[1], M. Mateo[2], E.W. Olszewski[3]

[1] *McMaster University, Hamilton, Ontario, Canada* [2] *Observatories of the Carnegie Institution of Washington, Pasadena, CA* [3] *Steward Observatory, University of Arizona, Tucson, AZ*

Abstract

The Magellanic Clouds remain an ideal place to study the properties of Cepheid variables. In this paper, we review historical and current work on Cepheids in LMC and SMC clusters, present new results for NGC 1866 and NGC 2164, and describe a new technique for automated selection of Cepheid variables using two-color photometry. We also emphasize the numerous advantages of high-precision radial velocities in the study of Magellanic Cloud variables.

1. Introduction and Historical Review

It has been over forty years since Thackeray (1951) first called attention to the Cepheid variables associated with the LMC cluster NGC 1866. Yet it has only been possible to study these stars with sensitive panoramic linear detectors in the last decade. In this paper, we will review the work to date and report new results for the young cluster NGC 2164. Furthermore, we highlight how current software developments are likely to influence work in the coming years and provide a glimpse of a highly sensitive, yet automated means of detection variables from photometry lists.

2. Modern Surveys

The first intensive studies of Cepheid variables in a Magellanic Cloud cluster were published by Arp and Thackeray (1967) and Arp (1967) for NGC 1866. Not long thereafter, Robertson (1974) published photoelectrically calibrated photometry of clusters listed in Table 1. Some periods were mentioned for probable Cepheid variables, but no lightcurve data were listed. More recently, Walker (1987) published BVI lightcurves for seven NGC 1866 Cepheids. In Table 1, Cepheid variables are listed in boldface and red supergiant variables in slant font.

Table 1
Variables Reported in Robertson (1974)

SMC	NGC 330	None
LMC	NGC 1818	*D3*
	NGC 1850	**D34, A86**
	NGC 1854	**A83**
	NGC 1866	No new variables
	NGC 2004	None
	NGC 2100	*B4, C32*
	NGC 2136	**HV 12230, 2868, 2870, B21**
	NGC 2157	None
	NGC 2164	None
	NGC 2214	**B1**

3. Current Status

To date, we have obtained BVRI CCD photometry of young clusters in the LMC and SMC during three (or more) observing seasons using the 1.0m Swope reflector at Las Campanas, Chile. (Some of the earliest epochs were obtained with the CTIO 0.9m or 4.0m telescopes). Typically, we have more than fifty BV pairs per cluster. These data were reduced using the photometric reduction routine DoPHOT (see Mateo and Schechter 1989) which is nearly ideal for the required reductions due to the minimal interaction required by the researcher. In all, more than 50 Gbytes of data have been reduced for this program.

In Table 2, we list the clusters which have been included in our survey or reported previously. This table contains only data for Cepheid variables. The columns N_C and N_F refer to the number of cluster and field Cepheids found, respectively. The distinction between these categories is not extremely well-defined — a case in point being the 10.8-day Cepheid found near NGC 2156. In all, some 60–70 Cepheids have been found in or near clusters. Note that this is a significant fraction of all Magellanic Cloud Cepheids with CCD or photoelectric photometry, recently estimated to be about 141 for the LMC and 186 for the SMC by Caldwell and Laney (1991). A '?' indicates that the data is reduced but that a careful search for variables has not yet been undertaken.

Table 2
Magellanic Cloud Cluster Cepheid Statistics

Object	Cluster	N_C	N_F	Comments
SMC	NGC 330	0	1	Balona 1992
	NGC 376	?	≥ 4	
LMC	NGC 1711	?	?	
	NGC 1755	2	0	
	NGC 1774	?	?	
	NGC 1818	0	0	
	NGC 1850	≥ 1	?	
	NGC 1854	1	0	
	NGC 1866	21	1	Welch et al. 1991
	NGC 2004	0	0	Balona 1992
	NGC 2010	4	1	
	NGC 2025	0	1	
	NGC 2031	14	0	Olszewski, Mateo and Madore 1991
	NGC 2041	0	0	
	NGC 2100	0	0	Balona 1992
	NGC 2134	2	0	
	NGC 2136	6	0	Olszewski, Mateo and Madore 1991
	NGC 2156	1	0	
	NGC 2157	3	0	Mateo, Olszewski, and Madore 1990
	NGC 2164	1	1	Welch et al. 1993
	NGC 2214	1	0	

4. New Results for NGC 1866

Photometric precision in Magellanic Cloud cluster work is limited by two factors: starlist differences from night-to-night and estimation of the underlying sky. A new routine written by Peter Stetson called ALLFRAME has recently been tested on NGC 1866 frames and promises to significantly improve our ability to work under the very crowded conditions in the cores of young Magellanic Cloud clusters. It produces photometry based on a single, optimized starlist and estimates sky 'under' the image of the star (rather than in an annulus around the star). In this test, 29 B and 27 V frames were reduced together. These frames were a subset of the frames used by Welch et al. (1991) which were selected for being centered on the cluster and representing many different observing nights.

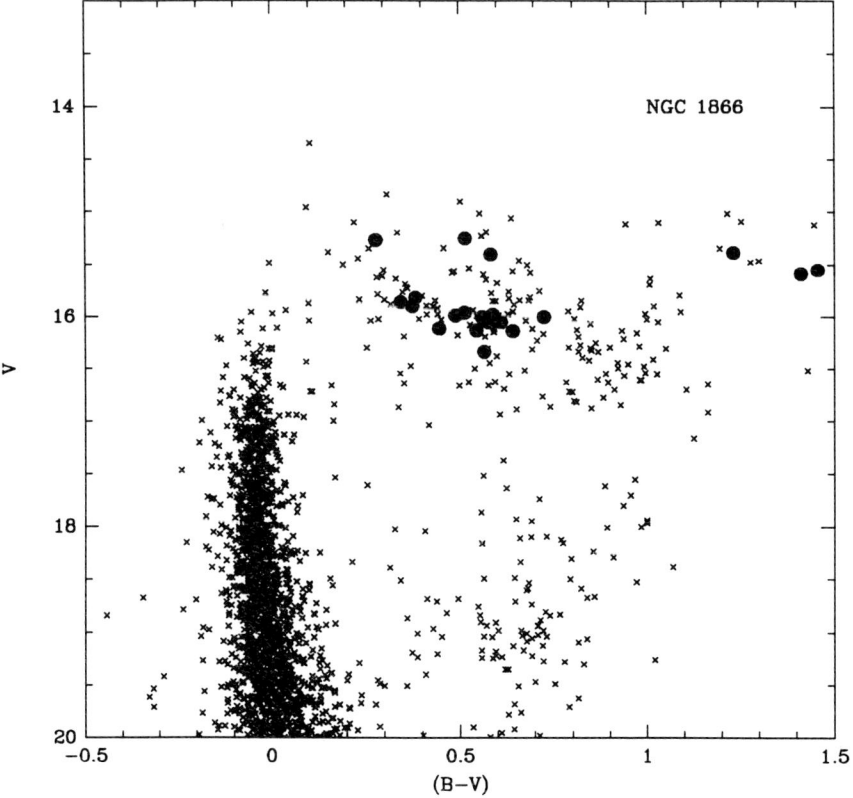

Figure 1 The color-magnitude diagram for NGC 1866 from 29 B and 27 V frames, using Peter Stetson's ALLFRAME. The filled circles are variable stars.

Our results can be summarized as follows:
- Periods for *all* reported Cepheids.
- Two additional low-amplitude Cepheids.
- One erratum: We 1 = V 7.
- Large number of apparently stable stars in the CIS.

A color-magnitude diagram for this field is shown in Figure 1, with variables indicated. B and V lightcurves for one of the newly discovered low-amplitude Cepheids are shown in

Figure 2. These results will be reported in more detail by Welch and Stetson (1993).

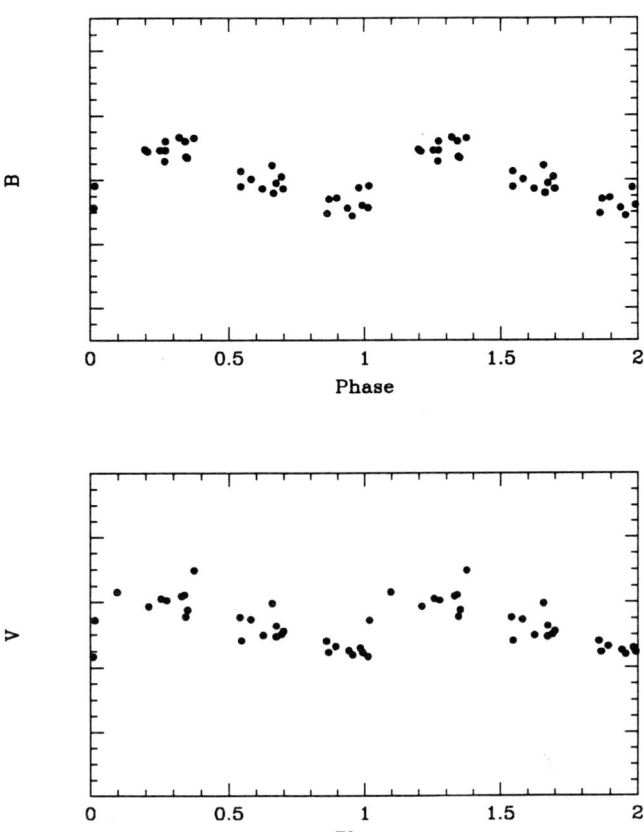

Figure 2 A previously unknown Cepheid in the central regions of NGC 1866, discovered with Peter Stetson's ALLFRAME. This variable is the bright, blue variable in Figure 1. The lightcurve is almost certainly being diluted by the light of an upper main sequence star. Major tick marks on the magnitude axis are separated by 0.2 mag.

5. Radial Velocities

In the last few years it has become possible to obtain high-dispersion spectra of Magellanic Cloud Cepheids from which, using cross-correlation techniques, radial velocities with errors of order 1 km s^{-1} can be extracted. Such velocity precision can be obtained for Cepheids of all periods. Among the uses for high-precision radial velocities are:

- Membership — the internal velocity dispersion of most young Magellanic Cloud clusters is ≤ 1 km s^{-1}, whereas the line-of-sight velocity dispersion of the massive disk population is about 5 km s^{-1}.

- Binarity — Several Cepheids are already known to be single-lined spectroscopic binaries.

- Period Determination — When crowded by main-sequence stars, the cross-correlation

peak is only minimally affected by photon noise. Hence there may be significantly greater dynamic range in the radial velocity curve than the lightcurve.
- Baade-Wesselink Analyses — Providing that light and color curves of sufficient quality are available, fundamental properties of Magellanic Cloud Cepheids can be directly compared with their Milky Way counterparts.
- Mode Identification — The radial velocity curve can be studied to extract mode information in the same way as the lightcurve. Under crowded conditions, the radial velocity curves are less effected.

A velocity curve for the 7.7156-day variable NGC 2157-2 is shown in Figure 3.

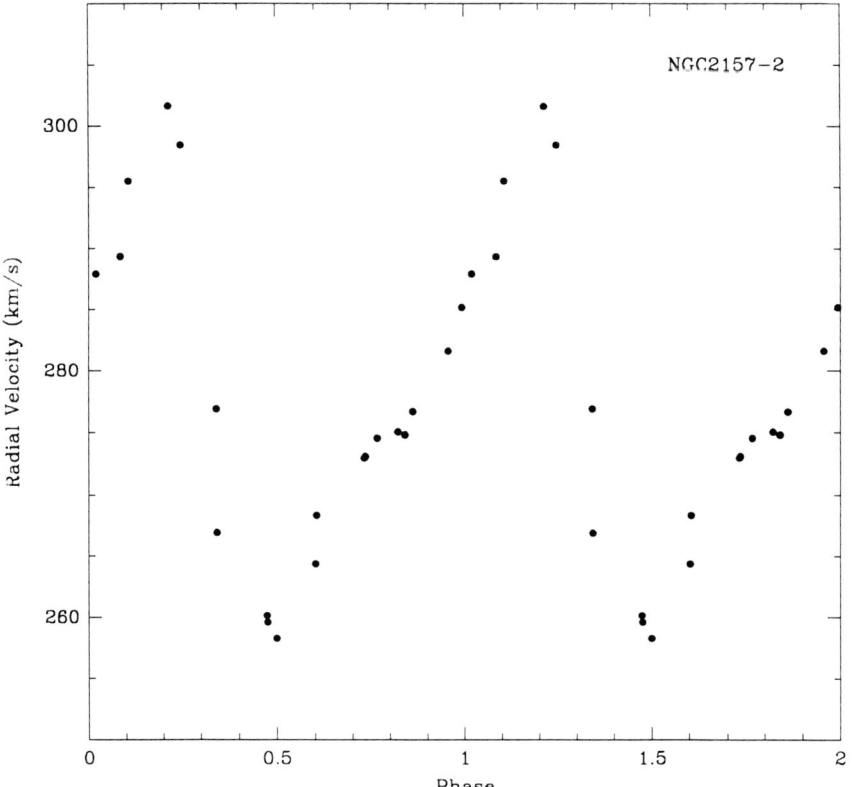

Figure 3 The radial velocity curve for NGC 2157-2, obtained using the 2D-Frutti plus echelle spectrograph on the 2.5m Dupont reflector at Las Campanas, Chile.

4. New Results for NGC 2164

We have recently completed a survey of the young cluster NGC 2164. The turnoff mass for this cluster has been estimated to be 7.7 $M_\odot$ by Elson, Fall, and Freeman (1989), so we expect the Cepheids to have longer fundamental periods than those in NGC 1866 where the turnoff mass is closer to 5 $M_\odot$.

Our survey made use of 64 B and V frames taken over the 1989, 1990, and 1991 observing seasons. Our results can be summarized as follows:

- A cluster member Cepheid — B11 in Robertson (1974) — which is a first overtone pulsator with $P_1 = 3.772$ days, confirmed to be a member with radial velocities.
- A new field first overtone pulsator with a period of 3.4262 days.
- Full phase coverage of the known 10.6878 day Cepheid HV 12078 which is sufficiently close to NGC 2156 that it may be a member.
- A measurement of the Fourier phase $\phi_{21} = 3.1$ radians for B11 which suggests that it is a so-called long-period s-Cepheid and confirms the assertion of Antonello, Poretti, and Reduzzi (1990) that such a distinct class of overtone pulsators exists.

A complete description of this work is contained in Welch et al. (1993).

5. Search Algorithms for Pulsating Stars

Search techniques for variables stars have evolved little with the years. The traditional technique of 'blinking' is still widely used with CCD frames. Other well-known algorithms are: 1) 'scatter' searches, where every star showing an unusual degree of photometric scatter is examined, and 2) 'brute force' searches, where every star of appropriate magnitude and color is examined. In an era of large detector format, large surveys, and short proprietary periods, it would be especially useful if a sensitive and selective technique for detecting pulsating variables from photometry lists were available.

One such technique has been identified which we call 'surface brightness change'. Simply put, the magnitude and color change of a star are correlated if the variation in flux is primarily the result of a change in surface brightness. This is not true of random photometric error. The most trivial demonstration of this technique involves combining two-color photometry from two epochs and examining the pattern of residuals. In Figure 4, we plot the V and (B–V) residuals for two sets of frames containing the SMC cluster NGC 376. Plotted in this way, random photometric errors result in an elliptical error distribution (since the V photometry appears on both axes). However, the objects which have changed their surface brightness scatter in a direction which is essentially uncontaminated by random errors. In this simple two-epoch comparison, probable Cepheid variables can be selected with changes in V mag as small as 0.10–0.15 mag. This compares to the 0.6–0.8 mag variation typically required for detection by blinking. Surface brightness change detection can be simply extended to an arbitrary number of epochs and the sensitivity increases correspondingly.

A full description of the surface brightness change algorithm (plus applications) will appear in Welch and Stetson (1993).

References:

Antonello, E., Poretti, E., and Reduzzi, L., 1990, Astron. Astrophys. 236, 138.
Arp, H., 1967, Astrophys. J. **149**, 91.
Arp, H., and Thackeray, A.D., 1967, Astrophys. J. **149**, 73.
Balona, L.A., 1992, Mon. Not. Royal Astron. Soc., in press.
Caldwell, J.A.R., and Laney, C.D., 1991, in: *IAU Symposium 148 The Magellanic Clouds*, eds. R. Haynes and D. Milne, (Kluwer: Dordrecht), 249.
Elson, R.A.W., Fall, S.M., and Freeman, K.C., 1989, Astrophys. J. **336**, 734.
Mateo, M., Olszewski, E.W., and Madore, B.F., 1990, Astrophys. J. Lett. **353**, L1.

Mateo, M., and Schechter, P., 1989, Proceedings of the 1st ESO-ECF Data Analysis Workshop, edited by P.J. Grosbøl, F. Murtagh, and R.H. Warmels, (European Southern Observatory, Garching)

Olszewski, E.W., Mateo, M., and Madore, B.F., 1991, in: *The Formation and Evolution of Star Clusters*, Astr. Soc. of Pacific Conf. Series **13**, 588.

Robertson, J.M., 1974, Astron. Astrophys. Suppl. **15**, 261.

Thackeray, A.D., 1951, Mon. Not. Royal Astron. Soc. **111**, 206. **353**, L1.

Walker, A.R., 1987, Mon. Not. Royal Astron. Soc. **225**, 627.

Welch, D.L., Mateo, M., Côté, P., Fischer, P., and Madore, B.F., 1991, Astron. J. **101**, 490.

Welch, D.L., Mateo, M., Olszewski, E.W., Fischer, P., Takamiya, M., 1993, Astron. J. (submitted).

Welch, D.L., and Stetson, P.B. 1993, in preparation.

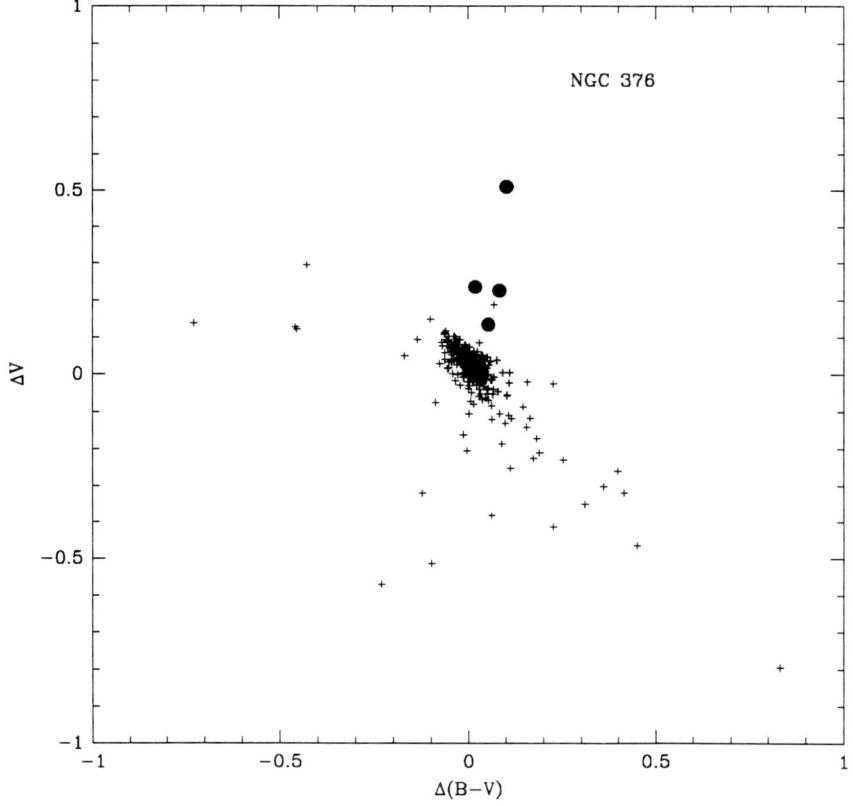

Figure 4 The magnitude and color residuals for the SMC cluster field taken on two consecutive nights. The five known Cepheids are indicated as filled circles (one of which appears in the thick of the zero-residual clump). The true variables are cleanly separated from random photometric scatter.

P. Moshalik: Do you have Fourier coefficients for 1H Cepheids, and how well do they fit the s-Cepheid progression of Antonello? Are there any discrepant 1H Cepheids?

D. Welch: I have Fourier coefficients for the 1H Cepheids in both NGC 1866 and NGC 2164. Their ϕ_{21} values lie on the s-Cepheid progression. We should have coefficients for the other cluster 1H Cepheids shortly.

D. Fernie: If Polaris (amplitude 0.02 or 0.03 mag) were in NGC 1866, would you have detected it as a Cepheid? What is the limiting amplitude of detectability?

D. Welch: There are really two aspects to this question which need to be addressed separately. First, the amount of noise introduced into the photometry of a given star will depend both on the mean seeing of the data set and the position of the star in the cluster. In the innermost regions of NGC 1866, Polaris would not be detected with our data. Second, in uncrowded conditions, Polaris could almost certainly be detected to be variable using the surface brightness change technique, but a large (prohibitive?) number of epochs might be required to establish the period.

J. Nemec: What are the SWB classes or ages of the clusters in your sample?

D. Welch: The SWB classes are I–III, although the best hunting is in cluster types II and III where a combination of richness and evolutionary timescale produces the largest number of stars in the Cepheid Instability Strip.

This work was supported in part by an Natural Sciences and Engineering Research Council (NSERC) Individual Operating Grant to DLW and was undertaken while he was an NSERC University Research Fellow. MM was partially supported by grant #HF-1007.01-90A awarded by the Space Telescope Science Institute which is operated by the Association of Universities for Research in Astronomy, Inc., for NASA under contract NAS5-26555. EWO was partially supported by NSF grants AST86-11405 and AST91-19343, and thanks Peter Strittmatter for additional support. DoPHOT was developed under grant AST 83-18504 from the National Science Foundation

CCD Observations of Forgotten Cepheids

David L. DuPuy[1], W. Todd Viar[1]
Raymond H. Bloomer[2], Bradley J. Ward[2]

[1]*Department of Physics and Astronomy*
Virginia Military Institute, Lexington, Virginia 24450
[2]*Department of Physics*
U.S. Air Force Academy, Colorado Springs, Colorado 80840

Abstract

CCD observations have been obtained of eight Cepheids which have few recent observations, or an uncertain period, or an uncertain finder chart. All CCD images were flat-fielded, with 4-min exposures (0.5-m telescope) or 6-min exposures (0.4-m telescope). The stars chosen for observation were CY Aqr, EV Aur, V395 Cas, V588 Cas, DW Per, HZ Per, MM Per, and SX Per.

Photometry shows clean, Cepheid light curves for six of these eight stars, with the results for MM Per shown as an example below. V588 Cas did not show a periodicity in our data set (only 9 points), and the results for EV Aur will be published separately. We obtained standard deviations of comparison-star differences for all useable stars in our CCD images, to aid future observers in choosing suitable comparisons.

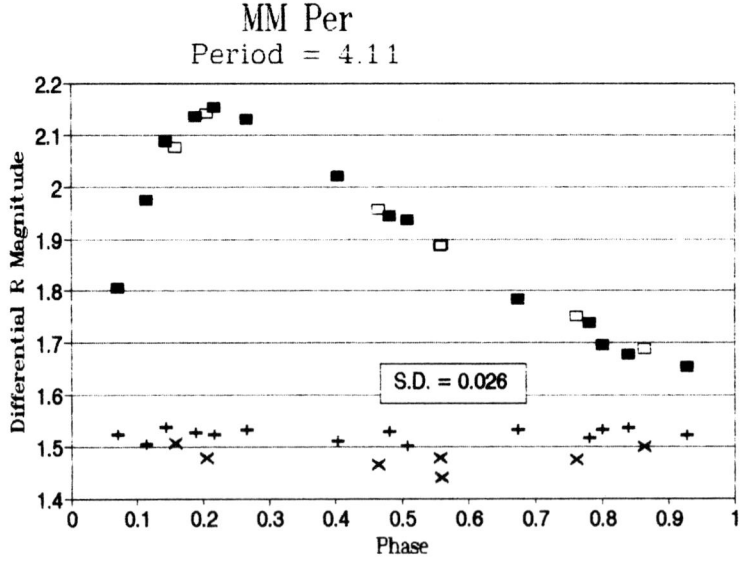

CCD Photometry of Faint Cepheids

Arne A. Henden
The Ohio State University, Columbus, OH

Abstract

The parameters and output products of a cepheid survey are presented.

A faint cepheid survey was begun seven years ago as a follow-on to the survey by Henden (1979,1980). The new survey studies those northern hemisphere cepheids fainter than V=12, with periods shorter than 5 days, and with little or no existing UBV photometry. The goals of the survey are: to complete a statistical sample; to generate uniform finding charts and comparison stars for each variable; to look for anomalies in periods and amplitudes; and to improve periods, epochs and coordinates for all variables.

Approximately 120 stars are included in the survey. These have now been observed using various CCD systems and telescopes, with at least four carefully chosen BVRI data points for each star. An attempt has been made to obtain data at pulsation phases 0.0, 0.25, 0.50 and 0.75 over consecutive cycles. Approximately 2000 BVRI data sets from 60 nights are contained in the survey. The initial paper highlights 36 stars for which sufficient published photometric data exists to permit a comparison. Berdnikov (1986,1987) has been especially active in performing a parallel survey with BVR photoelectric photometry, and our data sets are consistent after his R data has been transformed into the standard system.

The finding charts were generated from deep V CCD frames, and have a scale of 0.8 arcsec/pixel with a field width of 4 arcmin. The J2000.0 coordinates of all identified stars were obtained from the HST Guide Star Catalog. Periods were improved either using a theta-minimization program or by fitting Fourier coefficients. Photometry was taken in Johnson BV and Kron-Cousins RI, with calibration through the use of Landolt and KPNO VCAM standards. Light curves for all stars are given.

The first 36 survey stars show no major deviation from classical light curves. The periods for five of the stars (V526 Aql, HK Cas, V1033 Cyg, V1046 Cyg, and DW Per) have been improved. An earlier finding chart for IN Aur is in error. Of the remaining survey stars, about 60 have well-determined periods and will be examined next. The remainder have poor periods or have been classified as questionable cepheids and will be published last.

References:

Berdnikov, L. N., 1986, Variable Stars **22**, 369.
Berdnikov, L. N., 1987, Variable Stars **22**, 530.
Henden, A. A. 1979, Mon. Not. R. astr. Soc. **189**, 149.
Henden, A. A. 1980, Mon. Not. R. astr. Soc. **192**, 621.

First overtone pulsators among Cepheids

L. Mantegazza[1], E. Poretti[2]

[1] *Università degli Studi, Pavia, Italy*
[2] *Osservatorio Astronomico di Brera, Merate, Italy*

The Fourier decomposition has been successfully applied to several classes of pulsating variables. Antonello and Poretti (1986) and Antonello et al. (1990a) applied it to the Cepheids with $P < 8$ d. The latter authors redefined the $s-$Cepheids as Population I Cepheids that do not follow the Hertzsprung progression, but have a progression of their own. The same authors proposed a new denomination (Antonello et al., 1990b): C-a stars to indicate the Classical Cepheids and C-b stars to indicate the redefined $s-$Cepheids.

The new photometric data obtained at La Silla and Merate Observatories (Mantegazza and Poretti, 1992) increase the evidence of a separation of Cepheids into two well defined subclasses on the basis of the Fourier parameters of their light curves.

In the $\phi_{21} - P$ plane, the $s-$ and Classical Cepheids are characterized by two sequences well separated for P<5.5 d. In the period range 3 d< P <5.5 d, two different progressions are also present in the $\phi_{31} - P$ plane while a discriminating value $R_{21} = 0.20$ can be seen in the $R_{21} - P$ plane. In the $R_{31} - P$ plane the separation is well defined only for $P <$ 4.5 d, with a discriminating value $R_{31} = 0.08$.

Besides the identification of new first overtone pulsators located on the upper $s-$Cepheid sequence, we can point out that in the $\phi_{21} - P$ plane the increasing dispersion around 3 d strengthens the hypothesis that a resonance with a higher overtone abruptly stops the regular upper progression; consequently the Cepheids located on the lower sequence should also be first overtone pulsators. The common nature of all these stars finds another observational support in the relationships noticeable in the $R_{21} - P$, $\phi_{31} - P$, $R_{31} - P$ planes, now better established on the basis of the new available light curves.

References:

Antonello, E. and Poretti, E., 1986, Astron. Astrophys. **169**, 149.
Antonello, E., Poretti, E., Reduzzi, L., 1990a, Astron. Astrophys. **236**, 138.
Antonello, E., Poretti, E., Reduzzi, L., 1990b, in: *Confrontation between Stellar Pulsation and Evolution*, Astr. Soc. of Pacific Conf. Series **11**, p 209.
Mantegazza, L., Poretti, E., 1992, Astron. Astrophys., in press.

On Period Ratios and Resonance Sequences

J.O. Petersen

Copenhagen University Observatory, Denmark

Understanding Period Ratio Diagrams

Basic features of calibrations of period ratio diagrams in terms of mass or radius can be understood from two facts: (1) The pulsation Q-parameter for the fundamental mode of Cepheid-type models increases by a factor of about 3 from periods of about 0.05 days to 100 days, whereas Q for overtones vary much less. Therefore period ratios usually decrease with increasing pulsation period in a group of variable stars or in a model sequence (calibration curve). (2) Mass over radius (M/R) is a good pulsation parameter determining Q-values and therefore also period ratios (e.g. Cogan 1970). Using the pulsation criterion with constant Q, it is easy to show that this implies that periods of models with a fixed period ratio, e.g. 0.70, to a first approximation is proportional to both M and R. Thus all calibration curves in period ratio diagrams should move towards increasing period with increasing M or R. This is confirmed in numerous cases.

Resonance Sequences in Population I and II Variables

The center of the Hertzsprung progression, observed at about 10 days in classical Cepheids, corresponds to the 2:1 resonance between the fundamental mode and the second overtone. According to our analysis above, the period of stars at such resonances should also be proportional to the mass of the relevant stars. The population II Cepheids are believed to constitute a more or less homogeneous group of variables of mass about 0.7 solar masses. Hence, we predict the center of the type II progression to occur at a period of about $10 \cdot 0.7/5$ days $= 1.4$ days. Observationally this period is not well defined. But Petersen and Diethelm (1986) give 1.4-1.6 days, which supports the above interpretation. At the present colloquium Mantegazza and Poretti (Poster P50) give strong evidence for the presence of a similar 2:1 resonance sequence among the s-Cepheids believed to oscillate in the first overtone. The center period is 3.0 days. The corresponding resonance in the pop. II oscillators should occur at about $3.0 \cdot 0.7/4$ days $= 0.52$ days. This agrees well with the resonance proposed by Petersen (1984) in first overtone RR Lyrae variables. We conclude that the simple ideas described above give a unified understanding of population I and II variables.

References:

Cogan, B.C., 1970, Astrophys. J., **162**, 139.
Petersen, J.O., 1984, Astron. Astrophys. **139**, 496.
Petersen, J.O., Diethelm, R., 1986, Astron. Astrophys. **156**, 337.

Harmonics and Coupling-terms in the Pulsation of the Double-mode Cepheid TU Cas

L. Szabados

Konkoly Observatory, Hungary

Abstract

Preliminary results of the period analysis of the double-mode Cepheid TU Cassiopeiae are given. Up to now 29 frequencies have been identified: the frequency of the fundamental mode pulsation and the first overtone, and their linear combinations. The amplitudes of the frequency constituents show temporal variations.

First results of the analysis based on the largest homogeneous photometric data set on the double-mode Cepheid TU Cas are given. The UBV observations (more than 400 data points in each color) obtained during 14 years, especially the rich sample (160 observations on 86 nights) in a single observational season (1973/74) allowed the determination of a large number of frequencies present in the Fourier-spectrum of this bright beat Cepheid.

The following frequencies have been identified:

f_0, $2f_0$, $3f_0$, $4f_0$, $5f_0$, $6f_0$,

f_1, $2f_1$, $3f_1$,

f_0+f_1, $2f_0+2f_1$, $3f_0+3f_1$,

$2f_0+f_1$, $3f_0+f_1$, $4f_0+f_1$, $5f_0+f_1$, $6f_0+f_1$,

f_0+2f_1, $3f_0+2f_1$, $4f_0+2f_1$, $5f_0+2f_1$,

f_0+3f_1, $2f_0+3f_1$, $4f_0+3f_1$,

f_1-f_0, $2f_1-f_0$, $3f_1-f_0$,

$2f_0-f_1$, $3f_0-f_1$,

where $f_0=0.46734$ c/d is the frequency of the fundamental mode oscillation, while $f_1=0.65860$ c/d is that of the first overtone.

The preliminary analysis of the whole data set already revealed the temporal variation of the amplitudes. These changes can be studied when comparing with the results of Matthews et al. (personal communication).

The period analysis was performed with the help of the MUFRAN-package developed by Z. Kolláth. Thanks are due to him for putting this software at the author's disposal. The travel support from OTKA (Hungary) and IAU is gratefully ackowledged.

The Periods of X Cyg, T Mon, Y Oph, S Vul and SV Vul

A. M. Heiser

Dyer Observatory, Vanderbilt University, Nashville, Tennessee

Abstract

Photoelectric BV observations have been obtained with the 16-inch Vanderbilt APT over a number of recent years for five Cepheids. These data are being examined for periods, period changes, and for possible variations in the seasonal light curves.

The Vanderbilt University APT has been used since 1989 to collect differential BV data for a number of long-period Northern Hemisphere Cepheid variables. The primary purpose of this study is to obtain accurate light curves in order to compare them with other, previously acquired, photoelectric observations.

The analysis of the comparison and check star data for four of the programme stars (T Mon, Y Oph, S Vul and SV Vul) has shown that the average errors of the differential data during the total time period of our APT observations varied from 0.005 to 0.008 mag for the B data and from 0.006 to 0.009 mag for the V data. The check star for X Cyg, HD 196093 = 47 Cyg, was found to vary with an amplitude of 0.04 mag over about 80 days.

The periods that best fit just our APT data for SV Vul and S Vul are 45.036 and 68.500 days, respectively. A recent analysis of the period behaviour of SV Vul (Szabados 1991) has predicted a smaller period for this Cepheid in our observed time interval. S Vul is the faintest Cepheid in our present sample ($V \sim 8.8$ at light maximum) and the scatter in our light curve has not yet permitted a better period determination.

An analysis of a large portion of our APT observations of X Cyg and T Mon (Heiser and Cooper 1991) yielded periods of 16.377 and 27.010 days, respectively. Should these periods be confirmed using all our current APT data, then they would indicate period *decreases* for both these Cepheids. T Mon is a known binary system and Cyg has velocity variations which may be interpreted as being due to duplicity. A preliminary analysis of our APT data for Y Oph shows a period of 17.1269 days; in good agreement with the value determined by Szabados (1989).

References:

Heiser, A.M., Cooper, J.C., 1991, J. Tenn. Acad. Sci. 66, 69
Szabados, L., 1989, Comm. Konkoly Obs., Budapest, No. 94
Szabados, L., 1991, Comm. Konkoly Obs., Budapest, No. 96

A Spectroscopic Abundance Study of Dwarf Cepheid V1719 Cygni

Chulhee Kim[1], Kozo Sadakane[2]

[1]*Department of Earth Science Education, Chonbuk National University, KOREA,*
[2]*Astronomical Institute, Osaka Kyoiku University, JAPAN*

Abstract

Spectroscopic CCD observations were carried out for V1719 Cygni and the spectrum in the visual region is analysed relative to the Sun with a line-blanketed convective model atmosphere. Adopted atmospheric parameters are : an effective temperature $< T_{\text{eff}} > = 7000$ K, a surface gravity $log g = 3.4$. Although our result is dependent on microturbulent velocity and damping constant, it was found that Mg in V1719 Cygni is nearly solar, or underabundent by 0.2 to 0.3 dex according to the analysis of 5172.684 Å MgI line which is relatively free from blending. This is inconsistent with the previous photometric result where V1719 Cygni was known as an abnormally metal rich variable. Because the analysis was given to the single magnesium line which is not a good metallicity indicator and S/N ratio was low due to poor seeing condition, the investigation for iron lines in blue region is undertaken.

Asymmetry of metallic spectral lines in Cepheids

Michael Albrow, P.L. Cottrell

Mount John University Observatory, Department of Physics and Astronomy, University of Canterbury, Christchurch, New Zealand

Abstract

A program of high resolution spectroscopic observations of Cepheids has been carried out at Mt John University Observatory for several years. Radial velocities and asymmetries have been measured for selected metallic lines using the method of Wallerstein et al (1992).

The line profiles show the largest asymmetry at phases of maximum inward velocity. The asymmetry at phases of maximum outward velocity is smaller and sometimes in the same direction as for the inward velocity maxima. Enhanced asymmetry is also noticed at phases where the bump appears on the radial velocity curve.

To date our models are unable to predict such behaviour. It is important that a satisfactory explaination for these observations be found so that the accuracy of Baade-Wesselink radius solutions for such stars can be assessed.

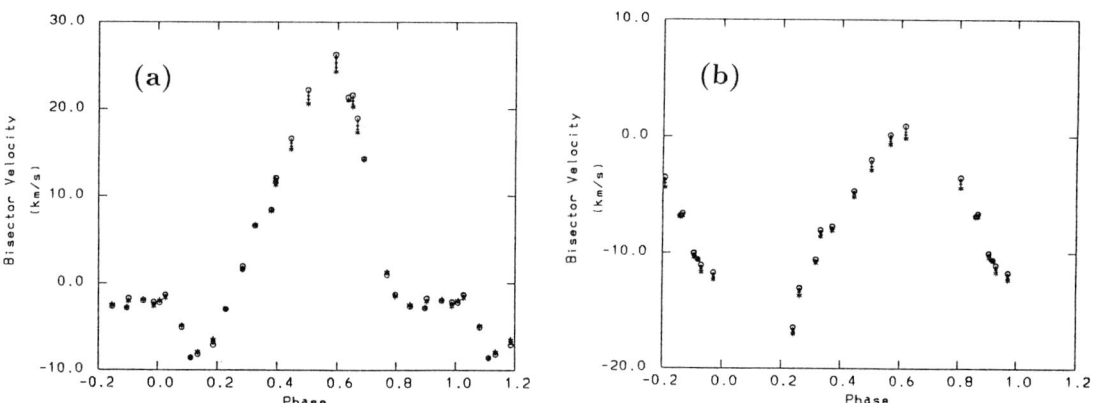

Figure 1 Radial velocities and asymmetries from the Fe I line at 6546Å for the Cepheids a) S Nor and b) Y Oph.

References:

Wallerstein,G., Jacobsen,T., Cottrell,P.L., Clark,M. & Albrow,M., 1992, M.N.R.A.S. in press.

RR Lyrae: Shock waves and atmospheric motions.

Agnès Lebre

DASGAL, Observatoire de Meudon, 92195 Meudon Cedex, France

Abstract

New high resolution spectroscopic observations of the star RR Lyrae are reported (resolution between 15,000 and 20,000). The spectra were taken in July 1990 with the 1.52m telescope of the Observatoire de Haute Provence, using the AURELIE spectrograph, and were centered around Hβ and NaD. They were distributed in time through almost an entire composite cycle ($P = 13.6$ hours). Photospheric FeI lines were analysed and their FWHM variations over one luminosity cycle are interpreted in the framework of the propagation of shock waves running through the low atmospheric layers.

From previous works (Gillet & Crowe, 1988 and 1989; Hill 1972; Fokin, 1992) striking features (the *Hump* and the *Bump*) on RR Lyrae light curves are (respectively) associated with two shock waves, generated during each pulsational cycle: the main strong shock, propagating outward (consequence of the κ-mechanism), and the early weak shock, resulting from the fast infall motion due to the large ballistic motion produced by the previous main shock.

From our new spectra we present for FeI $\lambda\lambda$ 4918.999Å and 4920.509Å photospheric lines the heliocentric radial velocity and the FWHM variations curves. On these latter, we clearly detect two peaks corresponding to important enlargments of the lines occuring at two phases in the pulsational cycle: a main and narrow peak occuring at the hump phase which is a first detection at the photospheric level of the outward propagating main shock; and a second wider peak occuring at the bump phase which already indicates the presence of a compression wave at the photospheric level. This result provides an important observational constraint to the theoretical developments devoted to RR Lyrae atmospheres that must take into account these enlargments observed on FeI lines confirming the presence at the photospheric level of the main and early shocks.

References:

Fokin A.B., 1992, M.N.R.A.S., submitted.
Gillet D. & Crowe R.A., 1988, Astron.& Astrophys., **199**, 242.
Gillet D. & Crowe R.A., 1989, Astron.& Astrophys., **225**, 445.
Hill S.J., 1972, Astrophys.J., **178**, 793.

Spectrophotometric determination of T_{eff}'s for six Cepheids with $P < 4$ d

E. Antonello[1], S. Fossati[2], L. Mantegazza[3]

[1] Osservatorio Astronomico di Brera, Merate, Italy
[2] Dipartimento di Fisica, Università di Milano, Italy
[3] Dipartimento di Fisica Nucleare e Teorica, Università di Pavia, Italy

The population I Cepheids with $1.8 < P < 7$ d are a non–homogeneous group in the sense that in this period range one can find fundamental, first overtone and double–mode pulsators. This implies that these three types of stars fall approximately in the same luminosity range. The knowledge of accurate physical parameters of these stars is therefore of paramount relevance in order to detect possible significant differences among different mode pulsators.

We have made spectrophotometric observations at La Silla Observatory of six Cepheids with $P < 4$ d (3 first-overtone and 3 fundamental mode pulsators). From a comparison with Kurucz's models of atmospheres (1979) we have derived their T_{eff}'s during the pulsational cycle. Due to the relevance of a correct estimate of the interstellar reddening in order to obtain accurate temperatures, we have independently determined the colour excesses of the studied objects by means of observations in the $uvby\beta$ system using the method proposed by Feltz and McNamara (1980).

The derived mean T_{eff}'s are 6350, 5850 and 6400 $°K$ for AZ Cen, BB Cen and BG Cru respectively (first overtone pulsators), and 6050, 6100 and 6150 $°K$ for EY Car, R TrA and UX Car respectively (fundamental mode pulsators).

These temperatures confirm the $\log T_{eff}$ vs.$(B-V)_0$ relationship derived by Teays and Smith (1986) and are in good agreement, for the 4 stars in common, with the temperatures obtained by Pel (1978) from Walraven photometry.

As a conclusion we can affirm on the basis of the data both of our stars and of those of the larger sample by Pel (1978) that fundamental, first-overtone and double mode pulsators coexist in the high temperature region of the instability strip, and that no segregation depending on the pulsation mode is apparent.

References:

Feltz, K.A., McNamara, D.H., 1980, Pub.Astr.Soc. Pacific, **92**, 609.
Kurucz, R.L., 1979, Astrophys. J. Suppl., **40**, 1.
Pel, J.W., 1978, Astron.Astrophys., **62**, 75.
Teays, T.J., Schmidt, E.G., 1986, in: *Stellar Pulsation*, eds. A.N. Cox, W.M. Sparks and S.G. Starrfield, Lecture Notes in Physics, p 173.

Spectroscopic Study of Cepheids in the Globular Cluster Omega Cen

Guillermo Gonzalez and George Wallerstein

Department of Astronomy, FM-20,
University of Washington, Seattle WA 98195

Abstract

Until now there have been few spectroscopic studies of cepheids in globular clusters. In this preliminary report we present sample spectra of the stars V1 and V29 in ω Cen. Eventually, we hope to use the abundance patterns, masses, and period changes of cepheids to better understand post-horizontal branch evolution in globular clusters.

1. Introduction

On the asymptotic giant branch (AGB) a star begins to experience thermal pulses due to the unstable nature of a double shell energy source, leading to "blue loops" on the HR diagram. Blue loops can explain the occurence of the W Vir and RV Tau variables in globular clusters. Lower luminosity BL Her variables are believed to be evolving from the hot side of the HB to the AGB, while at least some of the RV Tau variables are probably post-AGB stars. Globular clusters such as ω Cen with several cepheids and only one likely post-AGB star may provide the "rosetta stones" needed to understand post-HB stellar evolution. Since many globular clusters have well determined parameters including distance, age, metallicity, and reddening, the analysis of the composition and evolutionary status of the cepheids and related stars in globulars is more secure than the analysis of field stars.

2. Goals

We have undertaken this investigation with the following goals in mind: (1) to find out at what point in a star's post-HB life s-process and CNO elements reach the surface; (2) to calculate the amount of mixing between the envelope and the interior; (3) to calculate the mean neutron exposure of the s- process elements; (4) to look for possible fractionation effects; (5) to measure the masses of the cepheids from CCD multicolor photometry and spectroscopic data; and (6) to calculate the helium abundance from the He I emission lines.

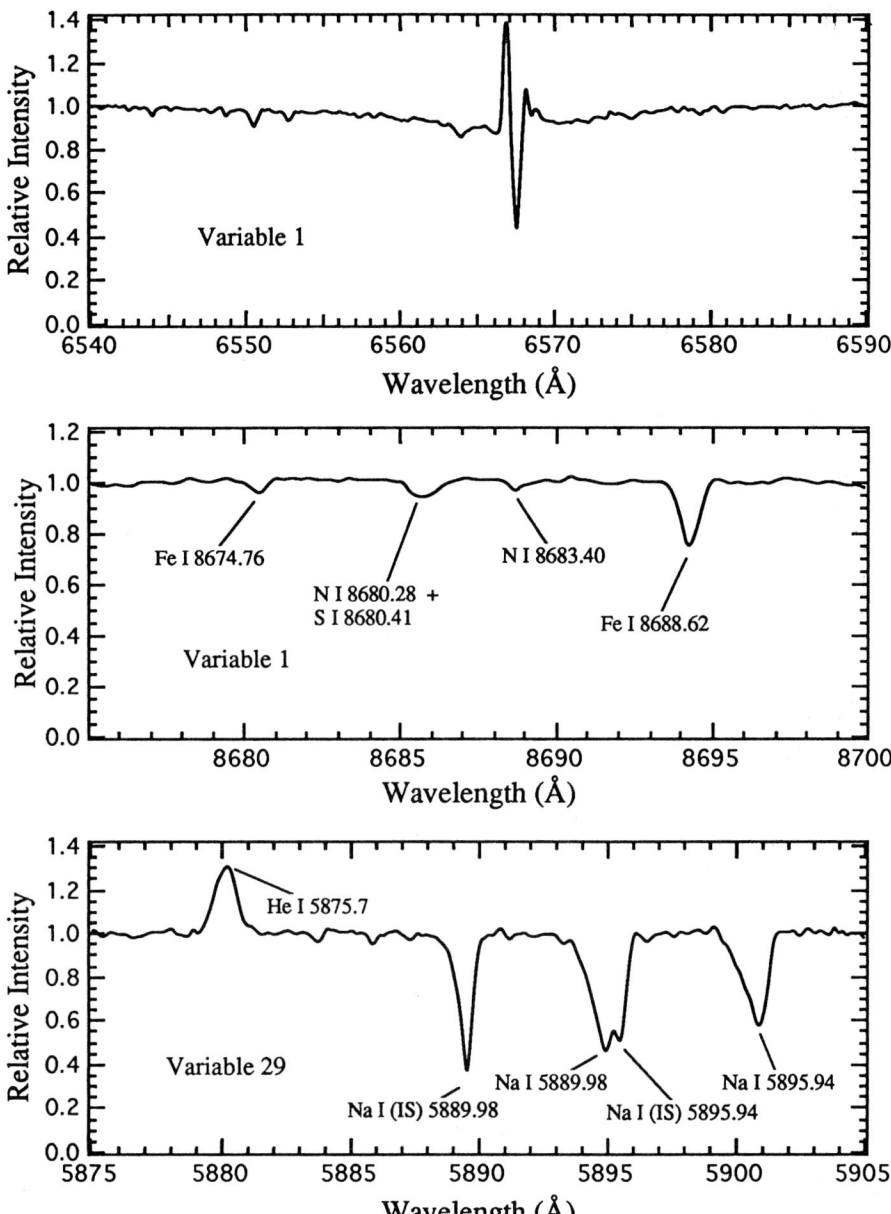

Figure 1. (Top) Spectrum of V1 obtained during rising light. The blue-shifted emission is produced by the shock in the atmosphere. The red-shifted absorption is produced by cooler infalling layers above the shock. (Middle) Infrared region of the spectrum of V1 showing some of the N I and S I lines. (Bottom) Spectrum of V29 showing one of the He I emission lines (visible during rising light) and the photospheric and interstellar Na I lines.

A Photometric and Spectrographic Study of SX Phœnicis

Chulhee Kim[1], D.H. McNamara[2], Kent A. Feltz Jr.[2], C.G. Christensen[2]

[1]*Department of Earth Science Education, Chonbuk National University, Korea,*
[2]*Department of Physics and Astronomy,
Brigham Young University, Provo, UT 84602*

Abstract

New simultaneous photometric ($uvby\beta$) and spectrographic observations of SX Phe are described. Analysis of the light variation frequency spectrum is performed and the oscillation modes are investigated. After separating the light curves and radial velocity curve, atmospheric and physical parameters corresponding to each different oscillation mode were determined. Intrinsic $(b-y)$, m_1, and c_1 values used in conjunction with a model-atmosphere grid yield a mean effective temperature $<T_{\text{eff}}> = 7590$K, a mean surface gravity, $<\log g> = 4.02$, and [Fe/H]= -0.47 for the first period. The pulsation theory and stellar model sequences yield bolometric magnitude of $2\overset{m}{.}6$, mass of 1.4 $\mathcal{M}_\odot$ and age of 2.0 Gyrs. It was found that these atmospheric parameters are almost independent upon different oscillation mode. The radial-velocity data indicates a mean radial velocity of -37km/s and a total velocity amplitude range of 38km/s.

1. Pulsating characteristics

By adopting the generalized least squares method proposed by Vanicek (1971) for frequency analysis, we obtained that f_1=18.1890(c/d) and f_2=23.3928(c/d). Also sixteen frequencies which are harmonics or combiantions of two main frequencies are confirmed. The amplitude and phase for all photometric indices corresponding to each frequency were computed and light curves were well reproduced. We separated the light curves according to each of period. To do this, we subtracted the component of curve corresponding to the fundamental or first harmonic with all their overtones and all other combination terms. The radial velocity data was analyzed by a similar procedure and five frequencies were confirmed.

In order to identify the mode of pulsation, we applied the method developed by Watson (1988). To do this the amplitude ratio (A_{B-V}/A_V)=0.351 and 0.301, and phase difference $(\phi_{B-V}-\phi_V)$=6.02 and 9.14 were computed for the first and second period by transforming $b-y$ to $B-V$ color indices. Then we applied model of (7400, 3.5) and (8000, 4.0) for Q=0.03 to identify the pulsation mode of SX Phe. We can see that the both modes are radial. We confirmed that this conclusion is supported by other methods.

2. Light curve analysis

In order to discuss the variation in the physical parameters of SX Phe, we have sorted the photometric data by phase into 20 equally spaced bins around the light cycle corresponding to two periods for all y, $b-y$, m_1, c_1, and β indices. Then normal points taken from smooth curves drawn through out the average points of each of the indices have been used in the data analysis. Intrinsic $(b-y)$ values were calculated with the aid of the Crawford calibration of A and F stars at normal points and we find an average color excess of $<E(b-y)>= -0\overset{m}{.}016$ and $-0\overset{m}{.}009$ for the first and second period respectively. According to the early A-star calibration relating [Fe/H] to m_1 by Crawford and Perry (1976), [Fe/H] changes from -0.44 to -0.53 and -0.43 to -0.54 for the first and second period for all phase interval.

By interpolating the $(b-y)_0$ and $(c_1)_0$ values in a grid of model atmospheres computed by Kurucz corresponding to [Fe/H] = -0.47, it was confirmed that the effective temperature of SX Phe varies from 7210K at light minimum to 8170K at light maximum and the surface gravity varies from 3.89 to 4.27 for the first period.

We applied the Surface–Brightness Method to determine the radius and distance of SX Phe but we failed due to a large phase shift between radial displacement curve and angular diameter variation. However we obtained 1.5 $R_\odot$ from the period – radius relation for dwarf cepheids. Also it was found that R_{min}=1.5 $R_\odot$ from the theoretical $(\log R, \log P_0)$ diagram for δ Scuti variables by Andreasen, et al. (1983).

3. Evolutionary status

With the aid of the evolutionary models of VandenBerg for $Z = 0.006$ and $Y = 0.25$ we have calculated the evolutionary track in the $\log P - \log T_{\text{eff}}$ plane. It was confirmed that the mass of SX Phe is 1.4 $\mathcal{M}_\odot$ and by referring to the models we find the age is 2.0 Gyr. Also by using Q_0=0.033 and R_{min}=1.5 $R_\odot$, we obtained the pulsating mass of 1.2 $\mathcal{M}_\odot$ for SX Phe. As a gravity mass, we obtained 1.5$\mathcal{M}_\odot$ by using our maximum surface gravity of 4.27 where the overestimation of 0.2 in $\log g$ for Stromgreen calibration was not taken into account.

According to the period–luminosity relation for fundamental mode pulsators, absolute magnitude of SX Phe is 2.8 mag close to the overall absolute magnitude of 3.0 mag for Population II dwarf cepheids. However both are much different from M_v=4.4 mag calculated from parallax. Furthermore $<R>= 0.7 R_\odot$ can be obtained by utilizing Stefan–Boltzman law by taking our T_{eff}=7590K for the first period and $M_b \sim M_v = 4\overset{m}{.}4$. All these inconsistent results are from uncertain parallax.

References:

Andreasen, G.K., Hejlesen, P.M., Peterson, J.O., 1983, A&A, 121, 241
Vanick, P., 1971, ApSS, 12, 10
Watson, R.D., 1988, ApSS, 140, 255

Investigation of the Double-mode Cepheid TU Cas: Atmospheric Parameters and Chemical Composition

S. M. Andrievsky, V. V. Kovtyukh, E. N. Makarenko, I. A. Usenko

Odessa State University, Department of Astronomy
Shevchenko Park, 270014 Odessa, Ukraine

Abstract

Atmospheric parameters and abundances of 25 elements were determined from two spectrograms of TU Cas (with dispersion 9 Å/mm) obtained in 1977 and 1990. We find $T_{\text{eff}} = 5860 - 6000 K$; $\log g = 1.0 - 1.5$; and $v_t = 2.9 - 5.3$ km/s.

The value of $[Fe/H] \sim -0.5$ testifies to the fact that TU Cas is deficient in metals compared with other double-mode Cepheids. More exact values of $[Fe/H]$ for 9 double-mode Cepheids in the Southern Hemisphere (Barrell 1982), V367 Sct (a member of the open cluster NGC 6649) and our value for TU Cas all show a good correlation with P_1/P_0. As the atmospheric Fe abundance increases, the ratio P_1/P_0 decreases.

Our measured values for abundances (in brackets) of various elements in the atmosphere of TU Cas are: **C** (8.26), **Na** (6.32), **Mg** (7.07), **Si** (6.85), **S** (6.96), **Ca** (5.88), **Sc** (2.71), **Ti** (4.34), **V** (3.35), **Cr** (5.29), **Mn** (4.92), **Fe** (6.94), **Co** (4.47), **Ni** (5.70), **Zn** (3.63), **Sr** (2.00), **Y** (2.17), **Zr** (2.21), **Ba** (1.41), **La** (1.08), **Ce** (0.96), **Nd** (0.82), **Sm** (0.42), **Eu** (1.04), and **Gd** (0.92).

References:

Barrell, S.L., 1982, MNRAS, 200, 127

The HF Precise Radial Velocity Programme at DAO

S. Yang[1], A. Larson[1], A.W. Irwin[1], C. Goodenough[1],
G.A.H. Walker[2], A. Walker[2], D. Bohlender[2]

[1] *University of Victoria, Canada,* [2] *University of British Columbia, Canada,*

Abstract

A programme to measure precise radial velocities of late-type stars is being carried out at the 1.22-m telescope of the Dominion Astrophysical Observatory (DAO). Wavelength-calibration fiducials are imposed directly on the stellar spectra by passing the starlight through a controlled hydrogen fluoride (HF) absorption cell placed in front of the coude spectrograph. Presently, the primary targets of the programme are bright G, K, and M giants. Preliminary results confirm the low-amplitude, radial-velocity (RV) variability of the yellow giants discovered at the Canada-France-Hawaii telescope using the HF technique. These yellow giants and additional bright candidates are now being continually monitored at DAO. Preliminary results also indicate that the "yellow giant" variability extends to the early-M giants. In addition to the RV variations, the data also yield information on the simultaneous variability of the Ca II $\lambda 8662$ line, T_{eff}, as well as the R - I index of the stars.

The HF Programme at DAO

Selected late-type giants have been reported to be low-amplitude, radial-velocity variables (Walker et al. 1989; Smith et al. 1987; Irwin et al. 1989; Murdoch et al. 1992). Photometric variations have also been reported (Percy & Fleming 1992). Preliminary results from the on-going programme at DAO to monitor late-type stars with the HF precise-radial-velocity technique (Campbell, Walker & Yang 1988) have confirmed the reported low-amplitude, radial-velocity variability of α Boo, α Tau, and ϵ Peg. Additional targets β UMi, γ Dra, and especially the M-type stars (β Peg, β And, μ Gem, R Lyr, g Her, α Ori, α Her) also appear to be variable.

References:

Campbell, B., Walker, G.A.H., Yang, S., 1988, ApJ, **331**, 902.
Irwin, A.W., Campbell, B., Morbey, C.L., Walker, G.A.H., Yang, S., 1989, PASP, **101**, 147.
Murdoch, K., Clark, M., Hearnshaw, J.B., 1992, MNRAS, **254**, 27.
Percy, J.R., Fleming, D.E.B., 1992, PASP, **104**, 96.
Smith, P.H., McMillan, S., Merline, W.J., 1987, ApJ, **317**, L79.
Walker, G.A.H., Yang, S., Campbell, B., Irwin, A.W., 1989, ApJ, **343**, L21.

The RV Variability of Yellow Giants

A.M. Larson[1], A.W. Irwin[1], S.L.S. Yang[1], C. Goodenough[1]
G.A.H. Walker[2], D.A. Bohlender[2], A.R. Walker[2]

[1] *University of Victoria, Canada*, [2] *University of British Columbia, Canada*

Abstract

The hydrogen fluoride (HF) absorption cell technique has been used at the Canada-France-Hawaii Telescope for over a decade to monitor the radial velocity variability of nearby, solar-type stars in a search for substellar companions. As a complement to this program, we have also been monitoring a select group of G and K subgiants, giants and supergiants, which have all proved variable. We present here a brief summary of our analysis.

Results

In Walker *et al* (1989) we reported on five K giants and one K supergiant included in the HF PRV program at the CFHT. With the addition of 6 more years of observations, we have been able to determine significant, long-term ($f_{period} < 0.01$, P > 100 days) periods in 5 of those stars plus an additional K giant, δ Sgr. Table I lists the best long-period sinusoidal solutions. Because of the limited sampling over short time spans, we are not able to rule out the possibility that these long periods are aliases of shorter periods. A program is currently underway at the Dominion Astrophysical Observatory to obtain better time coverage (Yang *et al*, this conference).

Table I. Best Long-Term Periodic Solutions (> 100 days)

Star	MK type[a]	γ(m/s) value	σ	K (m/s) value	σ	Epoch (days) value	σ	Period (days) value	σ
β Gem	K0 IIIb	-13.8	3.2	40.7	4.5	7148.4	9.2	583.5	4.7
α Boo	K1.5 III			To be covered in a separate paper					
δ Sgr	K2.5 IIIa	-6.3	15.0	93.2	18.0	7623.0	12.0	376.3	1.7
α Tau	K5 III	23.9	6.4	99.4	7.9	7614.0	2.2	123.1	0.2
α Hya	K3 II-III	-57.9	15.0	146.1	20.0	6752.6	16.0	794.8	13.0
ϵ Peg	K2 Ib			Multi-periodic					

[a] Garrison, R.F. 1992, Observer's Handbook of the Royal Astronomical Society of Canada

References:

Walker, G.A.H., Yang, S., Campbell, B., Irwin, A.W., 1989, ApJ, 343, L21
Yang, S., Larson, A.M., Irwin, A.W., Goodenough, C., Walker, G.A.H., Walker, A.R., Bohlender, D. 1992, these Proceedings

Extremely Low Amplitude Cepheids

R. Paul Butler

University of Maryland, College Park, USA

Abstract

Several photometric studies conducted in the 1970's indicate that at least half the stars in the Cepheid instability strip are stable at the level of 0.02 mag (Fernie & Hube 1971; Percy 1975; Fernie 1976; Percy, Baskerville, & Trevorrow 1979). A precision radial velocity survey of these "stable stars" is currently being conducted by the author. Radial velocity errors have been reduced to a few tens of meters per second with the use of an iodine absorption cell (Marcy & Butler 1992).

Extremely low amplitude (200 m/s) periodic radial velocity variations have been found for HR 7796, an F8Ib supergiant (Butler 1992). Two and a half complete cycles have been monitored over three observing runs. The period is found to be 11.87 days. Observations of the reference star HR 509 (G8 V) taken on the same nights, show no periodicities and a scatter of just 20 m/s. Both the period and the shape of the velocity curve appear similar to a normal F8Ib Cepheid. If HR 7796 is a Cepheid, it is the smallest amplitude Cepheid by more than an order of magnitude.

References:

Buter, R.P., 1992, Ap. J. Lett, **394**, L25.
Fernie, J.D., 1976, PASP, **88**, 116.
Fernie, J.D. & Hube, J.O., 1971, Ap. J., **168**, 437.
Marcy, G.W. & Butler R.P., 1992, PASP, **104**, 270.
Percy, J.R., 1975, I.A.U. Inf. Bull. Var. Stars, No. 983.
Percy, J.R., Baskerville I., & Trevorrow, D.W., 1979, PASP, **91**, 368.

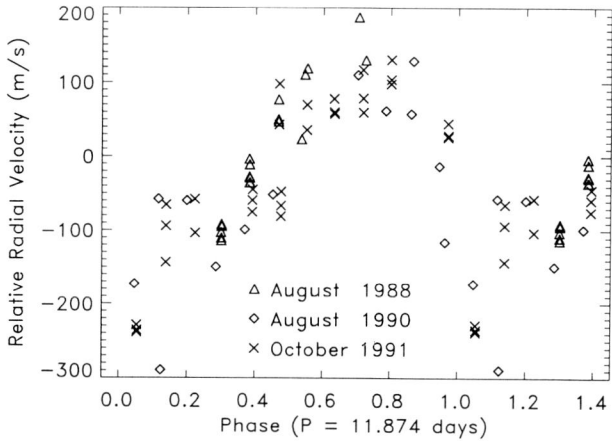

FIG. 5.—Velocity vs. pulsation phase for HR 7796. The period is 11.874 days, and the epoch is 2,447,386.29 for zero phase.

New Radial Velocities and the Barnes-Evans Method

Robert Hindsley[1] and R.A. Bell[2],

[1] U.S. Naval Observatory, Washington D.C.
[2] University of Maryland

Abstract

New radial velocities by Butler are used to recalculate the absolute magnitudes of Eta Aquilae and X Cygni using the Barnes-Evans method. The new velocities have little effect on the result for Eta Aquilae, but the absolute magnitude of X Cygni is found to be 0.40 mag fainter than that obtained earlier.

Butler (these Proceedings) has measured the radial velocities of four Cepheids (FF Aql, Delta Cep, Eta Aql, and X Cyg) using an iodine cell to give reference wavelengths for the spectral lines. With this technique the standard error of a single measurement is a few hundred metres per second. The ratio of the uncertainty in each velocity point to the full amplitude of the radial velocity variation now matches the corresponding ratio for the V magnitude light curve.

For Eta Aquilae, the Fourier decomposition of the radial velocities of Jacobsen and Wallerstein (1981, Publ. A.S.P, 93, 481) and Evans (1976, Ap. J. Suppl., 32, 399) were compared to the decomposition obtained by combining these data with the data of Butler. There is no difference in the derived coefficients, but the uncertainty in the coefficients is much reduced when the data of Butler are added. The relative precision of the amplitudes in the Fourier decomposition (ratio of each amplitude to its uncertainty) is not so good as for photometry, but this is due to the incomplete phase coverage of the radial velocity data of Butler.

Surface brightness techniques require integration of the radial velocity curve to obtain differences in radii in linear units, which then are matched to angular diameters derived from photometry, yielding the distance. Improved radial velocity data should yield more precise distance estimates. For Eta Aquilae the addition of Butler's data changes the radial velocity curve very little, and the distance is unchanged. But for X Cygni the data of Butler greatly improve the radial velocity curve, and an absolute magnitude of -5.06 is obtained, 0.40 mag fainter than the result found otherwise.

Ultraviolet Studies of Cepheids

Erika Böhm-Vitense[1,2]

[1] *University of Washington, Seattle, Washington,* [2] *Guest observer with the International Ultraviolet Explorer (IUE) Satellite*

Abstract

We discuss whether with new evolutionary tracks we still have a problem fitting the Cepheids and their evolved companions on the appropriate evolutionary tracks. We find that with the Bertelli et al. tracks with convective overshoot by one pressure scale hight the problem is essentially removed though somewhat more mixing would give still better fits.

Using results of recent nonlinear hydrodynamic calculations by Morgan we find that we also have no problem matching the observed pulsation periods of the Cepheids with those expected from their new evolutionary masses, provided that Cepheids with periods less than 9 days are overtone pulsators.

We investigate possible mass loss of Cepheids from UV studies of the companion spectrum of S Mus and from the ultraviolet spectra of the long period Cepheid l Carinae. For S Mus with a period of 9.6 days we derive an upper limit for the mass loss of $M < 10^{-9}$ $M_\odot$, if a standard velocity law is assumed for the wind. For l Carinae with a period of 35.5 days we find a probable mass loss of $M \sim 10^{-5\pm2}$ $M_\odot$.

1. Background

The Cepheid mass problem has been with us for a long time. Around 1980 standard evolution theory predicted for Cepheids with pulsation periods around 10 days an evolutionary mass around 7 $M_\odot$ for solar abundances, while pulsation theory and especially the bump Cepheids required masses between 4 and 5 $M_\odot$, see Cox 1980.

Ultraviolet observations with the IUE satellite revealed a number of Cepheid companions with temperatures $\leq$10,000 K, see Böhm-Vitense and Proffitt (1985). Several of them had radii larger than expected for main sequence stars. They appeared to be giants. Since evolution time scales for giants are very short they should then have masses which are only a few percent less than their Cepheid primaries. They should fit essentially on the same evolutionary track as the Cepheids. They were, however, observed to be much fainter and indicated masses of the order of the pulsational masses. In Table 1 we give the basic data for the Cepheids and their evolved companions as given by Böhm-Vitense and Proffitt 1985 in their Table 6C. We did apply a correction of $\Delta \log L = 0.2$ to all luminosities in order to match the Sandage and Tammann (1969) distance scale.

In Figure 1 we show the positions of the Cepheids and their companions in the HR diagram together with the newer evolutionary tracks of Becker (1981) with no convective overshoot. It is obvious that we had a problem, (except for AX Circini which seems to fit well).

Table 1
Data for Cepheids and companions used here

Star	log P	log T$_{eff}$ (Ceph)	log L (Ceph)	log T$_{eff}$ (comp)	log L (comp)	Δlog L
AX Cir	0.722	3.76	3.48(ov)	4.17	3.54	-0.06
AW Per	0.810	3.73	3.40(ov)	4.08	2.87	0.55
SV Per	1.046	3.72	3.50	4.08	3.11	0.39
SY Nor	1.102	3.69	3.36	4.04	2.75	0.61
RW Cam	1.215	3.67	3.45	4.08	2.99	0.46
KN Cen	1.532	?	4.27	4.40	3.40	0.87

The numbers in brackets give the deviations for the two sets of data.

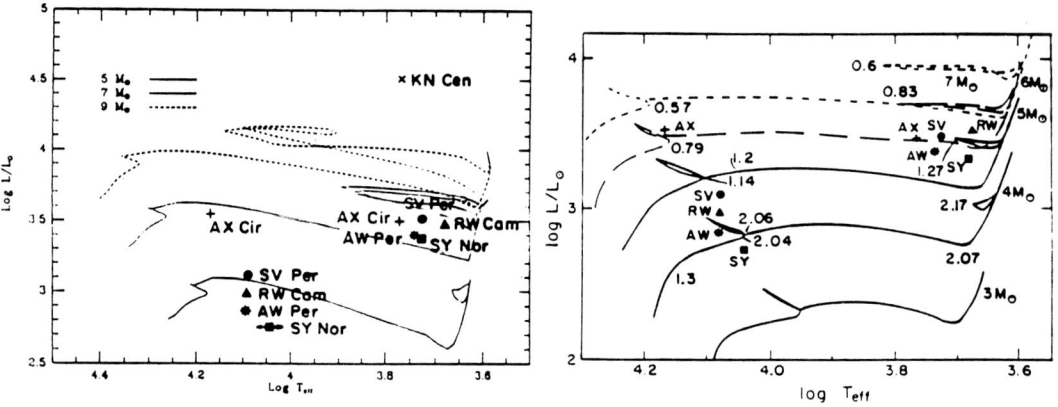

Figure 1 (left) Evolutionary tracks without convective overshoot are shown for 5, 7, and 9 M$_\odot$ stars according to Becker (1981). Also shown are the positions of the Cepheids and their "giant" companions according to Böhm-Vitense and Proffitt (1985) with L increased by 0.2 dex. Cepheids and evolved companions do not fit on one isochrone. AX Cir is the exception.

Figure 2 (right) Evolutionary tracks with convective overshoot by one pressure scale height are shown for 3, 4, 5, 6, and 7 M$_\odot$ stars according to Bertelli et al. (1986). Also shown are the positions of the Cepheids and their evolved companions. For these tracks the companions fit on the evolved part of the main sequence where evolution times are longer. The masses of the companions now come out to be about 10% smaller than those of the Cepheids (except for AX Cir). A minute increase in the overshoot length would place them on one isochrone.

2. New evolutionary tracks

Smaller Cepheid masses are one way to eliminate the mass problem for Cepheids. This requires that for a given mass the blue loops on which the Cepheids are believed to be found must have higher luminosities. Becker and Cox (1982) pointed out that increased mixing in the interiors of the main sequence progenitors would increase the luminosities of the blue loops. It would also explain the high nitrogen to carbon abundance ratios found for supergiants (Luck and Lambert 1981; Mena-Werth 1992). Bertelli, Bressan, Chiosi and Angerer (1986) followed up on that possibility and calculated a number of evolutionary tracks with overshoot above the convective core.

Maeder and Meynet (1988) calculated evolutionary tracks for overshoot by 0.3 pressure scale heights and some mass loss. In Figure 2 we show the Bertelli et al. evolutionary tracks for convective overshoot by one pressure scale height. We also show the positions of the Cepheid binaries with evolved companions. The evolved companions now fit on the evolved part of the main sequence where the evolution is still slower than on the giant branch. Their masses appear to be only 10 to 15% less than the Cepheid masses, which are about 4.8 ± 0.3 $M_\odot$. Considering the error limits the fit is quite good though a somewhat higher luminosity for the blue loops would give an even better match. (Tracks with the new OPAL opacities (Iglesias and Rogers 1991, give loops with somewhat lower not higher luminosities, see Stothers and Chin 1992.) The loops need to extend further to the blue than shown by the Bertelli et al. tracks in order to reach to the Cepheid positions. Alongi et al. (1991) showed, however, that this can be achieved by also considering overshoot at the bottom of the outer convection zone by about 0.5 to 0.7 pressure scale heights.

AX Circini is an exception. It must be a first crossing star if the Bertelli et al. tracks are correct, or it must follow the Becker track, which means it must have had a progenitor which had no additional mixing. Of course we do not know whether the mixing is due to convective overshoot or to a different mechanism.

N. Evans (this volume) reaches a different conclusion. She concludes that no evolutionary tracks match all the Cepheids with evolved companions. She emphasizes the discrepant results for AX Cir and SU Cyg (see below) and also BP Cir if the Bertelli et al. tracks are used. Three first crossing stars out of 10 are too many according to evolution times unless there is a special selection mechanism. All three stars have periods less than 4 days and they have rather high temperatures. Such high temperatures may not be reached by the blue loops for low mass Cepheids. We can then observe such short period Cepheids only for first crossing stars. If we see indeed first crossing stars, then they should still have solar nitrogen to carbon abundance ratios. It would be quite interesting to check.

Another possibility is, of course, that the excess mixing does not occur for all stars. In that case we probably have to look for another mixing mechanism than convective overshoot, which should be similar for all stars of a given mass and composition.

3. Pulsation properties

The next question is whether such Cepheids have the theoretically expected pulsation periods. Pulsation periods for solar abundance Cepheids with different combinations of L and T_{eff} were calculated recently by Buchler, Moskalik and Kovacs (1990) and also by Morgan in Seattle (1991). We use here the results by Morgan. In Figure 3 we show the relations between period and T_{eff} for different luminosities. Also shown are the positions of the Cepheids according to their luminosities and periods. Since the exact T_{eff} of these Cepheids are hard to determine because of the color contamination due to the bright companions we have plotted them at the best fitting values for T_{eff}. (In Figures 1 to 3 we also used these T_{eff}). For lower luminosities higher temperatures can be obtained. $\Delta \log L \sim -0.1$ would raise the temperature by about 350 K.

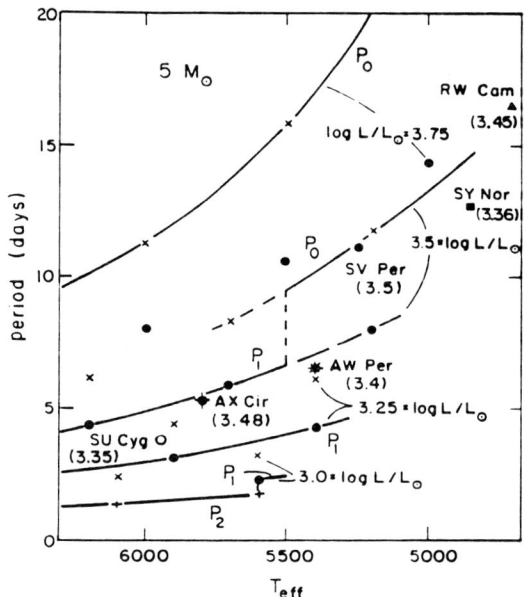

Figure 3 For given luminosities of the Cepheids the relation between periods and effective temperatures are shown according to non-linear hydrodynamic calculations by Morgan (1991). The fundamental periods P_o are indicated by x, the first overtone periods P_1 by dots and the second overtone P_2 by crosses. The prevailing modes are connected by solid lines. The dashed lines show the connections and the extensions for the non-prevailing modes. The Cepheid symbols are placed according to their luminosities and their periods. The log L of the Cepheids are given in brackets below the symbols. AW Per and AX Cir must be overtone pulsators to fit on this diagram with reasonable temperatures.

Morgan could not find any fundamental mode pulsators with periods less than 9.5

days. This agrees with our suggestion that for periods around 9 days a transition from fundamental mode pulsation to first overtone pulsation occurs for galactic Cepheids (Böhm-Vitense 1990). In Figure 3 the curves are drawn through the points that represent the prevailing mode of pulsation in nonlinear hydrodynamic modeling. We can fit AX Cir and AW Per on this diagram only if they are first overtone pulsators. As such they fit very well. It then seems that the discrepancy between evolutionary and pulsational masses has disappeared. If we adopt the Bertelli et al. tracks the Cepheid masses agree with the pulsation and bump masses.

One point of importance: If galactic Cepheids with periods less than 9 days are overtone pulsators we have to use different period luminosity relations for the two groups of stars as shown in Figure 4 taken from Morgan's thesis. For a given period and T_{eff} there is a shift in luminosity by about 0.2 dex. The overtone pulsators being the brighter ones as compared to the fundamental pulsators. For the fundamental mode pulsators the gradient of the period luminosity relation increases. This is important if we try to observe bright, long period Cepheids in other galaxies.

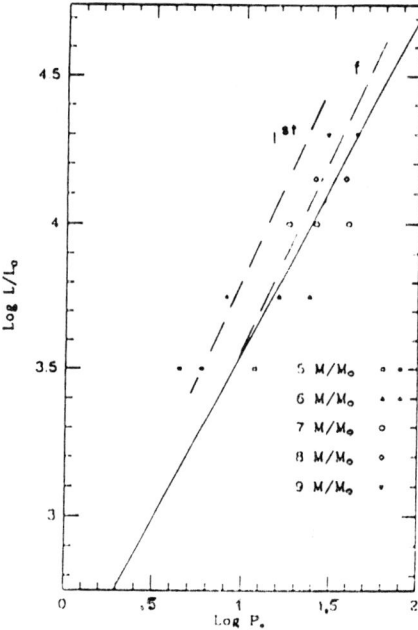

Figure 4 The relation between luminosities and periods is shown for stars of different masses and effective temperatures according to Morgan (1991). The open symbols refer to fundamental mode pulsators, the filled symbols to first overtone pulsators. The solid line gives the empirical relation according to Stothers and Carson. For a given period and T_{eff} the overtone pulsators have a luminosity larger by 0.2 dex as compared to the fundamental mode pulsators. For each mode of pulsation alone we will derive a steeper period luminosity relation than for the combined set of fundamental and first overtone pulsators as suggested schematically by the dashed lines added by us.

For AX Cir and AW Per, which we believe are overtone pulsators, we increased the luminosities by 0.2 dex to take this into account. This does not influence the luminosity difference between Cepheid and companion.

4. Dynamical mass determinations

Nancy Evans and I have tried very hard to determine dynamical masses for Cepheid binaries with main sequence companions. In Table 2 I have collected the results of our measurements. Within the error limits the results agree with the masses we just derived using the Bertelli et al. tracks except for SU Cyg. With its period of 3.8 days it has a larger mass than S Mus with a period of 9.6 days. We saw here that SU Cyg is expected to be an overtone pulsator. It fits on the evolutionary tracks only if it is at a first crossing. Both, the AX Cir and SU Cyg binaries, fit well on the blue loops of the Becker tracks. Possible explanations were discussed above.

Table 2

Star	Period	Mode	Mass Bertelli	Mass dynamic	Reference
S Mus	9.6	fund.	5.4 ± 0.2	5.3 ± 0.7	Böhm-Vitense et al. 1990
SU Cyg	3.8	1st	4.8 (6.0)	$> 5.8 \pm 0.4$	Evans and Bolton 1988
V636 Sco	6.0	1st	4.7 ± 0.2	4.8 ± 0.5?	Böhm-Vitense 1986

5. Mass loss of S Mus

Willson and Bowen (1986) suggested that pulsating stars may lose mass very efficiently. If so, they might have started out with larger masses. We have attempted to determine the mass loss for the Cepheid binary S Mus. If the Cepheid is losing mass the wind passes by the companion and causes absorption lines in the companion spectrum, see Figure 5. We (Rodrigues and Böhm-Vitense 1992) tried to study these absorption lines in the companion spectrum for S Mus. The available IUE spectra were taken at an orbital phase when the stars were at maximum separation, ($\Phi = 90°$). Liliya Rodrigues calculated the line profiles which would be expected for these wind absorption lines in the companion spectrum assuming a standard velocity law v(r) and $\Phi = 90°$. The largest absorption is due to particles near terminal velocities because of the long column density for this velocity. Near the center of the line the observed line profiles are confused by interstellar absorption and possibly Cepheid absorption lines or perhaps lines from S Mus B. We therefore studied mainly the broad, shortward shifted absorption component expected to be seen for such a passing cool Cepheid wind. The detailed structure of the profile is wiped out by the large turbulence expected to be present in the wind.

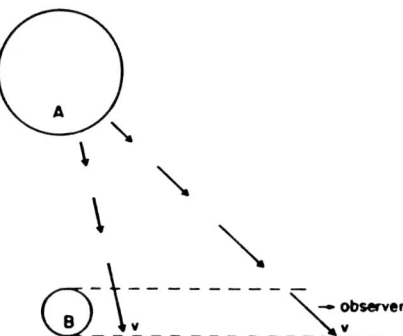

Figure 5 The geometry is shown for the S Mus system at the orbital phase of our observations. The Cepheid wind passes by the B star companion and causes absorption lines in its spectrum. The largest radial velocities are seen for the largest distance from the B star.

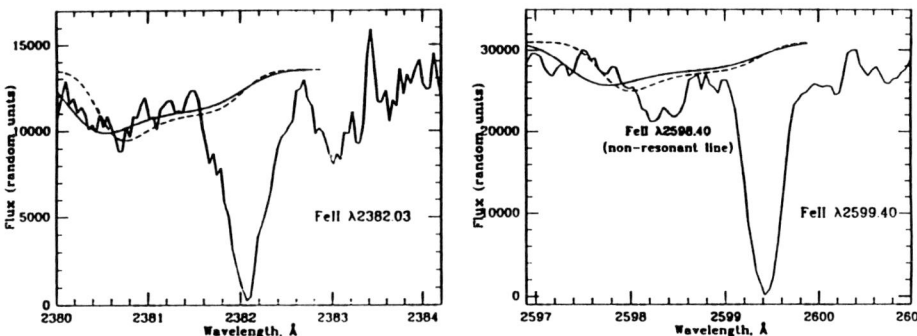

Figure 6 Observed and best matching theoretical line profiles for some Fe II lines in the B star companion spectrum are shown. The theoretical profiles correspond to a wind with a terminal velocity of 175 km/s and a turbulent velocity of 37 km/s. The depth of the shallow satellite lines determines the density in the wind.

For several Fe II lines such broad shallow blue shifted satellite lines were observed with the relative strengths as might be expected according to the oscillator strengths. In Figure 6 we show the comparison of observed Fe II line profiles with the best matching theoretical ones for a terminal wind velocity of 175 km/s and a turbulent velocity of 37 km/s. For these values and assuming that all Fe is present as Fe II we find a mass loss of about 10^{-10} $M_\odot$ per year. In order to check whether Fe could be present also as Fe III we looked at the Al III lines and found no sign of a blue satellite line. The Al III lines are expected to be weaker because of different abundances and different oscillator strengths. We therefore can conclude only that the amount of Fe III has to be less than about a factor 6 more than Fe II. This gives an upper limit for the mass loss of 6×10^{-10} $M_\odot$ per year. This upper limit also holds if the broad shallow line should be intrinsic to S Mus B.

6. Mass loss of *l* Carinae

It may be interesting to check whether we see evidence of stronger mass loss for Cepheids with longer periods. For *l* Carinae with a period of 35.5 days a large number of high resolution long wavelength (2000 Å-3000 Å) and low resolution (1200 Å-2000 Å) spectra are available. The Mg II lines are always in emission. They are highly variable during the pulsation period. The profiles and their variations can be understood if at the wavelengths for maximum emission, i.e. at about $\lambda_o \pm 75$ km/s, the emission occurs in an optically thick layer and shortly after maximum light in an optically thin layer. In addition there is some essentially constant absorption outside of this emitting layer. For this absorption we find at the wavelengths of maximum flux, i.e. at ± 75 km/s, an optical depth of about $\tau_\lambda = 0.6$ independent of phase. If the turbulent velocity is about 30 ± 5 km/s. and if Mg is all in the form of Mg II we derive a column density of $n_H = 10^{21\pm 2}$ cm^{-2}. In order to estimate how much mass is in this "shell" we have to know its volume, which means its radius.

We believe that we see spatially extended emission in the OI lines, which leads to an estimate for the radius of the surrounding shell or cloud of about 1000AU, i.e. the same order of magnitude as found by Mc Alary and Welch (1986) for some Cepheids with infrared excess.

Assuming a homogeneous sphere with the given column density and an outflow velocity of $v_r \sim 35$ km/s we estimate a mass loss of M $\sim 10^{-5\pm 2}$ $M_\odot$ per year. This is what would be expected according to Reimers' approximate relation for luminous stars.

It thus seems possible that mass loss may be of importance for longer period Cepheids. The uncertainties in our estimates (mainly of the turbulent velocity in the shell) are too large to reach a definite conclusion. Our studies seem to show a similar trend as seen by Deasy and Butler (1986) who found that short period Cepheids (P < 15 days) do not show excess infrared emission, while a few Cepheids with longer periods do. *l* Carinae does, however, not show excess infrared emission.

References:

Alongi, M., Bertelli, G., Bressan, A.,and Chiosi, C. 1991, Astr. Astrophys., **244**, 95.
Becker, S. and Cox, A. N. 1982, Ap.J., **260**, 707.
Becker, S. A. 1981, Ap. J. Suppl., **45**, 475.
Bertelli, G., Bressan, A., Chiosi, C., and Angerer, K. 1986, Astr. Astrophys. Suppl., **66**, 191.
Böhm-Vitense, E. and Proffitt, C. 1985, Ap. J., **296**, 175.
Böhm-Vitense, E., Clark, M., Cottrell, P. L. and Wallerstein, G. 1990, A.J., **99**, 353.
Böhm-Vitense, E., 1986, Ap. J., **303**, 262.

Böhm-Vitense, E. 1990, Ap. J., **324**, L27.
Böhm-Vitense, E. and Love, S. 1992, submitted to Ap. J.
Buchler, J. R., Moskalic, P. and Kovacs, G. 1990, Ap. J., **351**, 617.
Chiosi, C., Wood, P. R., Bertelli, G., Bressan, A., and Mateo, M. 1992, Ap. J., **385**, 205 .
Chiosi, C. and Wood, P. R. 1990, referenced by Chiosi, C. 1990, in ASP Conference Publ. 11, 158.
Cox, A.N . 1980, Ann. Rev. Astr. Astrophys., **18**, 15.
Deasy, H. P. and Butler, C. J. 1986, Nature, **320**, 726.
Evans, N. R. and Bolton, C. T. 1990, Ap. J., **356**, 630.
Iglesias, C. A. and Rogers, F. G. 1991, Ap. J., **371**, 408.
Maeder, A. and Meynet, G. 1988, Astr. Astrophys., **76**, 411.
Mc Alary, C. W. and Welch, D. L. 1986, A. J., **91**, 1209.
Morgan, S. 1991, Ph.D. Thesis, University of Washington, Seattle, WA.
Rodrigues, L. and Böhm-Vitense, E. 1992, Ap.J., Dec. 20.
Sandage, A. and Tammann, G. A. 1969, Ap. J., **157**, 683.
Stothers, R. B. and Chin, C. W. 1991, Ap. J., **381**, L67.
Willson, L. A. and Bowen, G. H. 1986, Ir. Astron. J., **17**, 2.

Acknowledgements

This research was supported by NASA grant NSG 5398, which is gratefully acknowledged. I am very much indebted to the staff of the IUE Observatory. Without their continuous help for obtaining and reducing the data this study would not have been possible.

Discussion

Comment by Nancy Evans: I am going to talk about several of the same stars as Erika, and although I disagree with many of the details of the analysis and interpretation, I think a lot of her study was driven by the difficulty of matching systems with a Cepheid and an evolved companion with current isochrones. I find I have a similar difficulty with these systems.

Comment by S. R. Sreenivasan: 1. Rotational mixing due to differential rotation is an alternative for producing core growth in stars to a uniform overshoot postulated for all stars. This way the extend of core growth is proportional to the vigor of differential rotation and does not occur for all stars as you require. 2. Mass loss can occur in these Cepheids because of the large outer (or subsurface) convection zones.*
It might also occur in the main sequence phase for stars whose mass is larger than $5M_\odot$ on the ZAMS if they have significant rotation. In that case surface rotational speed also undergoes evolutionary changes.
*due to both magnetic as well as non-magnetic (e.g. acoustic energy flux) causes.

X Sources in δ Scuti Stars: an Ultraviolet Study of 71 Tau

L.E. Pasinetti Fracassini[1], L. Pastori[2], F. De Nile[1], E. Poretti[2], E. Antonello[2]

[1]*Dipartimento di Fisica, Università degli Studi di Milano, Milano, Italy,*
[2]*Osservatorio Astronomico di Brera, Milano, Italy*

IUE observations of δ Scuti variables were planned to study the correlations between chromospheric activity and dynamics of pulsations, convection, rotation and to search for evidence of mass loss. So far we observed the following stars: ρ Pup, β Cas, o^1 Eri, K2 Boo, τ Peg, 69 Tau, 71 Tau and τ Cyg. Results and discussions on our survey may be found in Pasinetti Fracassini et al. (1990) and Fracassini et al. (1991).

Ultraviolet spectroscopic data (6 LWP and 3 SWP spectra) of 71 Tau were obtained with IUE in the year 1990, spanning an interval of 5^h35^m and covering about 1.5 cycles of the pulsation period. The period, derived from new photometric observations, is 4^h32^m with an amplitude of $0^m.028$. This variable is the most intense X-ray source in the Hyades cluster according to the results of Einstein Observatory.

The MgII lines exhibit an anomalous absorption profile; moreover some variations spanning for about 0.5 Å, confirmed also by the statistical analysis, occur during the pulsation phases; the features may also be interpreted as emission. Emissions were not detected in the highly ionized transition-region lines which are however underexposed in our spectra. The complete absence of chromospheric or transition region line emission was claimed by Zolcinski et al. (1982), who suggested that the apparent absence of MgII emissions may be ascribed to rotational smearing of the weak emission line in the core of the strong absorption line. CII emission, on the contrary, was detected by Walter et al. (1988); therefore the presence of chromospheric emission in 71 Tau is still an open problem. Finally, in our spectra no narrow absorption core was detected in the MgII k line; this feature was found by Zolcinski et al.(1982) and interpreted by the same authors as an interstellar absorption; probably the feature may be ascribed to the physical properties of the star.

References:

Fracassini,M., Pasinetti Fracassini,L.E., Pastori,L., Teays,T.J., and Mariani,A.: 1991, *Astronom.Astrophys.* **243**, 458.
Pasinetti Fracassini,L.E., Pastori,L., Schmidt,E.G., and Teays,T.J.: 1990, in "Confrontation between stellar pulsation and evolution", eds. C.Cacciari, G.Clementini, Bologna May 28-31 1990, p.230.
Walter,F.M.,Schrijver,C.J., and Boyd,W.: 1988, in "A Decade of UV Astronomy" with IUE, *ESA SP-281*, Vol.1, p.323.
Zolcinski,M.C.S., Anthiocos,S.K., Stern,R.A., and Walker,A.B.C.: 1982, *Astrophys.J.* **258**, 177.

A new pulsating white dwarf in the *Wide Field Camera*?

D. Wonnacott,

Rutherford Appleton Laboratory, Chilton, Didcot, Oxon., OX11 0QX, UK.

Abstract

Recent observations in the extreme ultraviolet have shown that the binary companion of the pulsating metallic-lined A8m: star IK Peg is a hot DA white dwarf. The EUV light curve is shown to be subject to too much photon noise to identify any periods, and simulations of the effect of a hotspot on the compact object indicate that it is unlikely that the observed photometric variability is *not* intrinsic to the A8m: star.

Given that there are more pulsating white dwarfs in the Galaxy than δ Scuti stars, this EUV detection warrants investigation to see if the pulsations ascribed to the Am star can, in fact, be attributed to hot spots on a rotating white dwarf or accretion from the ISM even though the magnitude difference is $\sim 7^m$.

Firstly, the *WFC* EUV light-curves for the two filter passbands (100 Å and 137 Å respectively for S1a and S2a) were subjected to a period analysis using the technique due to Scargle (1982). The periodograms of the data are very sensitive to the photon noise induced by the low count rates and short observation slots (~ 40–$80\,\mathrm{s}$). A comparison of the periodograms yields no coincident peaks making it unlikely that there is a real (detectable) period present.

As a further test, the times of the original observations of IK Peg (Kurtz 1978) were used to generate a series of 'observations' by sampling a single-frequency sine wave at those same times. As the measurements were nearly evenly spaced ($\Delta t \sim 0.^d 005 \sim 7.^m 2$), there is a strong possibility that a such a signal would be contaminated by beating with the observation period. These 'light-curves' were subjected to the same period analysis as above and compared with an identical analysis of the original y-band photometry. White dwarf variations were simulated in the range 2–25 minutes and for various phases relative to the first observation.

Examination of the power spectra of these light-curves, however, clearly shows that there is insufficient leakage of power from the white dwarf-like (high) frequencies to produce a dominant 23.9 cycles/day beat. This is true of all the high frequencies investigated. As the observed frequency is too low to be accounted for by a pulsating or rotating white dwarf, it can therefore *only* be attributed to the metallic-lined star and its apparently contradictory properties.

References:

Scargle J. D., *Ap. J.*, 1980. **263**, 835.
Kurtz D., 1978. *Ap. J.*, **221**, 869.

Recent Results on Binary Cepheids

Nancy Remage Evans[1,2]

[1]*Institute for Space and Terrestrial Science, York University,* [2]*IUE Guest Observer*

Abstract

IUE observations of hot companions of Cepheids have been used to determine the temperatures of the companions. For companions on the ZAMS, the temperatures can be used to determine the luminosity of the Cepheids from the inferred absolute magnitude of the companion. The accuracy of this method is comparable to the accuracy of calibration using Cepheids in clusters. An overtone pulsator (SU Cas) has been identified by this technique. The luminosity of the double mode pulsator Y Car agrees with that from the PLC, confirming that is a normal Population I Cepheid. The variation of the width of the instability strip (as a function of luminosity) in the HR diagram is confirmed, and overtone and double mode pulsators are shown to be near the blue edge of the instability strip. Comparison between variables and nonvariables in the HR diagram (from IUE and cluster studies) shows very little overlap. The shape of the observed instability region may be determined both by the boundary between the variables and nonvariables and also the location of the tips of the blue loops of evolutionary tracks. Finally, eight Cepheids with hot companions which are evolved beyond the ZAMS have been studied. Half the systems are matched well by evolutionary tracks with little or no core convective overshoot near the main sequence. At least two and possibly four systems, however, cannot be matched by current isochrones.

1. Cepheid Luminosity Determination

Cepheids which have hot companions can be "temperature resolved", that is, each component can be studied in detail in the wavelength region in which it dominates. Low resolution IUE spectra of the hot companions provide a well–determined spectral type, from which the absolute magnitude can be inferred. From this the absolute magnitude of the Cepheid can be determined. The details of the process are provided in Evans 1991 and 1992a. Previous discussions of IUE spectra of Cepheid companions are given by Böhm-Vitense 1985 and Böhm–Vitense and Proffitt 1985. Figure 1 shows the comparison of the results from IUE and the cluster Cepheids from Feast and Walker 1987. The IUE results (7 stars) have an rms deviation from the Feast and Walker period–luminosity–color (PLC) relation of 0.37 mag (for one observation), or a smaller deviation (0.33 mag) if the IUE results are used to determine the zero–point of the relation. Although the IUE results indicate a slightly fainter PLC than Feast and Walker, this result is preliminary because additional stars can be added to the IUE results and the analysis will be redone when the IUE data has been reprocessed.

Furthermore, we (Evans, Pitts, and Imhoff) have obtained spectra of comparable Pleiades stars, so that the calibration of the companions can be tied directly to them.

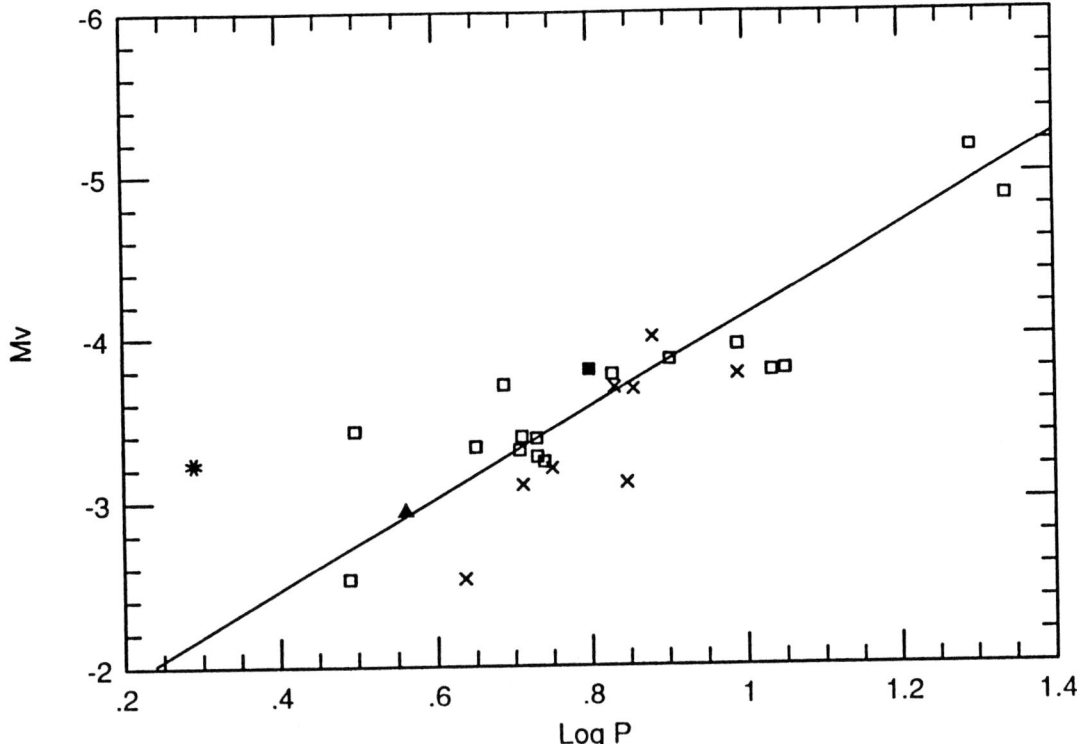

Figure 1 The period luminosity relation. Open and filled squares are the data from Feast and Walker for cluster Cepheids; the filled square is V367 Sct; other data are IUE determinations; the filled triangle is the double mode pulsator Y Car; the asterisk is SU Cas. The line is the period–luminosity relation from Feast and Walker.

In addition to the fundamental mode pulsators, we have observed a Cepheid we identify as an overtone pulsator (SU Cas, Evans 1991) and a double mode pulsator (Y Car, Evans 1992b). SU Cas is more luminous than the PLC relation would predict for a fundamental mode pulsator, and hence we conclude that it is pulsating in an overtone mode. This is independent evidence that this short period "s-Cepheid" (identified from the Fourier coefficients by Antonello, Poretti, and Reduzzi 1990) is actually an overtone pulsator. The luminosity of the double mode Cepheid Y Car is in good agreement with that predicted from the fundamental mode and the PLC, confirming that its properties are comparable to single mode pulsators.

An HR diagram can be plotted using the Cepheids with independent luminosities from the IUE determinations and the Feast and Walker cluster stars. We confirm the result of Fernie 1990 that the instability strip is narrower at lower luminosities than at high luminosities. In addition, the two double-mode pulsators Y Car and V637 Sct, and the overtone pulsator SU Cas fall on the blue edge of the instability strip. This is where pulsation calculations predict that overtone pulsation should be, but it is quite different from, for instance, the observed location of double mode pulsators found by Barrell 1981.

2. Comparison of Variables and Nonvariables

The same technique for determining Cepheid luminosities can also be used for nonvariable supergiants with hot companions. Recently we (Evans 1992c) have discussed the luminosities for two such systems, HD 183864 and HD 223047, also discussed by Ake 1988 and Parsons and Ake 1992. Figure 2 shows these results, the results from IUE luminosities for Cepheids, and the results for variables and nonvariables in clusters from Schmidt 1984. Both the IUE results and the Schmidt results have luminosities determined for variables and nonvariables using the same technique, and so the relation between the variables and the nonvariables should be accurate. As can be seen in Figure 2, Schmidt's conclusion that there is at most minimal overlap between Cepheids and nonvariables is still true. This does not, however, seem to be the case for the combined sample of Magellanic Cloud clusters (e.g. Mateo, et al. 1990). However, individual Magellanic Cloud clusters show a fairly clear separation between variables and nonvariables (Evans 1992c).

In Figure 2, the diagonal band in the HR diagram made up of both variable and nonvariable supergiants is what is expected from stars near the tips of the blue loops of the evolutionary tracks, as discussed by Schmidt. The hot edge of the blue loops is shown schematically by the solid line. (The solid point at (0.2, -2.4) is ignored in this discussion as a discordant point.) This line is quite well defined observationally, and can serve as a useful constraint on evolutionary calculations. What do we know about the boundaries of the instability region? On the red side, there are both variables and nonvariables for a large range of luminosity which define the boundary of the instability strip (shown schematically by the dashed line). On the blue side, on the other hand, only for high luminosities are there nonvariables as well as variables (dotted line). For lower Cepheid luminosities, there are no nonvariables hotter than the instability region. In fact, it looks in Figure 2 as though the location of the hottest Cepheids is determined more by the edges of the blue loops than by the stability/instability boundary. Since the location of the blue loop supergiant region affects the shape of the instability strip, it may even explain the narrowing of the strip at low luminosities, as discussed above. (The narrowing of the instability strip is more prominent in the IUE and Feast and Walker compilation than the IUE and Schmidt compilation.) To confirm this suggestion, it is desirable to have luminosities for additional nonvariable supergiants particularly on the hot side of the instability strip if any exist. The blue loop locus, however, is only one possible influence on the shape of the instability region (see, for instance Morgan, this conference).

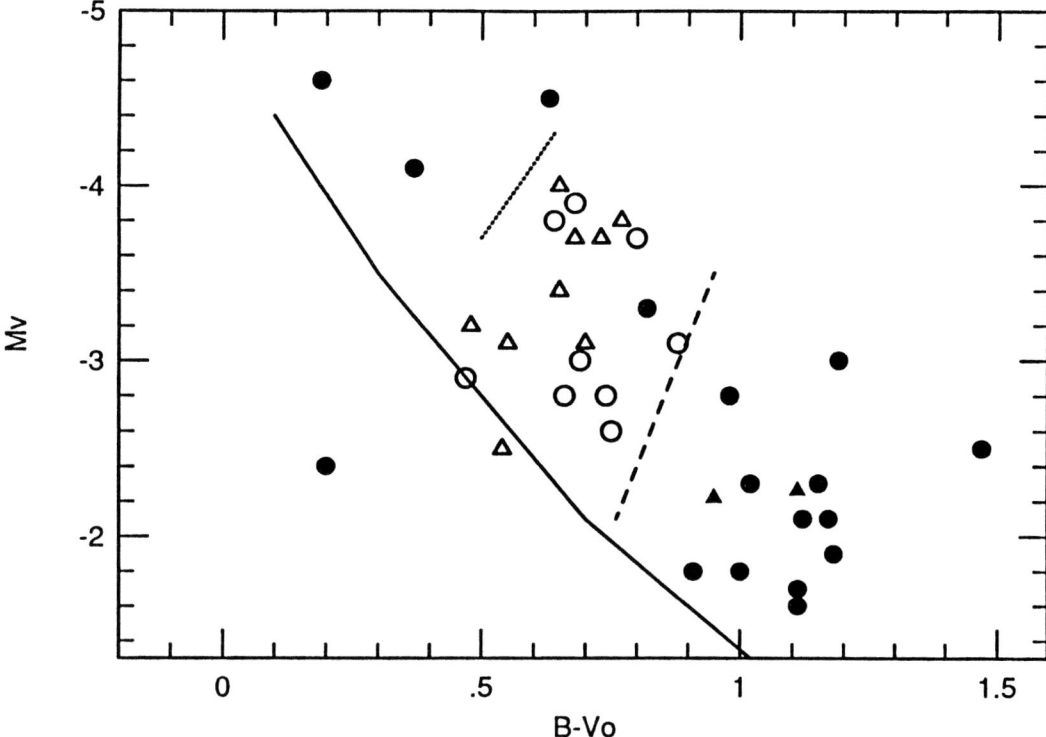

Figure 2. Comparison of Cepheids and nonvariable supergiants in the HR diagram. Triangles are from IUE observations, circles are from Schmidt; filled symbols are nonvariables, open symbols are variables. Lines show the schematic location of the tips of the blue loops (solid), the red edge of the instability strip (dashes) and the blue edge of the instability strip (dots) as determined from this sample.

3. Cepheids with Evolved Companions

A number of the Cepheid binary systems studied have hot companions which are evolved beyond the ZAMS. These have been treated differently than systems for which the large luminosity difference between the components confirms that the companions must be very close to the ZAMS. For systems with evolved companions, the luminosity of the Cepheid from the PLC has been used to place the components of the system on the HR diagram (Evans 1992d). These systems can then be compared

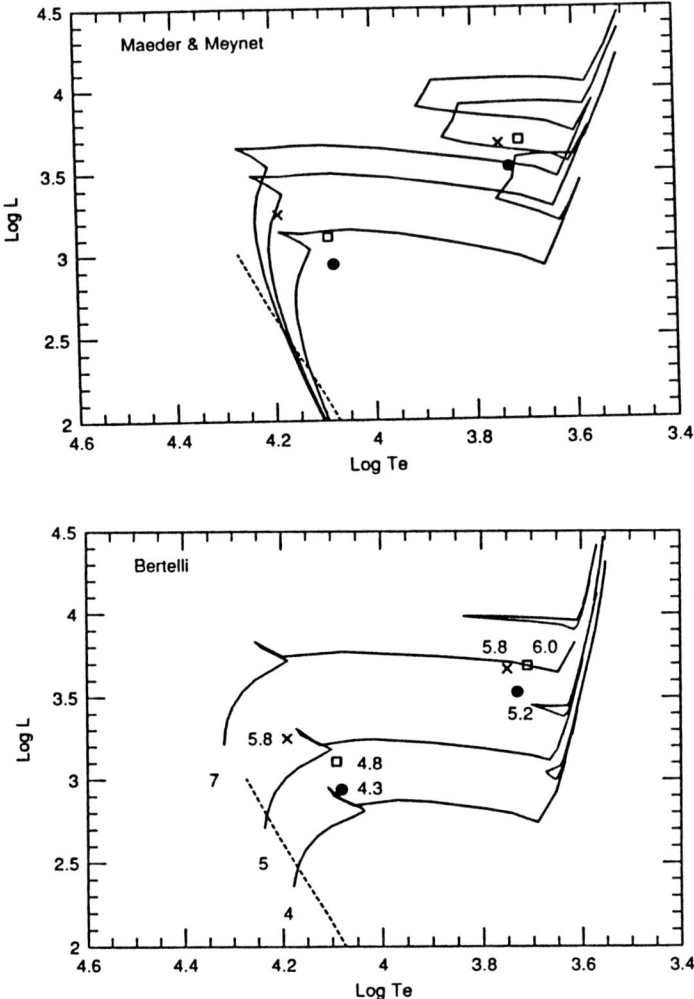

Figure 3 Theoretical HR diagrams. The symbols represent the stars in the following systems: SY Nor: x's; RW Cam: open squares; SV Per: filled circles. a. The isochrones from Meader and Meynet are shown for log age 7.7, 7.8, and 8.0 years. b. The evolutionary tracks from Bertelli, et al 1986 are shown for masses 4, 5, and 7 solar masses. Near each star is the mass estimated from these tracks, again in solar masses.

with evolutionary tracks and isochrones. One of the parameters which is especially uncertain in evolutionary calculations is the amount of core convective overshoot for stars near the main sequence. One of the purposes of this study is to examine groups of binary systems containing Cepheids with similar periods. Since the mass and luminosity should be simple functions of the period, Cepheids within a small period range should have similar masses, and presumably should be well matched by evolutionary tracks with similar input physics.

Three short period systems BP Cir (P = 2.4 days), AX Cir (P = 5.3 days), and V659 Cen (P = 5.6 days) agree well with isochrones with no convective overshoot, such as those of Stothers and Chin 1991. Because the luminosity difference between the Cepheids and companions is small for AX Cir and BP Cir, the companions must be near the end of their main sequence lifetimes, and the two stars in each system must have nearly the same mass. For V659 Cen, the luminosity difference is larger, and so, presumably, is the mass difference. As discussed by Evans, *et al.* 1992, for BP Cir this requires that the Cepheid be pulsating in the first overtone mode, making the fundamental mode 3.4 days. If a luminosity for the Cepheid is calculated assuming a period of 2.4 days, the companion is very close to the ZAMS, which disagrees with the small luminosity difference between the two stars. A period of 3.4 days makes both stars more luminous, and places the companion near the end of its main sequence lifetime. There are two other pieces of evidence that BP Cir is an overtone pulsator: the Fourier decomposition parameters (Antonello, *et al.* 1990), and the location of the Cepheid on the blue edge of the instability strip.

For the long period Cepheid SY Nor (P = 12.6), the two components fit an isochrone with a mild amount of overshoot such as that of Maeder and Meynet 1988 (Figure 3) better than one with no overshoot. This implies that the amount of convective overshoot *may* be a function of mass within the Cepheid mass range.

However, although these four systems can be understood within the framework of current evolutionary calculations, there are exceptions which are poorly fit by evolutionary tracks or isochrones. Figure 3 shows an example of the comparison of the SY Nor system with two systems with similar periods, RW Cam (P = 16.4 days) and SV Per (P = 11.1 days). Figures 3a and b respectively show isochrones based on the calculations of Maeder and Meynet 1988 with mild overshoot and some mass loss, and evolutionary tracks of Bertelli, *et al.*, 1986 with a large amount of overshoot. Note that recalculation of the Maeder–Meynet tracks (Meynet 1992) with new input physics, including the new opacities of Iglesias and Rogers 1991, removes the luminosity difference between the second and third crossing of the instability strip. That is, both branches of the blue loop lie very close to the second crossing in Figure 3b. Since the luminosities of the Cepheids in the SY Nor, RW Cam, and SV Per systems are all similar, the companions should fit a similar isochrone. Clearly in Figure 3a the companions RW Cam B and SV Per B are much too cool for the isochrone fitting the Cepheids. The RW Cam B and SV Per B temperatures are determined from well exposed IUE spectra, and the temperatures are in the temperature range where the IUE spectra are very sensitive to temperature, so the temperatures are well determined. (The SY Nor spectrum is not as well exposed.) A wide main sequence band is often taken as the signature of a large amount of convective overshoot. For this reason, we have also compared the same data with the Bertelli, *et al.* tracks, to see whether tracks with a large amount of overshoot (and calculations from a different code) result in better agreement. Figure 3b shows that the main sequence band is wide enough to include the companions for RW Cam and SV Per. However, the luminosity differences are not consistent with the tracks. The masses of each star estimated from the tracks are included in Figure 3b. The poor agreement with the SY Nor system is reflected in the fact that the stars have no mass difference at all with these tracks. For RW Cam and SV Per, on the other hand, the two stars differ by approximately a solar mass, which is too large for a blue loop star and a star at the end of its main sequence lifetime. The evolved nature of RW Cam B and SV Per B is confirmed by the fact that the IUE flux distributions match luminosity class III stars better than class V stars.

Two other systems *may* show the same inconsistency with evolutionary tracks, KN Cen (P = 34.0 days) and AW Per (P = 6.5 days). In both cases there are possible alternate explanations, which are, however, not entirely successful. Because KN Cen is the longest period star in the sample, any parameter such as overshoot which is a function of mass may have an extreme value in its case. The companion to AW Per may itself be a binary. While this would give the companion increased luminosity for its color, even the maximum correction for this effect would not result in a companion on the main sequence. In summary, two and possibly four of the eight systems with evolved companions studied are not well matched with current evolutionary tracks. Furthermore, the discrepant cases are spread out throughout the range of period, so a simple variation of some parameter as a function of mass is not a likely cause. Changes in evolutionary tracks because of rotation is an example of a possible cause.

4. Summary

Studies of the hot companions of Cepheids from IUE spectra have been used to determine the luminosities of Cepheids. Included in the sample are an overtone pulsator and a double mode pulsator, which are found on the blue edge of the instability strip in the HR diagram. A comparison between the location of variables and nonvariables (from the IUE studies and the cluster studies of Schmidt) shows very little overlap. Binary systems containing a Cepheid and a companion within the main sequence band but evolved beyond the ZAMS cannot be matched with current isochrones in all cases.

References:

Ake, T. 1988, in "New Directions in Spectrophotometry", eds. A. G. D. Philip, D. S. Hayes, and S. J. Adelman, (Schnectady: L. Davis Press), p. 27.

Antonello, E., Poretti, E. and Reduzzi, L. 1990, Astr. Ap., **236**, 138.

Barrell, S. L. 1981, M. N. R. A. S., **196**, 357.

Bertelli, G., Bressan, A., Chiosi, C., and Angerer, K. 1986, Astron. Ap. Suppl., **66**, 191.

Böhm-Vitense, E. 1985, Astrophys. J., **296**, 169.

Böhm-Vitense, E. and Proffitt, C. 1985, Astrophys. J., **296**, 175.

Evans, N. R. 1991, Astrophys. J., **372**, 597.

—. 1992a, Astrophys. J., **389** , 657.
—. 1992b, Astrophys. J., **385** , 680.
—. 1992c, preprint
—. 1992d, preprint
Evans, N. R., Arellano Ferro, A., and Udalska, 1992, Astron. J., **103**, 1638.
Feast, M. W. and Walker, A. R. 1987, Ann. Rev. Astr. Ap, **25**, 345.
Fernie, J. D. 1990, Astrophys. J., **354**, 295.
Iglesias, C. A. and Rogers, F. J. 1991, Astrophys. J., **371**, 408.
Maeder, A. and Meynet, G. 1988, Astron. Ap. Suppl., **76**, 411.
Mateo, M., Olszewski, E. W., and Madore, B. F. 1990, in A. S. P Conf. Ser., **11**, 214.
Meynet, G. 1992, private communication.
Parsons, S. B. and Ake, T. 1992, Bull. A. A. S. **24**, 769.
Schmidt, E. G. 1984, Astrophys. J., **287**, 261.
Stothers, R. B. and Chin, C. 1991, Astrophys. J., **381**, L67.

Question:
Turner: I have a comment and a question. First, one parameter which evolutionary models do not include is rotation, which can displace a star to cooler effective temperatures in the HR diagram. Also a B dwarf which is rapidly rotating and seen equator–on appears spectroscopically like a giant. This may help you explain the apparent anomalies you find for some of the Cepheid companions. Second, you and I disagree about the pulsation mode of SU Cas, and this seems to be tied to differences in the magnitude difference between the Cepheid and its companion. I am puzzled by the large ΔV you find for SU Cas, since it seems to contradict your K–line spectroscopic observations which implied a smaller ΔV.

Evans: Yes, rotation might explain the systems which fit evolutionary tracks poorly. For SU Cas, the magnitude difference between the two stars is much better determined from the IUE spectra than from the Ca II K line observations. For the K line spectra, the companion is only prominent for a small wavelength region in the line core, and any quantitative measure must take into account grating scattered light as well as the shape of the Cepheid line cores.

New Ways of Revealing Cepheid Binaries

L. Szabados

Konkoly Observatory, Hungary

Abstract

Two methods involving the observed amplitudes of radial velocity and *UBVR* light variations for classical Cepheids have been analysed, both being implicitly known: their principle is trivial but these methods had not yet been used systematically as indicators of duplicity.

The slope method is based on the alteration of the wavelength dependence of the light variation amplitude if either a blue or a red companion is added to the light of the Cepheid. The amplitude ratio (AR) method makes use of the fact that the companion reduces the amplitude of the light variation without observable effect on the pulsational radial velocity amplitude. This means that the ratio of these two amplitudes ($A_{rad.vel.}/A_B$) has a larger value for binary Cepheids as compared with the single pulsators.

Each method has been applied to more than 100 Cepheids, thus allowing to study how the uncontaminated parameters (amplitude ratio and slope) depend on the pulsation period. Binary Cepheids deviate from the regular pattern in these diagrams, and a number of new binaries can be discovered in this way. The effect of duplicity is revealed by both methods independently for VZ CMa, FM Cas, CR Cep, V402 Cyg, V1154 Cyg, V440 Per and DR Vel.

1. The "slope"-method

A new treatment of the amplitude ratios is suggested which makes use of four photometric amplitudes simultaneously. The relationship between the photometric amplitudes in *U, B, V* and *R* passbands as a function of the logarithm of the wavelength is roughly linear. The slope of this line is a good indicator of the companion if the color of the secondary star differs significantly from that of the Cepheid.

The values of the slope have been plotted in Figure 1 as a function of *log P*. The programme stars have been divided into four groups:
   - "small" amplitude Cepheids with known companion (circles),
   - other small amplitude Cepheids (diamonds),
   - "normal" amplitude Cepheids with known companion (asterisks),
   - other normal amplitude Cepheids (squares).

The blue companions strongly move the slope towards zero. As to normal amplitude Cepheids, the smallest absolute values of the slope almost always correspond to Cepheids belonging to known binary systems having a blue companion. The exceptions can be those Cepheids where the companion has not been detected yet. The red companions are usually much less luminous, therefore their effect is not as obvious as that of a blue secondary.

On the basis of this method the following Cepheids certainly belong to binary systems (because the other method to be discussed later on also indicates a companion): VZ CMa, FM Cas, CR Cep, V402 Cyg, V1154 Cyg, V440 Per and DR Vel. Their duplicity has not been reported previously. In addition, several stars have a value of the slope as if they had a blue companion without any spectroscopic confirmations: SY Aur, YZ Aur, SS Sct and RY Vel.

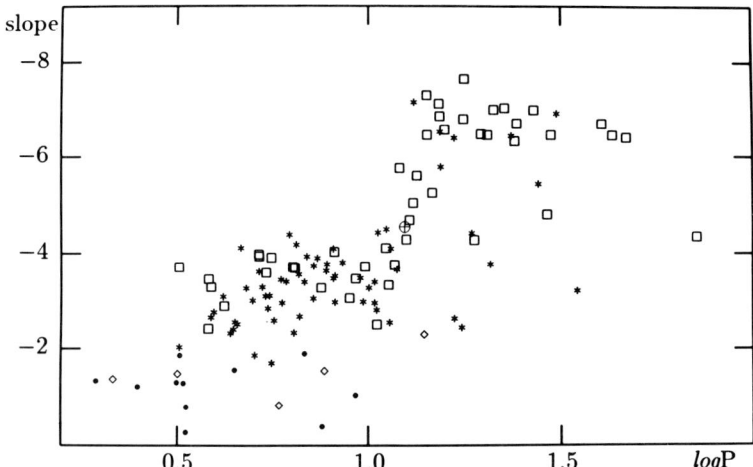

Figure 1 The period – slope relation.

2. The AR-method

The other method indicating the binary nature of the star makes use of both photometric and radial velocity data. The companion decreases the observable photometric amplitude but does not influence the observable pulsational radial velocity amplitude. As a consequence of this, the ratio of the amplitudes $AR = A_{rad.vel.}/A_B$ is larger for binary Cepheids. However, it cannot be assumed *a priori* that a single value of this amplitude ratio gives an adequate description for all Cepheids without companion.

The detailed study of the radial velocity amplitudes indicates that one cannot expect a single value of the amplitude ratio AR for the solitary Cepheids but an unusually large value of AR might well indicate the presence of a companion.

In addition to the seven stars newly classified as binary Cepheids (listed when discussing the "slope"-method), there are at least ten galactic Cepheids suspected in having a companion based on the large AR value: V1344 Aql, ER Car, XY Cas, BD Cas, BY Cas, V381 Cen, V419 Cen, MY Pup, EV Sct, EU Tau, and the Magellanic Cloud Cepheids HV834, HV1365 and HV2864. A thorough spectroscopic study of these Cepheids would be worthwhile.

Travel grants from the OTKA (Hungary) and IAU are gratefully acknowledged.

Orbital Solution and Physical Parameters of the Binary Cepheid AW Per

J. Vinko

JATE University, Szeged, and Konkoly Observatory, Budapest, Hungary

AW Per is a well-known binary Cepheid. Recently Welch & Evans (1989) determined a spectroscopic orbit. Evans (1989) pointed out that the spectral type of the companion is incompatible with the mass function of the system. We re-determined the orbit of AW Per using both photometric and spectroscopic data. The result of the simultanuous least-squares fit can be seen on Fig.1. The orbital elements are very close to the results of Welch & Evans. After correcting for the companion, the Cepheid's light and velocity curves were analyzed with the surface-brightness method. The details of the analysis will be published in MNRAS (Vinko, 1992). The mass of AW Per was found to be 6 solar masses, which confirms the mass-problem of the secondary. Observations are planned to continue at Konkoly Observatory.

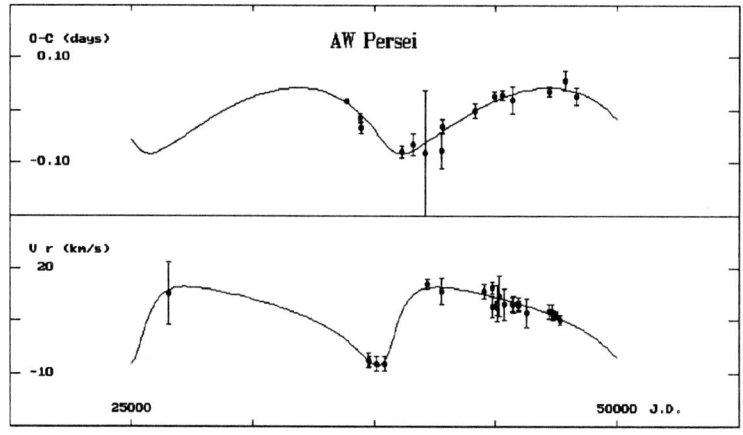

Figure 1. O-C and γ-velocity diagram of AW Per, assuming the results of the orbital solution.

References:

Evans,N.R., 1989, Astron.J. **97**, 1737.
Vinko, J., 1992, Mon.Not.Royal Astron.Soc. (in press)
Welch,D.L. & Evans,N.R., 1989, Astron.J. **97**, 1153.

The Periods of RR Lyrae

Arthur N. Cox

Los Alamos Astrophysics

Abstract

RR Lyrae (0.566 day period) exhibits the Blasko effect that suggests another natural mode with almost the same period as the accepted fundamental radial mode. This mode might be nonradial, but no one has done an extensive evaluation of this idea. An investigation requires a model that includes the deep composition structure where g-modes of low angular (observable) degree have weight and amplitude. An RR Lyrae model including the outer half of the mass and more than 99% of the radius, based on an asymptotic giant branch model from Hollowell (private communication), see below, was used for this study. It includes composition gradient ramps between the primordial surface hydrogen and helium and the almost pure helium shell and the one between this helium shell and the convective core that is burning helium.

Nonradial mode periods almost resonant with the radial fundamental mode period seem to occur for all low ℓ values. In addition to significant pulsation amplitudes in the composition gradient regions where the Brunt Väisälä frequency is large, these low degree and low radial order modes have near-surface amplitudes very similar to the low order radial modes. These modes are evanescent in the convective core. Classical κ and γ effects give enough driving in the very low mass surface layers, so that important deep radiative damping for these modes does not completely stabilize nonradial g-mode pulsations. The g_4, $\ell=1$ mode gives a double-mode RR Lyrae with Blasko effect.

A nonradial mode may not always be visible, depending on how rotation presents the nonspherical pulsations to the observer. Thus the Blasko effect might come and go, as observed for maybe 20% of all RR Lyrae variables. For many, the Blasko effect may not be observable, even when a nonradial mode is there.

700 Mass Shell RR Lyrae Model

Mass ($M_\odot$, g)	0.70, 1.392×10^{33}
Effective Temperature (K)	6800
Luminosity (erg/s)	2.272×10^{35}
Fundamental Mode Period (d)	0.566
Growth Rate (Δ KE/KE per period)	0.029
g_4 Mode Period (d)	0.570
Growth Rate (Δ KE/KE per period)	0.0007
Blasko Period (d)	81
Last Mass (g)	2.53×10^{24}
Envelope Mass (g)	6.96×10^{32}
Photosphere Radius (cm)	3.862×10^{11}
Central Ball Radius (cm)	4.536×10^9

The Blazhko Effect in RR Lyrae

Terry J. Teays

Computer Sciences Corporation/IUE Observatory

Abstract

The cause of the Blazhko effect, the long-term modulation of the light and radial velocity curves of some RR Lyr stars, is still not understood. The observational characteristics of the Blazhko effect are discussed in §1. Some preliminary results are presented from aomw recent of campaigns to observe RR Lyr, using the *International Ultraviolet Explorer* along with ground-based spectroscopy and photometry, throughout a pulsation cycle, at a variety of Blazhko phases. A set of ultraviolet light curves have been generated from low dispersion IUE spectra. In addition, the (visual) light curves from IUE's Fine Error Sensor are analyzed using the Fourier decomposition technique. The values of the parameters ϕ_{21} and R_{21} at different Blazhko phases of RR Lyr span the range of values found for non-Blazhko variables of similar period.

1. Characteristics of the Blazhko Effect

The first allusion to the Blazhko effect was in a paper which included some observations of RW Draconis (Blazhko 1907) published 85 years ago, so the phenomenon has been around for a long time, and yet we still do not have an adequate understanding of it. In brief, the effect is a long term modulation of the amplitude of the light and velocity curves of an RR Lyr star. Some additional details are given below, with special focus on the properties of RR Lyr itself (the brightest RR Lyr star to show the Blazhko effect). The amplitude variation in visual light is typically quite noticeable, for example, in RR Lyr there may be a difference as large as 0.3 magnitude (out of approximately one magnitude pulsational variation), between the extremes of the Blazhko cycle. In addition the shape of the light curve varies as a function of the Blazhko cycle. [A complete set of light curves can be seen in Walraven (1949).] In RR Lyr a "shoulder" appears during rising light, just prior to maximum, in the smaller amplitude Blazhko phases, but which is not present in the larger amplitude phases. Except for this particular feature, the radial velocity curve exhibits changes which are mirror images of those seen in the light curve. In addition, emission is seen in the hydrogen lines during some Blazhko phases, but not others.

Blazhko periods (P_B) generally run from about $20 - 100^d$, i.e. ≈ 100 times the fundamental pulsation period (Π). There does not appear to be any correlation between Π and P_B (Szeidl 1988). If one compares the amplitude of pulsation for a star showing Blazhko effect to ones with a similar period that does not show it, the

Blazhko variable's amplitude at maximum most closely resembles the non-Blazhko stars (Szeidl 1988).

The amplitude of the Blazhko effect itself is also modulated on an even longer time scale in a number of RR Lyr stars. RR Lyr itself shows a tertiary period of about 3.8-4.8 years. The Blazhko effect was especially weak during 1963, 1967, 1971, and 1975. For instance, the extensive photometric and spectroscopic study of rising light performed by Preston, Smak, and Paczyński (1965) showed that the Blazhko effect was virtually non-existent in RR Lyr in 1963. Spectroscopically, the variation in the hydrogen emission strength, however, was still observable during this time.

Estimates of the percentage of RR Lyr stars which show the Blazhko effect varies in the range of 15-35%. Given that the effect can all but disappear in RR Lyr at times, as discussed above, one must view these estimates with some caution. Several Bailey Type c variables (overtone pulsators) have been proposed as stars showing Blazhko effect, but it is much harder to study in these lower amplitude pulsators with sinusoidal light curves. In general the Blazhko effect is confined to the RR Lyr stars with shorter period; it is not seen in stars with periods longer than about $0.^d6$.

2. Proposed Explanations of the Blazhko Effect

It is the characteristics that I have listed in §1 that must be explained by any theory of the cause of the Blazhko phenomenon. There is space here to only summarize the two broad categories of working hypotheses, viz., the magnetic pulsator and various resonance interactions. The former is similar to the oblique magnetic pulsator models which have been used to explain rapidly oscillating Ap stars. In theses theories P_B is just the rotation period of the star. Support for a magnetic field modulated by the pulsation **and** Blazhko periods has been found in the polarization measurements of Romanov, Udovichenko, and Frolov (1987). Unfortunately, the original data were not published, and these results are in need of confirmation. I would strongly urge that observers attempt magnetic field measurements of RR Lyr at a variety of pulsation and Blazhko phases. The second general class of theories proposed are those which invoke resonance interactions with other modes. Examples of these scenarios can be found in Borkowski (1980), Moskalik (1986), and Cox (this conference). None of the proposed explanations mentioned above have met with universal acceptance, due largely to their failure to account for all of the observed phenomenon.

3. The Multi-wavelength Observing Campaigns

In order to provide a fairly extensive collection of observations of the Blazhko effect I organized two multi-site, multi-wavelength campaigns during the summers of 1990 and 1991. These campaigns included ground-based photometry and spectroscopy, which was obtained to coincide with observations conducted with the *International Ultraviolet Explorer* (IUE). The IUE observations covered a complete pulsation cycle, and were repeated at a variety of Blazhko phases. The participants in this project are:

T. J. Teays	(CSC-IUE)	J. T. Bonnell	(CSC-IUE)
T. G. Barnes III	(Texas)	J. M. Nemec	(WSU)
E. F. Milone	(Calgary)	E. F. Guinan	(Villanova)
E. G. Schmidt	(Nebraska/NSF)	J. Heath	(Texas/CalTech)
E. Poreti	(Milano-Merate Observatory)	D. G. Schleicher	(Lowell)
E. Dutchover	(Texas)	M. Frueh	(Texas)
D. Greenlaw	(Texas)	K. Venn	(Texas)

We are presently analyzing the large amount of data that were obtained during these two summers, but some preliminary results will be presented below.

a. Photometry

Ground-based photometry was obtained on a number of nights, and included (at different sites) UBVRHJK bandpasses. This photometry is being used to supplement the photometry provided by IUE's Fine Error Sensor (FES). The FES is an image dissector with an S-20 photocathode (effective wavlength = 5200 Å) which is used for target identification and tracking. It can serve as a good photometer under controlled conditions, except for the shortcoming that it provides no color information. We have used the FES to collect data during the times between taking spectra, when the spectral camera was being prepared for the next image. From these data we have constructed extensive light curves, which span a complete pulsation cycle (13.6 hours). The FES measurements have been converted into V magnitudes using the standard IUE calibration (Pérez et al. 1991). The one difficulty was in applying the correction for the (changing) color of the star. For the purposes of this paper we have made use of mean colors, calculated as a function of pulsation phase, from a model of RR Lyr by Bonnell. Figure 1 shows the FES light curves for RR Lyr at two different Blazhko phases, representing the extremes of amplitude that were observed. The Blazhko effect was clearly quite strong during the 1990 campaign!

b. IUE Observations

The 1990 campaign made use of IUE's long wavelength prime (LWP) camera in its low resolution mode, which has a wavlength range of $\approx$ 1910-3300Å. A contiguous pair of IUE shifts was used (i.e. 16 hours) in order to completely cover one pulsation cycle. Five such runs were conducted, in order to sample the Blazhko cycle in detail. The time required to read down an image on the LWP camera, and prepare the camera for the next exposure is 28 minutes. In order to increase the time resolution between subsequent spectra, we placed two spectra on the same IUE image, by offsetting them in the aperture. This technique requires additional effort beyond the usual procedures to extract the fluxes properly, but is necessary, to delineate the shape of

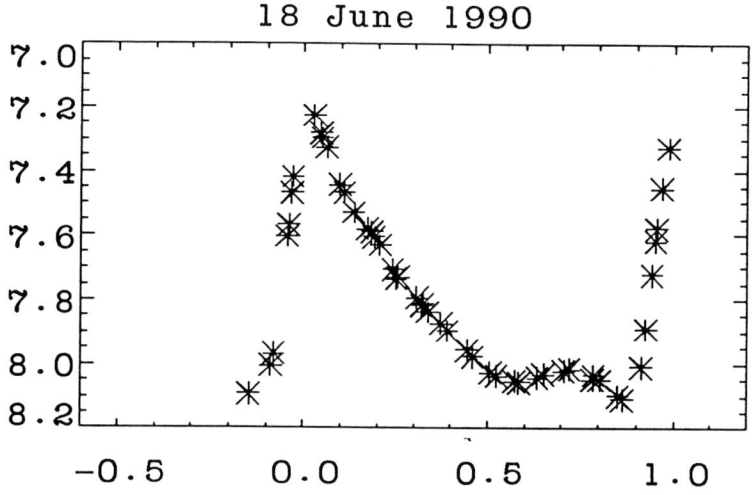

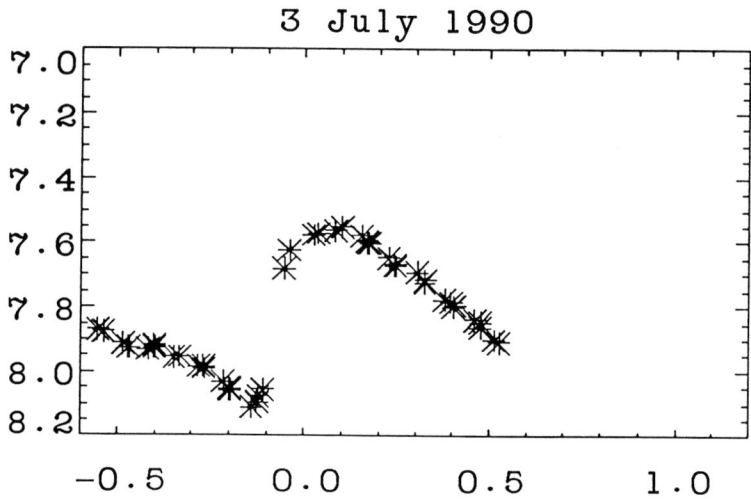

Figure 1. V light curves of RR Lyr at two different Blazhko phases, as measured by IUE's FES. These were selected to show the extremes of the variation in the light curve during the 1990 season.

the ultraviolet light curve, especially in the rapidly varying, rising light phases. From these spectra we have generated ultraviolet light curves at four wavelengths (2450, 2600, 2800, & 3000 Å) by simply binning the data. These data, when combined with the FES measurements, give the amplitude of the pulsation as a function of wavelength, from the visual through the ultraviolet, which can be directly compared to model predictions.

In 1991 we concentrated on using IUE's short wavelength prime (SWP) camera to obtain low dispersion spectra in the 1150-1975 Å range. The longer exposure times required in the SWP camera precluded us from obtaining the detailed ultraviolet light curves that we could get with the LWP camera, as well as from getting FES light curves with the necessary time resolution. Rather we concentrated on always getting a spectrum at both minimum and maximum light, so that the ultraviolet amplitude was well determined. We used four IUE shifts of eight hours duration to obtain these spectra, one of which was only two days later than a final pair of shifts that were used for obtaining detailed LWP and FES light curves, as in the 1990 campaign. This allowed us to have one time when we had complete waveleng coverage.

c. Ground-based Spectroscopy

During these campaigns spectrsocopy was obtained at McDonald, Dominion Astrophysical, and Palomar Observatories. These efforts, though hampered by the usual weather, instrument, and scheduling difficulties were successful on a number of nights. Reduction of these data is not complete, and we still have a great deal of work ahead of us to analyze the results, so in this review I will only mention the nature of the spectra taken (~ 100). Observations were concentrated on the region of the Hγ line or on the Ca II H and K lines. The former also include some metallic lines, which we can yield some radial velocity information. During the course of the pulsation cycle, one can see variable emission in the line core of Hγ.

4. Preliminary Results

The ultraviolet light curves obtained from the LWP camera have sufficient quality and phase coverage that we have been able to apply the standard Fourier decomposition techniques first introduced in Simon & Lee (1981). In addition, we have Fourier decomposed the V light curves obtained from the FES. In Figure 2 we plot the Fourier parameter ϕ_{21} vs. Π for RR Lyr at the different Blahzko phases which we sampled, and compare it to the sample of field RR Lyr stars used by Simon and Teays (1982). Similarly, Figure 3 shows the Simon & Lee parameter R_{21} plotted against period. What is seen in both cases is that the values for RR Lyr fall closely within the range of values for the non-Blazhko stars with similar pulsation period. In fact, **as RR Lyr goes through its Blazhko cycle its light curve recapitulates the appearance of all field RR Lyr stars of similar period**. We have also examined a period vs. amplitude diagram using the FES data, and confirm the results mentioned in Szeidl (1988), namely, that the closest match between Blazhko and non-Blazhko stars occurs

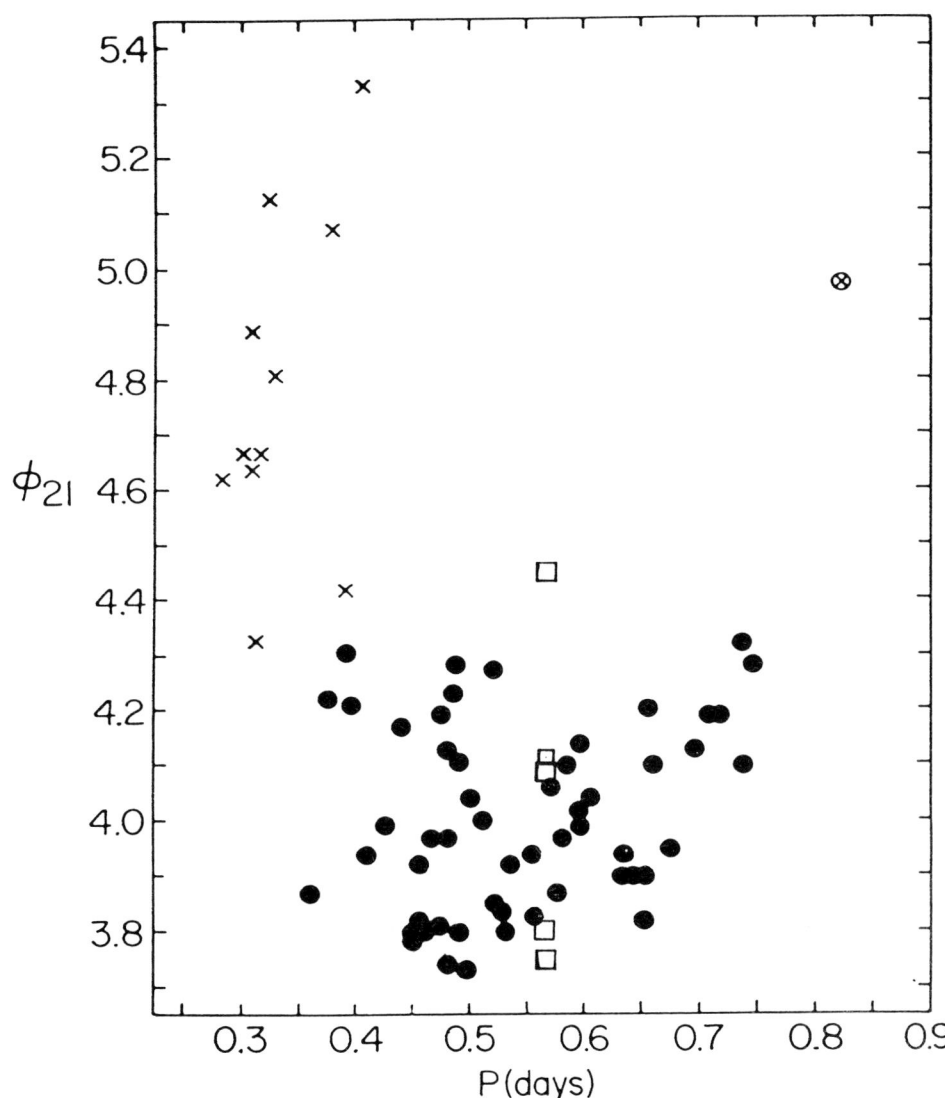

Figure 2. The Fourier decomposition parameter ϕ_{21} vs. period for non-Blazhko field RR Lyr stars (filled circles) and RR Lyr (open squares) at various Blazhko phases. The Fourier parameters are those defined by Simon & Teays (1982). (Crosses represent overtone pulsators)

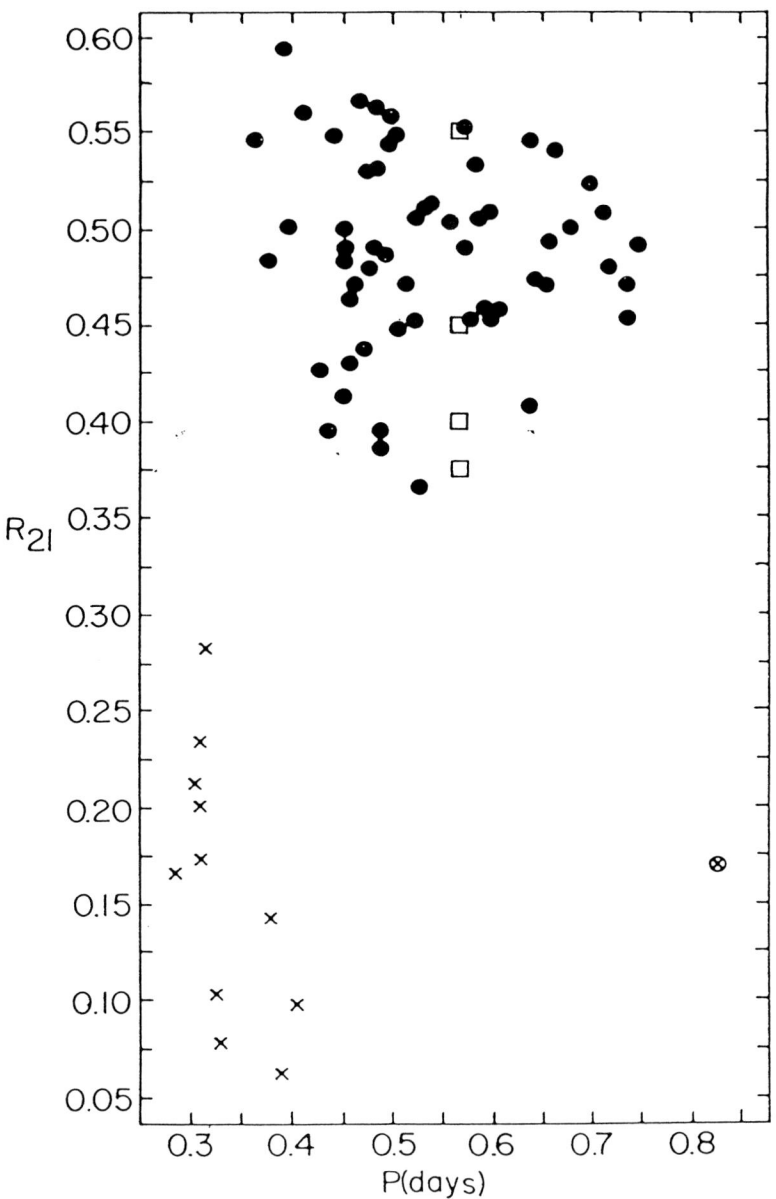

Figure 3. The Fourier decomposition parameter R_{21} plotted vs. period. The symbols are the same as in Figure 2.

at the Blazhko phase which corresponds to the largest amplitude.

Additional work is planned on these data, including seeing if one can determine phase shifts in maximum light as a function of wavelength, which can provide a useful test of pulsation models. So far, the only comparisons to models that we have performed is to the one-zone model of Grieco & Antonello (1990), the numerical values of which were provided to us by Antonello (private communication). The predicted and observed amplitudes match rather well, including the V, LWP and SWP wavelength regions. The closest match is for the largest amplitude Blazhko phase.

5. Concluding Remarks

This review has summarized the characteristics that must be explained by any theory of the Blazhko effect. Our intent in conducting the observing campaigns described in §3 was to provide a body of observations which cover complete pulsation cycles at various Blazhko phases, in order to make more detailed comparisons to theory possible. Clearly there is a great deal of analysis still to be done on these data, but it is also clear that (as usual) more studies are needed. The importance of intense coordinated campaigns for the study of phenomenon like the Blazhko effect is evident.

References:

Blazhko, S. 1907, *Astr. Nachr.*, **175**, 325.
Borkowski, K. J. 1980, *Space Sci. Rev.*, **27**, 511.
Cousens, A. 1983, *M.N.R.A.S.*, **200**, 807.
Cox, A. N. 1992, (this conference).
Grieco, A. & Antonello, E. 1990, in *Confrontation between Stellar Pulsation and Evolution*, ed. C. Cacciari & G. Clementini, (San Francisco: A.S.P.), p. 101.
Moskalik, P. 1986, *Acta Astron.*, **36**, 333.
Pérez, M., Loomis, C., Eaton, N., & Bradley, R. 1991, *Report to the IUE Three-Agency Coordination Committee*.
Preston, G. W., Smak, J., & Paczyński, B. 1965, *Ap. J. Suppl.*, **12**, 98.
Romanov, Yu. S., Udovichenko, S. N., & Frolov, M. S. 1987, *Sov. Astron. Lett.*, **13**, 29.
Simon, N. R. & Lee, A. S. 1981, *Ap. J.*, **248**, 291.
Simon, N. R. & Teays, T. J. 1982, *Ap. J.*, **261**, 586.
Szeidl, B., 1988, in *Multimode Stellar Pulsations*, eds. G. Kovács, L. Szabados, B. Szeidl (Budapest: Konkoly Observatory), p. 45.
Walraven, Th. 1949, *B. A. N.*, **11**, 17.

DISCUSSION

P. MOSKALIK: Is the scatter in ϕ_{21} vs. Period (Simon & Teays 1982) due to Blazhko effect?

T. TEAYS: No, the stars in Simon & Teays were carefully selected to be ones which did not show Blazhko effect.

S. SREENIVASAN: Could one not look for Zeeman splitting measurements of the magnetic field?

T. TEAYS: It is difficult, since one needs fairly high resolution, and RR Lyr stars are faint. A fall-back approach might be to compare the equivalent widths of lines which are magnetic proxies and those which are essentially unaffected by a magnetic field, as a function of phase.

G. MATHYS: About the magnetic field, one should be aware that what the Russian group measured through polarimetry in RR Lyr is the line-of-sight component of the field, which can only be detected if the field has a sufficient large-scale organization. If one determine the field from unpolarized line shape, what one gets is the field modulus, which is different. Still, it would be valuable to confirm the polarimetric detection. I believe that the magnitude limitation should not be taken seriously. Indeed, I have been measuring fields in 12th magnitude stars from spectra recorded in circular polarization at ESO. The only requirement is that the spectral lines should not be too broad.

T. TEAYS: I agree, and I urge observers to try and confirm the magnetic field measurements in RR Lyr as a function of pulsation and Blazhko phase, since a definite answer would go a long way towards determining the cause of the phenomenon.

E. ANTONELLO: I am amazed at the qualitative agreement between the model predictions and the observations in the far UV region. Since the one zone model takes into account simply the static model atmospheres, I think the shock effects, which are presumably larger in the UV region, are possible less strong than what is suspected.

T. TEAYS: It's true that the good agreement may be partly fortuitous. As I mentioned, I selected the best match to show today. Your one-zone model appears to be generally in agreement with observations from V to ≈ 1500 Å.

Moving Through The Instability Strip

Emilia Pisani Belserene
Former Director, Maria Mitchell Observatory

The purpose: To look at period changes in pulsating variables from the point of view of stellar evolution. Is there evidence of systematic, slow changes that might be caused by the changes in mean density during passage across the Instability Strip?

The data: $O-C$ diagrams for 67 RR Lyrae stars and Cepheids by student assistants at the Maria Mitchell Observatory, and for 88 northern Cepheids by L. Szabados

The method: Least-squares lines and parabolae (unless the $O-C$ diagram shows that the period has changed in both directions). The rate of change of period comes from the coefficient of the square term in the parabola. The principal feature of these analyses is that the rate is taken to be non-zero only if the parabola is significantly better than the linear fit, at the 2-sigma level.

The results: This table summarises the directions of period change.

	N	Inc	Dec	Both	Cnst	Inc	Dec	Both	Cnst	I/(I+D)
Pop. II										
RRab, $P < 0.4d$	3	0	1	2	0	0%	33%	67%	0%	
RRab, $P > 0.4d$	36	10	2	6	18	28%	6%	17%	50%	0.83
CWB, $P < 0.4d$	10	3	4	1	2	30%	40%	10%	20%	0.43
CWB, $P > 0.4d$	9	1	1	5	2	11%	11%	56%	22%	
Pop. I										
DCEP	79	20	17	14	28	25%	22%	18%	35%	0.54
DCEPS	11	1	0	7	3	9%	0%	64%	27%	
Pop. ?										
CEP	7	1	0	1	5	14%	0%	14%	71%	
Total	155	36	25	36	58	23%	16%	23%	37%	0.59

Discussion: Period change in both directions is most frequent among the W Vir stars and the s-Cepheids. The three shortest periods among the RR Lyrae stars, unusually short for RRab, are also quite unstable. For half of the other RR Lyrae stars, a constant period can be excluded at the 2-sigma level; increasing periods are favoured over decreasing by 5 to 1. Are we seeing these stars evolving? The e-folding times, typically a few million years, are consistent with stellar evolution on and immediately after the horizontal branch. Much of the other period behaviour is too noisy to be due to evolution. Its cause is undetermined.

RR Lyrae Variables in the Second-Parameter Globular Cluster NGC 7006

Amelia Wehlau[1] & James M. Nemec[2]

[1]*Dept. of Astronomy, University of Western ONtario, London ON N6A3K7 Canada*
[2]*Program in Astronomy, Washington State University, Pullman WA 99164 USA*

Abstract

The distant globular cluster NGC7006 was one of the first clusters studied for which the distribution of stars along the horizontal branch of its C-M diagram showed evidence for a "second parameter" in addition to metallicity. Studies of the more than 60 known RR Lyrae stars in this cluster should yield some statistically significant trends or correlations which might help to identify the second parameter. In the first stage of this study (Wehlau, Nemec, Hanlan $ Rich 1992, AJ, 103, 1583) photographic data from 1984 were combined with previously published data from the 1930's and 1950's and used to obtain period change rates for 46 variables. The median rate was found to fall one standard error below that predicted by the Yale evolutionary HB models. In addition, statistically significant evidence was found for a radial gradient in period change rates in the sense that rates for variables in the outer region of the cluster were more negative.

In the second stage of the investigation B and V magnitudes derived from CCD frames of the cluster are being used to obtain colors and to increase the number of variables for which good periods are known, in particular variables in the inner region of this centrally concentrated cluster.

The m_1 Index in RR Lyrae Stars

Eloy Rodríguez, Angel Rolland & Pilar López de Coca

*Instituto de Astrofísica de Andalucía,
Apartado 3004, 18080-Granada, Spain.*

Abstract

We have carried out simultaneous $uvby\beta$ photometry for several RR Lyrae stars. For each of these stars, the observed m_1 index variation along the pulsation cycle is compared with that expected variation from the $(\Delta m_1{}^*,\beta)$ grids of Rodríguez et al. (1991) for the corresponding temperatures, gravities and metallicities. The m_1 index variations are also calculated using the Kurucz's models. Good agreement is found

B,V Photometry of the Variable Star V9 in the Globular Cluster 47 Tuc

Michael Corwin[1] & Bruce Carney[2]

[1] *Dept. Astronomy, University of North Carolina, Charlotte NC 28223, U.S.A.*
[2] *Dept. Physics & Astronomy, Univ. North Carolina, Chapel Hill, NC 27599, U.S.A.*

Abstract

We present BV CCD photometry of the variable star V9 in the globular cluster 47 Tuc. V, B, and $(B-V)$ light curves are given. A colour-magnitude diagram based on four V and four B frames is given. V9's location on the diagram is considerably brighter and bluer than the edge of the red horizontal branch. Its radial velocity indicates that V9 is a member of the cluster.

The Stability of Cepheid Lightcurves

J.D. Fernie

David Dunlap Observatory, Richmond Hill, Ontario, Canada

Abstract

Published lightcurves of 25 to 40 day classical Cepheids and of BL Herculis stars have been examined closely for evidence of alternating deep and shallow minima, as predicted by Moskalik and Buchler (1991) and Buchler and Moskalik (1992). No such effect has been detected. Two stars, SZ Mon and EN Tra, which are sometimes taken to be classical Cepheids, do show this RV Tauri-like effect, but neither star is, in fact, likely to be a classical Cepheid.

Theoretical studies of pulsating stars by Moskalik and Buchler (1991) and by Buchler and Moskalik (1992) have predicted that BL Her stars in the approximate period range of 2.0 to 2.6 days should show alternating deep and shallow minima in their light- and velocity-curves. They predict the same effect for classical Cepheids in the 25 to 40 day range. If such an effect were present but unrealized by observers one would expect that lightcurves compiled from many different cycles would show excessive scatter in their minima. I have used existing data to search for this.

In the case of the classical Cepheids I have used the standard catalogues of Pel (1976), Moffett and Barnes (1984), and Coulson and Caldwell (1985) to search for stars with excessive scatter near minimum light. None were found. The BL Her stars had lightcurves compiled from many different sources covering many years, but again no unusual effects near minimum light were found.

Two stars, SZ Mon and EN Tra, were found to show RV Tauri-like behaviour. However, the galactic latitude of EN Tra would place it 900 pc below the galactic plane if it were a classsical Cepheid, while previous discussions (Stobie 1970; Lloyd Evans 1971) have already established that SZ Mon is not a classical Cepheid. Both stars show large far ir excesses (McAlary and Welch 1986), something not seen in classical Cepheids.

References:

Buchler, J.R., and Moskalik, P. 1992, ApJ, 391, 736.
Coulson, I.M., and Caldwell, J.A.R. 1985, So. Afr. Ast. Obs. Circ. No. 9, 1.
Lloyd Evans, T. 1971, The Observatory, 91, 159.
McAlary, C.W., and Welch, D.L. 1986, AJ, 91, 1209.
Moffett, T.J., and Barnes III, T.G. 1984, ApJS, 55, 389.
Moskalik, P., and Buchler, J.R. 1991, ApJ, 366, 300.
Pel, J.W. 1976, A&AS, 24, 413.
Stobie, R.S. 1970, MNRAS, 148, 1.

Period Variation of the Pop.II Cepheid AU Peg

J. Vinko

JATE University, Szeged, and Konkoly Observatory, Budapest, Hungary

Abstract

AU Peg, a short period BL Her-type variable, is unique among short period Cepheids. It is a member of a close binary system with an unseen companion. In t his paper the pulsational period variation is presented. It is shown that the period variation is very rapid and highly non-linear. It seems that there is no simple explanation of this phenomena by standard evolution or tidal interaction.

1. Introduction

The Population II Cepheid AU Peg is a metal-rich BL Her-type variable and a member of a close binary system which has the shortest orbital period among the known systems containing a Cepheid variable. Its binary nature was discovered from radial velo city measurements and the orbit was computed by Harris *et al.* (1979, 1984). The pulsating component is peculiar and it shows extremely rapid period change (Szabados 1977).

2. Period Variation

We have investigated the time dependence of the period variation from available photometric measurements. Constructing the O-C diagram, Fourier-spectra and phase dispersion minimization spectra of the individual light curves, seasonal periods were determined. The period variation is non-linear (Fig. 1) and there are at least two points where the rate of the period variation changed almost abruptly. From the most recent data (Szabados & Vinko 1992) it seems that the period increase stopped and the period began to decrease. It is of interest that this 'break' coincides with the appearance of a slight resonance between the orbital and the pulsational period ($P_{orb}/P_{pul} \sim 22$). Such a period variation is unprecedented among short period Cepheids. Since this star is located outside the instability strip (Harris *et al.* 1984) it has been suspected that the companion has a major effect on the pulsation of AU Peg. The presence of circumstellar matter (McAlary & Welch 1986) seems to strengthen this hypothesis. However, the period change is so complicated that it can hardly be explained by simple tidal interaction or stellar evolution alone. Continuous observations and detailed hydrodynamical modelling are desirable.

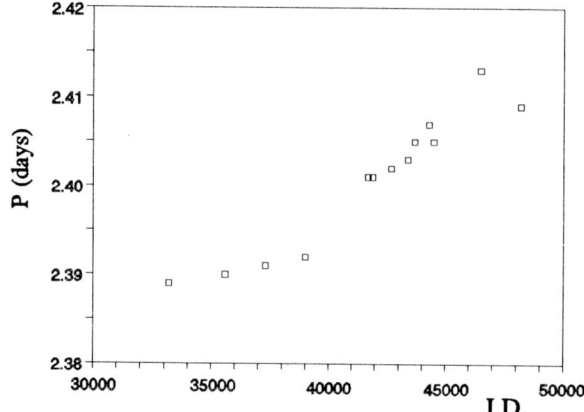

Fig. 1. Pulsational period variation of AU Peg. The error bars are smaller than the symbols.

References:

Harris, H.C., Olszewski, E. & Wallerstein, G. 1977, Astron.J. **84**, 1598.
Harris, H.C., Olszewski, E. & Wallerstein, G. 1984, Astron.J. **89**, 119.
McAlary, C.W. & Welch, D.L. 1986, Astron. J. **91**, 1209.
Szabados, L. 1977, Mitt. Sternw. Ung. Akad. Wiss. No.79.
Szabados, L. & Vinko, J. 1992, in preparation

Discussion:

D.WELCH: It is interesting that the three known binary Pop.II Cepheids are all metal rich. For AU Peg, Harris showed that the period change was consistent with the mass exchange.

J.VINKO: Yes, the mass transfer could cause many complications in the pulsation.

T.KREIDL: The Ap star ET And (a δ Sct pulsator) is in an eccentric, very close binary with even shorter P_{orb}. During a few years of observation we see no evidence of period change, so the duplicity may not be influencing any period change in AU Peg as well.

E.BÖHM-VITENSE: Binary Cepheids with $P_{orb} < 100$ days cannot exist because of the size of the stars. What does $P_{orb} = 50$ days tell you about the evolutionary state of this star?

C.WAELKENS: A Pop.II Cepheid must have been a red giant before, and it could not be in such a close system. Could this star be a Pop.I Cepheid caught during a first crossing?

J.VINKO: AU Peg star is definitely a Pop.II Cepheid because of its galactic position. I agree that AU Peg could not be a red giant in this system. It may be that AU Peg had a unique evolution due to the mass transfer between the components. But the evolutionary state of the metal-rich BL Her stars are quite uncertain.

Long-Term Brightness Changes in Cool Pulsating Variables

John R. Percy

Erindale Campus, University of Toronto, Mississauga, Ontario, Canada L5L 1C6

Abstract

Several types of cool pulsating variables show unexplained long-term changes in brightness, typically on time scales of 10 to 20 times the basic (pulsational) period. The visual and photoelectric programs of the American Association of Variable Star Observers (AAVSO) are well-suited for detecting and studying these changes. Some examples are given here, including yellow hypergiants, RV Tauri stars, small- and large-amplitude red giant and supergiant variables. The study of pulsating variables on long time scales provides "new perspectives" on their behavior.

Yellow hypergiants such as rho Cas and V509 Cas show complex pulsational variations on time scales of about a year. There are also variations on a time scale of a decade or more, which appear to be connected with the expansion of the photosphere and/or the ejection of a shell (Zsoldos and Percy 1991; Percy and Zsoldos 1992). These observations, especially when combined with the long-term spectroscopic data being accumulated by David Lambert and his collaborators at the University of Texas, should provide new information about the dynamical behavior of low-gravity, pulsationally-unstable atmospheres.

RV Tauri stars are defined as yellow pulsating supergiants showing alternating deep and shallow minima. In fact, there is a spectrum of behavior in different RV Tauri stars, ranging from strict alternating minima to almost random depths. About a third of RV Tauri stars show long-term variations in mean magnitude, and are designated as RVb stars. In the half-dozen well-studied RVb stars, the long period is about 15 times the shorter (pulsational) period. Figure 1 shows the long-term visual light curve of U Mon. The RVb variations have an unusually long period (2475 days) and a distinctly non-sinusoidal light curve reminiscent of that of an eclipsing binary. There are radial velocity variations of 30 km/s on the same time scale, but there appear to be no color variations accompanying the current RVb minimum, according to the results presented by Pollard et al. at this Colloquium. Note also that the amplitude of the RV Tauri variations appears to be smaller when the star is at long-term minimum. Figure 1 also demonstrates the AAVSO's newly-enhanced computer graphing capabilities.

Mira Variables. In a sample of 391 Mira variables, observed visually by the AAVSO for 75 years, about 15 showed long-term changes in mean magnitude (Percy et al. 1990). Almost all of these were S, N, C or R types, which suggests a relationship with the unique properties and/or evolution of these special types.

Small-Amplitude Red Variables are M giants pulsating with small amplitudes and periods of up to 200 days. A detailed study of one such star - EU Del - was published by Percy et al. (1989). We have now developed a simple autocorrelation method for analyzing the periodicity of large numbers of visual observations of these stars. Of the 10 stars studied, 8 show long-term variations on time scales of one to several years, an order of magnitude longer than the basic (pulsational) period, which is typically 20 to 100 days (Percy et al. 1992).

Supergiant Red Variables (SRc or Lc types) such as Betelgeuse show brightness variations on two time scales: a shorter one which is in approximate agreement with the fundamental radial period, and a second one which is about 10 times longer (Stothers and Leung 1971). In some cases, the long-term variations take the form of occasional large-amplitude cycles of the short-term variation (as in RS Cnc); in other cases, the long-term and short-term variations appear to be independent.

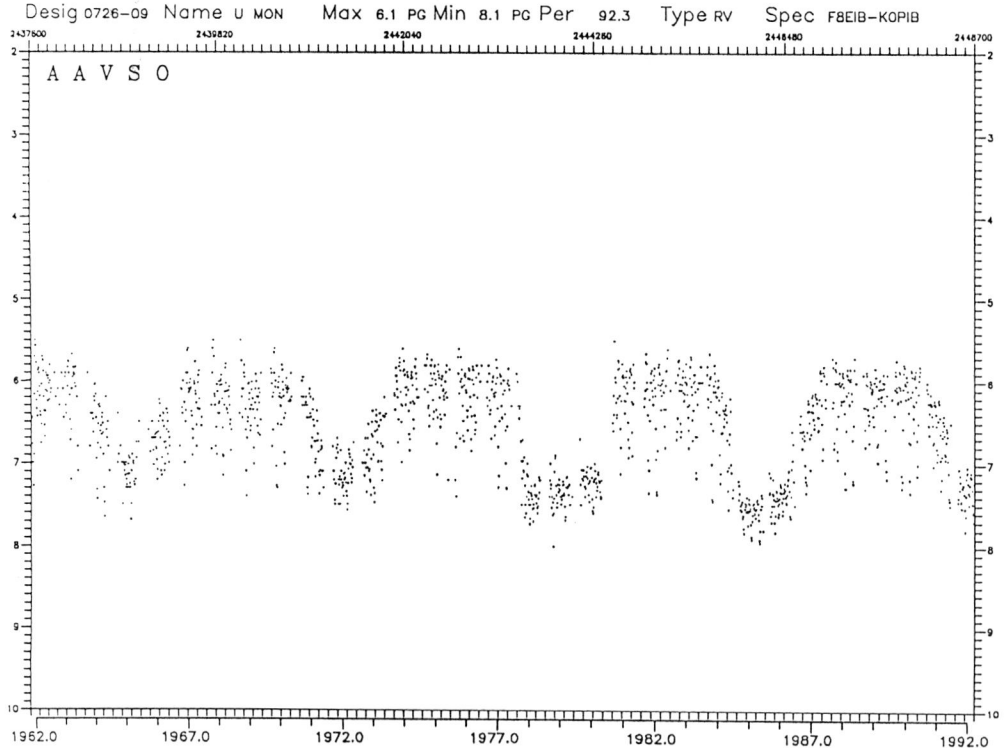

Figure 1 Five-day means of visual observations of U Mon, made by the AAVSO from 1967 to 1991. Note the 2475-day period, the unusual shape of the light curve, and the reduced

scatter (pulsation amplitude) at long-term minimum.

References:

Percy, J.R., Landis, H.J., Milton, R.E., 1989, Publ. Astron. Soc. Pacific **101**, 893.
Percy, J.R., Colivas, T., Sloan, W.B., Mattei, J.A., 1990, in em Confrontation between Stellar Pulsation and Evolution, ed. C. Cacciara and G. Clementini, Astron. Soc. Pacific Conf. Series **11**, 442.
Percy, J.R. and Zsoldos, E., 1992, Astron. Astrophys., in press.
Percy, J.R., Ralli, J., Sen, L.V., 1992, submitted to Publ. Astron. Soc. Pacific.
Stothers, R. and Leung, K.-C., 1971, Astron. Astrophys. **10**, 290.
Zsoldos, E. and Percy, J.R., 1991, Astron. Astrophys. **246**, 441.

Period Changes in the SX Phoenicis Star CY Aquarii

Michael D. Joner and John M. Powell

Brigham Young University, Provo, Utah, U.S.A.

Abstract

CY Aquarii is a Population II dwarf Cepheid that has been observed on a regular basis since 1930. We present evidence from unpublished times of maximum light from Brigham Young University observations secured in the mid-1970s to the present that indicate that the period of CY Aqr has changed in a discontinuous manner. Furthermore, we find that the period has changed significantly since the epoch of the last published investigation (1988). Finally, contrary to some past claims, examination of our high time resolution light curves yields no indication of multi-periodic pulsation in CY Aqr.

1. Data Interpretation

We utilize 84 unpublished times of maximum light secured between 1974 and 1991 at Brigham Young University along with about 500 light maxima from various literature sources to produce the O-C diagram for CY Aqr. Rolland *et al.* (1986) had suggested that the period for CY Aqr was changing in a continuous manner. In agreement with Percy (1975) and Kamper (1985), our observations indicate that the period has changed in a discontinuous manner over the years. The results of our period analysis are essentially indistinguishable from those of Coates *et al.* (1991) except for the fact that we find another sudden period change around 1989.4. The data were fit with a period of 0.061038302 days between 1968 and 1988. The new period is slightly longer at 0.061038663 days. The latest period change is solid evidence that the period cannot be changing in a continuous manner since both of the previous changes had resulted in a slightly shorter period. The last of the Coates *et al.* (1991) observations were gathered in 1988. Analysis of differential BV observations indicate that any multi-periodic pulsation for CY Aqr must currently be of an amplitude that is less than or comparable to the noise (< 0.01 mag) in the photometry.

References:

Coates, D.W., Barnes, T.G., Fernley, J.A., Frueh, M.L. & Sekiguchi, K., 1991, Delta Scuti Star Newsletter 4, 10

Kamper, B., 1985, IBVS, 2802

Percy, J.R., 1975, Astron. Astrophys. 43, 469

Rolland, A., Peña, J.H., Lopez de Coca, Peniche, R. & Gonzalez, S.F., 1986, Astron. Astrophys. 168, 125

Strömgren Photoelectric Photometry of the Dwarf Cepheid Stars DY Peg and BP Peg

J.H. Peña[1,2], R. Peniche[1,2] & R. Garrido[3]

[1] *Instituto de Astronomia-UNAM, Mexico*
[2] *Instituto Nacional de Astrofisica, Optica y Electronica, Mexico*
[3] *Instituto de Astrofisica de Andalucia, Granada, Espana*

Abstract

A study of short period high amplitude Dwarf Cepheid stars has been undertaken in order to discriminate between the two possible models, namely Bessel's (1969) proposal of low mass Pop. II or old Pop. I. The telecsope utilized was the 1.5 m at the SPM Observatory. The Danish spectrophotometer that allows the simultaneous acquisition of data in the $uvby$ filters and almost simultaneously in the $H\beta$ narrow and wide filters was attached. With this advantage of simultaneous observations, no phasing adjustment was needed, eliminating the risks of losing information due to amplitude varition explained either by multiple periodicity or by the Blazhko effect. We have restricted our discussion to the descending branch and light minimum phase interval between 0.175 to 0.725. For this interval the unreddened indices $(b-y)_0$ and $(c1)_0$ were calculated as in Nissen (1988), which will serve to determine the effective gravity and temperature variation of both stars, through the model atmsphere calibrations by Breger (1974), based on the ATLAS and Kurucz model atmospheres. Mean $\log g$ values of 4.0 and 4.33, and mean temperatures of 7500 and 7700 K were determined for DY Her and BP Peg, respectively. It can be concluded that both pulsating stars DY Her and BP Peg are Pop.I, normal Dwarf Cepheid stars.

References:

Bessell, M.S. 1969, ApJS, 18, 195
Breger, M. 1974, ApJ 192, 75
Breger, M., Campos, A.J. & Roby, S.W. 1978, PASP, 90, 754
Nissen, P.E. 1988, A&A, 199, 146

28 And: Simultaneous Strömgren Photometry

E. Rodríguez[1], A. Rolland[1], P. López de Coca[1], R. Garrido[1]
& E.E. Mendoza[2]

[1] *Instituto de Astrofísica de Andalucía, Apartado 3004, 18080 Granada, Spain*
[2] *Inst. de Astron., UNAM, Apartado Postal 70-264, CP-4510, México D.F., México*

Abstract

We have carried out simultaneous *uvby* photometry of the low amplitude δ Sct star 28 And. Analysis of the data, using the Fourier Transform method, establishes 28 And as a monoperiodic pulsator. Using the classical O-C method, it is found that the pulsation of this star can be well described by means of a linear ephemeris with a period of P=0.^{d}069304118 over the last twenty-four years. Amplitude variations are also shown to be present from season to season. The physical parameters of this star are determined and the nature of radial or nonradial pulsation is discussed on the basis on the derived phase shifts and amplitude ratios between Strömgren colours. The results indicate that 28 And pulsates in a nonradial mode with $\ell = 2$.

ρ Pup: A Monoperiodic Radially Pulsating δ Sct Star

Eloy Rodríguez, Angel Rolland & Pilar López de Coca

Instituto de Astrofísica de Andalucía,
Apartado 3004, 18080-Granada, Spain.

Abstract

ρ Pup is a δ Scuti type pulsator with relatively low amplitude ($\Delta V \sim 0.^m09$). Analysis of the data from different sources, using the Fourier Transform method, establishes ρ Pup as a monoperiodic pulsator. Using the classical O-C method, it is found that the pulsation of this star can be well described by means of a linear ephemeris with a period of P=0.^{d}140881372 over more than eighty years. The nature of radial or nonradial pulsation is discussed on the basis on the derived phase shifts and amplitude ratios between different colours. The results indicate that ρ Pup is a radial pulsator.

Index

active galactic nuclei (AGN)　　160
Ap stars (see rapidly oscillating Ap stars)

Baade, W.　　61
Baade-Wesselink methods (see Cepheids, δ Sct stars, RR Lyrae stars)
β CMa variable stars
　β Canis Majoris　　183
β Cep stars
　β Cephei　　184–185,186–187
　β Cep stars in LMC clusters 164–166
　β Cep stars in SMC　　164
　binarity　　171–179
　high-resolution spectrography　　173
　line-profile variations　　182
　location in HR diagram　　136
　non-radial pulsations　　182
　OPAL opacities　　163–170,221
　period changes and pulsation 171–179
　pulsation anomalies　　184–185
　pulsation mechanism　　164,171–179
　searches for β Cep stars　　172–173
BL Her stars (see Cepheids)
blue straggler stars　　31–52,347
B-type variable stars
　slowly pulsating B stars　　180–181

Cepheid variable stars
　anomalous Cepheids (ACs)　　31,37, 43–45,353
　asymmetry of metallic lines　　375
　ACs in Sculptor dwarf galaxy　43,358
　Baade-Wesselink methods　　73–76,84, 90,238,375
　Barnes-Evans method　　386
　BL Her stars 31,39–43,277,378, 422, 423–424
　binary Cepheids　　43,73,74,387–395, 406–407,408,423
　bump and beat masses　　237,239,268
　bump progression　　237,261,268
　C-a classification　　370
　C-b (s-Cepheids) classification　　370
　classical models of Cepheids 285–293
　colours of (anomalous) Cepheids 43–44,90
　diffusion in Cepheids　　244
　double-mode Cepheids　372,377,382
　evolution of Cepheids　　419
　extragalactic (anomalous) Cepheids:
　　in IC4182　　53–60
　　in LMC　　285–293,359–367
　　in NGC 6822　　91
　　in SMC　　43–45,285–293,349–356, 359–367
　　in Sculptor dwarf galaxy　　43–45
　　in Ursa Minor dwarf galaxy 43–45
　　in Virgo cluster galaxies　　81–89
　　in WLM (DDO221)　　92
　faint Cepheids　　369
　first-overtone Cepheids　　31–52,370, 377
　forgotten Cepheids　　368
　Fourier decomposition　268–269,370, 386
　Hβ photometry　　90
　Hertzsprung progression　　370,371
　individual Cepheids:
　　α Ursa Minoris　　77
　　AR Per　　25
　　AU Peg (Pop II)　　423
　　AV Ser　　333
　　AW Per (binary Cepheid)　　408
　　AZ Cen (first-overtone pulsator) 377
　　BB Cen (first-overtone pulsator) 377
　　BB Pup　　25
　　BB Sgr　　77,90
　　BD Cas (candidate binary)　　407
　　BG Cru (first-overtone pulsator) 377

BL Her	31
BY Cas (candidate binary)	407
CEa Cas	77
CEb Cas	77
CF Cas	77
CR Cep (candidate binary)	406–407
CS Vel	76–78
CV Mon	77,90
δ Cephei	386
DL Cas	77,90
DW Per	368,369
DR Vel (candidate binary)	406–407
η Aquilae	386
ER Car (candidate binary)	407
EU Tau (candidate binary)	407
EV Aur	368
EV Sct (candidate binary)	77,407
EY Car (fundamental mode pulsator)	377
FF Aql	386
FM Cas (candidate binary)	406–407
GY Sge	77
HK Cas	369
HR 7796	385
HZ Per	368
IN Aur	369
KN Cen (binary)	388
KQ Sco	77
MM Per	368
MY Pup (candidate binary)	407
QZ Nor	77
RS Boo	335
RS Pup	77
R TrA	377
RU Sct	77
RV Cet	333
RW Cam (binary)	388
RY Vel (candidate binary)	407
RZ Vel	77
S Mus (binary)	387,392,393
S Nor	77,90,375
SS Leo	25
SS Sct (candidate binary)	407
SU Cas	77,90
SU Cyg (binary)	393
S Vul	77,373
SV Per (binary)	388,390
SV Vul	77,90,373
SW Vel	77,90
SX Per	368
SY Aur (candidate binary)	407
SY Nor (binary)	388,390
SZ Tau	77,90
T Mon	77,373
TU Cas (double-mode)	372,382
TW Nor	76,77,78
U Car	77
U Sgr	77
UY Per	77
UX Car	377
V340 Nor	77
V367 Sct (in NGC 6649)	77,90,382
V381 Cen (candidate binary)	407
V395 Cas	368
V402 Cyg (candidate binary)	406–407
V419 Cen (candidate binary)	407
V440 Per (candidate binary)	406–407
V445 Oph	25
V526 Aql	369
V588 Cas	368
V636 Sco (binary)	392
V1033 Cyg	369
V1046 Cyg	369
V1154 Cyg (candidate binary)	406–407
V1344 Aql (candidate binary)	407
V Cen	77
VY Car	77
VZ CMa (candidate binary)	406–407

VZ Cnc	243	surface-brightness method	386
W Vir	61	ultraviolet studies of Cepheids	387–395
WZ Sgr	77, 90	W Vir stars	31–52,61,378,419
X Cygni	373,386	ZAMS-fitting distance scale	73–75,90
XY Cas (candidate binary)	407		
XZ Dra	333	deceleration parameter (q_0)	4
Y Oph	373,375	δ Scuti stars	
YZ Aur (candidate binary)	407	Baade-Wesselink radii	104
ζ Gem	77,90	diffusion in δ Sct stars	248–250
infrared photometry of Cepheids	61–71	dwarf Cepheids	374,419
kinematics of Galactic Cepheids	95–101	Fourier decomposition of δ Sct stars	104,144–146,147,148,149,150
light curves of Cepheids	244,422	identification in RR7 field	346
location of Cepheids in the H-R diagram	136	individual δ Sct stars:	
Los Alamos opacities	238	AI Vel	135,137,142
low-amplitude Cepheids	385	β Cas	137,396
luminosities of Cepheids	90,237	BI CMi	149
masses of Cepheids	268	DY Her	104
mass luminosity relation	240	DY Peg	104
mechanism for multimode pulsation	268	EH Lib	104
modelling Cepheid pulsations	267–274	ET And	424
period changes of Cepheids	419	γ Boo	137
period ratios of Cepheids	371	HD 18878	149
period ratio (Petersen) diagram	237,371	HD 93044	144–146
P-L relation	56,58,61–71,72–80,237	HD 224639	149
P-L-C relation	72–80,97	κ^2 Bootis	137
P-R relation	90	K2 Boo	396
P-R-mass relation	238	l Car (binary)	387,394
Pop I (classical) Cepheids	370–371, 377,424	o^1 Eri	396
Pop II Cepheids	31–52,371,378–379, 423–424	τ Cyg	396
radial velocities	385,386	τ Peg	137,147,396
resonance	237,371	θ^2 Tau	135,140,141
s-Cepheids	268,370,419	ρ Pup	396
stability of light curves	422	V1719 Cygni (dwarf Cepheid)	374
statistical parallaxes of Cepheids	73	X Caeli	149
		1 Mon	135,138
		4 CVn	136
		28 And	137
		44 Tau	149
		69 Tau	396
		71 Tau	396

large amplitude δ Sct stars 104
location in the instability strip
　　　　　　　　　　　125–126
luminosities 104
mode determinations 147,148
multiperiodic δ Sct stars 149
nonradial p- and/or g-modes 243
non-radial pulsation 135–143
non-linear radiation hydrodynamical
　calculations 242
spectral lines 396
spectroscopic abundances 374
SX Phe stars (see separate entry)
X-ray sources in δ Sct systems 396
D_n-σ relation 81
dwarf Cepheids (see δ Sct stars)
dwarf galaxies
　distances 32
　individual dwarf galaxies:
　　Ursa Minor dwarf galaxy 34,35
　　Sculptor dwarf galaxy 358
dynamical parallax 76

Faber-Jackson relation 49

galactic centre 26
globular clusters
　ages of globular clusters 3–14
　luminosity functions 9,81
　individual globular clusters:
　　NGC 104 (47 Tuc) 91–92
　　NGC 1851 91–92
　　NGC 2419 42
　　NGC 2808 280
　　NGC 3201 22,33,324
　　NGC 4147 324
　　NGC 4372 46
　　NGC 4590 (M68) 320,324,329,330
　　NGC 4833 4,324
　　NGC 5139 (ω Cen) 22–26,39,42,
　　　　46–49,297–299,322,378–379
　　NGC 5024 (M53) 316,320,324

NGC 5053 46–48,52
NGC 5272 (M3) 3,22,23,26,42,
　　　　295,316,320,324–332
NGC 5466 22,43,46–48
NGC 5904 (M5) 22–24,30,42,316,
　　　　320,324,329,330,332
NGC 6093 (M80) 42
NGC 6121 (M4) 22–27,33,335–336
NGC 6171 (M107) 22,33,52,316,
　　　　320,324
NGC 6205 (M13) 3,4,33,42,43,39
NGC 6254 (M10) 43
NGC 6266 (M62) 324
NGC 6273 (M19) 43
NGC 6341 (M92) 24–26
NGC 6388 338
NGC 6402 (M14) 39,41,43
NGC 6626 (M28) 43
NGC 6656 (M22) 4
NGC 6838 (M71) 46,47
NGC 6752 43
NGC 6864 280
NGC 7078 (M15) 22,23,26,30,
　　　　31–52,91,92,280,316,320,
　　　　322,324,329–332
NGC 7089 (M2) 4,39,43,91,92,295
in the Magellanic Clouds 15–20

HD 161796 (yellow supergiant star) 204
HD 161817 341,342
HD 196093 (= 47 Cyg) 373
Helium stars 217
Hertzsprung progression (see Cepheid
variables)
high-velocity HI clouds 346
Hipparcos satellite 31,49,51,125
HR Cam (high resolution camera) 82
HR 2680 (B5 V eclipsing binary) 188
HR 8762 (Be and shell star) 189
Hubble constant 3–4,10–12,58–59,64,
　　　　81–89

Hubble Space Telescope (HST)
   observations of Cepheids   89
   observ. of Cepheids in IC 4182 53–60
   observ. of Cepheids in M81   61–71
Hyades (see open clusters)

IC 1613 (galaxy)   63,65,67
IC 4182 (galaxy)   11,53–60,69
IUE satellite   410–418

λ Eri (Be variable) stars   136, 164
Large Magellanic Cloud (LMC)
  β Cep stars in LMC clusters 164–166
  Cepheids in the LMC:
    color excesses   79
    in NGC 1866   359–367
    P-L relation   73
    P-L-C relation   78
  distance modulus of LMC   17,20,
       63–68,91,92,352,353
  globular clusters in the LMC:
    NGC 1466   15,16,19
    NGC 1786   15,19,43
    NGC 1835   15,19
    NGC 1841   15,19
    NGC 2210   15,16,17
    NGC 2257   15,16,17,19
    Reticulum   15,16,17,19
  LVPs in the LMC   102,192–200
  Miras in the LMC   17
  star clusters (non-globular) in the LMC:
    NGC 1711   361
    NGC 1755   361
    NGC 1774   361
    NGC 1783   17
    NGC 1818   360,361
    NGC 1850   360,161
    NGC 1854   360,361
    NGC 1866   359–367
    NGC 2004   164–166,360,361
    NGC 2010   361
    NGC 2025   361
    NGC 2031   361
    NGC 2041   361
    NGC 2100   164–166,360,361
    NGC 2134   361
    NGC 2135   360,361
    NGC 2156   361
    NGC 2157   360,361,364
    NGC 2164   359–367
    NGC 2214   360,361
  number of variable stars in LMC 86
  RR Lyrae stars in the LMC   15–20, 160
late-type low-amplitude radial-velocity variables
  hydrogen fluoride technique 383,384
  individual stars:
    α Boo   283,383,384
    α Her   383
    α Hya   384
    α Ori (Betelgeuse)   383,426
    α Tau   383,384
    β And   383
    β Gem   384
    β Peg   383
    β UMi   383
    δ Sgr   384
    ϵ Peg   383,384
    γ Dra   383
    g Her   383
    μ Gem   383
    R Lyr   383
line-profile variations   147,148
long-period variable (LPV) stars
  dynamics of circumstellar shells 206
  extended pulsating atmospheres 281
  hydrogen emission lines   281
  hydrodynamical models   204–205
  LPVs in the Galaxy   192–200
  LPVs in LMC   102,192–200
  LPVs in M33   102
  LPVs in Per OB1   102

LPVs in SMC 192–200
LPVs in Virgo cluster galaxies 81–89
optical P-L relation 102
P-L relation in K 87–88
pulsation and mass loss 191
 (see Mira variable stars)
Los Alamos Opacity Library (LAOL)
 221–230,231–236,240,268,316
luminosity fluctuations 81

M31 (Andromeda galaxy) 9,63–67
M33 (Local Group galaxy) 63–67
M81 (galaxy) 61–71
M101 (galaxy) 63
Maia/ET And variables 136
Mira variable stars
 definition of a Mira variable star 192
 Hα emission line profiles 209,
  281–282
 hydrodynamic models 207
 implication of a P-L relation 103
 individual Mira variable stars:
  μ Gem 383
  o Ceti 201
  R Leo 201
  S Car 210–211
 in the LMC 17
 long-term brightness changes
  425–426
 masses 103
 non-linear models 208
 NLTE synthetic spectra 207
 photospheres 201–203
 radii 103
 S,N,C,R types 209,426
 time-dependent convection 208
 variable UV line emission 210–211
 (see long-period variable stars)

neutrinos 151
NGC 147 (galaxy) 69
NGC 185 (galaxy) 69

NGC 1783 16
NGC 2403 63,64
NGC 3109 67
NGC 4321 (Virgo galaxy) 83,84
NGC 4571 (Virgo galaxy) 83,84,85,86
NGC 6544 337
NGC 6642 337
NGC 6822 91,63,67
NGC 6822 (see globular clusters)
novae 81

Opacity Project (OP) 221–230,231–236,
 240,268
OPAL opacities 151,221–230,231–236,
 237,240,241,242,261,268,306
open (galactic) star clusters
 ejection of stars from open clusters
  346
 individual open clusters:
  C1814-191a 77
  Collinder 394 77
  Czerny 8 77
  Hyades 74
  King 4 77
  Lynga 6 77
  M25 77
  NGC 129 77
  NGC 1647 77
  NGC 2244 167
  NGC 3293 166
  NGC 5662 77
  NGC 6067 77
  NGC 6087 77
  NGC 6649 77
  NGC 6664 77
  NGC 7790 77
  NGC 4755 166
  Pleiades 74
  Trumpler 35 77

PG 1159-035 (pre-white dwarf) 107,
 117–119

Index   437

PG 1159 (pre-white dwarf) stars
                                    117–119
planetary nebulae   81,109
Pleiades (see open clusters)
PNN pulsating variable stars   136
Procyon (F5 IV star)   152
Proto-giant-planets   284

rapidly oscillating Ap (roAp) stars
  asteroseismology of roAp stars
                                  122–131
  chaos and roAp stars   133
  individual roAp stars:
    γ Equ   125
    HD 24712 (= HR 1217)   123–131,
                                    132
    HD 6532   127
    HD 60435   122–131
    HD 83368 (= HR 3831)   134
    HD 101065 (Pryzbylski's star) 123,
                              125,128,129
    HD 119027   125
    HD 134214   131–133
    HD 137949   132,133
    HD 166473   125
    HD 201601   132
    HD 203932   125
    HD 217522   129,133
    HD 218495   125
    HR 1217 (= HD 24712)   123–131,
                                    132
    HR 3831 (= HD 83368)   134
    10 Aql   125
    33 Lib   125
  limb darkening   128
  location in the HR diagram   136
  nonradial p- and/or g-modes   243
  obliquely rotating magnetic stars 134
R CoronæBorealis (RCB) stars
  echelle spectra of (RCB) stars   215
  individual RCB stars:
    HV12842 in LMC   215

    U Aqr   212
    RY Sgr   212
    SU Tau   215
    V854 Cen   214
    W Men in LMC   215
  RCB stars in the LMC   215
  pulsations and declines of RCBs
                                  212–213
  UV and Visual spectra of RCBs   214
resonance sequences   371
RR Lyrae stars
  absolute magnitudes   61
  Baade-Wesselink analysis   21–29,66,
                                 302,303
  binary RR Lyrae stars   332
  Blazhko effect   409,410–418
  chemical abundances   335–336
  convection   252–260,271
  double-mode RR Lyrae (RRd) stars:
    masses   241,324
    models of pulsation behaviour
                         256–260,275,276
    search for double-mode behaviour
                                    266
  mechanism for multimode pulsation
                                    268
  RRd stars in M15   322
  double-mode RR Lyrae with
    Blazhko effect   409
  evolutionary models   294–303,
                              309–311,312
  extragalactic RR Lyrae stars:
    in IC 1613   66
    in LMC   15–20,66,78
    in M31   66
    in M33   66
    in NGC 121 field of SMC   354
    in NGC 361 field of SMC   349–356
    in Lick field RR7 and NGC field
      SA57   340–348
    in Sculptor dwarf galaxy   358
    in SMC   93

Fourier decomposition of RR
 Lyrae light curves   269,315–323,
                      324,410–418
individual RR Lyrae stars:
  AC And (triple-mode)   279
  BB Vir (candidate binary) 330,331
  RR Lyr   409,410–418
  RW Dra   410
  VY Ser   21
infrared photometry of RR Lyrae
  stars   21–29,30,31–52,325–331,
                      332–334
instability strip   262
location in the HR diagram   136
masses of RR Lyraes   10,241,
           301–302,315–323,324
mass-[Fe/H] relationship   7
metal abundances   335–336
modelling RR Lyr pulsations 267–274
mode switching   254
new identifications   337,340–348
nonlinear models   265,266
luminosities 3–14,15–17,241,315–323
L-[Fe/H] relationship   8
Oosterhoff period effect   3–14,241
period-amplitude diagram   358
period ratio (Petersen) diagram 371
period changes   419
period-frequency distributions 353
P-L relationship   31–52,353
P-L-[Fe/H] relationship   31–52
P-[Fe/H] relationship 3–4,332–334
pulsation equation   7
RR Lyrae   409,410–418
Sandage effect   241,280
shock waves and atmospheric
  motions   374
similarity to the Sun and binary
  systems   160
spectrum of vibrational modes
                           263–264
T_{eff}-Color-[Fe/H] relationship   5

T_{eff}-[Fe/H] relationship for FBE   7
triple-mode RR Lyrae stars 278–279
RS Cnc (supergiant red variable) 426
runaway B stars   346
RV Tauri stars
  definition   425
  individual RV Tau stars:
    EN Tra (RV Tau-like)   422
    RU Cen   216
    SZ Mon (RV Tau-like)   422
    U Mon 210   425–426
  in globular clusters   378
  southern RV Tau stars   216

Small Magellanic Cloud (SMC)
  β Cep stars in SMC   164
  Cepheids in SMC:
    HV 829   93
    HV 834 (candidate binary) 407
    HV 1365 (candidate binary) 407
    HV 1871   351,352
    HV 2864 (candidate binary) 407
    HV 11449   351,352
  Cepheids in SMC field   43
  Cepheids in SMC star clusters
                           359–367
  distance modulus of SMC 61–71,93
  P-L-C zero point   73
  RR Lyrae stars in SMC 15–20,354,
                           349–356
  star clusters in SMC:
    NGC 121   16,17,19,349
    NGC 330   360,361
    NGC 361   17,19,349–356
    NGC 3 6   361,365,366
  surface-brightness technique 93
Sculptor dwarf galaxy   358
Small-amplitude red variables 426
  individual stars:
    EU Del   426
stellar associations
  Car OB2   77

Index

Mon OB2	77
Per OB1	88,102
Ruprecht	81,77
Sco OB anon	77
Vel OB1	77
Vel OB5	77
Vul OB1	77
Vul OB2	77

Sun
- diffusion in the Sun 245–248
- global oscillations 160
- solar convection zone 155
- solar oscillation frequencies 151–159
- nonradial p- and/or g-modes 243
- standard solar models 153

surface brightness method 381
Supergiant variable stars 136,426
supernovae 81
- Supernova 1937c 2–14,53–60,66
- Supernova 1987a 68

SX Phe stars
- definition 31
- number of known SX Phe stars 37
- P-A diagram 46
- P-L relationship 31,46–49
- P-L-[Fe/H] relationship 31,46–49
- individual SX Phe stars:
  - BL Cam 46,49
  - CY Aqr 46,150,368,428
  - DY Peg 46,49
  - KZ Hya 46,49
  - SX Phe 31,46,47,49,52,380–381

Type Ia Supernovae 53–60,11
Type II Supernovae 65
Trumpler 10 (star in Coma cluster) 342
Tully-Fisher method 49,64,81

V509 Cas (yellow hypergiant) 425

V1719 Cygni (dward Cepheid) 374
Virgo cluster of galaxies 81–89

weakly interacting massive particles (WIMPs) 152
white dwarfs (wds)
- DA wds 117–119,397
- DAV (ZZ Ceti) wds 107–115,120,136,397
- DB wds 116–119
- DSBV wds 107–115,136
- pulsation mechanisms
- D0V stars 109
- in binary systems 397
- individual white dwarfs:
  - GD 358 (DBV) 113
  - IK Peg companion (DA wd) 397
- nonradial p- and/or g-modes 243
- oscillations 107–115
- PG1159 stars 117–119
- pre-white dwarf evolution 108
- pulsation mechanisms 110–111
- observations with Whole Earth Telescope 117–119

Whole Earth Telescope 117–119
WLM galaxy (DDO 221) 91,92

yellow hypergiants
- definition 425
- individual stars:
  - ρ Cas 425
  - V509 Cas 425

zero-age main sequence (ZAMS) 74–76,90

ζ Oph stars
- location in HR diagram 136
- time-series spectroscopy 190

ZZ Ceti stars (see white dwarfs)

53 Per variables 136
89 Her (yellow supergiant star 204